Books are to be returned on or before the last date below.

The Maillard Reaction in Foods and Medicine

The Maillard Reaction in Foods and Medicine

Edited by

John O'Brien
School of Biological Sciences, University of Surrey, Guildford, UK

Harry E. Nursten
Department of Food Science and Technology, University of Reading, UK

M. James C. Crabbe
School of Animal and Microbial Sciences, University of Reading, UK

Jennifer M. Ames
Department of Food Science and Technology, University of Reading, UK

Proceedings of the 6th International Symposium on the Maillard Reaction held, by kind permission of the Treasurer, at the Royal College of Physicians, London, UK, 27–30 July 1997

The 6th International Symposium on the Maillard Reaction was officially sanctioned by the International Union of Food Science and Technology

Special Publication No. 223

ISBN 0-85404-733-6

A catalogue record for this book is available from the British Library

© The Royal Society of Chemistry 1998

All rights reserved.

Apart from any fair dealing for the purpose of research or private study, or criticism or review as permitted under the terms of the UK Copyright, Designs and Patents Act, 1988, this publication may not be reproduced, stored or transmitted, in any form or by any means, without the prior permission in writing of The Royal Society of Chemistry, or in the case of reprographic reproduction only in accordance with the terms of the licences issued by the Copyright Licensing Agency in the UK, or in accordance with the terms of the licences issued by the appropriate Reproduction Rights Organization outside the UK. Enquiries concerning reproduction outside the terms stated here should be sent to The Royal Society of Chemistry at the address printed on this page.

Published by The Royal Society of Chemistry,
Thomas Graham House, Science Park, Milton Road,
Cambridge CB4 4WF, UK

For further information see our web site at www.rsc.org

Printed and bound by Athenaeum Press Ltd, Gateshead, Tyne & Wear, UK.

Foreword

When I was asked to say a few words of introduction to the 6th International Symposium on the Maillard Reaction, I was aware of the nonenzymic glycation of proteins which I knew as the Maillard reaction. However, it came as a surprise that an international conference should be devoted to a single chemical reaction. After all, I have not heard of such a symposium on the Claisen rearrangement or the Reformatsky reaction. So what is so special about the Maillard reaction? I decided to do a little private research and looked through my early books on carbohydrates, some of my recent volumes on nutrition, and clinical texts including topics like diabetes and nephrology. In all of these I found plenty of reference to the Maillard reaction. For example, in a volume of *Advances in Carbohydrate Chemistry* (Vol. 3, 1948; W.W. Pigman and M.L. Wolfrom, eds), I found an article on the relationship between nitrogen content and unfermentable reducing substances in cane molasses ('Glucose and the unfermentable reducing substances in cane molasses', L. Sattler, pp. 113-128), where it was concluded that nitrogen in molasses is present as a condensation product of sugars with amino acids, i.e. the 'Maillard' reaction, with a reference to L.C. Maillard (*Comptes Rendus*, 1911, 154, 66-68). So the reaction is at least 86 years old. At the other extreme, I decided to search MEDLINE from 1992 to 1997 for the reaction by name only (i.e., not by 'glycation' or 'advanced glycation end-product', AGE) and found 192 references. I learned some interesting facts from these. For example, β-2 macroglobulin modified with AGE induces interleukin-6 from human macrophages and has a role in the pathogenesis of haemodialysis-associated amyloid deposits. The increase in the levels of 3-deoxyglucosone in diabetic rat plasma shows how the Maillard reaction leads to diabetic complications. This may contribute in the lens to senile and diabetic cataract. Products of the Maillard reaction were said to be genotoxic and, in coffee, were shown to have mutagenic properties. The cognitive enhancing drug Tenilsetam, which might be useful in the treatment of Alzheimer's disease, inhibits protein crosslinking by AGEs. NO-synthase is also inhibited by Maillard products. This could have interesting implications, as we have recently shown that NO stimulates glucose uptake into skeletal muscle and could, therefore, compensate against insulin resistance. Is this a further effect of glycation products in diabetic complications? A detailed review by Skog of the University of Lund (Department of Applied Nutrition and Food Chemistry) on 'Cooking procedures and food mutagens' claimed that creatine is a precursor of such compounds and that formation is inhibited by sugars; yet another example of the complexity of this chemical process. Thus, in principle and in practice, this conference dealt with a remarkable wealth of problems requiring a range of expertise.

You are dealing with chemical mechanisms, food preparation, food safety, human nutrition, public health and diseases that include diabetes, Alzheimer's, renal failure and cancer. There is an underlying trend in these diseases that involves protein deposits following glycation. It is interesting to compare these deposits to other forms of amyloid fibres and 'tangles'. In a recent paper, Booth *et al.* (*Nature*, 1997, 385, 787-793) described the interesting observation that specific single mutations of human lysozyme result in aggregation of the monomeric protein, forming amyloid-like fibres. In the same issue of *Nature* (pages 773-775), Max Perutz put forward the notion that some proteins may undergo a transformation from predominantly α-helical structure to β-sheets and thereby aggregate. He referred to these transformable proteins as 'chameleon proteins'. It is entirely possible that the prion proteins, so much in the news because of bovine spongiform encephalopathy (BSE) and its human equivalent, may well behave in this manner.

This conference exemplified the range of experts involved in studying the Maillard process and its consequences and, in particular, the 'division' between food scientists and clinical investigators. One of your roles must be, as I see it, to integrate the two sets of approaches. In a way your problem is no different from mine in my capacity as head of the Medical Research Council. The mission of the MRC is to "promote and support, by any means, high-quality basic, strategic and applied research and related postgraduate training in the biomedical and other sciences with the aim of maintaining and improving human health". I therefore must bring together the basic science and clinical communities to provide the necessary interactions to integrate studies on the structure of genes and gene products (proteins) with cellular observations and whole organism human physiology. I maintain that biomedical research is a continuum from molecules through cells to clinical and population studies. Perhaps in your case the Maillard reaction provides this continuum from basic chemistry through foods and proteins to medical problems. The papers in this book represent an interesting and highly varied programme, but it is the integration of the different sciences that is of particular value.

George K. Radda
Medical Research Council
London, UK

Contents

	Page
Foreword G.K. Radda	v

Plenary Lectures — 1

Role of Carbonyl Stress in Aging and Age-Related Diseases — 3
T.P. Degenhardt, E. Brinkmann-Frye, S.R. Thorpe and *J.W. Baynes*

Carcinogens in Cooked Foods: How do They Get There and Do They Have an Impact on Human Health? — 11
J.S. *Felton* and M.G. Knize

The Role of the Maillard Reaction in the Food Industry — 19
J.A. Mlotkiewicz

Amadoriase Isoenzymes (Fructosyl Amine: Oxygen Oxidoreductase EC 1.5.3) from *Aspergillus fumigatus* — 28
M. Takahashi, M. Pischetsrieder and *V.M. Monnier*

Oral Presentations

REACTION MECHANISMS — 35
Correlations Between Structure and Reactivity of Amadori Compounds: the Reactivity of Acyclic Forms — 37
M.S. Feather and V.V. Mossine

Study on the Formation and Decomposition of the Amadori Compound
N-(1-Deoxy-D-fructos-1-yl)-glycine in Maillard Model Systems — 43
I. Blank, S. Devaud and L.B. Fay

The Origin and Fate of α-Dicarbonyls Formed in Maillard Model Systems: Mechanistic Studies Using ^{13}C- and ^{15}N-Labelled Amino Acids — 51
V.A. Yaylayan and A. Keyhani

C_4, C_5, and C_6 3-Deoxyglycosones: Structures and Reactivity — 57
H. Weenen, J.G.M. van der Ven, L.M. van der Linde, J. van Duynhoven and A. Groenewegen

The Mechanism of Formation of 3-Methylcyclopent-2-en-2-olone — 65
H.E. Nursten

Fragmentation of Sugar Skeletons and Formation of Maillard Polymers — 69
R. Tressl, E. Kersten, G. Wondrag, D. Rewicki and R-P. Krüger

Recent Advances in the Analysis of Coloured Maillard Reaction Products — 76
J.M. Ames, R.G. Bailey and J. Mann

Determination of the Chemical Structure of Novel Coloured Compounds Generated During Maillard-Type Reactions — 82
T. Hofmann

Formation and the Structure of Melanoidins in Food and Model Systems — 89
S.M. Rogacheva, M.J. Kuntcheva, T.D. Obretenov and G. Vernin

Formation of N^ε-Carboxymethyllysine from Different Sugars and Glyoxal 94
A. Ruttkat and H.F. Erbersdobler

Maillard Compounds as Crosslinks in Heated ß-Casein–Glucose Systems 100
L. Pellegrino, M.A.J.S. van Boekel, H. Gruppen and P. Resmini

The Maillard Reaction of Ascorbic Acid with Amino Acids and Proteins –
Identification of Products 107
M. Pischetsrieder, B. Larisch and Th. Severin

Chemiluminescent Products of the Maillard Reaction: Studies on Model Systems 113
N. Suzuki, H. Hatate, I. Mizumoto and M. Namiki

FOOD TECHNOLOGY 119

The Use of Capillary Electrophoresis to Investigate the Effect of High Hydrostatic
Pressure on the Maillard Reaction 121
V.M. Hill, J.M. Ames, D.A. Ledward and L. Royle

Dehydroascorbic Acid Mediated Crosslinkage of Proteins Using Maillard Chemistry –
Relevance to Food Processing 127
J.A. Gerrard, S.E. Fayle, K.H. Sutton and A.J. Pratt

Linking Proteins Using the Maillard Reaction and the Implications for Food Processors 133
S. Hill and A.M. Easa

KINETICS AND ANALYTICAL ASPECTS 139

Modelling of the Maillard Reaction: Rate Constants for Individual Steps in the Reaction 141
B.L. Wedzicha and L.P. Leong

Molecular Modelling Study of the Mechanism of Acid-Catalysed Disaccharide
Hydrolysis: Implications for Nonenzymatic Browning Reactions 147
J. O'Brien and P. Bladon

Heating of Sugar-Casein Solutions: Isomerization and Maillard Reactions 154
M.A.J.S. van Boekel and C.M.J. Brands

Relationship of Ultraviolet and Visible Spectra in Maillard Reactions to CIELAB
Colour Space and Visual Appearance 160
D.B. MacDougall and M. Granov

Evaluation of the Acyclic State and the Effect of Solvent Type on Mutarotation
Kinetics and on Maillard Browning Rate of Glucose and Fructose 166
C.G.A. Davies, A. Kaanane, T.P. Labuza, A.J. Moscowitz and F. Guillaume

Influence of D-Glucose Configuration on the Kinetics of the Nonenzymatic
Browning Reaction 172
J. Häseler, B. Beyerlein and L.W. Kroh

Protein-Bound Maillard Compounds in Foods: Analytical and Technological Aspects 178
T. Henle, U. Schwarzenbolz, A.W. Walter and H. Klostermeyer

FLAVOUR CHEMISTRY 185

Formation of Maillard Aromas During Extrusion Cooking 187
C-T. Ho and W.E. Riha

Contents

Thermal Degradation of the Lacrymatory Precursor of Onion R. Kubec and J. Velíšek	193
The Interaction of Lipid-Derived Aldehydes with the Maillard Reaction in Meat D.S. Mottram and J.S. Elmore	198
Reaction Products of Glyoxal with Glycine J. Velíšek, T. Davídek and J. Davídek	204
Mechanistic Studies on the Formation of the Cracker-Like Aroma Compounds 2-Acetyltetrahydropyridine and 2-Acetyl-1-pyrroline by Maillard-Type Reactions P. Schieberle and T. Hofmann	209

TOXICOLOGY AND ANTIOXIDANTS — 217

Formation of Mutagenic Maillard Reaction Products P. Arvidsson, M.A.J.S. van Boekel, K. Skog and M. Jägerstad	219
Antioxidant Activity of Nonenzymatically Browned Proteins by Reaction with Lipid Oxidation Products F.J. Hidalgo, M. Alaiz and R. Zamora	225
Antioxidant and Prooxidant Activity of Xylose–Lysine Maillard Reaction Products G-C. Yen and M-L. Liu	231

HEALTH AND DISEASE — 237

A Novel AGE Crosslink Exhibiting Immunological Cross-reactivity with AGEs Formed *in Vivo* Y. Al-Abed and R. Bucala	239
Modeling of Protein and Aminophospholipid Glycation Using Low Molecular Weight Analogs. A Comparative Study O.K. Argirov, I.I. Kerina, J.I. Uzunova and M.D. Argirova	245
Protein Modifications by Glyoxal and Methylglyoxal During the Maillard Reaction of Higher Sugars M. Glomb and R.H. Nagaraj	250
The Relative Oxidation, Glycation and Crosslinking Activity of Glucose and Ascorbic Acid *in Vitro* B.J. Ortwerth, K-W. Lee, A. Coots and V. Mossine	256
Reaction Mechanisms Operating in 3-Deoxyglucosone-Protein Systems F. Hayase, N. Nagashima, T. Koyama, S. Sagara and Y. Takahashi	262
Effects of Vitamin E on Serum Glucose Levels and Nephropathy in Streptozotocin-Diabetic Rats M.J.C. Crabbe, B. Üstündağ, M. Cay, M.Nazıroğlu, N. Ilhan, I.H. Özercan and N. Dilsiz	268
Diabetic Cataract: Glycation of the Molecular Chaperone α-Crystallin and its Binding to the Membrane Protein MIP26 M.A.M. van Boekel and W.W. de Jong	273

Aminoguanidine Treatment Prevents the Depletion of Neurons Containing
Nitric Oxide Synthase in the Streptozotocin Diabetic Rat   279
E. Roufail, T. Soulis, M.E. Cooper and S.M. Rees

Modification of Low-Density Lipoprotein by Methylglyoxal Alters its Physicochemical
and Biological Properties   285
C.G. Schalkwijk, M.A. Vermeer, N. Verzijl, C.D.A. Stehouwer, J. te Koppele,
H.M.G. Princen and V.W.M. van Hinsbergh

Production and Metabolism of 3-Deoxyglucosone in Humans   291
S. Lal, B.J. Szwergold, M. Walker, W. Randall, F. Kappler, P.J. Beisswenger and T. Brown

The Role of 3-Deoxyglucosone and the Activity of its Degradative Pathways in the
Etiology of Diabetic Microvascular Disease   298
P.J. Beisswenger, S.Howell, R. Stevens, A. Siegel, S. Lal, W. Randall, B.J. Szwergold,
F. Kappler and T. Brown

Role of Maillard Reaction in Cataractogenesis and Altered α-Crystallin
Chaperone Function in Diabetes   304
E.C. Abraham, S. Swamy-Mruthinti, M. Cherian and H. Zhao

Immunochemical Approaches to AGE-Structures – Characterization of
Anti-AGE Antibodies   310
K. Ikeda, R. Nagai, T. Sakamoto, T.Higashi, Y. Jinnouchi, H. Sano, K. Matsumoto,
M. Yoshida, T. Araki, S. Ueda and S. Horiuchi

Immunolocalization of the Glycoxidation Product N^ϵ-(Carboxymethyl)lysine in
Normal and Inflamed Human Intestinal Tissues   316
E.D. Schleicher, K.E. Gempel, E. Wagner and A.G. Nerlich

Pharmacological Reversal of AGE-Related Protein Crosslinking with Agents that
Cleave α-Dicarbonyls   322
P. Ulrich

Role of Protein-Bound Carbonyl Groups in the Formation of Advanced Glycation
Endproducts   327
J. Liggins, N. Rodda, J. Iley and A. Furth

Role of Glucose Degradation Products in the Generation of Characteristic AGE
Fluorescence in Peritoneal Dialysis Fluid   333
A. Dawnay, A.P. Wieslander and D.J. Millar

Increased Formation of Pentosidine and N^ϵ-(Carboxymethyl)lysine in End Stage
Renal Disease: Role of Dialysis Clearance   339
M.A. Friedlander, Y.C. Wu, C.P. Randle, G.P. Baumgardner, P.B. DeOreo and V.M. Monnier

Reducing Sugars Induce Apoptosis in Pancreatic β-Cells by Provoking Oxidative
Stress via a Glycation Reaction   345
H. Kaneto, J. Fujii, T. Myint, N. Miyazawa, K.N. Islam, Y. Kawasaki,
K. Suzuki and N. Taniguchi

Low Density Lipoprotein Carboxymethylated *in Vitro* Does not Accelerate
Cholesteryl ester Synthesis in Mouse Peritoneal Macrophages   351
T. Sakurai, Y. Yamamoto, M. Shimoyama and M. Nakano

Contents

Pro-Inflammatory Cytokine Synthesis by Human Monocytes Induced by Proteins Minimally Modified by Methylglyoxal
E.A. Abordo and P.J. Thornalley ... 357

Detection of AGE-lipids *in Vivo*: Glycation and Carboxymethylation of Aminophospholipids in Red Cell Membranes
J.R. Requena, M.U. Ahmed, S. Reddy, C.W. Fountain, T.P. Degenhardt, A.J. Jenkins, B. Smyth, T.J. Lyons and S.R. Thorpe ... 363

Cellular Receptors for N^ε-Fructosyllysine: Potential Role in the Development of Diabetic Microangiopathy
S. Krantz, R.E. Brandt and R. Salazar ... 369

The Receptor for Advanced Glycation Endproducts Mediates the Chemotaxis of Rabbit Smooth Muscle Cells
T. Higashi, H. Sano, K. Matsumoto, T. Kanzaki, N. Morisaki, H. Rauvala, M. Shichiri and S. Horiuchi ... 374

Macrophage Scavenger Receptor Mediates the Endocytic Uptake of Advanced Glycation Endproducts (AGE)
S. Horiuchi, T. Higashi, H. Sano, K. Matsumoto, R. Nagai, H. Suzuki, T. Kodama and M. Shichiri ... 380

Insulin Accelerates the Endocytic Uptake and Degradation of Advanced Glycation Endproducts Mediated by the Macrophage Scavenger Receptor
H. Sano, T. Higashi, Y. Jinnouchi, R. Nagai, K. Matsumota, Z.W. Qin, K. Ikeda, Y. Ebina, H. Makino and S. Horiuchi ... 386

Endocytic Uptake of Age-Modified Low-Density Lipoprotein By Macrophages Leads to Cholesteryl Ester Accumulation *In Vitro*
Y. Jinnouchi, T. Higashi, K. Ikeda, H. Sano, R. Nagai, H. Hakamata, M. Yoshida, S. Ueda and S. Horiuchi ... 391

Poster Abstracts ... 397

Volatiles In Hydrolysed Vegetable Protein
M. D. Aaslyng and J. S. Elmore ... 399

Role of the Maillard Reaction Products in Affecting the Overall Antioxidant Properties of Processed Foods
M. Anese, L. Manzocco, M. C. Nicoli and C. R. Lerici ... 399

Separation of Coloured Components of Kecap Manis (an Indonesian Soy Sauce) by HPLC and Capillary Electrophoresis
A. Apriyantono, S. Marianti, L. Royle, R. G. Bailey and J. M. Ames ... 400

Determination of Dicarbonyl Compounds as Aminotriazines during the Maillard Reaction and *in Vivo* Detection in Aminoguanidine-Treated Rats
A. Araki, M. A. Glomb, M. Takahashi and V. M. Monnier ... 400

Glucosone and Glyoxal are Major Dicarbonyl Compounds Formed during Glucose Autoxidation: Role of Hydrogen Peroxide and Hydroxyl Radicals for Dicarbonyl Formation
A. Araki, J. Nourooz-Zadeh, M. A. Glomb, V. M. Monnier and S. P. Wolff ... 401

Commitment of the European Commission to Finance: "Optimisation of the Maillard Reaction – A Way to Improve Quality and Safety of Thermally Processed Foods" 401
A. Arnoldi (co-ordinator)

Involvement of Free Radicals in Pyrazine Formation in the Maillard Reaction 402
A. Arnoldi, A. D'Agostina and M. Negroni

New Coloured Compounds from Model Systems: Xylose-Lysine 402
A. Arnoldi, L. Scaglioni and E. Corain

Nonenzymic Browning Reaction Products Present in Pekmez (Concentrated Grape Syrup) 403
N. Bağdatlioğlu and Y. Hişil

Effective Inhibition of Free Radical Formation in the Maillard Reaction by Flavonoid Anion-Radicals 403
J. Čanadanović-Brunet, B. Lj. Milić and S. M. Djilas

Flavour and Oil-Stability Enhancement by Monitoring Maillard-Reaction Precursors in Peanut Kernels and Roasting Environments 404
R. Y. Y. Chiou

Effects of Advanced Glycation Endproducts on Vascular Permeability Increase as Induced by Histamine 404
C. M. S. Conde, F. Z. G. A. Cyrino, E. Michoud, D. Ruggiero-Lopez, N. Wiernsperger and E. Svensjö

Maillard Products: Interference in Carbonyl Assay of Oxidized Protein 405
P. J. Coussons, M. Bagga, S. Mulligan and J. V. Hunt

Development of Enzyme-Linked Immunosorbent Assays to Measure Advanced Glycation Endproducts in Human Serum 405
C. A. Dorrian, S. Cathcart, J. Clausen and M. H. Dominiczak

Antioxidative Activity of Maillard Reaction Products in Lipid Oxidation 406
S. M. Djilas, B. Lj. Milić and J. M. Čanadanović-Brunet

The Maillard Reaction in the Formation of Brown Pigments in Preserves 406
W. El-Gendy, H. Ismail and E. Salem

The Maillard Reaction in Meat Flavour Formation: The Role of Selected Precursors 407
L. J. Farmer, O. Paraskevas and T. D. J. Hagan

Identification of the Lactosylation Site of β-Lactoglobulin 407
V. Fogliano, S. M. Monti, A. Visconti, A. Ritieni and G. Randazzo

Serum 'Free' Pentosidine Levels and Urinary Excretion Predict Deteriorating Renal Function in Diabetic Nephropathy 408
M. A. Friedlander, R. A. Rodby, E. J. Lewis and D. Hricik

Production of an Antibody against Fructated Proteins and its Usefulness to Detect *in Vitro* and *in Vivo* Fructation 408
J. Fujii, N. Miyazawa, Y. Kawasaki, M. Theingi, A. Hoshi and N. Taniguchi

Cross-Linking of Proteins by Maillard Processes: Model Reaction of an Amadori Compound with N^{α}-Acetyl-L-Arginine 409
F. Gerum, M. O. Lederer and T. Severin

Contents

Structure and Biological Significance of Pentodilysine, a Novel Fluorescent Advanced Maillard Reaction Protein Crosslink
L. Graham, R. H. Nagaraj, R. Peters, L. M. Sayre and V. M. Monnier — 410

The Use of Electron Microscopy and X-Ray Diffraction to Study the Effects of Glycation on the Charge Distribution of Collagen and to Measure the Subsequent Swelling Behaviour of Corneal Tissue
J. C. Hadley, K. M. Meek and N. S. Malik — 411

Increased Formation of the Glycoxidation Product N^ϵ-Carboxymethyl-Lysine in Retinas from Diabetic Rats
H.-P. Hammes, K. Federlin, E. Wagner and E. Schleicher — 411

Factors Affecting Non-Enzymic Browning Reactions during the Aging of Port
P. Ho and M. C. M. Silva — 412

Characterization of Food Melanoidin
S. Homma, M. Murata, N. Terasawa and Y.-S. Lee — 412

Immunochemical Detection of Ascorbylated Proteins
B. Huber and M. Pischetsrieder — 413

Novel Observations of the Effect of Aminoguanidine on Previously Glycated Protein: A Pro-Oxidant Activity that Increases Proteolytic Susceptibility
J. V. Hunt and D. Schultz — 413

Effect of Pyrophosphate Buffer on 4-Hydroxy-5-Methyl-3-(2H)-Furanone Formation in Aqueous Model Systems
A. Jacquemier, L. Royle, J. K. Parker, C. M. Radcliffe and J. M. Ames — 414

The Use of an Electronic Nose for Detection of Volatile Aroma in Fried Meat Compared to GC-MS Analysis and Sensory Evaluation
M. Johansson and E. Tornberg — 414

Alteration of Skin Surface Protein with Dihydroxyacetone: A Useful Application of the Maillard Browning Reaction
J. A. Johnson and R. M. Fusaro — 415

Metabolism of Fructoselysine in the Kidney
F. Kappler, S. Djafroudi, H. Kayser, B. Su, S. Lal, W. C. Randall, M. A. Walker, A. Taylor, B. S. Szwergold, H. Erbersdobler and T. R. Brown — 415

The Effect of Glycated Proteins on the Expression of Cell-Adhesion Molecules of Human Endothelial Cells
N. Kashimura, M. Kitagawa, S. Noma, and S. Nishikawa — 416

Inhibitory Effect of Polei Tea Extract on the Formation of Advanced Glycation Endproducts *in Vivo*
N. Kinae, M. Matsuda, M. Shigeta and K. Shimoi — 416

Meat Surface Effects: Marinating Before Grilling Can Inhibit or Promote the Formation of Heterocyclic Amines
M. G. Knize, C. P. Salmon and J. S. Felton — 417

Flavour Formation in Soy Hydrolysates Prepared Using Enzymes
L. V. Kofod, M. Fischer and D. S. Mottram — 417

Role of Oligo- and Polymeric Carbohydrates in the Maillard Reaction   418
L. W. Kroh, J. Häseler and A. Hollnagel

Thermal Degradation of Lachrymatory Precursor of Onion   418
R. Kubec and J. Velíšek

Isolation and Characterisation of Melanoidins from Heat-Treated Fish Meat   419
M. J. Kuntcheva, S. M. Rogacheva and Tz. D. Obretenov

Autoxidation Mechanism of Reductones and its Significance in the Maillard Reaction   419
T. Kurata, N. Miyake and Y. Otsuka

Ascorbic Acid as the Principal Reactant Causing Browning in an Orange Juice Model System   420
T. P. Labuza, A. Kaanane and C. Davies

MALDI Mass Spectrometry in the Evaluation of Glycation Level of γ-Globulins in Healthy and Diabetic Subjects   420
A. Lapolla, R. Aronica, M. Battaglia, M. Garbeglio, D. Fedele, M. D'Alpaos, R. Seraglia B. and P. Traldi

The Maillard Reaction of Glucose and Ascorbic Acid with Guanosine under Oxidative Stress   421
B. Larisch, M. Pischetsrieder and Th. Severin

Amadori Products and a Pyrrole Derivative from Model Reactions of D-Glucose/3-Deoxyglucosone with Phosphatidylethanolamine   422
M. O. Lederer

Adverse Effect of Advanced Glycation Endproduct-Modified Laminin on Neurite Outgrowth and its Implications for Brain Aging   423
J. J. Li

Effect of Low Concentrations of Aminoguanidine on Formation of Advanced Glycation Endproducts *in Vitro*   424
J. Liggins, N. Rodda, V. Burnage, J. Iley and A. Furth

Maillard Reaction in Glucose-Glycine Systems Studied by Differential Scanning Calorimetry   424
L. Manzocco, P. Pittia and E. Maltini

Interaction between Maillard Reaction Products and Lipid Oxidation in Intermediate-Moisture Model Systems   425
D. Mastrocola, M. Munari, M. Cioroi and C. R. Lerici

Suppressive Effect of 4-Hydroxyanisole on Pyrazine Free Radical Formation in the Maillard Reaction   425
B. Lj. Milić, S. M. Djilas, J. Čanadanović-Brunet and N. B. Milić

Characterisation of the Fluorescence Associated with Human Serum Albumin Incubated with Glucose and Icodextrin-Based Peritoneal Dialysis Fluids   426
D. J. Millar, S. Turajlic, T. Henle and A. Dawnay

Advanced Glycation in Diabetic Embryopathy: Increase in 3-Deoxyglucosone in Rat Embryos in Hyperglycaemia in Vitro   426
H. S. Minhas, P. J. Thornalley, P. Wentzel and U. J. Eriksson

Reduced Susceptibility of Pyrraline-Modified Albumin to Lysosomal Proteolytic Enzymes   427
S. Miyata, B.-F. Liu, H. Shoda, T. Ohara, H. Yamada, K. Suzuki and M. Kasuga

Contents

Photo-Enhanced Modification of Human Skin Elastin in Actinic Elastosis by N^ϵ-Carboxymethyl-lysine, one of the Glycoxidation Products of the Maillard Reaction ... 427
K. Mizutari, T. Ono, K. Ikeda, K. Kayashima and S. Horiuchi

Identification and Antioxidative Activity of the Main Compounds from a Lactose-Lysine Maillard Model System ... 428
S. M. Monti, V. Fogliano, A. Ritieni and G. Randazzo

Effect of Model Melanoidins on Oxidative Stress of Neural Cells ... 428
G. S. Moon, W. Y. Lim and J. S. Kim

Analysis of Fluorescent Compounds Bound to Protein in Casein-Lactose Systems ... 429
F. J. Morales, C. Romero and S. Jiménez-Pérez

Metal-Binding Properties of Glycated Proteins and Amino Acids ... 429
V. V. Mossine and M. S. Feather

Cytotoxicity of Advanced Glycation Endproducts ... 430
G. Münch, B. Geiger, C. Loske, A. Simm and R. Schinzel

Determination of Advanced Glycation Endproducts in Serum by Fluorescence Spectroscopy and Competitive ELISA ... 430
G. Münch, R. Keis, U. Bahner, A. Heidland, H.-D. Lemke, H. Kayser and R. Schinzel

Advanced Glycation Endproducts as "Pacemakers" of β-Amyloid Deposition in Alzheimer's Disease ... 431
G. Münch, C. Loske, A. Neumann, J. Thome, R. Schinzel and P. Riederer

Browning; Does the Matrix Matter? ... 431
W. A. W. Mustapha, S. E. Hill, J. M. V. Blanshard and W. Derbyshire

N^ϵ-(Carboxymethyl)lysine Formation from Amadori Products by Hydroxyl Radicals ... 432
R. Nagai, K. Ikeda, T. Higashi, H. Sano, Y. Jinnouchi, K. Matsumoto and S. Horiuchi

Effects of an Organo-Germanium Compound (Ge-132) on Naturally Occurring Diabetes Mellitus Rats "OLETF" ... 432
K. Nakamura, Y. Fujita, T. Osawa and N. Kakimoto

Influence of pH and Metal-Ion Concentration on the Kinetics of Nonenzymic Browning ... 433
J. O'Brien

First Evidence for Accumulation of Protein-Bound and Protein-Free Pyrraline in Human Uraemic Plasma by Mass Spectrometry ... 433
H. Odani, T. Shinzato, Y. Matsumoto, J. Usami and K. Maeda

Oxidative Damage to Proteins and Abnormal Concentration of Pentosidine during Haemodialysis and after Renal Transplantation ... 434
P. Odetti, G. Gurreri, S. Garibaldi, L. Cosso, I. Aragno, S. Valentini, M. A. Pronzato and U. M. Marinari

A Novel Type of Advanced Glycation Endproduct Found in Diabetic Rats ... 434
T. Osawa, T. Oya, H. Kumon, Y. Morimitsu, H. Kobayashi, M. Akiba and N. Kakimoto

Effect of L-Ascorbate on the Oxidative Reaction of Lysine in the Formation of Collagen Cross-Links ... 435
M. Otsuka, M. Kuroyanagi, E. Shimamura and N. Arakawa

Different Responses of Retinal Microvascular Cells to Advanced Glycation Endproducts 435
C. Paget, N. Rellier, D. Ruggiero-Lopez, M. Lecomte, M. Lagarde and N. Wiernsperger

Mitochondrial Aminophospholipid Modification by Advanced Maillard Reaction is
Related to the Longevity of Mammalian Species 436
R. Pamplona, M. Portero-Otín, M. J. Bellmunt, J. Prat and J. R. Requena

The Inclusion of 5-Hydroxymethylfurfural in the Sulfite-Inhibited Maillard
Reaction Scheme 436
J. K. Parker, J. M. Ames and D. B. MacDougall

Identification and Inhibition of Glycation Cross-Links Impairing the Function of
Collagenous Tissues 437
R. G. Paul, T. J. Sims, N. C. Avery and A. J. Bailey

Structural Modifications and Bioavailablity of Starch Components upon the
Extent of the Maillard Reaction: An Enzymic Degradation and Solid
State ^{13}C CP MAS NMR Study 437
L. Pizzoferrato, M. Paci and G. Rotilio

Presence of Pyrraline in Human Urine and its Relationship with Glycaemic Control 438
M. Portero-Otín, R. Pamplona, M. J. Bellmunt and J. Prat

Fluorescent Products from Aminophospholipids and Glucose 438
J. Prat, M. J. Bellmunt, R. Pamplona and M. Portero-Otín

Diabetes and effects of collagen glycation in heart 439
T. J. Regan

Detection of Age-Lipids *in Vivo*: Carboxymethylation of Aminophospholipids in
LDL and Red Cell Membranes 439
J. R. Requena, M. U. Ahmed, S. Reddy, C. W. Fountain, T. P. Degenhardt, A. J. Jenkins,
C. Perez, T. J. Lyons, J. W. Baynes and S. R. Thorpe

Studies of the Formation of 2-Acetyl-1-Pyrroline in Model Systems and in Bread 440
L. L. Rogers, C. G. Chappell and J. A. Mlotkiewicz

Identification and Quantification of Class IV Caramels in Soft Drinks 440
L. Royle, J. M. Ames, L. Castle, H. E. Nursten and C. M. Radcliffe

Reaction of Metformin with Reducing Sugars and Dicarbonyl Compounds 441
D. Ruggiero-Lopez, M. Lecomte, N. Rellier, M. Lagarde and N. Wiernsperger

Determination of Heat Damage in Foods by Analysing N^ε-Carboxymethyl-Lysine
together with Available Lysine (Homoarginine Method) 441
A. Ruttkat, A. Steuernagel and H. F. Erbersdobler

Browning Reactions of Dehydroascorbic acid with Aspartame in Aqueous
Solutions and Beverages 442
H. Sakurai, K. Ishii, H. T. T. Nguyen, Z. Réblová and J. Pokorný

Effect of High Glucose Concentrations on Soluble Proteins from Mesophilic
and Thermophilic Bacteria 442
R. Schinzel and G. Münch

Studies on the Reaction of Glyoxal with Protein-Bound Arginine 443
U. Schwarzenbolz, T. Henle and H. Klostermeyer

Contents

Degradation Products Formed from Glucosamine in Water C.-K. Shu	443
Screening for Toxic Maillard Reaction Products in Meat Flavours and Bouillons K. Skog, A. Solyakov, P. Arvidsson and M. Jägerstad	444
Characterisation of a Novel AGE-Epitope Derived from Lysine and 3-Deoxyglucosone I. C. Skovsted, M. Christensen and S. B. Mortensen	444
Malondialdehyde Reactions with Collagen with Related Chemical Studies on Model Systems D. A. Slatter and A. J. Bailey	445
Influence of Maillard Browning on Cow and Buffalo Milk and Milk Proteins – A Comparative Study A. Srinivasan, A. Gopalan and S. Ramabadran	445
Nonenzymic Glycation of Phosphatidylethanolamine *in Vivo* K. Suyama and K. Watanabe	446
Overexpression of Aldehyde Reductase Will Protect PC 12 Cells from Cytotoxicity of Methylglyoxal or 3-Deoxyglucosone K. Suzuki and N. Taniguchi	446
Isolation, Purification and Characterization of Amadoriase Isoenzymes (Fructosylamine:Oxygen Oxidoreductase EC 1.5.3) from Aspergillus Sp. M. Takahashi, M. Pischetsrieder and V. M. Monnier	447
Nonenzymic Glycation as Risk Factor in Osteoarthritis J. M. teKoppele, J. de Groot, N. Verzijl and R. A. Bank	447
A Mass Spectrometric Investigation on the Products Arising from the Reaction of Glucose with Nucleotides P. Traldi, R. Seraglia, M. D'Alpaos, A. Lapolla, R. Aronica, M. Battaglia, M. Garbeglio and D. Fedele	448
Antibody Titres against Oxidatively Modified Proteins Reveal an Increased Oxidative Stress in Diabetic Rats N. Traverso, S. Menini, L. Cosso, P. Odetti, D. Cottalasso, M. A. Pronzato, E. Albano and U. M. Marinari	448
Antioxidant Reduction of Lens Damage in Model Diabetic Cataract J. R. Trevithick, F. Kilic and J. Caulfield	449
Studies of the Mutagenic Action of Maillard Reaction Products from Triose Reductone and Amino Acids Y.-K. Tseng and H. Omura	449
Digestibility and the Peptide Patterns of Lysozyme Modified by Glucose H. Umetsu and N. van Chuyen	450
The Study of Renin-Angiotensin-Aldosterone in Experimental Diabetes Mellitus Rats B. Üstündağ, M. Cay, M. Naziroğlu, N. Ilhan, N. Dilsiz and M. J. C. Crabbe	450
Kinetic Modelling of the Maillard Reaction between Glucose and Glycine M. A. J. S. van Boekel	451
Improvement of Diabetes Mellitus Complications by Dietary Antioxidants N. van Chuyen, J. V. Jale, K. Shinada, N. Kemmotsu and H. Arai	451

α-Acetyl-N-Heterocycles from Glutamic Acid and Carbohydrates   452
J. G. M. van der Ven and H. Weenen

**Maillard Reaction of free and Nucleic Acid-Bound 2-Deoxy-D-Ribose and
D-Ribose with ω-Amino Acids**   452
G. Wondrak, R. Tressl and D. Rewicki

Modification of Protein Structure following Reaction with Epoxyalkenals 453
R. Zamora and F. J. Hidalgo

Radiation-Induced Maillard Reactions   453
A. Zegota and S. Bachman

Subject Index   455

Scientific Committee

Dr J.M. Ames, University of Reading, UK
Dr J.W. Baynes, University of South Carolina, USA
Professor M.J. Crabbe, University of Reading, UK
Dr F. Hayase, Meiji University, Japan
Dr S. Horiuchi, Kumamoto University, Japan
Professor H. Kato, Otsuma Women's University, Japan
Dr T.P. Labuza, University of Minnesota, USA
Dr V.M. Monnier, Case Western Reserve University, USA
Prof. H.E. Nursten, University of Reading, UK
Dr J. O'Brien, University of Surrey, UK (Chair)
Dr G. Reineccius, University of Minnesota, USA
Dr E. Schleicher, Eberhard-Karls-Universität, Tübingen, Germany
Prof. B.L. Wedzicha, University of Leeds, UK

Organizing Committee

Dr J.M. Ames
Prof. M.J.C. Crabbe
Prof. H.E. Nursten
Dr J. O'Brien

The organizers wish to thank the following student and other helpers for their assistance in running a successful conference:

N. Boudaud, I. Boulajoun, J. Boyd, M. Bristow, C. Davies, O. de Peyer, D. Drake, C. Duckham, S. Elmore, H. Hepburne-Scott, V. Hill, A. Jacquemier, G. Marakis, K. Masters, J. Parker, L. Royle, D. Schuette and K. Weissman

Corporate Sponsors

Generous financial support from the following companies is gratefully acknowledged:

Ajinomoto Company
Associated British Foods
British American Tobacco Company Ltd
Coca Cola Europe Ltd
Dalgety Plc
Firmenich SA
Givaudan-Roure Research Ltd
Guinness Plc
Nestle SA
The Procter & Gamble Company
Scotia Pharmaceuticals
Unilever Plc
United Biscuits Plc
Yakult UK Ltd

Plenary Lectures

Role of Carbonyl Stress in Aging and Age-Related Diseases

T.P. Degenhardt, E. Brinkmann-Frye, S.R. Thorpe and J.W. Baynes

DEPARTMENTS OF CHEMISTRY & BIOCHEMISTRY, AND OPHTHALMOLOGY, UNIVERSITY OF SOUTH CAROLINA, COLUMBIA SC 29208, USA

Summary

The browning of proteins *via* the Maillard reaction *in vivo* involves nonenzymatic autoxidation and/or enzyme-catalysed oxidation of carbohydrates, lipids and amino acids. Carbohydrates also contribute to non-oxidative browning of proteins by rearrangement and elimination pathways which generate deoxydicarbonyl compounds, such as deoxyglucosone and methylglyoxal. A common feature of both the oxidative and non-oxidative reactions is the formation of reactive carbonyl compounds, suggesting that carbonyl stress, as well as oxidative stress, is involved in chemical modifications of proteins during the Maillard reaction. Oxidative stress may enhance the damage produced by carbonyl stress, both by providing an additional source of carbonyls and by limiting the rate of their detoxification.

Introduction: The AGE Hypothesis and Oxidative Stress

A prevailing theme in diabetes research is that diabetic complications may not be a direct consequence of insulin deficiency or resistance, but an indirect result of chemical or metabolic sequelae of hyperglycemia. Fifteen years ago, glycation of protein was considered a reasonable source of diabetic complications. Increased glycation of hemoglobin, albumin and immunoglobulins was invoked, at one time or another, to explain the hematological abnormalities, altered pharmacokinetics and increased susceptibility to infection among diabetic patients.[1] The glycation hypothesis gradually evolved into the AGE hypothesis which focused on chemical modifications and crosslinking of protein occurring subsequent to glycation.[2] The AGE hypothesis offered a reasonable explanation for the gradual development of chronic complications in diabetes. Long-lived proteins, such as collagens and crystallins, were recognized as important target proteins which accumulated AGEs in affected tissues. Carbohydrates other than glucose, such as ascorbate, fructose and ribose, were also identified as possible precursors of AGEs in diabetes, and inhibition of AGE formation, rather than glycation, was recognized as an important goal for pharmaceutical management of diabetes.[3]

With the realization that transition metal ions and air catalysed the browning of proteins by glucose and the observation that the best characterized AGEs in tissue proteins, carboxymethyllysine (CML) and pentosidine, were products of combined glycation and oxidation reactions of hexoses or ascorbate, the AGE hypothesis evolved to accommodate a role for oxidative stress.[4] Multiple autoxidative mechanisms are involved in the formation of AGEs, as illustrated in Figure 1 for formation of CML from glucose including: 1) the Wolff pathway of *autoxidative glycosylation*,[5] initiated by autoxidation of glucose to form reactive intermediates, including arabinose and glyoxal;[6] 2) the Namiki pathway, involving cleavage of the Schiff base adduct, forming protein-bound aldehydes and glycolaldehyde and glyoxal;[7] and 3) the Hodge pathway proceeding by oxidative cleavage of the Amadori adduct.[8] There is evidence that the Wolff and Namiki pathways prevail in phosphate buffer *in vitro*, but that the Hodge pathway may be important *in vivo*.[9,10] Other carbohydrates, such as ascorbate, fructose or glycolytic intermediates may also be the primary source(s) of CML and other AGEs *in vivo*. The complexity of the Maillard reaction in biological systems is illustrated in Figure 1 by the multiple sources of CML that have been identified, at least *in vitro*, including pathways of carbohydrate autoxidation, as well as those involving lipid peroxidation[11] and oxidation of serine by metal ion independent pathways involving HOCl formed by the myeloperoxidase reaction.[12] CML may be described as an AGE, but its origin in tissues is not readily defined.

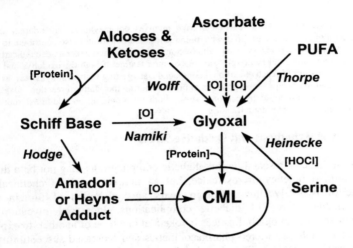

Figure 1. *Pathways for formation of CML in vitro. Not all of these pathways are known to occur in vivo, but CML and other AGEs are formed in vitro by a variety of routes from multiple precursors.* References: Heinecke,[12] Hodge,[8] Namiki,[7] Thorpe,[11] and Wolff[5]

Generalized *versus* Tissue-Specific Oxidative Stress

Despite the multiple pathways of the Maillard reaction, anaerobic conditions or inhibition of reactive oxygen production prevent the formation of CML and crosslinking of collagen by aldoses, ketoses, ascorbate, PUFA and amino acids *in vitro*. Because antioxidant defenses may have a role in limiting Maillard reactions *in vivo*, there is increasing interest in the use of antioxidant therapies to decrease oxidative stress, thereby retarding the formation of AGEs and the development of diabetic complications. However, the status of oxidative stress in diabetes is controversial. The increased levels of glycoxidation products in skin collagen and urine in diabetes are adequately explained by the increase in glucose alone, without invoking an increase in oxidative stress.[13] Age-dependent increases in amino acid oxidation products, *ortho*-tyrosine and methionine sulfoxide, in skin collagen are also unaffected by diabetes.[14] In contrast, there are numerous reports that lipid peroxidation products are increased in plasma in diabetes, although many of these have noted that the increase in lipid peroxidation is more closely associated with the presence of vascular complications, rather than with diabetes itself.[15] A limitation of these studies is that they focus on extracellular proteins or fluids, such as collagen, blood and urine, rather than on the intracellular milieu in tissues in which complications develop. Oxidation products in extracellular proteins and fluids may reflect the detoxification potential of the liver, rather than the status of intracellular oxidative stress in tissues. Only when intracellular oxidative damage to tissues has progressed to an overtly clinical stage, may increased levels of oxidation products be detectable in plasma. Thus, a fundamental relationship between diabetes and oxidative stress, if it exists, may not become apparent by analysis of blood or urine until complications have progressed to an advanced stage.

The dissociation between complications, such as nephropathy and retinopathy, and differences in their time of onset suggest that tissue-specific differences in antioxidant protection may modulate the development of complications. Set-points of antioxidant defenses or oxidative stress may not be altered by diabetes, but those patients or tissues most susceptible to complications may have lower levels of antioxidant defenses. Oxidative stress, like genetic susceptibility to hypertension, may be an independent risk factor for complications. There is growing evidence that high glucose causes oxidative damage to cells in culture,[16] and it is therefore important to assess the status of oxidative stress in affected tissues and cells *in vivo*, such as vascular endothelia, retinal pericytes and renal mesangial cells. Oxidative stress may be both nonenzymatic and metabolic in origin, so that it is also important to identify biomarkers of specific sources of oxidative stress in tissues.

Non-Oxidative Routes of the Maillard Reaction *in vivo*.

While the role of oxidative stress in development of diabetic complications is under intense study, it is clear that the Maillard reaction can proceed quite efficiently through dicarbonyl intermediates formed by a number of routes in the total absence of metal ions and air (Figure 2). Dicarbonyl compounds, such as dehydroascorbate (DHA), 3-deoxyglucosone (3DG)[17] and methylglyoxal (MGO),[18] are present in tissues at micromolar concentrations, are increased in blood of diabetic patients, and brown proteins efficiently under anaerobic conditions. Pentoses and tetroses also brown and crosslink proteins rapidly, even under rigorously anaerobic, antioxidative conditions,[19] apparently by rapid rearrangement and hydrolysis of Amadori adducts to form deoxyglycosones. While DHA is a product of the oxidation of ascorbate, 3DG and MGO are formed from glucose or its metabolites by *non-oxidative* routes. 3DG, and probably its isomer 1-deoxyglucosone, are formed by rearrangement and hydrolysis of Amadori adducts of glucose to protein[8] or by mixed enzymatic and nonenzymatic processes involving phosphorylation of fructose and β-elimination of phosphate from fructose-3-phosphate.[20] MGO is formed by elimination of phosphate from the metabolic intermediates, glyceraldehyde-3-phosphate (G3P) or dihydroxyacetone phosphate (DHAP), but is also a product of amino acid metabolism.[18] The efficient browning of proteins by 3DG and MGO under anaerobic conditions emphasizes that oxidation is not essential for Maillard reactions *in vivo*, although oxygen and oxidation reactions appear to be rate-limiting for browning and crosslinking of proteins by aldohexoses, ketohexoses and ascorbate.

Non-oxidative browning can be distinguished from oxidative browning *in vitro* by the formation of CML. Although ribose and 3DG brown proteins at essentially the same rate under both oxidative and antioxidative conditions, CML is formed only under oxidative conditions.[19] This requirement for oxygen in formation of CML and pentosidine should be considered a good approximation, rather than an absolute. They could be formed from precursors during work-up of samples for analysis, or by intermolecular (substrate-level) redox reactions which could generate oxidized precursors, even under antioxidative conditions. The role for oxidation in the browning of protein *via* the Maillard reaction is obviously dependent on both precursor and pathway. For six-carbon aldoses and ketoses, oxidation increases the rate of browning reactions. For ascorbate, oxygen is essential for browning, while for smaller sugars and dicarbonyl compounds, it is non-essential.

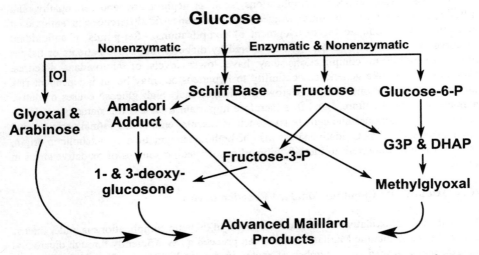

Figure 2. *Alternative pathways for formation of AGEs from glucose via dicarbonyl sugars. G3P, glyceraldehyde-3-phosphate; DHAP, dihydroxyacetone phosphate*

While it is relatively easy to assess the role of oxidation in the formation of specific AGEs *in vitro*, the distinction between glycoxidation products and non-oxidative AGEs *in vivo* is not always obvious. Consider that the carbons in GO, a product of the oxidation of glucose,[6] are in a +1 oxidation state, compared to that of 0 for carbons in glucose. In contrast, MGO, with an average carbon oxidation state of 0, the same as glucose, is not a product of oxidation of glucose - it is formed from triose phosphates during *anaerobic* glycolysis.[18] Would CML, a product of reaction of GO with protein, be considered a

Table 1. *Levels of AGEs in human tissue proteins* [a]

Marker		Lens protein	Skin collagen	Reference
CML	(mmol/mol Lys)	3.9	1.7	13,22
CEL	"	4.6	0.52	21
GOLD	"	0.15	0.04	Unpublished
MOLD	"	0.77	0.4	Unpublished
Pentosidine	"	0.006	0.03	13,22
ortho-Tyrosine	(mmol/mol Phe)	-	0.023	14
MetSO	(mmol/mol Met)	-	185	14

[a] Estimated average levels of compounds in human tissue proteins at ~80 years of age, based on data from our laboratory. GOLD, glyoxal lysine dimer; MOLD, methylglyoxal lysine dimer; MetSO, methionine sulfoxide

glycoxidation product, while its homolog N^ε-[(1-carboxy)ethyl]-lysine[21] (CEL), which is derived from MGO, be considered a non-oxidative Maillard product? CML and CEL are present at similar concentrations in proteins (Table 1), both increase with age in lens proteins and skin collagen, and their concentrations in the lens correlate strongly with one another.[21] They also increase to similar extents in skin collagen of diabetic, compared with age-matched control animals (unpublished). Since both compounds are formed from dicarbonyl intermediates and some dicarbonyls are more oxidized than glucose while others are not, it may be more appropriate to focus on the dicarbonyls themselves, rather than on their origin. A similar argument can be made regarding the imidazolium salt crosslinks, glyoxal- and methylglyoxal-lysine dimer (GOLD and MOLD), and imidazolones formed by reaction of GO, compared with MGO, with arginine residues in protein. Perhaps, the major derangement in diabetes is in the metabolism of dicarbonyl compounds, regardless of their source.

Although dicarbonyl compounds may be derived from carbohydrates by either glycoxidative or non-oxidative pathways, they are detoxified by common antioxidant-related mechanisms, such as the glyoxalase pathway, aldehyde reductases or aldehyde dehydrogenases, yielding, for example, D-lactate from MGO or 3-deoxyfructose from 3DG. The efficiency of these pathways depends on cellular levels of reduced glutathione (GSH) and/or NADPH. High levels of dicarbonyls in tissues may result from overloaded or compromised antioxidant defense systems. For example, excessive consumption of NADPH by aldose reductase may lead to a decrease in GSH. Thus, increases in MGO and 3DG might result from oxidative stress, even though these compounds are not formed from glucose by oxidative reactions. For this reason, it may not be too useful to classify Maillard reaction damage *in vivo* into glycoxidative and non-oxidative categories. A more relevant consideration is that increases in the rate of formation or steady-state concentration of dicarbonyl compounds, regardless of their origin, is consistent with decreased antioxidant defenses and a resultant increase in levels of AGEs in tissue proteins.

AGEs, ALEs or AMPs - a Current Problem?

Although CML and CEL are formed from carbohydrates and were originally classified as AGEs, these same compounds have now been detected as products of lipid peroxidation reactions. They may be formed directly from GO and MGO, or from other precursors, during metal-catalysed oxidation of PUFA in the presence of protein. GO and MGO are formed simultaneously with other carbonyl compounds more characteristic of lipid peroxidation reactions, such as malondialdehyde (MDA) and 4-hydroxynonenal (HNE) (MDA may also be formed during oxidative degradation of deoxyribose). The formation of CML and CEL, as well as MDA and HNE adducts to protein, during lipid peroxidation reactions is inhibited by AGE inhibitors, such as aminoguanidine[3] and pyridoxamine[23] (Thorpe, unpublished). Under these circumstances, it seems more appropriate to use the term, Advanced Maillard Product (AMP), to describe the broad range of products formed by carbonyl-amine chemistry during the Maillard reaction *in vivo* (Table 2). AMPs may be divided into AGEs and Advanced Lipoxidation End-products (ALEs) and other sub-classes. While the formation of AGEs may or may not require oxygen, the formation of ALEs *via* lipid peroxidation (lipoxidation) is clearly an oxygen-dependent process.

The classification of AMPs as AGEs or ALEs is sometimes straightforward. For example, pentosidine and 3DG-arginine imidazolones appear to be formed exclusively from carbohydrates, and HNE adducts only during lipid peroxidation. The origin of the major AMPs (CML and CEL) and crosslinks (GOLD and MOLD) and the role of oxidative stress in their formation *in vivo* is less certain. Regardless of their source and mechanism of formation, however, all AMPs are products of the chemistry of reactive carbonyl compounds, including enols, enals, and α-amino, α-hydroxy and α-keto aldehydes and ketones. Reactive carbonyl chemistry is the *modus operandi* of the Maillard reaction.

Table 2. *Classes and members of the family of Advanced Maillard Products* (AMPs)

Advanced Glycation End-Products (AGEs)	Mixed AGEs and ALEs	Advanced Lipoxidation End-Products (ALEs)
Pyrraline	CML, CEL	HNE-Cys, His
Pentosidine	CM-PE, CM-PS[25]	& Lys Adducts
Crosslines	GOLD, MOLD Imidazolium Salts	
AGE-X$_1$	MDA-Lys Adduct & Crosslink	
3DG-Arg Imidazolone	MGO-Arg Imidazolone	

CM-PE, carboxymethyl phosphatidyl ethanolamine; CM-PS, carboxymethyl phosphatidyl serine

During the past 20 years, we have learned that that biochemistry and molecular and cell biology proceed in a background of un-catalysed, nonenzymatic, spontaneous organic reactions and that the control of Maillard chemistry is essential for survival. Understanding this chemistry and the complementary roles of carbonyl and oxidative stress in Maillard reaction damage to tissues is essential in the search for pharmaceutical agents to intervene in diseases associated with accumulation of AMPs in tissues. These diseases now include not only diabetes, but also atherosclerosis, end-stage renal disease, rheumatoid arthritis and neurodegenerative diseases.[24] While the pathogenesis of these diseases is diverse in origin and includes a range of genetic and environmental factors, the Maillard reaction appears to be a common element in the pathology of age-related diseases. Controlling the chemistry of this reaction should have a broad and beneficial effect on our health and welfare.

Abbreviations

AGEs: Advanced Glycation End-products, ALEs: Advanced Lipoxidation End-products, AMPs: Advanced Maillard Products, CEL: N$^\varepsilon$-[(1-carboxy)ethyl]lysine CML: N$^\varepsilon$-(carboxymethyl)lysine, CM-PE: carboxymethyl phosphatidyl ethanolamine, CM-PS: carboxymethyl phosphatidyl serine, 3-DG: 3-deoxyglucosone, DHA: dehydroascorbate, DHAP: dihydroxyacetone phosphate, G3P: glyceraldehyde-3-phosphate, GO: glyoxal, GOLD: glyoxal lysine dimer, HNE: 4-hydroxynonenal, MDA: malondialdehyde, MetSO: nethionine sulfoxide, MGO: methylglyoxal, MOLD: methylglyoxal lysine dimer, *o*-tyrosine: *otho*-tyrosine, PUFA: polyunsaturated fatty acids,

References

1. L. Kennedy and J.W. Baynes, Nonenzymatic glycosylation and the chronic complications of diabetes: an overview, *Diabetologia*, 1984, **26,** 93-98.
2. H. Vlassara, R. Bucala and L. Striker, Pathogenic effects of advanced glycosylation: biochemical, biologic and clinical implications for diabetes and aging, *Lab. Invest.,* 1994, **70,** 138-151.
3. M. Brownlee, H. Vlassara, A. Kooney, P. Ulrich and A. Cerami, Aminoguanidine prevents diabetes-induced arterial wall protein crosslinking, *Science*, 1986, **232,** 1629-1632.
4. J.W. Baynes, The role of oxidative stress in the development of complications in diabetes, *Diabetes*, **40,** 405-412.
5. S.P. Wolff, Z.Y. Jiang and J.V. Hunt, Protein glycation and oxidative stress in diabetes mellitus and ageing. *Free Radic. Biol. Med.*, 1991, **10,** 339-52.
6. K.J. Wells-Knecht, D.V. Zyzak, J.E. Litchfield, S.R. Thorpe and J.W. Baynes, Mechanism of autoxidative glycosylation: Identification of glyoxal and arabinose as intermediates in autoxidative modification of proteins by glucose, *Biochemistry*, 1995, **34,** 3702-3709.
7. T. Hayashi and M. Namiki, Role of sugar fragmentation in the Maillard reaction, in 'Amino-Carbonyl Reactions in Food and Biological Systems,' M. Fujimaki, M. Namiki and H. Kato (eds), Elsevier Press, Amsterdam, 1986, p. 29-38.
8. J.E. Hodge, Dehydrated foods: chemistry of browning reactions in model systems, *J. Agric. Food Chem.*, 1953, **1,** 928-943.
9. M.C. Wells-Knecht, S.R. Thorpe, and J.W. Baynes, Pathways of formation of glycoxida- tion products during glycation of collagen, *Biochemistry*, 1995, **34,**15134-15141.
10. M.A. Glomb and V.M. Monnier, Mechanism of protein modification by glyoxal and glycol-aldehyde, reactive intermediates of the Maillard reaction, *J. Biol. Chem.*, 1995, **270,**10017-10026.
11. M. Fu, J.R. Requena, A.J. Jenkins, T.J. Lyons, J.W. Baynes and S.R. Thorpe, The advanced glycation end-product, N^{ε}-(carboxymethyl)lysine (CML), is a product of both lipid peroxidation and glycoxidation reactions, *J. Biol. Chem.*, 1996, **271,** 9982-9986.
12. M.M. Anderson, J.R. Requena, S.L. Hazen, M.X. Fu, S.R. Thorpe and J.W. Heinecke, A pathway for the generation of advanced glycosylation end products by the myeloperoxidase system of activated macrophages, *Circulation*, 1997, **8,** I-37 (Abst.).
13. D.G. Dyer, J.A. Dunn, S.R. Thorpe, K.E. Bailie, T.J. Lyons, D.R. McCance and J.W. Baynes, Accumulation of Maillard reaction products in skin collagen in diabetes and aging, *J. Clin. Invest.*, 1993, **91,** 2463-2469.
14. M.C. Wells-Knecht, T.J. Lyons, D.R. McCance, S.R. Thorpe and J.W. Baynes, Age-dependent accumulation of *ortho*-tyrosine and methionine sulfoxide in human skin collagen is not increased in diabetes: evidence against a generalized increase in oxidative stress in diabetes, *J. Clin. Invest.*, 1997, **100,** 839-846.
15. J.W. Baynes and S.R. Thorpe, The role of oxidative stress in diabetic complications. *Curr. Opin. Endocrinol. & Diabetes*, 1996, **3,** 277-284.
16. D. Giugliano, A Ceriello and G. Paolisso, Oxidative stress and diabetic vascular complications, 1996, *Diabetes Care*, 1996, **19,** 257-265.

17. T. Niwa, T. Katsuzaki, S. Miyazaki, T. Miyazaki, Y. Ishizaki, F. Hayase, N. Tatemichi and Y. Takei, Immunohistochemical detection of imidazolone, a novel advanced glycation end product, in kidneys and aortas of diabetic patients, *J. Clin. Invest.*, 1997, **99**, 1272-1280.
18. P.J. Thornalley, Advanced glycation and the development of diabetic complications. Unifying the involvement of glucose, methylglyoxal and oxidative stress, *Endocrinol. Metab.*, 1996, **3**, 149-166.
19. J.E. Litchfield, The Role of Carbohydrate Autoxidation in the Maillard Reaction, PhD Thesis, University of South Carolina, 1996.
20. S. Lal, B.S. Szwergold, A.H. Taylor, W.C. Randall, F. Kappler, K. Wells-Knecht, J.W. Baynes and T.R. Brown, Metabolism of fructose-3-phosphate in the diabetic rat lens, *Arch. Biochem. Biophys.*, 1995, **318**, 191-199.
21. M.U. Ahmed, E. Brinkmann Frye, T.P. Degenhardt, S.R. Thorpe and J.W. Baynes, N^ε-(Carboxyethyl)lysine, a product of chemical modification of protein by methylglyoxal, increases with age in human lens protein, *Biochem. J.*, 1997, **324**, 565-570.
22. T.J. Lyons, G. Silvestri, J.A. Dunn, D.G. Dyer and J.W. Baynes, Role of glycation in modification of lens crystallins in diabetic and non-diabetic senile cataracts, *Diabetes*, 1991, **40**, 1010-1015.
23. A.A. Booth, R.G. Khalifah, P. Todd and B.G. Hudson, In vitro kinetic studies of formation of antigenic advanced glycation end products (AGEs). Novel inhibition of post-Amadori glycation pathways, *J. Biol. Chem.*, 1997, **272**, 5430-5437.
24. S.R. Thorpe and J.W. Baynes, Role of the Maillard reaction in diabetes mellitus and diseases of aging. *Drugs & Aging*, 1996, **9**, 69-77.
25. J.R. Requena, M.U. Ahmed, C.W. Fountain, T.P. Degenhardt, S. Reddy, C. Perez, T.J. Lyons, A.J. Jenkins, J.W. Baynes and S.R. Thorpe, Carboxymethylethanolamine: a biomarker of phospholipid modification during the Maillard reaction *in vivo*, *J. Biol. Chem.*, 1997, **272**, 17473-17479.

Carcinogens in Cooked Foods: How Do They Get There and Do They Have an Impact on Human Health?

James S. Felton and Mark G. Knize

MOLECULAR AND STRUCTURAL BIOLOGY DIVISION, BIOLOGY AND BIOTECHNOLOGY RESEARCH PROGRAM, LAWRENCE LIVERMORE NATIONAL LABORATORY, UNIVERSITY OF CALIFORNIA, LIVERMORE, CA, 94551 USA

Summary

Fifty years ago, skin painting experiments on mice showed that extracts from heated animal muscle caused cancer. Today we know of more than 20 heterocyclic amines (HAs) formed through the Maillard reaction during the cooking of beef, pork, lamb, chicken and fish muscle. These compounds cause mutations in bacteria, mammalian cells and animals, and cancer in animals. Analytical methods today are sufficient to measure these compounds at concentrations of 0.1 ppb and practical enough for examining numerous foods cooked by different methods, and to varying degrees of doneness. There are widespread exposures to low amounts of HAs in the Western diet. In general, frying and grilling food well-done result in the highest yield of HAs. Fast-food hamburgers are very low in total heterocyclic amine content, usually below 1 ppb. In contrast, restaurant prepared meats have about ten-fold higher amounts of HAs. For chicken cooked well-done over an open flame, up to 500 ppb of 2-amino-1-methy-6-phenylimidazo[4,5-b]pyridine (PhIP) has been found. Human risk is based on exposure and potency. Current results suggest that reducing exposure to HAs is feasible and should lower the human risk to those consuming such foods.

Introduction

Diet has been associated with varying cancer rates in human populations for many years, yet the causes of the observed variation in cancer patterns have not been adequately explained.[1] Coupled with the role of diet on human cancer incidence is the more recent evidence that mutations form critical changes in oncogenes and tumor suppressor genes that control and monitor cell division. These changes are directly involved in the initiation, promotion, and progression of cancer.[2] Heterocyclic amines (HAs), which are potent genotoxins in bacteria, mammalian cells in culture, and in rodents *in vivo*,[3] are formed from the cooking of foods derived from muscle.[4] The observed higher cancer rates at organ sites such as breast and colon in populations consuming a 'Western diet' high in muscle meats, have been attributed to high fat or low fiber in their diets. The mutagenic potency of these HAs in the *Salmonella* mutation assay (Ames test) is greater than for any other class of chemical.[5] The types of mutation induced by these compounds has been studied by a number of laboratories and is inherent in the particular strains of *Salmonella* sensitive to the mutagens. Our laboratory has put extensive effort into understanding the exact lesions caused by HAs by sequencing the mutants in *Salmonella* and Chinese hamster ovary cells (CHO). HAs induce primarily frameshift mutations in *Salmonella* and base-pair substitutions in CHO. For both systems there are very specific lesions seen repeated frequently and they most often occur because of an adduct forming on the C-8 of guanine. These covalent interactions with DNA are exclusively formed through the activated amino group of the HA.

HAs initiate tumors in rodents at multiple sites,[6-8] many of which are frequent tumor sites for humans on a Western diet. One HA, PhIP, causes mammary gland and colon tumors in the rat. Recently, it was reported by Shirai *et al.*[9] that PhIP also induces carcinomas in the prostate of the male rat. Estimating human exposure to these food carcinogens by quantifying the amounts present in foods is a major goal of our laboratory.

Accurate knowledge of dietary intakes of these compounds is needed to begin any type of risk assessment for human exposure to HAs. Exposure information can be integrated with the rodent carcinogenic potency assessment for extrapolating to estimates of human cancer risk. Finally, quantitative chemical analysis is also useful for devising cooking methods and preparation strategies for reducing formation of these compounds in foods.

The cooking process is directly related to the formation of HAs, with cooking time and temperature being important determinants in both the qualitative and the quantitative formation of these compounds in foods.[10-12] Higher temperatures and longer cooking times favor the formation of HAs. A number of studies have shown the precursors of HAs to be amino acids, such as phenylalanine, threonine, and alanine; biochemicals specific to muscle, such as creatine or creatinine; and sugars.[13] Because of the requirement for temperatures in the 200°-300°C range, HAs are frequently formed in muscle meats when they have been fried, broiled, or grilled. Lower temperature cooking such as stewing, boiling, and baking do not result in the formation of HAs. Although foods appear to be the major source of human exposure to heterocyclic amines, their presence in food smoke,[14,15] cigarette smoke[15,17] and outdoor air[18] has been reported.

The carcinogenic HAs formed during cooking have stable multi-ring (2-3) aromatic structures. There can be 2 to 4 heterocyclic atoms (primarily nitrogen) in the rings and an exocyclic amino group is always present. This amino group is the primary site for oxidation and subsequent conjugation required for either detoxification and excretion or activation and binding to macromolecules such as DNA and proteins. The structures of these compounds commonly detected in foods are shown in Figure 1. Additional mutagenic heterocyclic amines have been isolated from foods as well, and the whole group of compounds is discussed in two recent reviews.[19,20] Since time and temperature of cooking influence the final concentration of HAs in food, the analysis of a large number of food samples is needed to determine the concentrations of these carcinogens in the human diet.

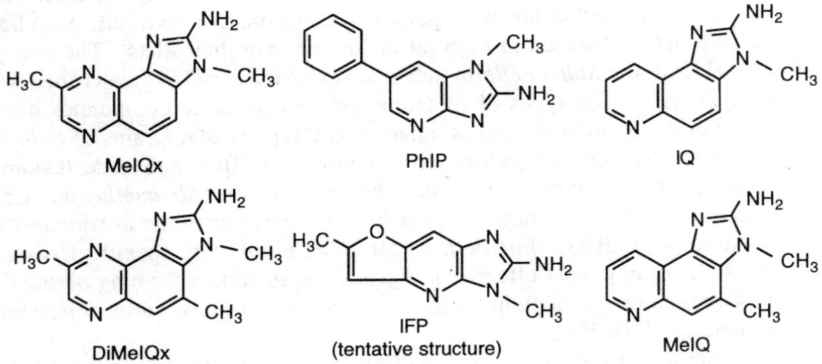

Figure 1. *Structures and common names of heterocyclic amines*

Several factors make the chemical analysis of carcinogens in foods a difficult problem. HAs are present in foods at low nanogram per gram levels. Many compounds are formed under the same reaction (heating) conditions, so the number of compounds to be quantified requires that the extraction(s), chromatographic separation, and detection be general enough to detect all the carcinogens present in the sample. Sample matrix differs with food type, thus the analysis method must be rugged enough to not be affected by these variables.

Gross[21] developed a solid-phase extraction and liquid chromatography method to purify and analyse heterocyclic amines in foods with the primary goal of high sample throughput, low cost, and high analytical sensitivity. The key to this method is the coupling of liquid/liquid extraction to a cation exchange resin column (propylsulfonic acid silica) and the ability to concentrate the dilute sample present in the large volume of organic solvent used in the first extraction step. The cation exchange properties of the propylsulfonic acid silica column allows selective washing and elution to further purify the sample.

Materials and Methods

Cooked food samples were purchased in grocery stores or restaurants and the edible part of the food was analysed by solid-phase extraction and HPLC. For marinating experiments, whole chicken breasts were cooked on a propane grill for 10, 20, 30, or 40 min with the skin and bone removed. The marinade, which was derived from a published recipe, consisted of brown sugar, olive oil, lemon juice, cider vinegar, mustard, garlic and salt.

Results

Table 1 shows that only MeIQx and DiMeIQx were detected in measurable amounts in one industrially cooked sample. Tikkanen[22] analysed industrially processed foods from Finland. Her results showed that the majority of flame-broiled fish, chicken, and pork samples had detectable heterocyclic amines (0.03 to 5.5 ng/g). In her work, meat patties commercially processed did not have detectable levels of HAs.

Table 1. *Heterocyclic amines (ng/g) in industrially and restaurant-cooked meat products*

Food product	IQ	MeIQ	MeIQx	DiMeIQx	PhIP
Industrially cooked					
Beef jerky	nd	nd	nd	nd	nd
Pork rinds A	nd	nd	0.42	0.1	nd
Pork rinds B	nd	nd	nd	nd	nd
Chili/beef	nd	nd	nd	nd	nd
Brown gravy	nd	nd	nd	nd	nd
Polish sausage	nd	nd	nd	nd	nd
Meatballs	nd	nd	nd	nd	nd
Restaurant-cooked					
Blackened Pork	nd	nd	0.53	nd	3.0
Chicken Fajita	nd	nd	0.54	nd	6.4
Hamburger	nd	nd	0.89	nd	11
Blackened Beef	nd	nd	0.48	nd	1.0
Top Sirloin Steak	nd	nd	0.87	nd	13

nd, Not Detected.

In contrast to industrially processed foods, restaurant-cooked meats (Table 1) had much more PhIP, up to 13 ng/g, and detectable levels of MeIQx averaging less than 1 ng/g. Traditional restaurant food preparation gave much higher levels of HAs than the fast foods[23]. This can be explained by the short cooking times used in fast-food restaurants, which do not allow the HAs to form from the precursors. Also, we have found that ground meat in the form of patties produces less of the heterocyclic amines than intact muscle meat (steaks) cooked under the same conditions.

The Maillard reaction and heterocyclic amine formation
The import effect of sugars and the Maillard reaction was shown recently when using a sucrose-containing marinade.[24] Table 2 shows results from the analysis of chicken breast meat samples grilled for 10, 20, 30, or 40 min with or without marinating before grilling. MeIQx showed an increase, particularly at the longer cooking times, when marinated meat was compared with non-marinated controls.

In the same experiment, PhIP was reduced from 92 to 99% after marinating and grilling for 20, 30, or 40 minutes. It appears that the marinade ingredients may directly inhibit the PhIP formation reactions or alter the pH of these reactions or, alternatively, may push the formation reactions in favor of MeIQx leaving less substrate available for PhIP formation. PhIP clearly requires higher temperatures of activation so insufficient substrate cannot be ruled out.[25]

Table 2. *A sucrose-containing marinade increases MeIQx and reduces PhIP in grilled chicken*

Grilling time	Marinade MeIQx, ng/g	Control MeIQx, ng/g	Ratio, marinade to control, MeIQx	Marinade PhIP, ng/g	Control PhIP, ng/g	Ratio, marinade to control, PhIP
10	0	0	1	0.04	0.68	.06
20	1.0	0.74	1.4	0.64	54	.01
30	3.5	0.35	10	6.3	156	.04
40	13	0.19	68	26.8	328	.08

In a dry-heated model system containing amino acids, creatinine, and glucose or sucrose, we observed that MeIQx was increased 17-fold when sucrose was used compared with reactions containing the natural meat sugar glucose. This supports our work showing sucrose to increase MeIQx in marinated chicken. The sugar does not affect PhIP formation in our model system, but another mutagen, IFP,[26] is increased two-fold when sucrose is used instead of glucose. Involvement of the Maillard reaction of sugars reacting with amino groups of amino acids in foods is supported by these studies. Model reactions are a useful way to study carcinogen-forming Maillard reactions while avoiding much of the complexity and variability of muscle foods.

Reducing the formation of heterocyclic amines
Many studies have shown the importance of cooking practices on the formation of HAs. Some have looked at time and temperature effects on mutagenic activity,[27,28] and others have actually measured HA concentrations in the food.[11,12,29] These are especially difficult tasks to do reproducibly, as the foods 'doneness' is difficult to quantify. In our experience, measuring internal temperatures with thermocouples or depending on surface appearance are not reliable indicators of doneness. The best indicator of HA formation is weight loss. Weight loss over 20% usually results in the highest PhIP

concentrations.[12,28,29] Reducing the cooking temperature is a very practical way to lower HA content of the food; however, internal meat temperatures below 70°C should be avoided because of potential bacterial contamination.

Another method for reducing HAs in foods is precooking using a microwave oven.[30] Beef patties receiving microwave treatment for various times up to 2 minutes before cooking exhibited a 3-9 fold reduction in mutagenicity and HAs following 'well-done' cooking. Analysis of beef patties after microwave treatment revealed the reason for the reduction in HAs. The precursors for HA formation (creatine, creatinine, amino acids such as phenylalanine and threonine, and glucose) were reduced in the patties by up to 30%. These precursors end up in the liquid that accumulates in the bottom of the microwave dish and, thus, are not available for the reactions to take place in the meat while it is cooking following microwaving. A reaction of two substrates in which each component is reduced by 30% could give a 10% yield of product.

Schemes for reducing mutagenic activity or the specific HA by adding substances to ground meat has been reported in the literature. Creatinase treatment was used to reduce the availability of creatine,[31] and food additives such as antioxidants,[32] or glucose or lactose[33] were shown to lower mutagenic activity. The work with sugars contradicts our results with marinades and in the model system on the formation of the potent mutagen MeIQx, but there may be other explanations for the lower mutagenic activity reported.

Discussion

The analysis of foods for HAs is important because the exposure estimates for these potent carcinogens/mutagens, which are widely consumed in the Western diet, must be correlated with epidemiology efforts to relate cause and effect, primarily on cancer etiology in humans. Exposures appear to vary over a 1000-fold range depending on diet and variations in cooking and preparation. Individuals may also vary in their susceptibility to the chemicals based on differences in metabolic activation and repair of DNA damage. This area of research provides a unique opportunity in cancer etiology, the ability to evaluate carcinogen exposure and effects of exposure in human populations. The variability in the formation reactions of these compounds also provides the opportunity for intervention to reduce exposure to HAs. The manipulation of sugar types and amounts provides an opportunity to greatly affect HA formation and could be an important consideration in the production of meat-based food flavors.

Risk estimates based on animal cancer data and human exposure to the same compounds are subject to large errors due to extrapolation, but do give some understanding of the degree of importance of exposure to HAs. Layton et al.[34] based a study on animal cancer data and surveys for food intake and concluded that the average exposure to cooked food containing HAs gave a 10^{-4} risk for lifetime cancer. Based on differences in diet and cooking preference this could change by 100-fold in either direction from the average. In addition, Lang and his co-workers found that genetic differences in metabolic activation can have a significant effect on the colon cancer rates of consumers of well-done meat, showing that individual susceptibility is important for individual risk. Finally, Turteltaub et al.[35] showed that humans bind more MeIQx per unit dose in the colon (in the form of adducts) than do rodents, suggesting the human is more sensitive.[35] This finding could increase the human risk estimates calculated by Layton up to 10-fold. In summary, the actual risk for humans exposed to heterocyclic amines is not known at this time, but continual exposure for a lifetime to these potent carcinogens can and should be reduced by simple food preparation changes.

Acknowledgments

This work was performed under the auspices of the US Department of Energy by Lawrence Livermore National Laboratory under contract No. W-7405-Eng-48 and supported by NCI grant CA55861.

References

1. R. Doll and R. Peto, The causes of cancer: Quantitative estimates of avoidable risks of cancer in the United States today, J. Nat. Cancer Inst.,1981, **66**, 1191-1308
2. B. Vogelstein and W. W. Kinzler, Carcinogens leave fingerprints. Nature, 1992, **355**, 209-210.
3. J.S. Felton and M.G. Knize, Heterocyclic-amine mutagens/carcinogens in foods, *Chemical carcinogenisis and Mutagenesis I*, ,Springer-Verlag, Berlin-Heidelberg, 1990, p. 471-502
4. J.S. Felton and M.G. Knize, Occurrence, identification, and bacterial mutagenicity of heterocyclic amines in cooked food. Mutat. Res., 1991, **259**, 205-218
5. J.S. Felton, M.G. Knize, F.A. Dolbeare, R. Wu, Mutagenic activity of heterocyclic amines in cooked foods, Env. Health Persp., 1994, **102**, 201-204
6. H. Ohgaki, H. Hasegawa, M. Suanaga, S. Sato, S. Takayama and T. Sugimura Carcinogenicity in mice of a mutagenic compound, 2-amino-3,8-dimethylimidazo[4,5-f]quinoxaline (MeIQx) from cooked foods, Carcinogenesis, 1987, **8**, 665-668
7. N.Ito, R. Hasegawa, K. Imaida, S. Tamano, A. Hagiwara, M. Hirose and T. Shirai, Carcinogenicity of 2-amino-1-methy-6-phenylimidazo[4,5-b]pyridine (PhIP) in the rat, Mutation Res., 1997, **376**, 107-114
8. H. Esumi, H. Ohgaki, E. Kohzen, S. Takayama and T. Sugimura, Induction of lymphoma in CDF1 mice by the food mutagen 2-amino-1-methy-6-phenylimidazo[4,5-b]pyridine, Jpn. J. Cancer Res., 1989, **80**, 1176-1178
9. T. Shirai, M. Sano, S. Tamano, S. Takahashi, M. Hirose, M. Futakuchi, R. Hasegawa, K. Imaida, K-I. Matsumoto, K. Wakabayashi, T. Sugimura, N. Ito, The prostate: A target for the carcinogenicity of 2-amino-1-methyl-6-phenylimidazo[4,5-b]pyridine derived from cooked foods. *Cancer Res.* 1997, **57**, 195-198
10. M.G. Knize, B.D. Andresen, S.K. Healy, N.H. Shen, P.R. Lewis, L.F. Bjeldanes, F.T. Hatch and J.S. Felton, Effect of temperature, patty thickness and fat content on the production of mutagens in fried ground beef. *Food Chem. Toxic.* 1985, **23**, 1035-1040
11. G.A. Gross and A. Grüter, Quantitation of mutagenic/carcinogenic heterocyclic aromatic amines in food products. *J. Chromatogr.* 1992, **592**, 271-278
12. K. Skog, G. Steineck, K. Augustsson and M. Jägerstad, Effect of cooking temperature on the formation of heterocyclic amines in fried meat products and pan residues. Carcinogenesis 1995, **16**, 861-867
13. K. Skog, Cooking procedures and food mutagens: A literature review. Food Chem. Toxic. 1993, **31**, 655-675
14. S. Vainiotalo, K. Matveinen, A. Reunamen, GC/MS Determination of the mutagenic heterocyclic amines MeIQx and DiMeIQx in cooking fumes. Fresenius J. Anal. Chem. 1993, **345**, 462-466
15. H.P. Thiebaud, M.G. Knize, P.A. Kuzmicky, J.S.Felton, D.P. Hsieh, Mutagenicity and chemical analysis of fumes from cooking meat. J. Agric. Food Chem. 1994, **42**, 1502-1510

16. S. Manabe, K. Tohyama, O. Wada, T. Aramaki, Detection of a carcinogen, 2-amino-1-methyl-6-phenylimidazo[4,5-b]pyridine (PhIP), in cigarette smoke condensate. Carcinogenesis 1991, **12**, 1945-1947
17. K. Wakabayashi, I-S. Kim,R. Kurosaka, Z. Yamaizumi, H. Ushiyams, M. Takahashi, A. Koyota, A. Tada, H. Nukaya. S. Goto, T. Sugimura, M. Nagao, In *Heterocyclic amines in cooked foods: possible human carcinogens*; Adamson, R.H.: Gustafsson, J-A.: Ito, N.: Nagao, M.; Sugimura, T.; Wakabayashi, K.; Yamazoe, Y., Eds.; Princeton Scientific Publishing, Princeton, NJ, USA, **1995**, p. 39.
18. S. Manabe, N. Kurihara, O. Wada, S. Izumikawa, K. Asakuna, M. Morita, Detection of a carcinogen, 2-amino-1-methyl-6-phenylimidazo[4,5-b]pyridine, in airborne particles and diesel-exhaust particles. *Environ. Pollut.* 1993, **80**, 281-286
19. S. Robbana-Barnat, M. Rabache, E. Rialland and J. Fradin, Heterocyclic amines: Occurrence and prevention in cooked food, Environ. Health Persp. 1996,**104**, 280-288
20. B. Stavric, Biological significance of trace levels of mutagenic heterocyclic aromatic amines (HAAs) in the human diet: a critical review. Food Chem. Toxic. 1994, **32**, 977-994
21. G.A. Gross, Simple methods for quantifying mutagenic heterocyclic amines in food products. Carcinogenesis, 1990,**11**, 1597-160.
22. L. M. Tikkanen, T.M. Sauri and K.J. Latva-Kala, Screening of heat-processed Finnish foods for the mutagens 2-amino-3,4,8-dimethylimidazo[4,5-f]quinoxaline, 2-amino-3,8-dimethylimidazo[4,5-f]quinoxaline, and 2-amino-1-methyl-6-phenylimidazo[4,5-b]pyridine. Food Chem. Toxic. 1993, **31**, 717-721
23. M.G. Knize, R. Sinha, N. Rothman, E.D. Brown, C.P. Salmon, O.A. Levander, P.L.Cunningham and J.S. Felton Fast-food meat products have relatively low heterocyclic amine content. Food Chem.Toxic. 1995, **33**, 545-551
24. C.P. Salmon, M.G. Knize and J.S. Felton, Marinating before grilling greatly effects heterocyclic amine carcinogen production in chicken. Food Chem. Toxic. 1997 35, 433-441.
25. M.G. Knize, , F.A. Dolbeare, K.L. Carroll, D.H. Moore II, and J.S. Felton, Effect of cooking time and temperature on the heterocyclic amine content of fried-beef patties. Fd. Chem. Toxic. 1994, 32, 595-603
26. M. G. Knize, M. Roper, N. H. Shen and J.S. Felton, Proposed Structures for an amino-dimethylimidazo-furopyridine Mutagen in Cooked Meats. Carcinogenesis, 1990, **11**, 2259-2262
27. B. Commoner, A.J. Vithayathil, P. Dolara, S. Nair, P. Madyastha, G.C. Cuca, Formation of mutagens in beef and beef extract during cooking. Science 1978, **201**, 913-916
28. T.Sugimura, M. Nagao, T. Kawachi, M. Honda, T. Yahagi, Y. Seino, S. Sato, ; N. Matsukura, T. Matsushima,A. Shirai, M. Sawamura, H. Matsumoto, Mutagen-carcinogens in foods with special reference to highly mutagenic pyrolytic products in broiled foods. In *Origins of Human Cancer*; H.H.; Watson, J.D.; Winsten, J.A., Eds.; Cold Spring Harbor, New York, 1977, 1561-1577
29. R. Sinha, N. Rothman, E. Brown, O. Levander, C.P. Salmon, M.G. Knize and J.S. Felton, High concentrations of the carcinogen 2-amino-1-methyl-6-imidazo[4,5-b]pyridine (PhIP) occur in chicken but are dependent on the cooking method. Cancer Res. 1995, **55**, 4516-4519
30. J.S. Felton, E. Fultz, F.A. Dolbeare and M.G. Knize, Reduction of heterocyclic amine mutagens/carcinogens in fried beef patties by microwave pretreatment. Food Chem. Toxic. 1994, **32**, 897-903

31. R. Viske and P.E. Joner, Mutagenicity, creatine, and nutrient contents of pan fried meat from various animal species. Acta Vet. Scand. 1993, **34**, 1-7
32. Y.Y. Wang, L.L. Vuolo, N.E. Springarn and J.H.Weisburger, Formation of mutagens in cooked foods, V. The mutagen reducing effect of soy protein concentrates and antioxidants during frying of beef. Cancer Lett. 1982, **16**, 179-186
33. K. Skog, M. Jägerstad and A. Laser Reuterswärd, Inhibitory effect of carbohydrates on the formation of mutagens in fried beef patties. Food Chem. Toxic. 1992, **30**, 681-688
34. D.W. Layton, K.T. Bogen, M.G. Knize, F.T. Hatch, V.M. Johnson and J.S. Felton, Cancer risk of heterocyclic amines in cooked foods: An analysis and implications for research. Carcinogenesis. 1995, **16**, 39-52
35. K.W. Turteltaub, R.J. Mauthe, K.H. Dingley, J.S. Vogel, R.C. Franz, R.C. Garner and N.Shen, Mutation Research, 1997, **376**, 243-252

The Role of the Maillard Reaction in the Food Industry

Jerzy A Mlotkiewicz

DFI, FOOD INGREDIENTS DEVELOPMENT CENTRE, STATION ROAD, CAMBRIDGE CB1 2JN, UK

Summary
In the food industry, the role of flavour and colour both desirable and undesirable, is key in the manufacture of products of consistent organoleptic quality. The generation of flavour and colour in a variety of processes, especially those utilizing heat, owes much to the Maillard reaction.

Introduction

Since the beginning of time, through the power of the sun, lightning and eventually man's discovery of how to make and control fire, the means for cooking existed. The use of heating would have improved the eating quality of food in terms of flavour and digestibility and it would also have become apparent fairly quickly that cooked food could be stored for longer than the raw material. Early biblical writings describe the benefits of fire.[1]

Culinary practices have developed over the centuries into art forms. However, in scientific terms, the earliest reported studies of nonenzymatic browning reactions were initiated by Louis Camille Maillard who heated amino acids in solution with high levels of glucose as he investigated the biosynthesis of proteins.[2] In 1951, the first general review of the Maillard reaction was published, linking the reactions between sugars and nitrogenous compounds to food problems.[3] Two years later, a review was published by Hodge[4] which addressed model systems and the mechanisms of reaction. This review became the foundation for modern mechanistic studies and Hodge's scheme is still used widely today to describe the reaction.

Chemistry of the Maillard Reaction

The complexity of the chemical pathways in the Maillard reaction is mirrored in the diversity of research work undertaken and reported in the years following Hodge's publication. Regular reviews and several conference proceedings have been published which describe some elegant model studies.[5-9] The bulk of the studies in the literature have focused on flavour generation with particular attention being given to identification of the chemical classes formed at defined stages of the Maillard reaction. The kinetics of the reaction have received much less attention and knowledge of the chemistry of colour formation is sparse compared with flavour formation.

The Maillard reaction takes place in 3 major stages and is dependent upon factors such as pH, time, temperature, concentration of reactants and reactant type.

Early stage
This involves the condensation of an amino acid with a reducing sugar to form Amadori or Heyns rearrangement products via an N-substituted glycosylamine. It is also considered that the N-glycosylamine can degrade to fission products via free radicals without forming the Amadori or Heyns rearrangement products.[10]

Advanced stage
The degradation of the Amadori and Heyns rearrangement products occurs via four possible routes involving deoxyosones, fission or Strecker degradation.[11-12] A complex series of reactions including dehydration, elimination, cyclization, fission and fragmentation result in a pool of flavour intermediates and flavour compounds Following the degradation pathways as illustrated schematically in Figure 1, key intermediates and flavour chemicals can be identified. One of the most important pathways is the Strecker degradation in which amino acids react with dicarbonyls to generate a wealth of reactive intermediates.

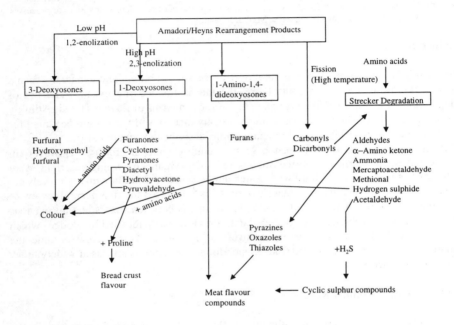

Figure 1. *Pathways of formation of key flavour intermediates and products in the Maillard reaction*

Final Stage
The final stage of the Maillard reaction is characterized by the formation of brown nitrogenous polymers and co-polymers. While the development of colour is an important feature of the reaction, relatively little is known about the chemical nature of the compounds responsible;[13] colour compounds can be grouped into two general classes - low molecular weight colour compounds, which comprise two to four linked rings, and the melanoidins, which have much higher molecular weights.[14-15] Colour development increases with increasing temperature, with time of heating, with increasing pH and by intermediate moisture content (a_w = 0.3-0.7).[16] Generally, browning occurs slowly in dry systems at low temperatures and is relatively slow in high-moisture foods.[17] Colour generation is enhanced at pH > 7.

The Role of the Maillard Reaction in the Food Industry

In exploiting the Maillard reaction, the key target for industry is to understand and harness the reaction pathways enabling improvement of existing products and the development of new products. While it would be easy to assume that this means the generation of flavour and colour, not all Maillard products endow positive characteristics to foods and ingredients. The positive contributions of the Maillard reaction are flavour generation and colour development. The negative aspects are off-flavour development, flavour loss, discoloration and loss of nutritional value. In applying the Maillard reaction, there are challenges that are common to the food industry, independent of the type of product. These challenges can be classified as follows: maintenance of raw material quality; maintenance of controlled processes for food production; maintenance of product quality; extension of product shelf-life.

Processes

Thermal processes utilized in the food industry add value to raw materials and are applied to produce texture, colour, flavour and to sterilize/pasteurize the materials enabling longer shelf-life and improving product safety. A selection of these processes can be examined in terms of the role of the Maillard reaction in overcoming or magnifying the challenges outlined above.

Extrusion

Extrusion is a process widely used in the food industry to prepare dry petfoods, animal feeds, snack foods, ready-to-eat cereals, textured vegetable proteins and heat-processed flours. Such processing conditions should favour the Maillard reaction but major problems include insufficient time to generate flavour and flavour flash-off at the die.[18-19] Post-extrusion coating with flavour blends may create problems due to rapid chew-out (the rapid loss of coated flavour due to mastication). Flavour encapsulation is an alternative approach, but can result in poor flavour release.[20] Recent developments in protein encapsulation appear to provide selective protection for flavours in the extrusion process.

Generation of pyrazines and other toasted notes are suitable for ready-to-eat cereals but not where bland products are required as in the heat processing of flour. With the rise in vegetarianism and the growing culture of reduced meat consumption, the requirement for meat replacement has led to a focus on the extrusion of cereal proteins to produce textured chunks. Consumer demands point to the need not only for meat-like texture but also meat-like flavour. The challenge to the food industry is the generation, *in situ*, of high-quality stable meat flavours in the extrusion process which retain flavour quality during subsequent cooking and chewing. The use of precursor systems, such as hydrolysed vegetable proteins, autolysed yeasts, sugars and amino acids, to generate flavour is common and specific configurations, temperatures, pressures, shear and dwell times are selected to influence the flavours generated. Little work has been published on the generation of colour in the extruder,[21] though it was noted that increasing die temperature and pH and decreasing moisture content resulted in increased colour formation.[22]

Microwave cooking

The use of the microwave oven in the kitchen has been limited because of the lack of flavour and colour development.[19] Flavour, added or generated, is lost due to moisture migration, and the surface conditions of high moisture and low temperature do not support the Maillard reaction.[23] Comparison of flavour generation in conventional and microwave-cooked foods showed both chemical and sensory differences. Studies showed that more carbohydrate decomposition occurred and more off-notes were formed.[24-25] Generation of flavour in the microwave requires the use of quick-acting precursor systems that produce satisfactory flavours within short cooking times.

Colour generation is also a major problem in the design of microwaveable foods but this has been overcome in part by the development of susceptor packaging and reactive coatings.

Baking
The Maillard reaction contributes significantly to the flavour and colour of baked products. Bread, in various forms, is one of the most popular staple foods in the world. Studies have concluded that the crumb fraction contains predominantly fat-degradation products (alcohols, aldehydes and ketones) while the crust contains mostly Maillard reaction products, with furans, pyrroles, pyrrolines and cyclic sulphur compounds predominating.[26] 2-Acetyl pyrroline has been described as a character impact compound of wheat bread crust and a number of heterocyclic compounds have been proposed as being key to bread flavour.[27] The mechanism of formation of these compounds has been investigated and reviewed.[28-29] In the baking process, high temperature and low moisture are critical as well as the careful selection of raw materials, wheat variety being of major importance.

Canning
The sterilization of canned meat products involves very specific temperature and pressure conditions, usually in a continuous batch process. The cooking conditions are not conducive to the formation of roast-type notes, so a different chemistry profile is produced. Investigation of the heating of model systems containing water, glucose and serine or phenylalanine under different conditions ('soup' cooking, roasting and autoclaving) indicated that autoclaving produced similar compounds to roasting.[30] However, in a real meat system, the Maillard reaction does occur but lipid oxidation and sugar degradation are also important reactions, introducing a reaction 'soup' chemically more complex than a model system. Optimization of the ingredients is vital as the dominant pathways are governed by the choice of raw materials.

Retort-processed flavours
The Maillard reaction has been exploited to produce meat-like commercial flavours from a range of raw materials and ingredients, including amino acids, hydrolysed vegetable proteins, carbohydrates, autolysed yeasts and others.[31-33] One of the challenges in using this approach is that aqueous systems generally are retorted for a period of hours to produce flavours. In many cases, flavour of the resultant product is either weak because a sufficiently high temperature has not been achieved or the flavour character produced is not specific enough. To overcome these drawbacks, ingredients enabling higher boiling points can be used and specific character can be added to the system by the use of topnote flavour systems. Processed flavours prepared in a sealed system, under pressure, have a stronger flavour character.

Product Quality/Shelf Life

Control of process parameters is paramount to addressing the issues of product quality and shelf-life. However, storage stability is a key challenge in preventing off-flavour development, discoloration and reduction in nutritional value. Stale off-flavour and colour can develop during storage of whey protein concentrate following spray drying.[34-35] Milk components can undergo degradation during sterilization and spray drying[17] and depletion of lysine occurs due to the ε-amino group being susceptible to glycation resulting in the blocking of lysine. This reaction can happen on storage under mild conditions over a period of time. It can also happen during processing if the temperature is allowed to rise excessively. Prevention of such problems involves careful selection of raw materials, definition and monitoring of process parameters and ensuring that packaging and storage conditions are appropriate.

Industrial Examples

Canned products
A comparison was made between a standard recipe canned product containing low fat, medium protein and high moisture and the same product where the formulation was altered through addition of selected raw materials which would react with the meat components of the recipe to produce flavour. Both were processed under similar parameters of temperature, pressure and time. The headspace volatiles of the altered recipe showed a more complex profile than the standard sample. The volatiles can be identified, the volatile profiles can be manipulated by varying the ingredients, but all the factors responsible for the overall profile are not understood because of the complexity of the recipe.[36]

Ruminant Feed
Dietary polyunsaturated fats undergo hydrogenation to saturated fats by microbial action in the rumen, resulting in loss of desired nutritional benefits. Dietary proteins can be crosslinked using an encapsulating aldehyde such as formaldehyde to generate a protective coating which is stable to the high pH in the rumen. However, the aldehydes in question may harm the digestive tract of the animal or produce undesirable metabolites. Whey protein concentrate and lactose have been used to produce crosslinking through the Maillard reaction resulting in encapsulation of lipid material which can by-pass the rumen; the crosslinked material is hydrolysed in the low pH of the abomasum, releasing the lipid for absorption.[37] The potential benefit is the production of milk and meat having lower saturated fat and increased unsaturated fat concentrations.

Processed wheat flour
During the processing of wheat flour for use in delicately flavoured products, e.g. baby food, the generation of malty flavour and off-colour needs to be avoided. These are caused by the degradation of flour proteins to give amino acids and peptides by the action of native proteases and subsequent reaction of these materials with reducing sugars. To avoid the problem, the temperature of the process must be raised rapidly to inactivate the proteases.

Sugar spot
A major problem in recent years was the enhanced incidence of brown blemishes, called 'sugar spot', during the processing of potatoes in the manufacture of potato crisps (chips). This led to product rejection and waste. The cause was the localized degradation of starch to give high reducing sugar concentrations. Upon frying, the sugars reacted with the proteins in the potato. Studies showed that the problem was exacerbated during storage of the potatoes and therefore a rapid spot test was developed to identify problem potatoes in the field, allowing selection of potatoes suitable for storage. The cause of the problem was found to be related to the presence of a virus in the potato tuber and agronomical measures were taken to reduce the incidence.

Colour generation
Addition of amino acids and sugars as flavour precursors to a food formulation creates a reaction environment in which the precursors are diluted by the bulk of the ingredients. Investigation of enhancement of the reactivity of the precursor species by linking them chemically at a very early stage of the Maillard reaction without progressing down the reaction cascade involved the rapid drying of solutions of a series of amino acid/sugar pairs in a spray drier under low-temperature conditions. However, instead of colourless coupled reactants, a large array of coloured materials was produced dependent on the amino acids and sugars used.

Conclusions

The examples given highlight both the positive and negative aspects of the Maillard reaction in the food industry. The Dalgety petfood and agriculture companies each utilize some six hundred raw materials. In the case of ingredients for human consumption, Dalgety handles more than two and a half thousand raw materials. Processing of ingredients to generate products is key to reducing the effect of variability and process control is critical. To gain control of the process, it is necessary to understand the system. Researchers have spent years studying simple model systems, identifying the reaction pathways and products of reaction. Industrialists manipulate complex formulations and observe the results using analysis and hindsight. There is often a 'suck-it-and-see' approach. The future can best be addressed by a series of questions.
1. How do we design and monitor better model systems that are more closely related to industrial formulations? This can only be achieved by closer cooperation between academe and industry
2. How can Maillard reactants be 'protected' from premature reaction or reaction during long-term storage? The development of flavour-delivery/release systems is extremely important to the food industry and there is a need to look towards researchers involved not only in food research but in other areas of technology.
3. How can reactant systems be developed for use in extrusion and microwave processes?

References

1. Isaiah 44:15-16.
2. L.C. Maillard and M.A. Gautier, Action des acides amines sur les sucres: formation des melanoidines par voie methodique, *Compt. Rend. Acad. Sci.*, 1912, **154**, 66-68.

3. J.P. Danehy and W.W. Pigman, Reactions between sugars and nitrogenous compounds and their relationship to certain food problems, *Adv. Food Res.*, 1951, **3**, 241-290.
4. J.E. Hodge, Chemistry of browning reactions in model systems, *J. Agric. Food. Chem.*, 1953, **1**, 928-943.
5. G.R. Waller and M.S.Feather (eds), 'The Maillard Reaction in Foods and Nutrition', American Chemical Society, Washington DC, 1983.
6. T.H. Parliment, R.J. McGorrin and C-T. Ho, American Chemical Society (eds), 'Thermal Generation of Aromas', Washington DC, 1989.
7. T.H. Parliment, M.J. Morello and R.J.McGorrin (eds), 'Thermally Generated Flavours: Maillard, Microwave and Extrusion Processes' American Chemical Society, Washington DC, 1994.
8. T.P. Labuza, G.A. Reineccius, V.M.Monnier, J. O'Brien and J.W. Baynes (eds), 'Maillard Reactions in Chemistry, Food and Health,' The Royal Society of Chemistry, Cambridge, 1994.
9. A.J. Taylor and D.S. Mottram (eds), 'Flavour Science: Recent Developments', The Royal Society of Chemistry, Cambridge, 1996.
10. M. Namiki, Chemistry of Maillard reactions: recent studies on the browning reaction mechanism and the development of antioxidants and mutagens, *Adv. Food Res.*, 1988, **32**, 115-184.
11. F. Ledl, in, 'The Maillard Reaction in Food Processing, Human Nutrition and Physiology, P.A. Finot, H U. Aeschbacher, R.F. Hurrell and R. Liardon (eds), Birkhauser Verlag, Basel, 1990, 19-42.
12. F. Ledl and E. Schleicher, New aspects of the Maillard reaction in foods and in the human body, *Angew. Chem. Int. Ed. Engl.*, 1990, **29**, 565-594.
13. H.E. Nursten and R. O'Reilly, Coloured compounds formed by the interaction of glycine and xylose, in 'The Maillard Reaction in Foods and Nutrition', G.R. Waller and M.S. Feather (eds), American Chemical Society, Washington DC, 1983, 103-121.
14. H. Kato and H. Tsuchida, Estimation of melanoidin structure by pyrolysis and oxidation, Prog. Food Nutr. Sci., Pergamon Press, Oxford, 1981, **5,**147-156.
15. J. M. Ames, A. Apriyantono and A. Arnoldi, Low molecular weight coloured compounds formed in xylose-lysine model systems, *Food Chem.*, 1993, **46**, 121-127.
16. J.M. Ames, L. Bates and D.B. MacDougall, Colour development in an intermediate moisture Maillard model system, in 'Maillard Reactions in Chemistry, Food and Health', T.P. Labuza, G.A. Reineccius, V.M.Monnier, J. O'Brien and J.W. Baynes (eds), The Royal Society of Chemistry, Cambridge, 1994, 120-125.
17. J.M. O'Brien and P.A.Morrissey, The Maillard reaction in milk products, *Bull.Int.Dairy Federation*, 1989, **238**, 53-61.
18. J.A. Maga and C.H. Kim, Protein-generated extrusion flavours, in 'Thermal Generation of Aromas', T.H. Parliment, R.J. McGorrin and C-T. Ho (eds), American Chemical Society, Washington DC, 1989, 494-503.
19. I. Katz, Maillard, microwave and extrusion cooking: generation of aromas, in 'Thermally Generated Flavours: Maillard, Microwave and Extrusion Processes', T.H. Parliment, M.J. Morello and R.J.McGorrin (eds), American Chemical Society, Washington DC, 1994, 2-6.

20. R. Villota and J.G. Hawkes, Flavouring in extrusion: an overview, in 'Thermally Generated Flavours: Maillard, Microwave and Extrusion Processes', T.H. Parliment, M.J. Morello and R.J.McGorrin (eds), American Chemical Society, Washington DC, 1994, 280-295.
21. A. Noguchi, K. Mosso, C. Aymanrod, J. Jeunink and J.C. Cheftel, Maillard reactions during extrusion cooking of protein-enriched biscuits, *Lebensm. Wiss. u. Technol.*, 1982, **15**, 105-110.
22. S. Sgaramella and J.M. Ames, The development and control of colour in extrusion cooked foods, *Food Chem.*, 1993, **46**, 129-132.
23. T. van Eijk, Flavor and flavorings in microwave foods: an overview, in 'Thermally Generated Flavours: Maillard, Microwave and Extrusion Processes', T.H. Parliment, M.J. Morello and R.J.McGorrin (eds), American Chemical Society, Washington DC, 1994, 395-404.
24. H.C.H. Yeo and T. Shibamoto, Chemical comparison of flavours in microwaved and conventionally heated foods, *Trends Food Sci. Technol.*, 1991, **2**, 329-332.
25. T.H. Parliment, Comparison of thermal and microwave mediated Maillard reactions, in 'Food Flavours, Ingredients and Composition', G. Charalambous (ed), Elsevier Science Publishers, Amsterdam, 1993, 657-662.
26. W. Baltes and C. Song, New aroma compounds in wheat bread, in 'Thermally Generated Flavours: Maillard, Microwave and Extrusion Processes', T.H. Parliment, M.J. Morello and R.J.McGorrin (eds), American Chemical Society, Washington DC, 1994, 192-205.
27. P. Schieberle and W. Grosch, Bread flavour, in 'Thermal Generation of Aromas', T.H. Parliment, R.J. McGorrin and C-T. Ho, American Chemical Society (eds), Washington DC, 1989, 258-267.
28. P. Schieberle, Formation of 2-acetyl-1-pyrroline and other important flavour compounds in wheat bread crust, in 'Thermal Generation of Aromas', T.H. Parliment, R.J. McGorrin and C-T. Ho, American Chemical Society (eds), Washington DC, 1989, 268-275.
29. N.G. De Kimpe, W.S. Dhooge, Y. Shi, M.A. Keppens and M.M. Boelens, On the Hodge mechanism of the formation of the bread flavour component 6-acetyl-1,2,3,4-tetrahydropyridine from proline and sugars, *J.Agric.Food Chem.*, 1994, **42**, 1739-1742.
30. W. Baltes, J. Kunert-Kirchhoff and G. Reese, Model reactions on generation of thermal aroma compounds, in 'Thermal Generation of Aromas', T.H. Parliment, R.J. McGorrin and C-T. Ho, American Chemical Society (eds), Washington DC, 1989, 143-155.
31. G.A.M. van den Ouweland, E.P. Demole and P. Enggist, Process meat flavor development and the Maillard reaction, in 'Thermal Generation of Aromas', T.H. Parliment, R.J. McGorrin and C-T. Ho, American Chemical Society (eds), Washington DC, 1989, 433-441.
32. D.S. Mottram and F.B. Whitfield, Aroma volatiles from meatlike Maillard systems, in 'Thermally Generated Flavours: Maillard, Microwave and Extrusion Processes, T.H. Parliment, M.J. Morello and R.J.McGorrin (eds), American Chemical Society, Washington DC, 1994, 180-191.

33. M. Rubin, Process flavours, *The World of Ingredients*, C&S Publishers, 1996, 28-30.
34. C.V. Morr and E.Y.W. Ha, Off-flavors of whey protein concentrates: a literature review, *Int. Dairy J.*, 1991, **1**, 1-11.
35. J.E. Hodge, F.D. Mills and B.E. Fisher, Compounds of browned flavor derived from sugar-amine reactions, *Cereal Sci. Today*, 1972, **17**, 34-40.
36. Dalgety internal project report, private communication.
37. T. Richardson, Method to produce unsaturated milk fat and meat from ruminant animals, *U S Patent*, 1992, 5,143,737.

Amadoriase Isoenzymes (Fructosyl Amine: Oxygen Oxidoreductase EC 1.5.3) from *Aspergillus fumigatus*

Motoko Takahashi, Monika Pischetsrieder and Vincent M. Monnier

INSTITUTE OF PATHOLOGY, SCHOOL OF MEDICINE, CASE WESTERN RESERVE UNIVERSITY, CLEVELAND, OHIO 44106, USA

Summary
Four 'amadoriase' enzyme fractions that oxidatively degrade glycated low molecular weight amines and amino acids forming hydrogen peroxide and glucosone were isolated from an *Aspergillus fumigatus* soil strain selected using fructosyl adamantanamine as sole carbon source. The enzymes were purified to homogeneity using a combination of ion exchange, hydroxyapatite, gel filtration and Mono Q column chromatography. Molecular masses of amadoriase enzymes Ia, Ib and Ic were 51 kDa; amadoriase II was 49 kDa. FAD was identified in all enzymes based on its typical absorption spectrum. The N-terminal sequence was identical for enzymes Ia and Ib except that the first 5 amino acids were truncated. The sequence of enzyme II was different. These data show the presence of two distinct Amadoriase enzymes in the *Aspergillus fumigatus* soil strain selected on fructosyl adamantanamine and induced by fructosyl propylamine. In contrast to previously described enzymes, these novel Amadoriase enzymes can deglycate both glycated amines and amino acids. To identify primary structure of the enzymes, we isolated a cDNA clone from a cDNA library of *Aspergillus fumigatus* induced with fructosyl propylamine. Northern blotting analysis revealed that amadoriase II was induced by fructosyl propylamine in a dose-dependent manner.

Introduction

It has been suggested that increased levels of Amadori product may be responsible for much of the glucotoxicity underlying the pathogenesis of diabetic complications. Based on this premise we began to search for novel ways to prevent selectively the effects of protein glycation by attempting to deglycate proteins enzymatically,[1] with the ultimate goal of utilizing such enzymes in transgenic models of hyperglycemia. We found and purified a deglycating enzyme in a *Pseudomonas* sp. soil strain selected using glycated ε-aminocaproic acid as sole carbon source.[2,3] However, the enzyme obtained cleaved glycated substrates at the N-alkyl instead of the ketoamine bond. We synthesized a highly sterically inhibited Amadori product using adamantanamine, and utilized this substrate as sole carbon source to select soil organisms. Two molecularly distinct isoenzymes that form glucosone and H_2O_2 from glycated substrate were obtained from an *Aspergillus fumigatus*. Since such products are potentially highly injurious molecules, we hypothesized that such enzymes in diabetes might contribute to the development of complications if present *in vivo*.[4]

Materials and Methods

Materials
α-Glucosone was a gift from Dr. Milton Feather (Department of Biochemistry, University of Missouri). Adamantanamine and *o*-phenylenediamine (OPD) were purchased from Aldrich (Milwaukee, WI). Glucosone triazine was a gift from Dr. Marcus Glomb. Fructosyl adamantanamine, glycated BSA, fructosyl propylamine and its reduced form were

prepared as described.[4] All other materials were of analytical grade.

Microorganism and culture media
An *Aspergillus fumigatus* soil strain was isolated from soil by selection on minimal medium (3 g KH_2PO_4, 7 g K_2HPO_4, 0.1 g $(NH_4)_2SO_4$, 1 mM $MgSO_4$ in 100 mL), containing fructosyl adamantanamine as sole carbon source (5.5 mg/mL). Batch cultures were grown in nutrient broth containing 1 mg/mL fructosyl propylamine. Incubation was carried out under aerobic conditions at 37 °C for 3 days.

Assays for enzymatic activity
Three different assays were developed to measure the activity of the enzyme.[4]
1) *Glucosone formation.* For purification, the enzyme activity was monitored by the release of glucosone measured by a colorimetric reaction with OPD[6] using fructosyl propylamine as a substrate. Synthesized glucosone was used as a standard.
2) *H_2O_2 assay.* Hydrogen peroxide was quantitated by the quinone dye assay according to Sakai *et al.*[7]
3) *Oxygen consumption.* Oxygen consumption was determined using a YSI-Beckman glucometer II equipped with a Clarke-type oxygen electrode as described before.[2]

Purification of the enzyme
Washed mycelia (180 g wet weight) were suspended and homogenized in 900 mL of homogenizing buffer (20 mM sodium phosphate buffer, pH 7.4, 1 mM phenylmethylsulfonyl fluoride, 1 mM EDTA, 2 mM dithiothreitol). The homogenate was centrifuged to remove unbroken cells and cell debris and fractionated between 45% and 70% ammonium sulfate. The fraction was chromatographed on DEAE Sepharose, hydroxyapatite, Sephacryl S-200, and a Mono Q column. Fractions containing enzyme activity were used as the purified preparation of amadoriases.[4] Kinetic constants were calculated from the data by least-squares linear regression analysis.

Cloning and expression of amadoriase II cDNA
Screening of the cDNA library was carried out as follows. After transferring plaques onto nitrocellulose filters, plaques presenting amadoriase II epitopes by isopropyl-β-D-thiogalactopyranoside (IPTG) induction were detected by anti-amadoriase II antiserum. Positive clones were converted to phagemids carrying cDNA inserts between *Eco*RI and *Xho*I sites of pBluescript SK(-) by helper phage superinfection. The plasmid carrying amadoriase II cDNA was transformed into *E. coli* XL1-Blue MRF'. The transformant was cultured in the presence of 10 mM IPTG and the cell extract was subjected to enzyme assays and immunoblot analysis.

Northern blotting
Northern blot analysis was carried out with total RNA (10 µg) from *Aspergillus fumigatus* cultured in the presence of various concentrations of fructosyl propylamine. Random primer ^{32}P-labeled amadoriase II cDNA was used as a probe.

Results

Search for an Amadori-product degrading microorganism and identification of deglycating activity

Soil specimens were screened for a microorganism that can degrade Amadori products by using culture media with glycated substrates as sole carbon source. From the medium enriched with 5.5 mg/mL of fructosyl adamantanamine, a microorganism was isolated and identified as an *Aspergillus fumigatus*. The Amadori product degrading activity was ascertained and the products were identified as glucosone and free amine by TLC and GC/MS.[4] Oxygen consumption during the reaction was detected polarographically using an oxygen electrode and H_2O_2 generation was detected by the quinone dye assay (data not shown). Thus, all the data confirm that the Amadori product is degraded by an oxidative cleavage of the ketoamine bond, in which oxygen is used as an electron acceptor and free amine and glucosone are released (Figure 1).

Purification and characterization of four enzyme fractions

Since glucosone could be easily assayed by the OPD method, it was used for routine assays in the course of purification. Four amadoriase enzyme fractions were isolated from *Aspergillus fumigatus*, which had been selected using fructosyl adamantanamine and grown in the presence of fructosyl propylamine.[4] The apparent molecular masses of amadoriases Ia, Ib and Ic were found to be 51 kDa and that of amadoriase II was 49 kDa. From the results of Sephacryl S-200 gel filtration, the apparent molecular masses of native amadoriases Ia, Ib and Ic were calculated to be 40 kDa and II was 55 kDa. The results indicated that all four enzymes were monomers. The absorption spectrum of amadoriase enzymes indicated the presence of flavine as prosthetic group (data not shown). To identify the proteins, N-terminal sequencing was carried out using a gas-phase protein sequencer. On close inspection, it turns out that all the enzymes have the consensus sequence for the ADP-binding $\beta\alpha\beta$-fold common to all FAD and NAD enzymes.[8] The sequences of all four enzymes are shown in Table 1.

Substrate specificity and kinetic constants of the purified amadoriases Ia, Ib, Ic and II

Substrate specificity was examined for each purified enzyme (Table 2). All four enzymes were active toward N^ε-fructosyl N^α-*t*-Boc-lysine as well as fructosyl propylamine. However, the relative activity of amadoriases Ia, Ib and Ic toward fructosyl adamantanamine was low compared with that for fructosyl propylamine. They had no activity against N^α-fructosyl N^ε-acetyl lysine. On the contrary, amadoriase II was active toward fructosyl adamantanamine *and* N^α-fructosyl N^ε-acetyl lysine. None of them were active toward reduced fructosyl propylamine (10 mM), glycated BSA (2 mg/mL), glycated poly-L-lysine (0.02%) or glucose (10 mM). Apparent kinetic constants for N^ε-fructosyl N^α-*t*-Boc-lysine and fructosyl adamantanamine are shown in Table 3. The stoichiometry of the reaction catalysed by the enzymes was determined using fructosyl propylamine as substrate.

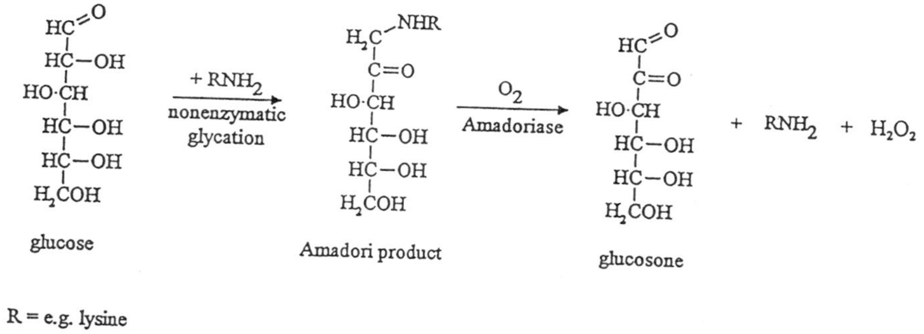

Figure 1. Proposed reaction mechanism of amadoriase

R = e.g. lysine

Table 1. N-terminal sequence of amadoriases

Protein	Sequences	
Ia: 1	A P S I L S T E S S I(C/T) V I G A G T W G	20
Ib: 1	A P S I L S T E S S I I V I G A G T W G	20
Ic: 1	S T E S S I I V I G A G T W G(C) (S) T A L	20
II: 1	A V T K S S S L L I V G A G T W G T S T	20

Table 2. Substrate specificity of amadoriases

Substrates (10 mM)	Specific Activity (U/mg)			
	Amadoriase Ia	Ib	Ic	II
Fructosyl propylamine	4.7 (100%[a])	4.7 (100)	4.3 (100)	2.5 (100)
Fructosyl adamantanamine	0.49 (10)	0.49 (10)	0.49 (11)	3.0 (121)
N^ϵ-fructosyl N^α-t-Boc-lysine	4.3 (92)	4.4 (94)	4.5 (104)	4.8 (193)
N^α-fructosyl N^ϵ-acetyl-lysine	N.D.[b]	0.16 (3)	0.12 (3)	3.7 (151)

[a] The activity against fructosyl propylamine was taken as 100% for each enzyme.
[b] N.D., not detected.

Table 3. *Kinetic constants of amadoriases*

	N^ε-fructosyl N^α-t-Boc-lysine			Fructosyl adamantanamine		
	K_m (mM)	k_{cat} (min^{-1})	k_{cat}/K_m (M^{-1} min^{-1})	K_m (mM)	k_{cat} (min^{-1})	k_{cat}/K_m (M^{-1} min^{-1})
Ia	3.0 ± 0.3	250 ± 13	0.8 × 10^5	14.4 ± 0.3	48 ± 1.3	3.3 × 10^3
Ib	3.1 ± 0.5	320 ± 32	1.0 × 10^5	14.7 ± 0.9	52 ± 1.7	3.5 × 10^3
Ic	3.3 ± 0.2	320 ± 12	1.0 × 10^5	14.7 ± 0.9	52 ± 1.7	3.5 × 10^3
II	1.6 ± 0.1	330 ± 5.8	2.0 × 10^5	3.4 ± 0.4	135 ± 6.5	4.0 × 10^4

Values of the parameters are means ± SD (n = 5).

Cloning and expression of cDNA of amadoriase II
Library constructed in Uni-ZAP XR was screened with antiserum against amadoriase II of *Aspergillus fumigatus*. By screening 5 × 10^4 plaques, we obtained six positive clones. Sequence analysis of one of these clones was as described previously.[5] The amadoriase II coding sequence was considered to have 1314 bp coding for a novel protein of 438 amino acids (nucleotide positions 174-1487). To confirm the identification of the cDNA, pBluescript II containing amadoriase II cDNA was introduced into *E. coli* XL1-Blue MRF' and the enzyme was expressed as a β-galactosidase fusion protein. The transformant exhibited significant amadoriase activity, and immunoblot analysis indicated the amadoriase II protein was present in the transformant.[5] The results confirmed that the cloned cDNA encodes 49 kDa amadoriase II.

Northern blotting
Using amadoriase II cDNA as a hybridization probe, Northern blotting analysis was performed for an *Aspergillus fumigatus* which was cultured in the presence of various concentrations of fructosyl propylamine. Amadoriase II was induced in a dose-dependent manner.[5]

Discussion

We have isolated and purified four amadoriase enzyme fractions from *Aspergillus fumigatus*. One of the four enzymes described above (amadoriase II) was different form the other three based on its N-terminal amino acid sequence and enzymatic properties. Amadoriase Ia and Ib appear identical and amadoriase Ic have Michaelis-Menten properties and substrate specificities similar to amadoriase Ia and Ib. The N-terminal amino acid sequence of enzyme Ic is similar to that of enzymes Ia and Ib except for the truncation of the first five amino acids which may have resulted from partial proteolytic breakdown during isolation. Thus, taken together, the data suggest the presence of two distinct amadoriase enzymes in the isolated Aspergillus strain. These amadoriase enzymes cleave glycated amino acids and glycated alkylamine under consumption of oxygen and release of H_2O_2 and glucosone in a reaction involving FAD as a cofactor. However, they do not deglycate larger molecules such as glycated BSA and glycated poly-L-lysine. It is considered to be due to steric inhibition of the substrates.

The novel amadoriase enzymes described in this paper join the growing family of fructosyl amino acid oxidase enzymes that can deglycate low molecular weight substrates leading to regeneration of the free amine while producing H_2O_2 and glucosone.[7, 9-12] The first fully characterized enzyme was described by Horiuchi et al.[9] in Corynebacterium sp., followed by a similar enzyme from Aspergillus sp. by the same group.[10] Both enzymes have activity against many glycated amino acids, but, in contrast to the enzymes described in this work, have no activity against glycated alkyl amines. GENZYME scientists[11] described the presence of deglycating enzymes in Klebsiella and Corynebacterium, the fungi Acremonium and Fusarium, and the yeast genus Debaryomyces. Molecular weights of these enzymes are undetermined. Sakai et al.[7] isolated a fructosyl lysine oxidase (FLOD) from the fungus Fusarium oxysporum using N^ε-fructosyl N^α-Z-lysine as sole nitrogen source. The monomeric enzyme has, like our enzymes a molecular mass of 50kDa. In contrast to the two other enzymes of Horiuchi, FOLD had covalently bound FAD. While this work was in progress, Yoshida et al.[12] described a fructosyl amino acid oxidase from Aspergillus terreus and Penicillium janthinellum that exhibits high homology with amadoriase II. Their Km values for ε-fructosyl lysine are 10 times smaller than that of amadoriase II.

In spite of the failure to induce soil organisms to produce enzymes with activity against glycated proteins, the discovery of inducible amadoriase isoenzymes in our fungus, together with the reports above describing the presence of amadoriase enzymes in several genetically unrelated organisms, such as prokaryotes, fungi and yeast, raises the important question of the evolutionary significance of deglycating enzymes. Except for the Pseudomonas enzyme described earlier,[2, 3] all amadoriase enzymes that regenerate the intact amine also generate the highly reactive molecules glucosone and H_2O_2. Since glucosone itself is basically toxic to these Aspergillus, Aspergillus overexpression amadoriase might be coexpression other enzyme such as 'aldehyde reductase'. The Aspergillus is able to survive by utilizing fructose as a carbon source which is produced from glucosone by the action of aldehyde reductase.

In view of the growing data that directly implicate H_2O_2, oxidative stress and Maillard reaction intermediates in diabetes,[13, 14] the question thus arises as to whether amadoriase enzymes occur in human tissues. The existence of an enzyme such as amadoriase suggests a possible role of flavin as an electron acceptor in oxidation of Amadori products, by a mechanism similar to metal-catalyzed oxidation of Amadori products that also generate glucosone and H_2O_2 in vivo. Although the BLAST search detected no significant homology with between amadoriase and any human protein in the database, the possible occurrence of amadoriase homologues in human tissue could have considerable implications for the development of diabetic complications.

References

1. C. Gerhardinger, S. Taneda, M. S. Marion and V. M. Monnier, Isolation, purification, and characterization of an Amadori product binding protein from a *Psudomonas* sp. soil strain, *J. Biol. Chem.*, 1994, **269**, 27297-27302.
2. C. Gerhardinger, M. S. Marion, A. Rovner, M. Glomb and V. M. Monnier, Novel degradation pathway of glycated amino acids into free fructosamine by a *Pseudomonas* sp. soil strain extract, *J. Biol. Chem.*, 1995, **270**, 218-224.
3. A. K. Saxena, P. Saxena and V. M. Monnier, Purification and characterization of a membrane-bound deglycating enzyme (1-deoxyfructosyl alkyl amino acid oxidase, EC 1.5.3) from a *Pseudomonas* sp. soil strain, *J. Biol. Chem.*, 1996, **271**, 32803-32809.
4. M. Takahashi, M. Pischetsrieder and V. M. Monnier, Isolation, purification, and characterization of amadoriase isoenzymes (fructosyl amine-oxygen oxidoreductase EC 1.5.3) from *Aspergillus* sp., *J. Biol. Chem.*, 1997, **272**, 3237-3443.
5. M. Takahashi, M. Pischetsrieder and V. M. Monnier, Molecular cloning and expression of amdoriase isoenzyme (fructosyl amine:oxygen oxidoreductase, EC 1.5.3) from *Aspergillus fumigatus*, *J. Biol. Chem.*, 1997, **272**, 12505-12507.
6. S. Kawakishi, R. Z. Cheng, S. Saito and K. Uchida, Biomimic Oxidation of glycated protein and Amadori product, in 'Maillard Reactions in Chemistry, Food, and Health' (T. P. Labuza, G. A. Reineccius, V. M. Monnier, J. O'Brien and J. W. Baynes, eds.), Royal Society of Chemistry, Cambridge, 1994, p. 281-285.
7. Y. Sakai, N. Yoshida, A. Isogai, Y. Tani and N. Kato, Purification and properties of fructosyl lysine oxidase from *Fusarium oxysporum* S-1F4, *Biosci. Biotech. Biochem.*, 1995, **59**, 487-491.
8. R. K. Wierenga, P. Terpstra and W. G. J. Hol, Prediction of the occurrence of the ADP-binding $\beta\alpha\beta$-fold in proteins, using an amino acid sequence fingerprint, *J. Mol. Biol.*, 1986, **187**, 101-107.
9. T. Horiuchi, T. Kurokawa and N. Saito, Purification and properties of fructosyl-amino acid oxidase from *Corynebacterium* sp. 2-4-1, *Agric. Biol. Chem.*, 1989, **53**, 103-110.
10. T. Horiuchi and T. Kurokawa, Purification and properties of fructosylamine oxidase from *Aspergillus* sp. 1005, *Agric. Biol. Chem.*, 1991, **55**, 333-338.
11. GENZYME LIMITED, European patent # 0526150A1, appl # 92306844.9, 1993.
12. N. Yoshida, Y. Sakai, A. Isogai, H. Fukuya, M. Yagi, Y. Tani and N. Kato, Primary structures of fungal fructosyl amino acid oxidases and their application to the measurement of glycated proteins, *Eur. J. Biochem.*, 1996, **242**, 499- 505.
13. J. W. Baynes, Role of oxidative stress in development of complications in diabetes, *Diabetes*, 1991, **40**, 405-412.
14. A. H. Elgawish, M. Glomb, M. Friedlander and V. M. Monnier, Involvement of hydrogen peroxide in collagen cross-linking by high glucose *in vitro* and *in vivo*, *J. Biol. Chem.*, 1996, **271**, 12961-12971.

Reaction Mechanisms

Correlations between Structure and Reactivity of Amadori Compounds: the Reactivity of Acyclic Forms

Milton S. Feather and Valeri V. Mossine

DEPARTMENT OF BIOCHEMISTRY, UNIVERSITY OF MISSOURI, COLUMBIA, MO 65211, USA

Summary
Several Amadori compounds were prepared and solution (NMR) and, in some cases, solid-state structures (X-ray) were determined. Several of the compounds studied contain sugar units which exist predominantly in acyclic forms. Such compounds were used as model Amadori compounds and rates of reactivity with respect to $C(1)H_2$ proton exchange, the ability to generate oxygen free radicals, and the type of dicarbonyl intermediates produced were evaluated. The results of the experiments enable direct, although qualitative, correlations to be made between the population of acyclic forms in solutions of Amadori compounds, the rates of their reversible enolization, and their reactivity in redox and degradation reactions.

Introduction

The available data on the solution properties and structures of Amadori compounds (ACs) is not extensive, but suggests that the carbohydrate portion is a major structural determinant. For the relatively small number of ACs, largely derived from fructose, that have been examined in solution, the major tautomeric form appears to be the β-pyranose, with lesser amounts of the α- and β-furanose forms present. In addition, only small amounts of the α-pyranose are present and less than 2% of the total population is present as the acyclic form.[1,2] The structure of the sugar residue is thought to be important in determining reactivity, including susceptibility to degradative oxidation and dehydration. It is also generally thought that acyclic forms of sugars, which are thermodynamically unstable, represent the principal reactive species in oxidation and dehydration reactions associated with the Maillard reaction. This rationale has been accepted over the years,[3,4] even though direct experimental data are not available to support this assumption, primarily because acyclic forms of sugars, or derivatives thereof, are not available for study (nearly all sugars exist in solution as stable cyclic forms). In the past, we have been interested in the detection of dicarbonyl intermediates that are produced when ACs undergo degradation in solution.[5,6]

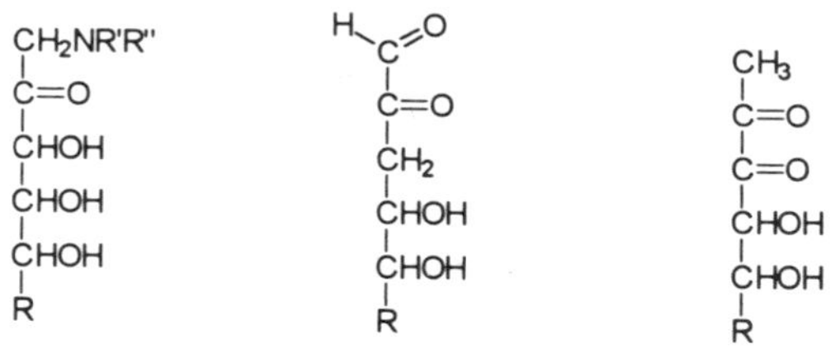

Figure 1. *Dicarbonyl compounds produced from Amadori compounds*

Based on the structures of end-products, there appear to be several intermediates produced that correspond to sugar, minus one molecule of water, and which are shown in Figure 1. The production of such intermediates is thought to proceed via 1,2- (for *1*) and 2,3- (for *2*) enolization of acyclic forms of Amadori compounds.

Acyclic forms of Amadori compounds

Using two glycine-derived ACs ['fructoseglycine' (*Fru-Gly*) and 'difructoseglycine' (*(Fru)$_2$-Gly*)], we were able to demonstrate that both gave rise to both intermediates *1* and *2*. However, *(Fru)$_2$-Gly* gave much higher yields of *1*, and the ratio of *1:2*, produced from the two ACs were quite different. The data collected clearly showed that *(Fru)$_2$-Gly* (the N,N-disubstituted AC), is considerably more reactive than the monosubstituted material. We have now prepared and examined the solution properties of these compounds in some detail by NMR spectroscopy and, in addition, have crystallized both of these compounds and determined their crystal structures by X-ray diffraction. The solid-state structure of 'fructoseglycine' (not shown) shows that it exists as the β-pyranose form.[7] In solution at 20°C, however, the molecule exists in the following tautomeric forms: ~ 70% as the β-pyranose form, ~ 5% as the α-pyranose form, with the remainder as a mixture of furanose forms.[2] Approximately 0.8% is in the acyclic form. The X-ray structure of 'difructoseglycine'[8] is shown in Figure 2. The data clearly show that the two sugar residues of this molecule exist in two different forms, one

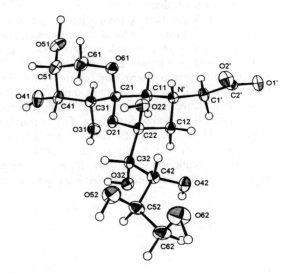

Figure 2. *ORTEP drawing of the crystal structure of difructoseglycine* (Reproduced with permission from Harwood Academic Publishers, GmbH)

in the β-pyranose form and the other in an acyclic form. One possible reason for such an unusual and unexpected structure is that the expected bis-β-pyranose form is less stable than this *spiro*-bicyclic hemiketal tautomer, which is the structure found. The latter may be stabilized by an additional anomeric effect, which is known to influence sugar conformations. NMR data for *(Fru)$_2$-Gly* show[8] that, in solution, one of the sugar residues of 'difructoseglycine' is also in the acyclic form, thus indicating that the solution structure parallels

Reaction Mechanisms

the solid-state structure of the material. This unusual structure may explain the high rate of reactivity of disubstituted compared with monosubstituted molecules, but the implications of this, in terms of the Maillard reaction remain to be elucidated.

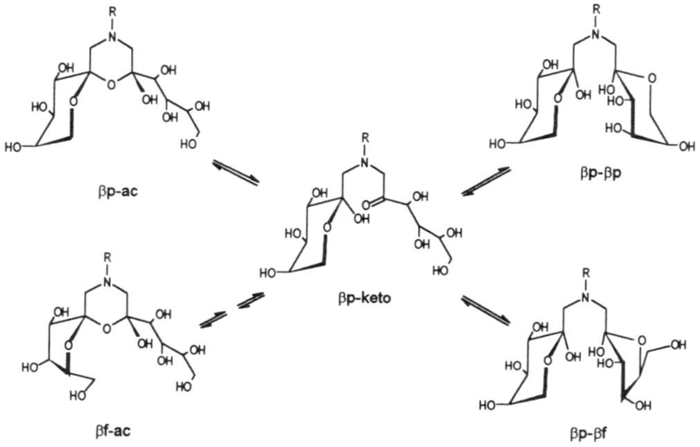

Figure 3. *Major tautomeric forms of N,N-difructoseamino acids in aqueous solutions*

As a result of these findings, we also prepared and purified some Amadori compounds derived from lysine and examined these materials in a similar manner. For systems such as these, the tautomeric equilibrium is very complex, with numerous combinations of isomers possible. The major forms are shown in Figure 3. Based on NMR data, the tautomeric forms found are shown in Table 1. It is quite clear that the presence of two sugar residues on one nitrogen atom ('N,N-difructoseamino acids') leads to the formation of large amounts of acyclic forms both in solution and in the solid state, as is shown in Table 1, and that such compounds would be expected to have an unusually high reactivity. In such N,N'-difructoseamino acid such as N^α,N^ϵ-$(Fru)_2$-Lys, the fructosyl residues tautomerize totally independently of one another.

Pentose-derived ACs were also prepared, along with some derived from hexoses, other than glucose, as well as from some disaccharides. The tautomeric composition of these, as determined by NMR are shown in Table 2. It is noteworthy that, for pentose-derived ACs,

Table 1. *Tautomeric composition of difructoseamino acids (0.3M) in D_2O solutions at (25 °C)*

Amadori compound	Tautomeric forms (% of total difructoseamine population)								
	βp-ac	βf-ac	βp-βp	βp-βf	βp-αf	βp-αp	βf-βf	βf-αf	others
$(Fru)_2$-Gly	54	3.5	17	10	7.5	2	1.5	1.5	3
N^ϵ,N^ϵ-$(Fru)_2$-N^α-Fm-Lys	84	4.6	4.3	3	2.2	-	-	-	2
	Tautomeric forms (% of total fructose population)					
N^α,N^ϵ-$(Fru)_2$-Lys	α-pyr	β-pyr	α-fur	β-fur	keto	hydr
	6	68	12	11	1.5	1.5

appreciable amounts of acyclic forms are also present in solution; keto forms significantly exceed hydrated acyclic pentulose anomers. 'Xyluloseglycine' was also crystallized and its X-ray structure obtained, which clearly shows that this material, in the solid state, exists in the acyclic *keto* form. In general, the anomeric equilibrium patterns for ketoseamino acids are essentially the same as for the parent ketoses.

Table 2. *Solution composition of some Amadori compounds and the parent ketoses[9] (in parentheses) in D_2O, as measured by NMR spectroscopy* (T = 20 - 30°)

Amadori Compound	(Parent sugar)	Anomeric form present (% of total population)					
		α-pyr	β-pyr	α-fur	β-fur	keto	hydr
Rbu-Gly	(Ribulose)	-	-	48 (63)	28 (20)	23 (17)	0.9 (-)
Xlu-Gly	(Xylulose)	-	-	32 (18)	40 (62)	27 (20)	1.1 (-)
Fru-Gly	(Fructose)	5 (2)	69 (70)	13 (6)	12 (25)	0.8 (0.7)	-
Tag-Val	(Tagatose)	67 (71)	21(18)	9 (7.5)	3 (2.5)	- (0.3)	-
Lct-Gly	(Lactulose)	2 (-)	66 (62)	17 (7.5)	15 (29)	- (1.6)	-
Mlt-Gly	(Maltulose)	2 (-)	64 (64)	18 (12)	16 (23)	- (1.5)	-
Clb-Gly	(Cellobiulose)	2 (-)	67 (61)	16 (10)	15 (29)	-	-

Reactivity of Amadori compounds

We have recently[6] established the relative reactivities of some ACs towards non-oxidative degradation which resulted in release of osones *1* and *2*. Our most recent experiments have involved the use of aminoguanidine, which rapidly reacts with dicarbonyl compounds as they are produced in solution and converts them into relatively stable triazine derivatives.[10,11] This technique allows the accumulation of dicarbonyl compounds as the triazine derivative during the course of a reaction and, enables an indirect assessment of the rate of production of the intermediate. The order of the AC degradation rates in 50 mM phosphate buffer (pH 7.0) at 37°C was *(Fru)$_2$-Gly* >> *Fru-Gly* > *N$^\varepsilon$-Fru-N$^\alpha$-Fm-Lys*; over a period of 2 weeks, the ratio of yields for trapped *1* was approximately 6 : 1 : 0.6, respectively.

The same order of reactivity was observed when the relative rates of superoxide anion formation by the same compounds were evaluated under aerobic conditions in the superoxide dismutase - cytochrome *c* assay (Table 3). In these experiments,[12] the catalytic influence of transition metal ions was minimized by using chelex-treated reagent solutions, in addition to 1 mM of the chelator DETAPAC present in the reaction mixtures. We have found that the disubstituted compounds serve to generate superoxide much more rapidly than the monosubstituted compounds.

In order to evaluate the relative rates of 1,2-enolization of fructose derived ACs, which differ in the amounts of acyclic forms available in aqueous solution, we measured rates of proton exchange at fructosyl carbon-1 in deuterium oxide solution by NMR. The results are shown in Table 3. Under such conditions, no measurable proton exchange at the carbon-3, which would be caused by 2,3-enolization, was detected. The relative rates of the C-1 proton exchange were found to be much higher for N,N-difructoseamino acids as well. Our preliminary experiments on ACs also confirm the catalytic role that phosphate and other nucleophiles play in reactions of reversible enolization.

The results of the experiments described above enables the establishment of direct, although qualitative, correlations between the population of acyclic forms in solutions of Amadori compounds, the rates of their enolization, and their reactivity in redox and degradation reactions.

Table 3. *Relative rates of superoxide radical generation by Amadori compounds in 50 mM P_i, pH 7, T = 37 °C; and 1,2-enolization rates of ACs in unbuffered D_2O at pD= 4 - 5, T = 37 °C*

Amadori compound	Relative rate of superoxide formation*	Rel. rate of H↔D exch.*	H↔D (T1/2)
Fru-Gly	1	1	30 h
N^ϵ-Fru-N^α-Fm-Lys	0.7	0.25	5 d
N^α,N^ϵ-(Fru)$_2$-Lys	2	0.8	1.5 d
(Fru)$_2$-Gly	7.7	7.4	4 h
N^ϵ,N^ϵ-(Fru)$_2$-N^α-Fm-Lys	15	~70	25 min
Fru-Gly / 50 mM P_i, pH 7		15	2 h

* Rates were calculated relative to the value for Fru-Gly (= 1)

Discussion

It is interesting to note that, for monosubstituted ACs, the solution compositions do not significantly differ from those of the parent sugars, i.e. fructoseglycine exists in roughly the same ratios of anomers and ring forms as D-fructose. This appears to hold for other sugars as well. It is the disubstituted compounds that have an unexpected solution structure. The data collected herein clearly show that the acyclic form of an AC constitutes a highly reactive form of the molecule, and confirms earlier views to this effect. Such structures, if they exist in any significant amount in solution, will have a great effect on the reactivity of AC. Based on the data presently available, a 'disubstituted' AC, such as difructoseglycine or N^ϵ,N^ϵ-difructoselysine, as well as AC derived from pentose sugars all can be expected to contain significant amounts of acyclic forms in solution, which explains the high reactivity of these materials. These conclusions are supported by the data on reversible enolization and superoxide generation. The food-related and physiological implications of these findings are more difficult to assess. Clearly, if disubstituted AC are produced during food processing and *in vivo*, this will be cause for us to view the Maillard reaction differently. For example, hypotheses as to how protein crosslinks are formed are often postulated on the assumption that a monosubstituted AC is the reactive component. The fact that disubstituted AC have, to our knowledge, not been reported as reactants/intermediates, suggests that they may not be present in large amounts (if at all), in either food products nor as *in vivo* reactants. The fact that they have not been reported, however, does not mean that they are not present. It is also noteworthy that, in most cases, protein crosslinks, when isolated, represent an exceedingly small amount of the total mass of the preparation. It is, therefore, conceivable that disubstituted AC could play a role in such reactions.

References

1. H. Röper, S. Röper, K. Heyns, and B. Meyer, N.m.r. spectroscopy of N-(1-deoxy-D-fructos-1-yl)-L-amino acids ("fructose-amino acids"), *Carbohydr. Res.*, 1983, **116**, 183-195.
2. V.V. Mossine, G.V. Glinsky, and M.S. Feather, The preparation and characterization of some Amadori compounds (1-amino-1-deoxy-D-fructose derivatives) derived from a series of aliphatic ω-amino acids. *Carbohydr. Res.*, 1994, **262**, 257 - 270.
3. E.F.L.J. Anet, 3-Deoxyglycosuloses (3-deoxyglycosones) and the degradation of carbohydrates. *Adv. Carbohydr. Chem.*, 1964, **19**, 181-218.
4. T.M. Reynolds, Chemistry of nonenzymic browning. 1. The reaction between aldoses and amines, *Adv. Food Res.*, 1963, **12**, 1-52.
5. K.J. Knecht, M.S. Feather and J.W. Baynes, Detection of 3-deoxyfructose and 3-deoxyglucosone in human urine and plasma: evidence for intermediate stages of the Maillard reaction in vivo, *Arch. Biochem. Biophys.*, 1992, **294**, 130-137.
6. J. Hirsch, V.V. Mossine and M.S. Feather, The detection of some dicarbonyl intermediates arising from the degradation of Amadori compounds (The Maillard reaction), *Carbohydr. Res.*, 1995, **273**, 171 - 177.
7. V.V. Mossine, G.V. Glinsky, C.L. Barnes and M.S. Feather, Crystal structure of an Amadori compound, N-(1-deoxy-β-D-fructopyranosyl-1-yl)-glycine ("D-fructose-glycine"), *Carbohydr. Res.*, 1995, **266**, 5 - 14.
8. V.V. Mossine, C.L. Barnes, G.V. Glinsky and M.S. Feather, Interaction between two C(1)-X-C'(1) branched ketoses: observation of an unprecedented crystalline *spiro*-bicyclic hemiketal tautomer of N,N-di(1-deoxy-D-fructos-1-yl)-glycine ("difructose glycine") having open chain carbohydrate, *Carbohydr. Lett.*, 1995, **1**, 355-362.
9. S.J. Angyal, The composition of reducing sugars in solution, *Adv. Carbohydr. Chem. Biochem.*, 1984, **42**, 15-68.
10. J. Hirsch, E. Petrakova and M.S. Feather, The reaction of some 1-deoxy-2,3-dicarbonyl hexose derivatives with aminoguanidine (guanylhydrazine), *J. Carbohydr. Chem.*, 1995, **14**, 1179 - 1186.
11. J. Hirsch, C.L. Barnes and M.S. Feather, X-Ray structures of a 3-amino-5- and a 3-amino-6-substituted triazine, produced as a result of a reaction of 3-deoxy-D-erythro-hexos-2-ulose (3-deoxyglucosone) with aminoguanidine, *J. Carbohydr. Chem.*, 1992, **11**, 891-901.
12. V.V. Mossine, M. Linetsky, G.V. Glinsky, B.J. Ortwerth and M.S. Feather, Manuscript submitted for publication in *Chem. Res. Toxicol.*

Study on the Formation and Decomposition of the Amadori Compound N-(1-Deoxy-D-Fructos-1-yl)-Glycine in Maillard Model Systems

Imre Blank, Stéphanie Devaud and Laurent B. Fay

NESTEC LTD, NESTLÉ RESEARCH CENTRE, VERS-CHEZ-LES-BLANC, PO BOX 44, 1000 LAUSANNE 26, SWITZERLAND

Summary
The formation and degradation of N-(1-deoxy-D-fructos-1-yl)-glycine (DFG) were studied in unbuffered aqueous solutions at 90°C over a period of 2 hours. The pH was kept constant at 6 or 7 during the reaction. Both the formation and decomposition of DFG was faster at pH 7 than at pH 6. Yields of DFG were 13 mol% after a reaction period of 2 h. Almost 60 % of DFG was degraded at pH 7 after 1 h, whereas 55 % was still unreacted at pH 6 after 2 h of heating. 4-Hydroxy-2,5-dimethyl-3(2H)-furanone (furaneol) was the most flavour-active compound generated. After 2 h of reaction, DFG yielded 110 and 190 mg/mol furaneol at pH 6 and 7, respectively. The reaction of glucose and glycine resulted in 2 and 27 mg furaneol per mol glucose at pH 6 and 7, respectively. These results indicate that furaneol is preferably generated from DFG under conditions favouring 2,3-enolization.

Introduction

Amadori compounds are N-(1-deoxyketos-1-yl)-amine derivatives formed during the Maillard reaction.[1] N-(1-Deoxy-D-fructos-1-yl)-glycine (DFG) is formed from D-glucose and glycine in the initial stage of the Maillard reaction by Amadori rearrangement of the corresponding N-substituted glycosylamine.[2,3] Amadori compounds are key intermediates in the Maillard reaction and are responsible for changes in colour, flavour, and nutritive value of foods.[4] DFG has been found in foods such as malt,[5] soy sauce,[6] dried vegetables,[7] roasted cocoa[8] and roasted meat.[9]

Amadori compounds decompose via the enediol or aminoenol intermediates and form 1-deoxy-2,3-hexodiuloses and 3-deoxy-2-hexosuloses (3-deoxyosones) by 2,3- and 1,2-enolization, respectively.[10] These reactive α-dicarbonyls show a tendency to further enolization, dehydration, and carbon-carbon fission reactions and can form polymeric colorants and volatile compounds. 4-Hydroxy-2,5-dimethyl-3(2H)-furanone (furaneol) is a well-known flavour-active degradation product of hexoses.[11]

The aim of this work was to study the formation and decomposition of DFG at pH 6 and 7 in model systems. Isotope dilution FAB-MS/MS was applied as a quantification method using ^{13}C-labelled DFG as internal standard.[12] The formation of furaneol was monitored by isotope dilution GC-MS using $^{13}C_2$-furaneol as internal standard.[13]

Materials and Methods

Glycine and 4-hydroxy-2,5-dimethyl-3(2H)-furanone (furaneol) were obtained from Fluka (Buchs, Switzerland); D-(+)-glucose from Merck (Darmstadt, Germany); and [1-^{13}C]glycine from Tracer Tech. Inc. (Sommerville, Massachusetts, USA).

DFG and ^{13}C-DFG (i.e., N-(1-deoxy-D-fructos-1-yl)-[1-^{13}C]glycine), were synthesized from D-glucose and glycine or [1-^{13}C]glycine as reported by Staempfli *et al.*[12] 4-Hydroxy-

2(or 5)-[^{13}C]methyl-5(or 2)-methyl-3(2H)-[2(or 5)-^{13}C]furanone (^{13}C$_2$-furaneol) was prepared as recently described by Blank and co-workers.[13]

Sample preparation[12]
Aqueous solutions of DFG or equimolar amounts of glucose and glycine were heated at 90°C for 15, 30, 60 or 120 minutes. Experiments were performed in duplicate at constant pH (6 or 7) using a pH-stat device (Metrohm, Herisau, Switzerland).

Quantification of DFG[12]
Isotope dilution FAB-MS/MS was performed on a TSQ 700 triple quadrupole mass spectrometer (Finnigan, Bremen, Germany). Aliquots of the sample and the internal standard (^{13}C-DFG) were mixed directly on the probe tip with a thioglycerol/glycerol (1+1, v/v) matrix. Measurements were performed in triplicate.

Quantification of furaneol[13]
The internal standard (^{13}C$_2$-furaneol) was added to the reaction mixture, the pH adjusted to 4.0, and the solution saturated with NaCl. Neutral compounds were extracted with diethyl ether (2 x 20 mL). The organic phase was dried over Na$_2$SO$_4$ at +4°C and concentrated to 0.5-1 mL. Measurements were performed in triplicate.

The MAT 8430 mass spectrometer (Finnigan) was connected to an HP-5890 gas chromatograph (Hewlett-Packard). The interface was kept at 220°C. The ion source, working in EI mode at 70 eV, was held at ~ 180°C. The samples were injected via a cold on-column injector onto a DB-FFAP capillary column (J&W Scientific, Folsom, CA, USA) using the chromatographic conditions previously described.[13] Calibration curves were established with standard mixtures containing defined amounts of labelled and unlabelled furaneol in different ratios by measuring the ions at m/z 130 and m/z 128, respectively.

Results and Discussion

Degradation of DFG
Continuous degradation of DFG was observed at both pHs (Figure 1). DFG decomposed at a moderate rate at pH 6, but more than 50 % was still unreacted after 2 h. The rate of decomposition was considerably increased at pH 7: more than 50 % was already degraded after 1 h.

The lower decomposition rate of DFG at pH 6 compared with that at pH 7 can be explained by the type of enolization favoured. Decomposition at pH 6 proceeds via the reversible 1,2-enolization, forming the relatively stable 3-deoxyglucosone. At pH 7, however, DFG mainly undergoes 2,3-enolization which, according to Simon and Heubach,[14] is an irreversible reaction, leading to 1-deoxy-2,3-hexodiulose.

The pH of the reaction medium is an important parameter controlling the enolization pathway.[10] In general, 1,2-enolization is promoted under acidic conditions by protonation of the nitrogen atom. Neutral and alkaline conditions lead to partially unprotonated Amadori compounds favouring 2,3-enolization. As reported by Ledl and co-workers,[15] 20-times more 1-deoxy-2,3-hexodiulose than 3-deoxyhexosone was produced at pH 7. By contrast, 40-times more 3-deoxyglucosone was found at pH 5 than at pH 7. Furthermore, a more rapid fragmentation of these reactive α-dicarbonyls was observed with increasing pH.

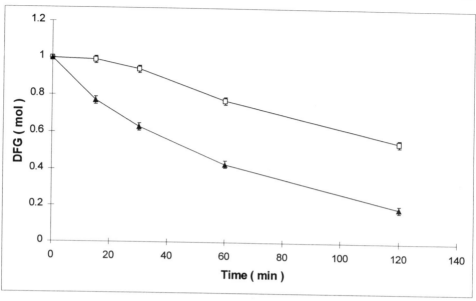

Figure 1. *Thermal degradation of N-(1-deoxy-D-fructos-1-yl)-glycine (DFG) at 90°C and pH 6 (□) or pH 7 (▲)*

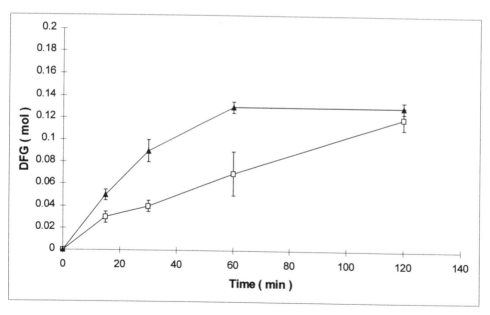

Figure 2. *Formation of N-(1-deoxy-D-fructos-1-yl)-glycine (DFG) by heating glucose and glycine at 90°C and pH 6 (□) or pH 7 (▲)*

While 3-deoxyglucosones have long been isolated in the pure form,[16] the existence of 1-deoxy-2,3-hexodiuloses, already postulated by Hodge in 1953,[2] was only recently verified.[17] These relatively unstable Maillard intermediates could only be detected as quinoxaline derivatives.

Formation of DFG

The formation of DFG from glucose and glycine at pH 6 or 7 was studied using the same reaction conditions as reported above. As shown in Figure 2, DFG was continuously formed at both pHs. The reaction yield was low, ~ 10 mol% after 2 h. DFG formation was favoured at pH 7: nearly twice as much DFG was generated after heating for 15, 30 or 60 min compared with the reaction at pH 6. At the end of the reaction, however, approximately the same amount of DFG (0.12 mol) was formed at both pHs. The saturation of the curve obtained for DFG at pH 7 indicates that the rates of formation and decomposition were nearly equal after 1 h of reaction. Therefore, the curves in Figure 2 represent the sum of formation and decomposition of DFG, i.e. formation of DFG increased with pH, but the product formed was degraded at a rate that also increased with pH and time.

A similar rate of formation of DFG was described by Potman and van Wijk.[18] Highest yields were obtained in the absence of phosphate due to its catalytic effect on the decomposition of Amadori compounds. However, inorganic phosphate also acts as a base catalyst in the Amadori rearrangement and increases the rate of conversion of the starting materials following first-order kinetics.

The effect of pH on the formation of DFG in the initial phase of the Maillard reaction may be explained by the favoured conformation of the sugar. The amino acid reacts with the β-anomeric carbon atom of the carbohydrate in the cyclic pyranose conformation, which is the most abundant form of aldohexoses in solution.[19] Also, the resulting N-substituted aldosylamines are mainly present in the β-pyranoid form. Furthermore, 1-epimerization of D-glucose is much faster at pH 7 than at pH 6, thus favouring the formation of the thermodynamically more stable β-anomer.[20] Consequently, β-D-glucose as the favoured conformer dominates in the model reaction at pH 7 and accelerates the reaction with glycine to form N-glucosylglycine which, in turn, rearranges to DFG.

Formation of furaneol

Furaneol was identified as one of the most odour-active volatile components present in the aroma extract of the Maillard model systems analysed (data not shown). Despite several studies dealing with the formation of volatile compounds through thermolysis of DFG,[9,21-22] this is the first time that furaneol has been reported as a degradation product of DFG. In general, furaneol is known as a pyrolysis product of 1-deoxy-1-piperidino-D-fructose[23] and the Amadori compounds from histidine,[22] methionine,[24] proline,[25] and valine and alanine.[26]

Formation of furaneol strongly depends on the experimental conditions. As shown in Figures 3 and 4, furaneol was continuously formed with time. The reaction of glucose and glycine at pH 7 resulted in 27 mg furaneol per mol glucose (Figure 3). In general, yields were higher at pH 7 than at pH 6, particularly when using the Amadori compound (Figure 4). Best results were obtained by heating DFG at pH 7 for 120 min: 191 mg/mol furaneol, which corresponds to a yield of ~ 0.15 mol%. However, furaneol is a potent odorant having threshold values of 0.1-0.2 mg/L water (nasal) and 0.03 mg/L water (retronasal).[27]

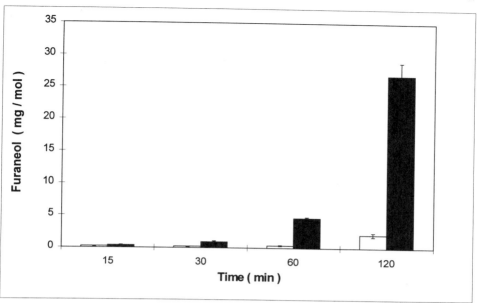

Figure 3. *Formation of 4-hydroxy-2,5-dimethyl-3(2H)-furanone (furaneol) from glucose and glycine at pH 6 (□) or pH 7 (■)*

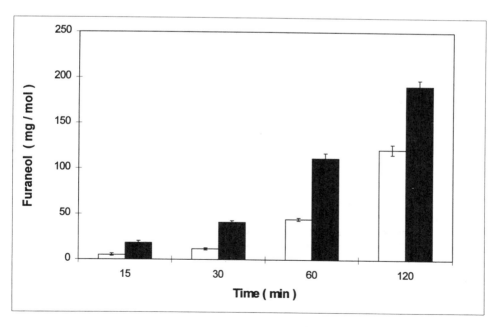

Figure 4. *Formation of 4-hydroxy-2,5-dimethyl-3(2H)-furanone (furaneol) from N-(1-deoxy-D-fructos-1-yl)-glycine (DFG) at pH 6 (□) or pH 7 (■)*

The formation of furaneol from DFG (Figure 5) can be explained on the basis of the work published by Hodge et al.[2,28] and Schieberle.[11] The reactive 1-deoxy-2,3-hexodiulose is formed by 2,3-enolization of DFG followed by β-elimination of the amino acid from a postulated intermediary tautomeric 3-ketose.[2] The resonating reductone structure of the 1-deoxy-2,3-hexodiulose is the basis for an allylic elimination of water at C_6 of the 4,5-enediol intermediate, thus resulting in acetylformoine with a second terminal methyl group.[28]

Furaneol can then be formed from acetylformoine by reduction of the open-chain form either by disproportionation or by reduction with further enoloxo compounds followed by subsequent enolization and elimination of water.[11] The direct formation of furaneol from acetylformoine has recently been confirmed by labelling experiments.[29] $^{13}CH_3$-Labelled furaneol was identified as the main reaction product obtained by thermally induced degradation of $^{13}CH_3$-labelled acetylformoine.

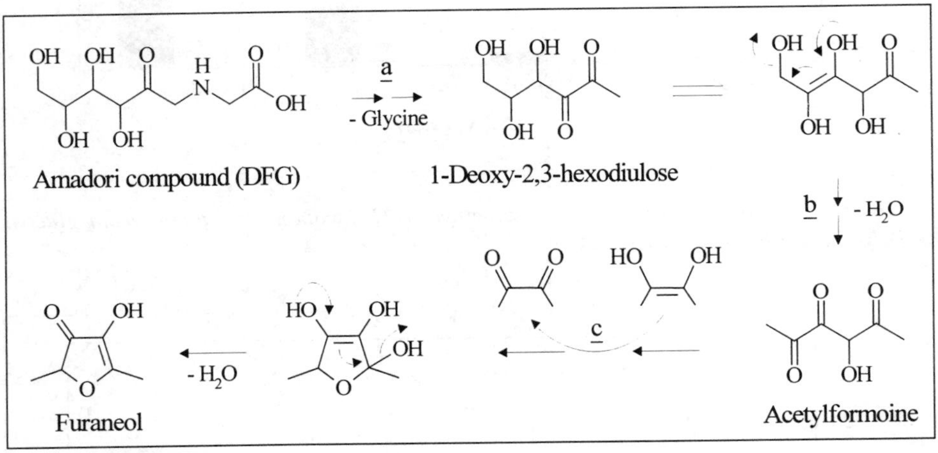

Figure 5. *Schematic formation of furaneol from N-(1-deoxy-D-fructos-1-yl)-glycine (DFG) by (a) 2,3-enolization, (b) allylic elimination of water, and (c) reduction of acetylformoine (shown in the open-chain form: 4-hydroxy-2,3,5-hexanetrione) (for explanation see text)*

Conclusion

Formation and decomposition of Amadori compounds is accelerated at pH 7 compared with pH 6. This may explain higher browning rates and more intense flavours generated under neutral and alkaline conditions. Furaneol is preferably formed from the Amadori compound DFG via acetylformoine as an active intermediate. Significant amounts of furaneol were produced after 1-2 h of reaction when sufficient amounts of DFG had been generated.

Acknowledgements
We are grateful to Dr. A. Stämpfli for the FAB-MS/MS measurements and Mrs. S. Metairon for expert technical assistance. We also thank Dr. R. Badoud for valuable discussions and Dr. E. Prior for linguistic proof-reading of the manuscript.

References

1. F. Ledl and E. Schleicher, New aspects of the Maillard reaction in foods and in the human body, *Angew. Chem.* (Int. Ed. Engl.), 1990, **29**, 565-594.
2. J. E. Hodge, Dehydrated foods: Chemistry of browning reactions in model systems, *J. Agric. Food Chem.*, 1953, **1**, 928-943.
3. J. E. Hodge, The Amadori rearrangement, *Advan. Carbohyd. Res.*, 1955, **10**, 169-205.
4. J. Mauron, The Maillard reaction in food: a critical review from the nutritional standpoint, *Prog. Food Nutr. Sci.*, 1981, **5**, 5-35.
5. R. Wittmann and K. Eichner, Detection of Maillard products in malts, beers, and brewing colorants, *Z. Lebensm. Unters. Forsch.*, 1989, **188**, 212-220 (in German; *Chem. Abstr.* 1989, **111**, 55721b).
6. H. Hashiba, Isolation and identification of Amadori compounds from soy sauce, *Agric. Biol. Chem.*, 1978, **42**, 763-768.
7. M. Reutter and K. Eichner, Separation and determination of Amadori compounds by HPLC and post-column reaction, *Z. Lebensm. Unters. Forsch.*, 1989, **188**, 28-35 (in German; *Chem. Abstr.* 1989, **110**, 211066d).
8. M. Heinzler and K. Eichner, Behavior of Amadori compounds during cocoa processing. Part I. Formation and decomposition, *Z. Lebensm. Unters. Forsch.*, 1991, **192**, 24-29 (in German; *Chem. Abstr.* 1991, **115**, 27981a).
9. G. A. M. van der Ouwerland, H. G. Peer and S. B. Tjan, Occurrence of Amadori and Heyns rearrangement products in processed foods and their role in flavor formation, in 'Flavor of foods and beverages', G. Charalambous and G. E. Inglett (eds), Academic Press, New York, 1978, pp. 131-143.
10. E. F. L. J. Anet, 3-Deoxyglycosuloses (3-deoxyglycosones) and the degradation of carbohydrates, *Adv. Carbohyd. Chem.*, 1964, **19**, 181-218.
11. P. Schieberle, Formation of furaneol in heat-processed foods, in 'Flavor precursors - Thermal and enzymatic conversions', R. Teranishi, G. R. Takeoka, M. Güntert (eds), ACS Symp. Ser. 490, Am. Chem. Soc., Washington, 1992, pp. 164-175.
12. A. A. Staempfli, I. Blank, R. Fumeaux and L. B. Fay, Study on the decomposition of the Amadori compound N-(1-deoxy-D-fructos-1-yl)-glycine in model systems: quantification by fast atom bombardment tandem mass spectrometry, *Biol. Mass Spectrom.*, 1994, **23**, 642-646.
13. I. Blank, L. B. Fay, F. J. Lakner and M. Schlosser, Determination of 4-hydroxy-2,5-dimethyl-3(2H)-furanone and 2(or 5)-ethyl-4-hydroxy-5(or 2)-methyl-3(2H)-furanone in pentose sugar-based Maillard model systems by isotope dilution assays, *J. Agric. Food Chem.*, 1997, **45**, 2642-2648.
14. H. Simon and G. Heubach, Alicyclic, open-chain, N-containing reductones by the reaction between secondary amino salts and monosaccharides, *Chem. Ber.*, 1965, **98**, 3703-3711 (in German; *Chem. Abstr.* 1966, **64**, 5193b).
15. W. Nedvidek, F. Ledl and P. Fischer, Detection of 5-hydroxymethyl-2-methyl-3(2H)-furanone and of α-dicarbonyl compounds in reaction mixtures of hexoses and pentoses with different amines, *Z. Lebensm. Unters. Forsch.*, 1992, **194**, 222-228.
16. E. F. L. J. Anet, Degradation of carbohydrates. I. Isolation of 3-deoxyhexosones, *Aust. J. Chem.*, 1960, **13**, 396-403.
17. J. Becker, F. Ledl and T. Severin, Formation of 1-deoxy-D-*erythro*-2,3-hexodiulose from Amadori compounds, *Carbohyd. Res.*, 1988, **177**, 240-243.

18. R. P. Potman and T. A. van Wijk, Mechanistic studies of the Maillard Reaction with emphasis on phosphate-mediated catalysis, in 'Thermal generation of aromas', T. H. Parliment, R. J. McGorrin and C.-T. Ho (eds), ACS Symp. Ser. 409, Am. Chem. Soc., Washington, 1989, pp. 182-195.
19. G. Westphal and L. Kroh, Mechanism of the "early Phase" of the Maillard reaction. Part I. Influence of the structure of the carbohydrate and the amino acid on the formation of the N-glycoside, *Nahrung*, 1985, **29**, 757-764 (in German; *Chem. Abstr.* 1985, **103**, 121808y).
20. J. Vogel, G. Westphal and C. Pippig, Mutarotation of D-glucose dependence on the reaction environment, *Nahrung*, 1988, **32**, 709-714 (in German; *Chem. Abstr.* 1989, **110**, 8498d).
21. E. J. Birch, J. Lelievre and E. L. Richards, Thermal analysis of 1-deoxy-1-glycine-D-fructose and 1-β-alanino-1-deoxy-D-fructose, *Carbohyd. Res.*, 1980, **83**, 263-272.
22. V. Yaylayan and A. Keyhani, Pyrolysis/GC/MS analysis of non-volatile flavor precursors: Amadori compounds, in 'Contribution of low- and non-volatile materials to flavor', W. Pickenhagen, C.-T. Ho, A. M. Spanier (eds), Allured Publ. Corp., Carol Stream, Illinois, 1996, pp. 13-26.
23. F. D. Mills, B. G. Baker and J. E. Hodge, Amadori compounds as nonvolatile flavor precursors in processed foods, *J. Agric. Food Chem.*, 1969, **17**, 723-727.
24. G. Vernin, J. Metzger, C. Boniface, M.-H. Murello, A. Siouffi, J.-L. Larice and C. Párkányi, Kinetics and thermal degradation of the fructose-methionine Amadori intermediates, *Carbohyd. Res.*, 1992, **230**, 15-29.
25. F. D. Mills and J. E. Hodge, Amadori compounds: Vacuum thermolysis of 1-deoxy-1-L-prolino-D-fructose, *Carbohyd. Res.*, 1976, **51**, 9-21.
26. H. Shigematsu, S. Shibata, T. Kurata, H. Kato and M. Fujimaki, Thermal degradation products of several Amadori compounds, *Agric. Biol. Chem.*, 1977, **41**, 2377-2385.
27. A. O. Pittet, P. Rittersbacher and R. Muralidhara, Flavor properties of compounds related to maltol and isomaltol, *J. Agric. Food Chem.*, 1970, **18**, 929-933.
28. J. E. Hodge, B. E. Fischer and E. C. Nelson, Dicarbonyls, reductones, and heterocyclics produced by reactions of reducing sugars with secondary amine salts, *Am. Soc. Brew Chem. Proc.*, 1963, 84-92.
29. M.-O. Kim and W. Baltes, On the role of 2,3-dihydro-3,5-dihydroxy-6-methyl-4(*H*)-pyran-4-one in the Maillard reaction, *J. Agric. Food Chem.*, 1996, **44**, 282-289.

The Origin and Fate of α-Dicarbonyls Formed in Maillard Model Systems: Mechanistic Studies Using ^{13}C- and ^{15}N-Labelled Amino Acids

V. A. Yaylayan and A. Keyhani

DEPARTMENT OF FOOD SCIENCE AND AGRICULTURAL CHEMISTRY, MCGILL UNIVERSITY, 21111 LAKESHORE, STE. ANNE DE BELLEVUE, QUEBEC, CANADA, H9X 3V9

Summary
The origin of reactive α-dicarbonyl species formed during the Maillard reaction can be identified through labelling studies using ^{13}C-labelled sugars and amino acids. Such studies could be performed conveniently by Py-GC-MS. Analysis of model systems consisting of short-chain dicarbonyls, D-glucose and labelled glycine; indicated that α-ketoaldehydes produced by retro-aldol reactions of sugar derivatives could undergo chain elongation through incorporation of amino acid carbons, by two processes; (1) aldol condensation with Strecker aldehydes or (2) aldol condensation with the α-carbon of the amino acids followed by deamination. The resulting α-carboxydicarbonyl, in turn, can either undergo decarboxylation to produce dicarbonyls that incorporate the side chain of the amino acids or undergo further additions of amino acids to produce pyrazinones.

Introduction

α-Dicarbonyl compounds such as pyruvaldehyde, glyoxal, 2,3-pentanedione and other dicarbonyls are implicated in the formation of different heterocyclic compounds, especially pyrazines. They are formed in Maillard model systems through carbohydrate degradation mechanisms from Amadori products and deoxyglucosones. α-Dicarbonyl compounds could be detected as their stable derivatives upon reaction with o-diaminobenzenes. The initial sugar dicarbonyls have been suggested[1] to undergo aldol condensation with Strecker aldehydes to form various dicarbonyl compounds incorporating amino acid carbons. Keyhani and Yaylayan[2] identified an aldol condensation reaction of α-ketoaldehydes with glycine that leads also to the incorporation of the amino acid carbons into the dicarbonyl thus effecting a chain-elongation process. Pyrolysis coupled with gas chromatography-mass spectrometry (Py-GC-MS) is ideally suited to conducting model studies in the pyrolysis probe and to separate and identify the products formed in a relatively short period of time.[2,3] The use of the Py-GC-MS technique developed in our laboratory, as an integrated reaction, separation and identification system, to perform such studies, not only accelerates the time required to accomplish the labelling studies but also reduces the high cost associated with expensive labelled starting materials. Application of this technique allowed, for the first time, identification of the origin of C_2, C_3 and C_4 dicarbonyl fragments and of all the carbon atoms incorporated in the products formed in a glycine-D-glucose model system. The labelling studies have also helped to identify a new chemical transformation of α-dicarbonyls, affected by the amino acid, that leads to the addition of the C-2 atom of the amino acid to the α-dicarbonyl compounds; this process is in direct competition with Strecker-type interactions. In addition, the role of the Amadori compound in generating products containing the six carbon atoms of intact sugar was also elucidated.[3]

Materials and Methods

All reagents and chemicals were purchased from Aldrich Chemical Company (Milwaukee, WI) or from ICON Services Inc. (Summit, New Jersey).

Pyrolysis-GC-MS Analysis: A Hewlett-Packard GC-mass-selective detector (5890 GC-5971B MSD) interfaced to a CDS pyroprobe 2000 unit was used for the Py-GC-MS analysis. Solid samples (1-4 mg) of amino acid-glucose or amino acid-dicarbonyl compound, were introduced inside a quartz tube (0.3 mm thickness), plugged with quartz wool, and inserted inside the coil probe. The Pyroprobe was set at the desired temperature (250 °C) at a heating rate of 50 °C/ms and with a THT (total heating time) of 20 s. The GC column flow rate was 0.8 mL/min for a split ratio of 92:1 and a septum purge rate of 3 mL/min. The pyroprobe interface temperature was set at 250 °C. The capillary direct MS interface temperature was 180 °C; ion-source temperature was 280 °C. The ionization voltage was 70 eV, and the electron multiplier was 1682 V. The mass range analysed was 30-300 amu. The column was a fused-silica DB-5 column (30m length x 0.25 mm i.d. x 25 µm film thickness; Supelco, Inc.). The column initial temperature was -5 °C for 2 min and was increased to 50 °C at a rate of 30 °C/min; immediately, the temperature was further increased to 250 °C at a rate of 8 °C/min and kept at 250 °C for 5 minutes.

Results and Discussion

Reactive α-dicarbonyl intermediates in Maillard model systems could be detected and identified by post-reaction derivatization with o-diaminobenzene. A more convenient technique was developed in our laboratories that uses the formation of selected pyrazines and pyrazinones as *in-situ* derivatization reactions that trap two reactive dicarbonyls as stable pyrazines or a single dicarbonyl as pyrazinone, which could be detected readily by GC-MS. The retro-synthetic scheme shown in Figure 1 indicates, for example, that a tetrasubstituted pyrazine could be formed either by condensation of two α-diketones followed by oxidation or by the condensation of an α-diketone and an α-aldehydoketone to form dihydropyrazine that eventually condenses with an aldehyde followed by dehydration. On the other hand, pyrazinone incorporates one dicarbonyl intermediate.[2] In this approach, selected model systems with labelled reactants should be used to prevent unambiguous results. Model systems consisting of glyoxal, pyruvaldehyde, 2,3-butanedione and labelled glucoses and labelled amino acids were analyzed for the presence of pyrazines and pyrazinones using Py-GC-MS. Label incorporation was used to identify reaction pathways leading to the formation of α-dicarbonyl compounds. However, although reactive dicarbonyls are known to be formed through retro-aldol reactions of sugar intermediates, studies using labelled amino acids[2,3] indicated the participation of the C-2 atom and the side-chain of amino acids in a process that leads to carbon-chain elongation of sugar-derived α-dicarbonyls, either through aldol condensation as Strecker aldehydes or as intact amino acids.

Glycine system - formation of pyrazines.

The glyoxal-glycine model system can generate valuable information regarding the formation of chain-elongated sugar-derived dicarbonyls. Glyoxal cannot undergo aldol condensation either with itself or with formaldehyde (Strecker aldehyde of glycine) to generate longer-chain dicarbonyls. Any dicarbonyl containing more than two carbon atoms, detected directly or observed indirectly as part of pyrazine or pyrazinone structures, should arise mainly from the glycine interaction with glyoxal. Table 1 lists selected pyrazines, pyrazinones and dicarbonyls observed in different model systems containing [^{13}C-2]-glycine.

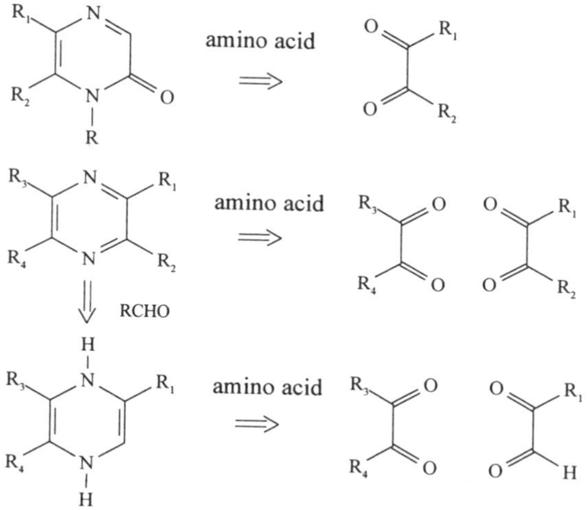

Figure 1. Retro-synthetic scheme for the formation of pyrazines and pyrazinones

To rationalize the observed ^{13}C-2 incorporation patterns in the listed compounds, a lesser known chemical property of amino acids was used, which involves the reactivity of their α-carbons. Amino acids, in addition to their reactivity as N-nucleophiles, are also known to react as C-nucleophiles through their α-carbons. Under basic conditions, free glycine or its cobalt or copper complexes can react with acetaldehyde or benzaldehyde through aldol condensation to generate threonine[4] or β-phenylserine, respectively,[5] as shown in Figure 2. This reaction is used industrially for large-scale production of threonine. It seems such interactions play a major role in Maillard reactions as an alkylation mechanism involving α-aldehydocarbonyls and amino acids.

Formation of pyrazine and methylpyrazine in glyoxal-glycine system could be rationalized as involving Strecker reaction and interaction of dihydropyrazine with formaldehyde. However, the formation of ^{13}C-labelled di, tri and tetra-substituted pyrazines and free pyruvaldehyde and 2,3-butanedione, should involve an aldol-type condensation of the α-carbon of glycine with glyoxal. All of the pyrazine derivatives listed in Table 1 could be formed by the Strecker reaction of glycine with glyoxal, pyruvaldehyde, 2,3-butanedione or 2-ketobutanal. The formation of these dicarbonyls could be explained by a reaction mechanism as illustrated in Figure 3.

Table 1. *Number and extent (%) of incorporation of carbon-13 atoms in selected α-dicarbonyls, pyrazines and pyrazinones formed in model systems containing [^{13}C-2]-glycine[a]*

Compound	[^{13}C-2]-glycine				
Pyrazines[b]	Glucose[c]	Glyoxal[c]	Glyc[c]	Py	Bd
pyrazine	0	0	nd	nd	nd
methylpyrazine	0(90),1(10)	0(5), 1(95)	t	nd	nd
2,5-dimethylpyrazine	0(85),1(10),2(5)	1(15), 2(85)	t	0	nd
2,3-dimethylpyrazine	0(35),1(65)	1(10), 2(90)	t	nd	nd
trimethylpyrazine	1	1(5), 2(20), 3(75)	0(35), 1(35), 2(15), 3(15)	1	t
tetramethylpyrazine	0(20),1(35), 2(40), 3(5)	1(5),2(10), 3(35),4(50)	t	t	0
2-ethyl-6-methylpyrazine	nd	1(10),2(5),3(85)	t	nd	nd
vinylpyrazine	nd	0	t	nd	nd
Pyrazinones[b]					
1,6-dimethylpyrazinone	2(40),3(60)	2(15),3(85)	2(25), 3(75)	nd	nd
1,5,6-trimethylpyrazinone	2(30),3(70)	2(5) ,3(15), 4(80)	2(20), 3(55), 4(25)	3	2
α-Dicrabonyls					
pyruvaldehyde	nd	1	nd	0	nd
2,3-butanedione	0(35),1(65)	2	nd	1	0

[a] Py, pyruvaldehyde; Bd, 2,3-butanedione, Glyc, glycoladehyde; nd; not detected, t, trace. [b] All pyrazines and pyrazinones incorporate (100 %) two nitrogen atoms and all pyrazinones incorporate (100 %) one ^{13}C-1 atom of glycine. [c] values in paranthesis indicate incorporation (%) of labells.

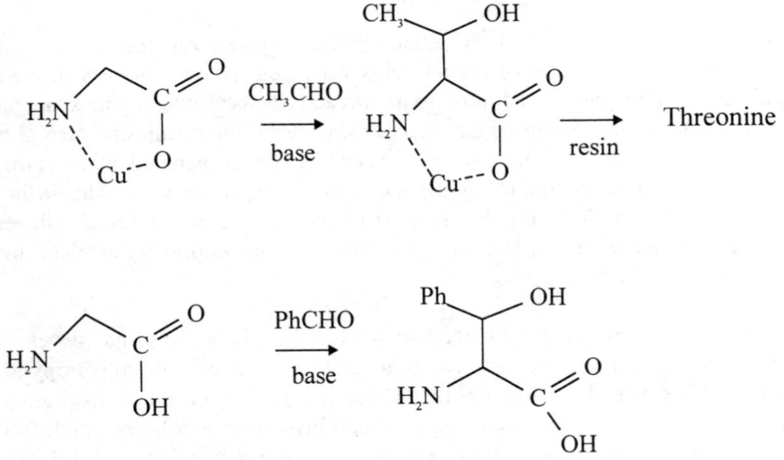

Figure 2. *Reaction of the α-carbon of glycine with aldehydes*[4,5]

Reaction Mechanisms

Glycine undergoes an aldol addition with glyoxal followed by elimination of ammonia to produce 3-carboxypyruvaldeyde (I) a key intermediate that can either undergo multiple additions with glycine to produce pyrazinones (see Figure 4), or decarboxylates to produce pyruvaldehyde (a) with the C-2 atom of glycine as the methyl group. Pyruvaldehyde, in turn, can undergo a similar process to produce 2,3-butanedione (b) with two C-2 atoms of glycine incorporated as methyl groups. 4-Amino-3-hydroxy-2-butanone formed after the second glycine reaction, can undergo β-elimination to produce 2-ketobutanal (c) after an hydrolysis step. The incorporation pattern of ^{13}C-2 atoms into the pyrazines and dicarbonyls detected in the glycine-glyoxal model system and listed in Table 1 is consistent with the above mechanism. The glycoladehyde model system, compared with glyoxal, generates only trace amounts of similar pyrazines and less intense pyrazinone peaks. Glycoladehyde should undergo oxidation[1,6] to glyoxal before it can generate pyrazines and pyrazinones.

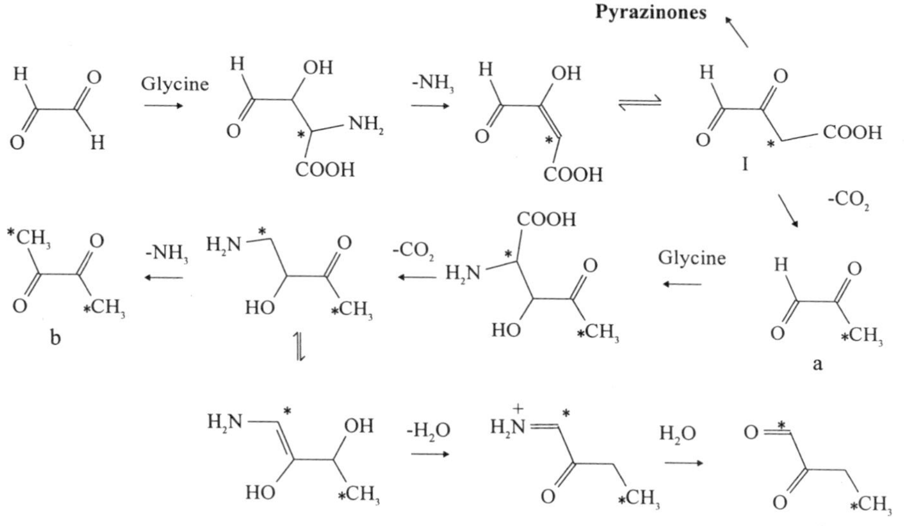

Figure 3. *Chain-elongation reactions of glyoxal (asterisks indicate ^{13}C-2 atom of glycine)*

Glycine system - formation of pyrazinones.

The mechanism of formation of pyrazinones and fused-benzopyrazinones (quinoxalinones) from different dicarbonyls and glycine has been published.[2,3] Figure 4 (pathway A) illustrates the mechanism of formation of dimethyl pyrazinone from glyoxal (85%) and trimethylpyrazinone from pyruvaldehyde (15%), both pyrazinones require the incorporation of three C-2 atoms of glycine.

The incorporation of four ^{13}C-2 atoms of glycines into trimethylpyrazinone (80%) could be explained by decarboxylation of intermediate I and formation of [^{13}C-3]-pyruvaldehyde. Formation of pyrazinones with only two ^{13}C-2 atoms of glycine (dipeptide pathway B[7]) indicates an aldol condensation process that can generate pyruvaldehyde and 2,3-butanedione.

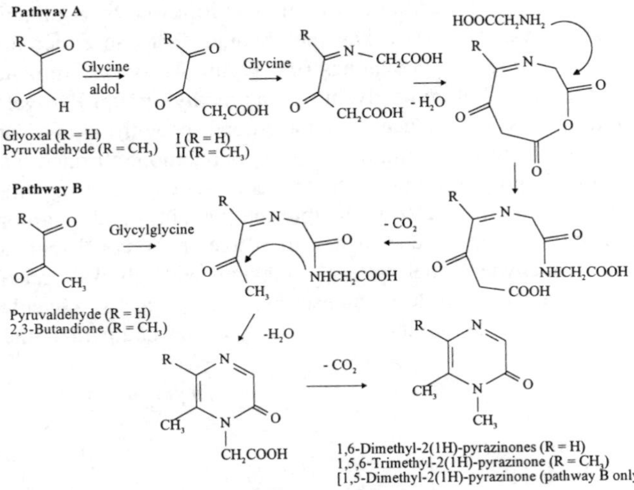

Figure 4. *Mechanism of formation of pyrazinones*

Conclusion

Similar studies performed using [^{13}C-2]-labelled L-alanine, L-valine and L-leucine confirmed the generality of carbon-chain elongation of sugar-derived dicarbonyls through aldol type condensation of the α-carbon of amino acids. Confirmation of the nucleophilic character of the α-carbon of amino acids under Maillard reaction conditions can further elucidate the role of amino acids in Maillard reactions.

Acknowledgments

V. Y. acknowledges funding for this research by the Natural Sciences and Engineering Research Council (NSERC) of Canada.

References

1. H. Weenen and W. Apeldoorn, Carbohydrate cleavage in Maillard reaction, in 'Flavour Science: Recent developments', A. J. Taylor and D. S. Mottram, (eds), The Royal Society of Chemistry, Cambridge, 1996, 211-216.
2. A. Keyhani and V. Yaylayan, Elucidation of the mechanism of pyrazinone formation in glycine model systems using labelled sugars and amino acids, *J. Agric. Food Chem.*, 1996, **44**, 2511-2516.
3. A. Keyhani and V. Yaylayan, Glycine specific novel Maillard reaction products: 5-hydroxy-1,3-dimethyl-2-[1H]quinoxalinone and related compounds, *J. Agric. Food Chem.*, 1997, **45**, 697-702.
4. T. Kaneko, Y. Izumi, I. Chibata and T. Itoh (eds), 'Synthetic production and utilization of amino acids', Wiley, New York, 1974, p. 197.
5. D. H. Barton, and W.D Ollis, (eds), 'Comprehensive Organic Chemistry', Pergamon Press, Oxford, 1979, Vol. 2, p. 825.
6. V. A. Yaylayan, Classification of the Maillard reaction: A conceptual approach, *Trends Food Sci. Technol.*, 1997, **8**, 13-18.
7. Y-C. Oh, C-K. Shu and C-T, Ho, Formation of novel 2(1H)-pyrazinones as peptide specific Maillard reaction products, *J. Agric. Food Chem.*, 1992, **40**, 118-121.

C_4, C_5, and C_6 3-Deoxyglycosones : Structures and Reactivity

Hugo Weenen,[1] Jos G. M. van der Ven,[1] Leendert M. van der Linde,[1] John van Duynhoven[2] and Anneke Groenewegen[2]

[1] QUEST INTERNATIONAL, HUIZERSTRAATWEG 28, 1400 CA NAARDEN, THE NETHERLANDS; [2] UNILEVER RESEARCH LABORATORY, OLIVIER VAN NOORTLAAN 120, 1234 AB, VLAARDINGEN, THE NETHERLANDS

Summary
Because of the importance of deoxyglycosones as reactive intermediates in the Maillard reaction, C_4, C_5 and C_6 3-deoxyglycosones were prepared, their structures were elucidated, and their reactivity studied. Whereas 3-deoxyglucosone consists exclusively of bicyclic (hemi)acetal/(hemi)ketal structures, 3-deoxypentosone consists of monocyclic isomers with 63% hydrates (*gem*-diols) present; 3-deoxytetrosone was found to occur as monocyclic and non-cyclic hydrates, with a free carbonyl present in 27% of the isomers. The reactivity (instability and browning) is in the order 3-deoxytetrosone > 3-deoxypentosone > 3-deoxyglucosone.

Introduction

Deoxyosones are reactive Maillard intermediates that play an important role in flavour formation, and have been postulated as intermediates in the formation of various cyclization products, such as 4-hydroxy-2,5-dimethyl-3(2*H*)-furanone, 4-hydroxy-5-methyl-3(2*H*)-furanone, furfural and maltol, as well as secondary products derived from these cyclization products and amino acids. In addition, deoxyglycosones are precursors of sugar fragmentation products.[1,2] The formation of deoxyglycosones from Amadori rearrangement products (ARPs) has been demonstrated by Beck *et al.* using 1,2-diaminobenzene as a trapping agent.[3,4]

In the past, we elucidated the structure of 3-deoxyglucosone in water using NMR, its degradation in acid and alkaline media, its function as a flavour precursor in pyrazine formation and its behaviour in fragmentation.[1,2,5] We reported that 3-deoxyglucosone occurs as a mixture of 5 major, and at least 6 minor isomers, all having bicyclic structures with no free carbonyl. When reacted with asparagine, 3-deoxyglucosone was found to be a rather selective precursor of dimethylpyrazine, pointing to retro-aldolization as a major cleavage mechanism. Unexpectedly, 3-deoxyglucosone was a less efficient precursor of methylated pyrazines than glucose or fructose, suggesting that other more efficient cleavage precursors play a role in pyrazine formation as well.

Studies on carbohydrate fragmentation[5] confirmed that 3-deoxyglucosone is a rather selective precursor of pyruvaldehyde, but also showed that normal aldoses and ARPs are quite efficient fragmentation precursors as well. Interestingly, xylose was also found to be a relatively efficient precursor of pyruvaldehyde. We tentatively explained this, by pointing out that C_5 3-deoxyosones are expected to be less stable than C_6 3-deoxyosones, because the extra carbon of C_6 3-deoxyosones allows the formation of more stable bicyclic isomers. This raised our interest in the structure and reactivity of C_5 and C_4 3-deoxyosones. The synthesis, structural analysis, and reactivity of the 3-deoxyosones prepared from D-xylose and D-erythrose, 3-deoxy-D-*glycero*-pentos-2-ulose or D-3-deoxypentosone (3-dPone) and 4-hydroxy-2-oxobutanal or 3-deoxytetrosone (3-dTone), respectively, are reported here, and compared with 3-deoxyglucosone (3-dGone).

D-Xylose (or wood sugar) is one of the four naturally occurring D-pentoses and is commercially available. It is used in the production of process flavours because it is readily available and is a precursor to a range of interesting furanoid flavour substances. D-Erythrose is one of the two naturally occurring D-tetroses. In nature it is only found as erythrose 4-phosphate, an intermediate in enzymatic transformations of carbohydrates.[6] It is easily prepared through oxidative glycol cleavage of 4,6-acetal protected D-glucose.[6,7]

Because D-tetroses (and L-tetroses) all give the same 3-deoxyosone, we use the name 3-deoxytetrosone throughout this paper for the 3-deoxyosone prepared from D-erythrose. Similarly, because all the D-pentoses (ribose, arabinose, xylose, and lyxose) give the same D-3-deoxyosone, we use the name 3-deoxypentosone for the 3-deoxyosone derived from D-xylose, rather than D-3-deoxyxylosone.

Materials and Methods

Synthesis

D-Erythrose was synthesized by periodate oxidation[7] of 4,6-O-benzylidene-α,β-D-glucopyranose,[8] followed by hydrogenolysis of the benzylidene acetal using 5% Pd/C in EtOH-EtOAc (6:1, v/v, containing 0.25% HOAc). The 3-deoxyosones (Figures 1-3) were synthesized according to the method of El Khadem et al.[9] with an additional purification using column chromatography (Kieselgel 60, pentane-EtOAc-MeOH [2:2:1, v/v/v]). Fractions were monitored on TLC and visualized by charring after spraying with 50% H_2SO_4/MeOH. The R_f-values for 3-dGone (1), 3-dPone (2) and 3-dTone (3) are 0.27, 0.46, and 0.68, respectively. Since these compounds are present as a mixture of many isomers they tend to tail on TLC. 3-dGone, 3-dPone and 3-dTone were obtained in 24%, 40% and 29% yield, respectively. The compounds were stored at -20 °C under an inert atmosphere.

Structural analysis

For structural analysis, ^{1}H- and ^{13}C-NMR spectra for 2 and 3 were obtained in D_2O at 298°K, using a Bruker AMX 600 MHz spectrometer operating at 600.13 MHz for ^{1}H and 150.9 MHz for ^{13}C. In addition, spectra for 3 were obtained in D_2O at 298 °K on a Jeol EX 400 MHz spectrometer operating at 399.65 MHz for ^{1}H and 100.4 MHz for ^{13}C. The sodium salt of deuterated 3-(trimethylsilyl)propionic acid was used as the reference compound. The stability and reactivity studies of the deoxyosones were performed on the latter spectrometer. The structure elucidation of **1** has been reported previously.[1]

Stability and reactivity

The stability of the 3-deoxyosones was determined by ^{1}H-NMR measurements of samples stored at room temperature in phosphate buffer (25 mM, pH 7), containing 1% sodium benzoate as an antimicrobial agent, and methanol as the internal standard.

Acid-treated samples were obtained by heating the deoxyosones at 60°C in D_2O containing trifluoroacetic acid-d_1 (TFA) (1 M) and methanol as the internal standard. At regular intervals, the samples were analysed by measuring their ^{1}H-NMR spectra.

Alkali-treated samples were obtained by keeping the 3-deoxyosones at room temperature in D_2O containing sodium deuteroxide (100 mM) and methanol as the internal standard for 1 h, followed by spectral analysis using ^{1}H-NMR.

Browning

The browning of aldoses, 3-deoxyosones, 3-deoxyosone + alanine mixtures, and the aldose + alanine mixtures (equimolar concentrations of 33 mM) was followed at 70°C, in phosphate buffer (220 mM, pH 8.0) at 420 nm using an Unicam 8700 series UV/VIS spectrometer. If necessary, samples were diluted in order to keep the A_{420} below 1.

Results and Discussion

Structures of 3-deoxypentosone isomers

Detailed NMR analysis of 3-deoxypentosone showed that, although this compound contains one CHOH less than 3-deoxyglucosone, it still occurs as a complex mixture of several isomers. The assignment of structures 2a-d was first of all based on short-range connectivities observed in homo- and heteronuclear 2D experiments. To allow differentiation between 2a-d, long-range ^{1}H-^{13}C couplings (as they become manifest in HMBC experiments), were used. ROESY (through-space interactions) was used to resolve ambiguities in the assignments. A more elaborate interpretation of the NMR data will be published elsewhere. The spectral data of these six isomers are in agreement with the two anomers of the pyranoid structure 2a, the two anomers of the furanoid structure 2b, and the two anomers of the furanoid structure 2d, as shown in Figure 1. The available experimental short- and long-range J-coupling information cannot unambiguously discriminate between 2c and 2d. The coupling constants are not in agreement with values observed previously in bicyclic systems.[10] However, 2 has electronegative substituents which can strongly influence J-couplings. Additional relevant information comes from the observation that the

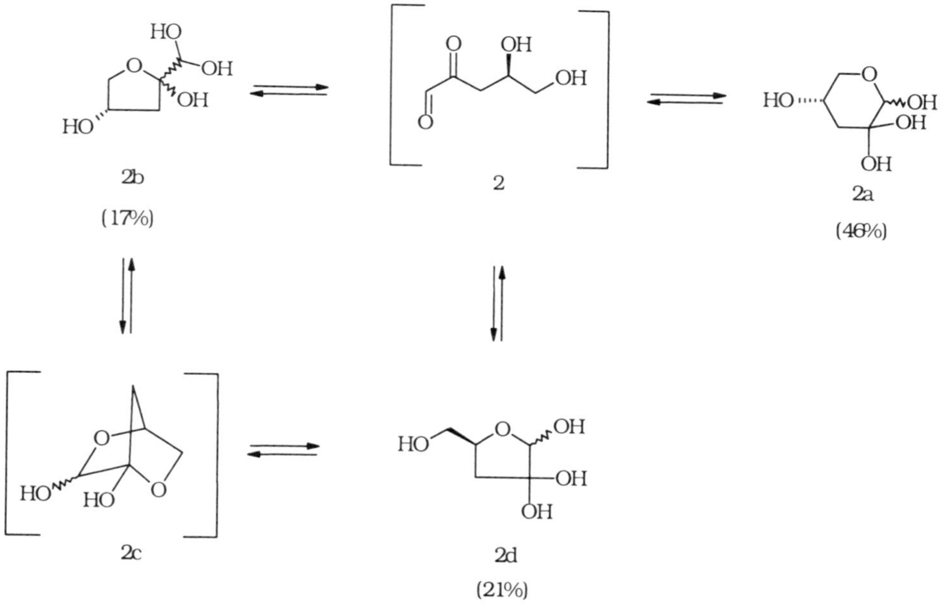

Figure 1. *Structures of main isomers of 3-deoxypentosone*

OCH$_2$ protons in these two anomers are almost degenerate, and the ROE interactions, which both support structure 2d for these isomers.

The proposed structures are similar to what has been reported for pentos-2-uloses or pentosones,[11] which are reported to occur not as bicyclic isomers, but as furanose and pyranose endocyclic hydrates, similar to 2a and 2d.

Structures of C₃ hydroxycarbonyls

Since 3-deoxypentosone was found to contain significant concentrations of hydrated isomers, the occurrence of these *gem*-idols in C_3 model compounds was studied. For this purpose, the NMR-spectra of hydroxyacetone, 1,3-dihydroxyacetone, and D,L-glyceraldehyde were recorded. Of these compounds the free carbonyl fractions are ~98%, ~80% and ~5% for hydroxyacetone, 1,3-dihydroxyacetone and D,L-glyceraldehyde respectively, as determined by ^{1}H-NMR in D_2O.

Apparently electronegative substituents at the α-carbon increase the reactivity of the carbonyl, as the *gem*-diol content for 3-C carbonyl containing compounds is in the order: dihydroxyacetone > hydroxyacetone > acetone. Aldehydic carbonyls are more reactive than ketone carbonyls, which explains the very high *gem*-diol content of glyceraldehyde (95%).

Structures of 3-deoxytetrosone isomers

Figure 2 shows the structures of 3-dTone. NMR-analysis showed that the linear dicarbonyl structure (3) is not present. Instead, the compounds 3a, 3b, 3c, and 3d, were present in ~12%, ~3%, ~15%, and ~70%, respectively. 3-Deoxytetrosone has only four carbons, and can, therefore, not form the relatively stable bicyclic structures, as in the case of 3-deoxyglucosone and 3-deoxypentosone. The structure of 3a clearly confirms the general rule that aldehyde carbonyls form hydrates easier than ketone carbonyls. Of the three 3-deoxyosones studied by us so far, 3-deoxytetrosone is the only one with a significant free carbonyl content (27%).

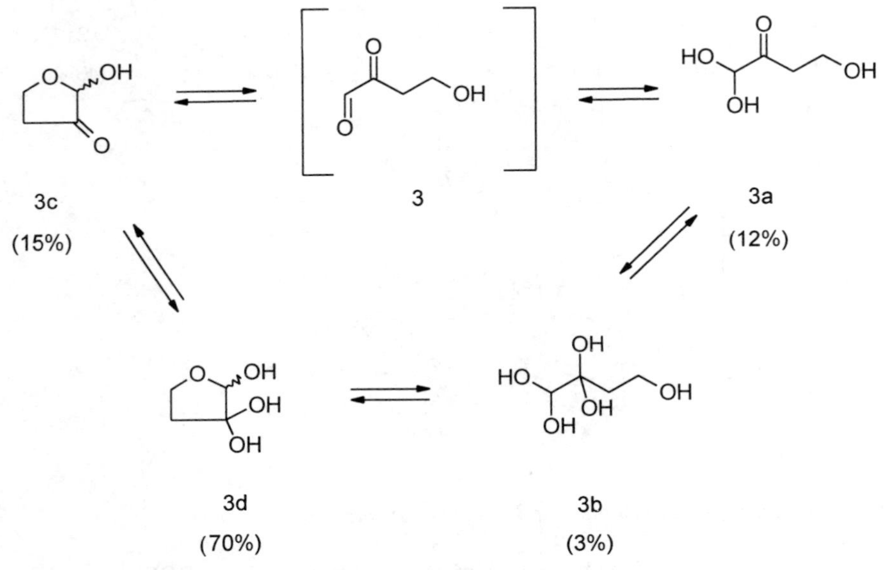

Figure 2. *Observed structures of 3-deoxytetrosone in D_2O*

Stability

Table 1 shows the results of stability tests in neutral, acidic, or alkaline aqueous solutions. NMR measurements showed that the stability of the 3-deoxyosones at room temperature in aqueous buffer (pH 7.0) is: 3-deoxytetrosone < 3-deoxypentosone < 3-deoxyglucosone. During storage, H-D exchange occurs at the α-position of the carbonyl. Browning of the samples was negligible.

3-Deoxytetrosone is least stable: after 1 hour at room temperature, the decay of the compound is detectable using NMR, and within 3-4 hours 50% of 3-deoxytetrosone had disappeared. After 24 hours, 3-deoxytetrosone had completely decomposed, but products could not be identified. The NMR spectrum resembled the spectrum of the alkali-treated 3-deoxytetrosone.

For 3-deoxypentosone, the decay is much slower. After 2 days at room temperature, ~50% of the sample had decomposed, again into compounds that were also detected during alkali treatment. After 20 days of storage some starting material was still present in the samples.

Finally, 3-deoxyglucosone is the most stable compound of the three 3-deoxyosones investigated. After 20 days of storage in D_2O at room temperature, only H-D exchange was detected (~50% after 1 day, ~100% after 6 days) and no significant decomposition. Apparently, the bicyclic structures that predominate in the case of 3-deoxyglucosone are relatively stable, and strongly effect the relative stabilities of the deoxyosones.

Table 1. *Half-lives for 3-deoxyosones during stability tests*

3-deoxyosone	Stability		
	neutral[a]	acidic[b]	alkaline[c]
3-deoxyglucosone	> 20 d	15 h	< 1 h
3-deoxypentosone	2 d	5 h	< 1 h
3-deoxytetrosone	3-4 h	3 h	< 1 h

[a] in phosphate buffer (25 mM, pH 7), [b] in 1 M CF_3COOD, [c] in 100 mM NaOD

Acid treatment

In all acid-treated samples, a brown precipitate formed. Previous experiments showed that 5-(hydroxymethyl)furfural is formed when 1 is treated with dilute acid (Figure 3). Thus, it was expected that 3-deoxypentosone would give furfural, and 3-deoxytetrosone would give 3-hydroxyfuran.

After 3 hours of heating at 60°C, ~50% of the starting 3-deoxytetrosone had disappeared. Apart from, polymeric material only formic acid could be identified. Apparently the primary breakdown product(s) are not stable under the reaction conditions.

For 3-deoxypentosone, 5 hours were needed to obtain a 50-60% degradation of the starting material. Indeed, furfural was formed, as predicted, in addition to polymers and a small amount of formic acid. During the first 3 hours of heating the formation of 2-hydroxy-2,6-dihydropyran-3-one (Figure. 3b) was seen (~5 mol%). However, during prolonged heating, this pyran disappeared. The maximum yield of furfural (after 15 hours) was about 50 mol%.

Acid degradation of 3-deoxyglucosone gives 5-(hydroxymethyl)furfural and formic acid as previously reported. After 15 hours, about 50% of the starting material had degraded.

Alkali treatment

Formation of the so-called metasaccharinic acids(Figure 3) from an internal Cannizzaro reaction of the 3-deoxyosones, and from carbohydrates is well documented.[12-14] When the C_4- and C_5-3-deoxyosones were treated with dilute aqueous base rapid browning occurred, but meta-saccharinic acids could not be detected, in contrast to 3-deoxyglucosone which readily forms metasaccharinic acid upon alkaline treatment, as previously reported.[1]

Browning

The degrees of browning of the 3-deoxyosones with or without alanine and of the corresponding aldoses with or without alanine are shown in Table 2. The initial browning rates for aldoses are higher with alanine than without alanine, as expected. The difference between initial browning rates for glucose with or without alanine is surprisingly small. Possibly phosphate catalysis is more important than amino acid catalysis for glucose under the conditions used. The relative order of browning is as follows:

$$D,L\text{-glyceraldehyde} > D\text{-erythrose} > D\text{-xylose} > D\text{-glucose}$$

D,L-Glyceraldehyde has a free carbonyl content of about 5%, but also the hydrate easily forms the free aldehyde which explains its high reactivity. The reactivity of the C_4, C_5, and C_6 aldoses also seems related to the presence of the corresponding aldehyde, or the ease with which a free aldehyde can be formed. The initial browning rate for the 3-deoxyosones is as follows:

$$3\text{-deoxytetrosone} > 3\text{-deoxypentosone} > 3\text{-deoxyglucosone} > \text{pyruvaldehyde}$$

Table 2. *Relative browning of 3-deoxyosones and the corresponding aldoses*[a]

Aldose / 3-Deoxyosone	Without alanine[b]		With alanine[b]	
	after 15 min	after 7 h	after 15 min	after 7 h
D-glucose	1	1	1	1.3
D-xylose	0	1.6	2.3	5.2
D-erythrose	1.8	4.8	15	14
D,L-glyceraldehyde	3.5	13	300	42
3-deoxyglucosone	38	29	51	57
3-deoxypentosone	103	64	195	102
3-deoxytetrosone	155	33	588	76
pyruvaldehyde	19	17	87	43

[a] *The values for browning of glucose without alanine were taken as 1,* [b] *O.D. at 420 nm / O.D. at 420 nm for glucose. Samples were heated for 15 min. or 7 h. in phosphate buffer (220 mM, pH 8.0) at 70 °C*

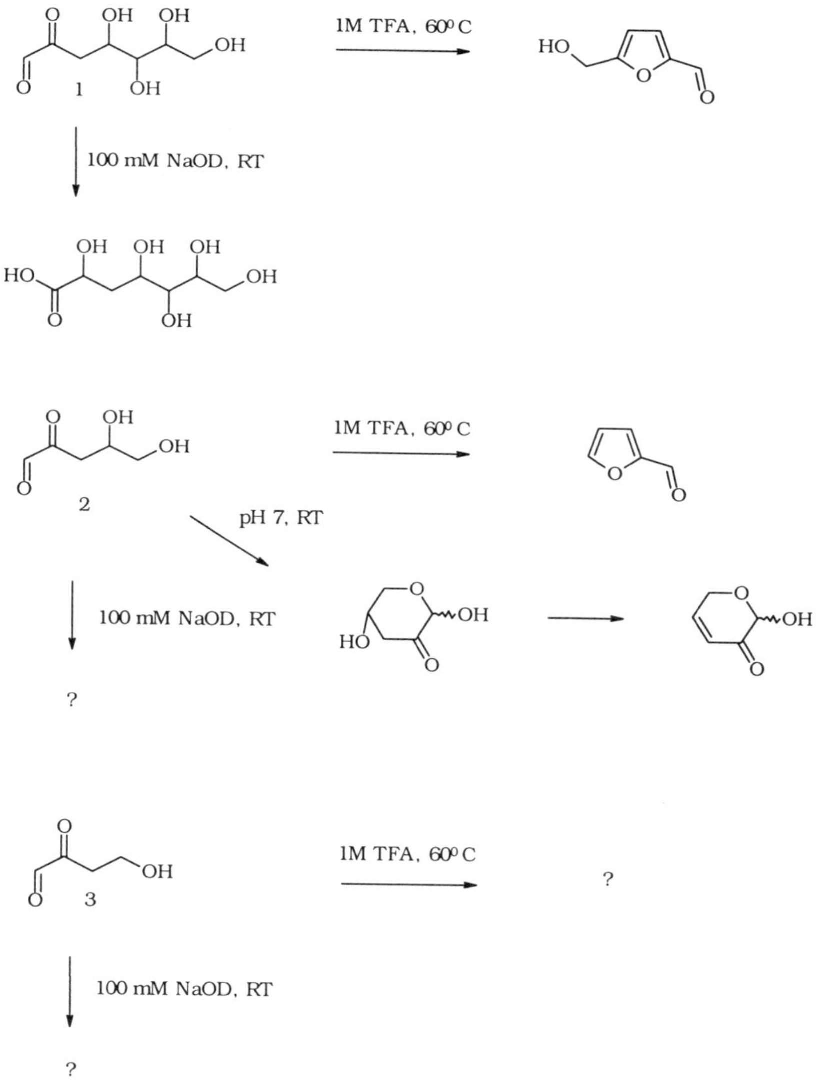

Figure 3. *Acid/base degradation products from 3-deoxyosones 1, 2, and 3*

The relative initial browning rates for the 3-deoxyosones are in the same order as their stabilities. There appears to be a discrepancy between the relative browning of the osones after 15 minutes and 7 hours: After 15 minutes 3-deoxytetrosone shows the highest browning, whereas after 7 hours 3-deoxypentosone shows the highest browning. Possibly the melanoidins formed from 3-deoxytetrosone decompose to less absorbing species upon heating, as suggested by the decreasing O.D. at 420 nm for 3-deoxytetrosone, but not for the other osones, after about 3-4 hours.

Pyruvaldehyde showed unexpectedly low relative browning. Its structure in aqueous solution has to the best of our knowledge not been reported, but is certainly complex, and lacks a free aldehyde moiety (NMR observations), and unlike the deoxyglycosones with 4 or more carbons, it cannot form furans via intramolecular condensation. Possibly furans are more efficient browning precursors, as suggested by some of the coloured condensation products which have been isolated from Maillard reaction mixtures.[15]

References

1. H. Weenen, and S.B. Tjan, Analysis, structure, and reactivity of 3-deoxyglucosone, in 'Flavor precursors, thermal and enzymatic conversions', R. Teranishi, G. R. Takeoka, and M. Guntert, Eds, ACS, Washington, DC, USA, 1992, 217-231.
2. H. Weenen, S.B. Tjan, 3-deoxyglucosone as flavour precursor, in 'Trends in flavour research', H. Maarse, and D. G. van der Heij, Eds, Elsevier, Amsterdam, The Netherlands, 1994, 327-337.
3. J. Beck, F. Ledl, and T. Severin, Formation of 1-Deoxy-D-*erythro*-2,3-hexodiulose from Amadori compounds, *Carbohydr. Res.*, 1988, **177** 240-243.
4. J. Beck, F. Ledl, and T. Severin, Formation of glucosyl-deoxyosones from Amadori compounds of maltose, *Z. Lebensm. Unters. Forsch.*, 1989, **188**, 118-121.
5. H. Weenen, and W. Apeldoorn, Carbohydrate cleavage in the Maillard reaction, in 'Flavour science, recent developments', A. J. Taylor, and D. S. Mottram, Eds, The Royal Society of Chemistry, London, UK, 1997, 211-216.
6. R. Schaffer, Occurrence, properties, and preparation of naturally occurring monosaccharides (including 6-deoxy sugars), in 'The carbohydrates', W. Pigman, and D. Horton, Eds, Academic Press, New York, 1972, 69-111.
7. A.S. Perlin, D-Erythrose, *Methods. Carbohydr. Chem.*, 1962, **1**, 64-70.
8. P. Brigl, and H. Grüner, Neue Benzal- und Benzoyl-Derivate der Glucose. Ber. 65 (1932) 1428-1434.
9. H. El Khadem, D. Horton, M.H. Meskreki, and A.M. Nashed, New route for the synthesis of 3-deoxyaldos-2-uloses, *Carbohydr. Res.*, 1971, **17**, 183-192.
10. J.L. Marshall, S.R. Walter, The complete proton nuclear magnetic resonance analysis of norcamphor, *J. Am. Chem. Soc.*, 1974, **96**, 6358-6362.
11. T. Vuorinen, A.S. Serianni, ^{13}C-Substituted pentos-2-uloses: synthesis and analysis by ^{1}H- and ^{13}C-n.m.r. spectroscopy, *Carbohyd. Res.*, 1990, **207**, 185-210.
12. E.F.L.J. Anet, 3-Deoxyglycosyloses (3-deoxyglycosones) and the degradation of carbohydrates, *Adv. Carbohydr. Chem.*, 1964, **19**, 165-194.
13. W. Pigman, and E.F.L.J. Anet, Mutarotations and actions of acids and bases, in 'The carbohydrates', W. Pigman, and D. Horton, Eds, Academic Press, New York, 1980, 1013-1099.
14. O. Theander, Acids and other oxidation products, in 'The carbohydrates', W. Pigman, and D. Horton, Eds, Academic Press, New York, 1980, 1013-1099.
15. J.M. Ames, A. Apriyantono, and A. Arnoldi, Low molecular weight coloured compounds formed in xylose-lysine model systems, *Food Chem.*, 1993, **46** 121-127.

The Mechanism of Formation of 3-Methylcyclopent-2-en-2-olone

Harry E. Nursten

DEPARTMENT OF FOOD SCIENCE AND TECHNOLOGY, THE UNIVERSITY OF READING, WHITEKNIGHTS, PO BOX 226, READING RG6 6AP, UK

Summary
A revised mechanism is advanced for the formation of 3-methylcyclopent-2-en-2-olone. It is based on a ketol addition between two molecules of acetol, followed by dehydration, tautomerization, cyclization through aldol addition, a second dehydration, and a final isomerization. 3-Methylcyclopent-2-en-2-olone possesses no less than twelve tautomeric forms. The formation of acetol from sugars has been demonstrated, but its mechanism is not entirely clear.

Introduction

Shaw *et al.*[1] proposed that 3-methylcyclopent-2-en-2-olone (cyclotene) is formed by the dehydration of 2 molecules of hydroxyacetone (acetol) between the hydroxy group of one and the methyl group of the other, followed by a second dehydration between the methylene hydrogens of the remaining hydroxymethyl group and the carbonyl group near the other end of the molecule (see Figure 1). The second stage would be a normal ketol addition followed by dehydration, but the first stage does not seem realistic, as was pointed out previously.[2]

Cyclotene is important not only because of its flavour, but also because of its interesting chemistry. It is that which is to be explored in this paper.

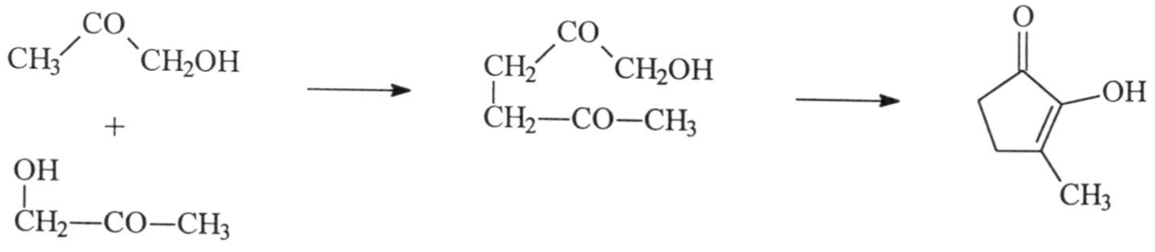

Figure 1. *Formation of 3-methylcyclopent-2-en-2-olone according to Shaw et al.*[1]

Discussion

Cyclotene is a labile molecule. It has 12 tautomers, as shown in Figure 2. From this, it can be seen that there is only one diketo form, three 3-methyl-2-olones with double bonds in the 2-, 3-, and 4-position, respectively, three 5-methyl-2-olones with double bonds in the 2-, 3-, and 4-position, respectively, and five 3-methyl-1,2-diols with double bonds in the 1,3-, 1,4-, 2,4-, 2,5-, and 3,5-positions. Some of the tautomerism in Figure 2 involves vinylogy. Because the double bond is not conjugated with either a keto group or another double bond, the 3-methyl- and the 5-methyl-3-penten-2-olone would be expected to be

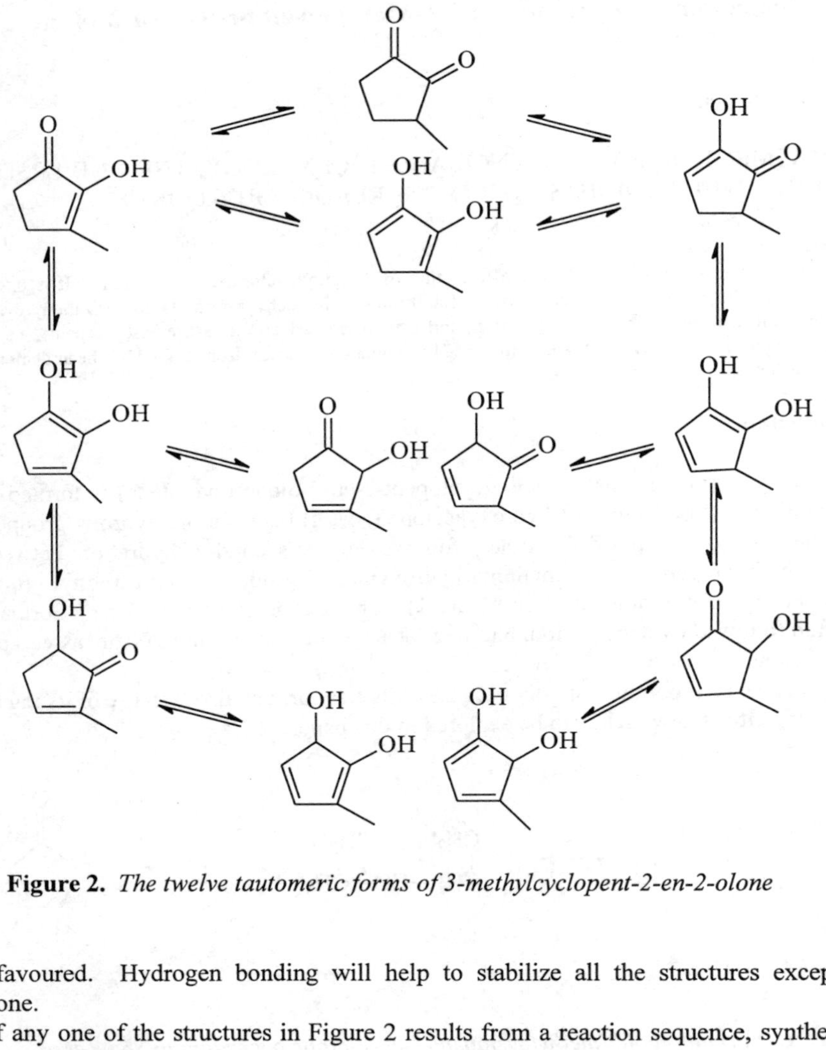

Figure 2. *The twelve tautomeric forms of 3-methylcyclopent-2-en-2-olone*

less favoured. Hydrogen bonding will help to stabilize all the structures except the diketone.

If any one of the structures in Figure 2 results from a reaction sequence, synthesis of cyclotene has been achieved.

Armed with the above information, it is worth returning to the consideration of acetol as the precursor. The aldol condensation has long been an integral part of the Maillard reaction scheme.[3] If one were to take place between the hydroxymethyl group of one molecule with the keto group of the other, the result would be as shown in Figure 3. The product would be able to lose water and the vinyl alcohol would tautomerize to the aldehyde. Another aldol condensation could then take place between the aldehyde group and the methyl group at the other end of the molecule. The resultant cyclopentane could dehydrate to 3-methylcyclopent-4-en-2-olone, one of the 12 tautomers of cyclotene. This is clearly a more acceptable sequence of reactions than that of Shaw et al.[1]

A mechanism relatively closely related to Figure 3 was outlined by Scarpellino and Soukup[4], who suggested that the first molecule of hydroxyacetone undergoes two

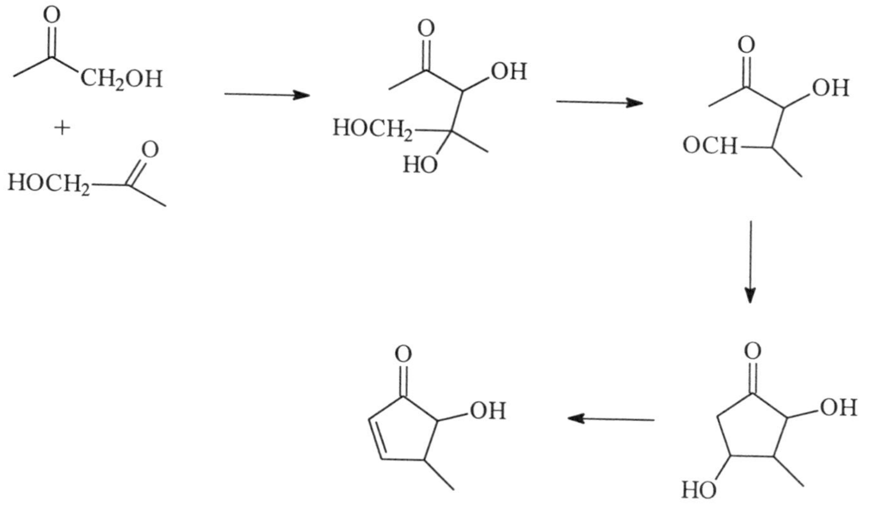

Figure 3. *Proposed mechanism for the formation of 3-methylcyclopent-2-en-2-olone*

tautomeric changes to 2-hydroxypropional before aldol condensation with the methyl group of the second molecule.

Figure 3 uses $C_3 + C_3$. $C_4 + C_2$ would also be possible, as in Figure 4. However, 2-oxobutanal is not known as a Maillard product, whereas the isomeric butane-2,3-dione is, but then the location of the oxygens is incompatible with the cyclotene structure.

That leaves the remaining question: from where does acetol originate? There is no doubt that many sugars, including glucose and xylose, are converted at least in part into acetol on treatment in boiling aqueous sodium hydrogen carbonate.[5] Acetol was also obtained by Weenen and Tjan[6] on analysis of the headspace of the reaction of fructose or glucose with asparagine in glycerol. Both pyruvaldehyde and dihydroxyacetone are well recognized Maillard reaction intermediates. Both are in an oxidation state one level higher than acetol and so a reduction would be required. Hydrolysis of diacetylformoin, another recognized Maillard reaction intermediate, is not likely to go as required to acetol and pyruvic acid, but rather to 2 molecules of pyruvaldehyde. Hayami[7] found that acetol was the main C_3 fragment when glucose was treated with 40% aqueous phosphate buffer at pH 6.7 and he proposed that 1-deoxyglucosone isomerizes to the β-diketose, which cleaves hydrolytically to acetol and glyceric acid. Since both 1-^{14}C-xylose and 5-^{14}C-arabinose lead to exclusively methyl-labelled acetol with about half the activity of the starting material, a symmetrical intermediate must be involved, which Hayami[8] suggested to be the 3-pentoketose. Thus, overall, the mechanism of the formation of acetol during the Maillard reaction is only gradually becoming apparent.

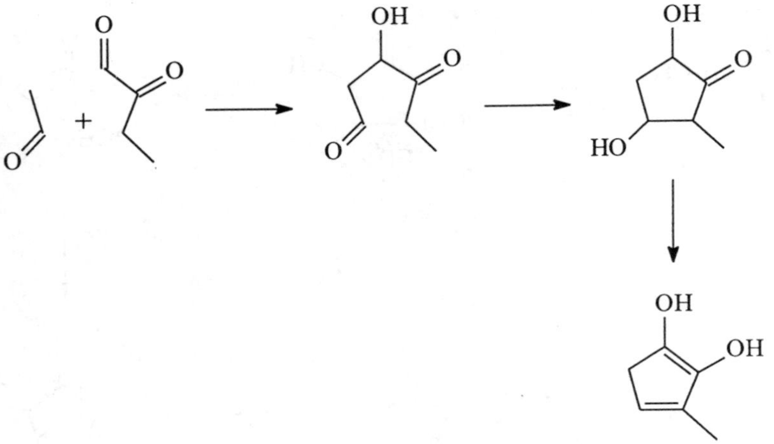

Figure 4. *Possible formation of 3-methylcyclopent-2-en-olone from C_2 + C_3 fragments*

References

1. P.E. Shaw, J.H. Tatum and R.E. Berry, Base-catalyzed fructose degradation and its relation to non-enzymic browning, *J. Agric. Food Chem.*, 1968, **16**, 979-982.
2. H.E. Nursten, Key mechanistic problems posed by the Maillard reaction, in 'Maillard Reaction in Food Processing, Human Nutrition and Physiology', P.A. Finot, H.U. Aeschbacher, R.F. Hurrell and R. Liardon (eds), Birkhäuser, Basel, 1990, pp. 145-153.
3. J.E. Hodge, Chemistry of browning reactions in model systems, *J. Agric. Food Chem.*, 1953, **1**, 928-943.
4. R. Scarpellino and R J. Soukup, Key flavors from heat reactions of food ingredients, in 'Flavor Science: Sensible Principles and Techniques', T.E. Acree and R. Teranishi (eds), American Chemical Society, Eashington, DC, 1993, pp. 309-335.
5. O. Baudisch and H.J. Deuel, Studies on acetol. I.A new test for carbohydrates, *J. Amer. Chem. Soc.,* 1922, **44**, 1585-1587.
6. H. Weenen and S.B. Tjan, 3-Deoxyglucosone as flavour precursor, in 'Trends in Flavour Research'. H. Maarse and D.G. van der Heij (eds), Elsevier, Amsterdam, 1994, pp. 327-337.
7. J. Hayami, Studies on the chemical decomposition of simple sugars. XII. Mechanism of the acetol formation, *Bull. Chem. Soc. Japan,* 1961, **34**, 927-932.
8. J. Hayami, Studies on the chemical decomposition of simple sugars. XI. Acetolformation from ^{14}C-labeled pentoses, *Bull. Chem. Soc. Japan,* 1961, **34**, 924-926.

Fragmentation of Sugar Skeletons and Formation of Maillard Polymers

Roland Tressl, Evelyn Kersten, Georg Wondrak, Dieter Rewicki,[1] and Ralph-Peter Krüger[2]

TECHNISCHE UNIVERSITÄT BERLIN, INSTITUT FÜR BIOTECHNOLOGIE, SEESTR. 13, 13353 BERLIN, GERMANY; [1]FREIE UNIVERSITÄT BERLIN, INSTITUT FÜR ORGANISCHE CHEMIE, TAKUSTR. 3, 14195 BERLIN, GERMANY; [2]INSTITUT FÜR ANGEWANDTE CHEMIE ADLERSHOF e.V., RUDOWER CHAUSSEE 5, 12484 BERLIN, GERMANY

Summary
Labelling experiments involving the reaction of ^{13}C-labelled hexoses, pentoses, and D-lactose with 4-aminobutyric acid (GABA) are described. The distribution of the label was investigated by MS and gave an insight into the formation pathways leading to complementary labelled compounds from hexoses and pentoses with intact carbon skeletons and indicated distinct fragmentations of the sugar skeletons into C_5- and C_4-compounds (furans, N-alkylpyrrolemethanols, N-alkyl-2-formylpyrroles and N-alkylpyrroles). These compounds undergo polycondensations to melanoidin-like macromolecules under mild reaction conditions. In a series of model experiments, different types of polymers were investigated, and individual oligomers were characterized by $^1H/^{13}C$-NMR spectroscopy and FAB-/MALDI-TOF-MS. We postulate that these polycondensation reactions represent the most important driving force in the Maillard reaction.

Introduction

In the Maillard reaction, the formation of macromolecular compounds, generally referred to as melanoidins,[1] is predominant (> 95%, w/w). Low molecular weight coloured condensation products, isolated in Maillard model systems[2,3] do not necessarily represent melanoidin-like substructures. The structure of melanoidins as well as the structure of the low molecular weight compounds, and the factors, that channel the Maillard reaction irreversibly into melanoidins, are hitherto still obscure. In an attempt to clarify the nature of melanoidins, we investigated model reactions that produced melanoidin-like polymers.

Materials and Methods

The labelling experiments in hexose(pentose)-GABA(isoleucine) systems have been described in detail.[4,5] 2-Deoxy-D-ribose and D-lactose have been studied analogously, the details will be described elsewhere.[9,10] 2-Hydroxymethyl (or 2-[^{13}C]Hydroxymethyl)-N-methylpyrrole was prepared from the corresponding aldehydes by $NaBH_4$-reduction in methanol, and the compound, dissolved in trichloromethane, polymerizes spontaneously.[6] Mixtures (1:1, v/v) of N-methylpyrrole and 2-formyl-N-methylpyrrole (or N-methylpyrrole and furfural) in methanol, after addition of catalytic amounts of 1 N HCl, yield polycondensation products during 10 min at 20°C. Oligomers were separated by TLC on 1-mm-silica gel (petroleum ether-ethylacetate 3:1, 1% NH_3). MALDI-TOF-MS (polarity positive, flight path reflection, 20 kV acceleration voltage, nitrogen laser 337 nm): 0.5 µL of the oligomer/polymer mixture (1 mg/mL in $CHCl_3$) and 0.5 µL of a matrix solution (25 mg of 2,5-dihydroxybenzoic acid or 2,4,6-trihydroxyacetophenone/mL ethanol) were mixed on the sample slide, and the solvents were evaporated.

Results and Discussion

On the basis of ^{13}C- and ^{2}H-labelling experiments, we proposed an extended Maillard reaction scheme of hexoses[5] demonstrating the formation of products from intact sugar skeletons as well as from sugar fragments. A new pathway to pyrroles via a 3-deoxy-2,4-diketose intermediate was established (β-dicarbonyl route; see Figure 1, **C**), which competes with the well-known 3,4-dideoxyaldoketose route via **B**. Unexpected results of earlier labelling experiments[7] are easily explained by this new pathway. In the case of pentoses (Figure 1), the quantitative contribution of the two competing pathways (40:60) could be calculated from the observed isotopomeric pyrrolealdehydes.

In the case of hexoses at pH < 5, some of the C_6 compounds formed (e.g., pyrraline from lysine) show crosslinking activity towards nucleophiles, but appear to be slow to polymerize. Based on our previous browning experiments,[8] which indicated the higher reactivity of pentoses, 2-deoxypentoses (free or nucleic-acid-bound) and tetroses, we focused on C_5 and C_4 fragmentation products as precursors of melanoidins derived from hexoses. The results of detailed ^{13}C-labelling experiments[4,5,9] (including complementary labelled sugars) are summarized in Figure 1. Three different fragmentation reactions (α-dicarbonyl -, retroaldol -, and vinylogous retroaldol cleavage) transform the hexose Schiff's base or the hexose intermediates **A - C** into several products, which are related to the pentose, 2-deoxypentose and tetrose systems.

The [1-^{13}C]-D-lactose-GABA labelling experiment indicated a disaccharide-specific fragmentation pathway, which is revealed by the extensive formation of [$^{13}CH_2OH$]furfuryl alcohol.[10] According to Figure 1, only unlabelled furfuryl alcohol is formed from [1-^{13}C]-hexoses via 2-deoxypentoses after loss of the label by α-dicarbonyl cleavage. In contrast, furfuryl alcohol generated from D-lactose is predominantly (74%) singly labelled. To explain this result, the formation route shown in Figure 2 is postulated: due to the blocked β-dicarbonyl route (by 4´-glycosylation), obviously, a vinylogous retroaldol cleavage of intermediate **D** is favoured. Subsequent β-elimination of galactose from the activated allylic position and reduction-cyclization generates furfuryl alcohol with the expected isotopic pattern.

We attributed a high polymerization potential to the endproducts shown in Figure 1 and 2, even under mild reaction conditions (20ºC, H^+-catalysis). This was tested using N-methyl pyrrole derivatives (and their ^{13}C-labelled analogues) as simple synthetic model compounds.[6] In fact, complete polymerization was observed in certain cases. On the basis of MALDI-TOF-MS analysis of the polymeric products and using FAB-MS and $^1H/^{13}C$-NMR for structure elucidation of selected oligomers, two types of regular polymers were identified. The structures of these polymers and the reaction sequences involved are given in Figures 3 and 4.

Type I polymers (Figure 3) were formed from 2-hydroxymethyl-N-methylpyrroles (or furfuryl alcohol) by repetitive electrophilic substitution and subsequent deformylation and dehydration/dehydrogenation, respectively. After exposure to air, the polymers showed a brown to black colour and a strong yellow fluorescence in $CHCl_3$ (λ_{ex} = 385 nm, λ_{em} = 493 nm). The colour may be due either to the chromophore of the methine-bridged polypyrroles or to delocalized radicals. The oligomers P_2 and P_3 have been identified in the 2-deoxyribose-methylamine Maillard system.[6] Interestingly, similar linear oligomers of type I have been detected among the reaction products of lipid peroxidation products with amines.[11]

Type II polymers (Figure 4) were formed by polycondensation of N-methylpyrrole with N-methyl-2-formylpyrrole (or furfural) and subsequent dehydration/hydrogenation in the case of even-numbered oligomers. Model reactions of N-(2-methoxycarbonylethyl)pyrrole and N-(2-methoxycarbonylethyl)-2-formylpyrrole result in corresponding structures of up to about 14 pyrrole moieties. In type II polymers, branched structures are possible.

Of course, the described type I and type II polymers formed in parallel in complex Maillard systems will, presumably, represent distinct domains of native melanoidins, because the formation of regular homopolymers is rather unlikely. On the other hand, the oligo- and polymerizations described are reasonable processes, by which simple Maillard endproducts are incorporated irreversibly into macromolecules, representing the most important driving force for the conversion of sugar-amine systems into melanoidins.

Reaction Mechanisms

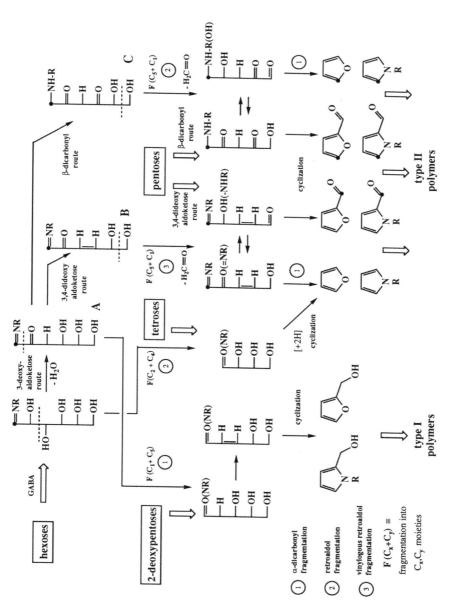

Figure 1. Fragmentation of [1-^{13}C]-D-glucose to C5- and C4- compounds suitable for polycondensation reactions to type I / type II polymers

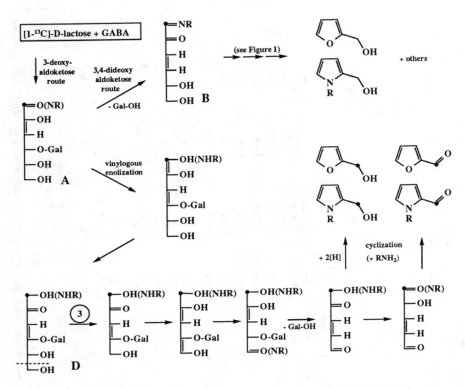

Figure 2. *Fragmentation routes in D-lactose-4-aminobutyric acid (GABA) Maillard systems*

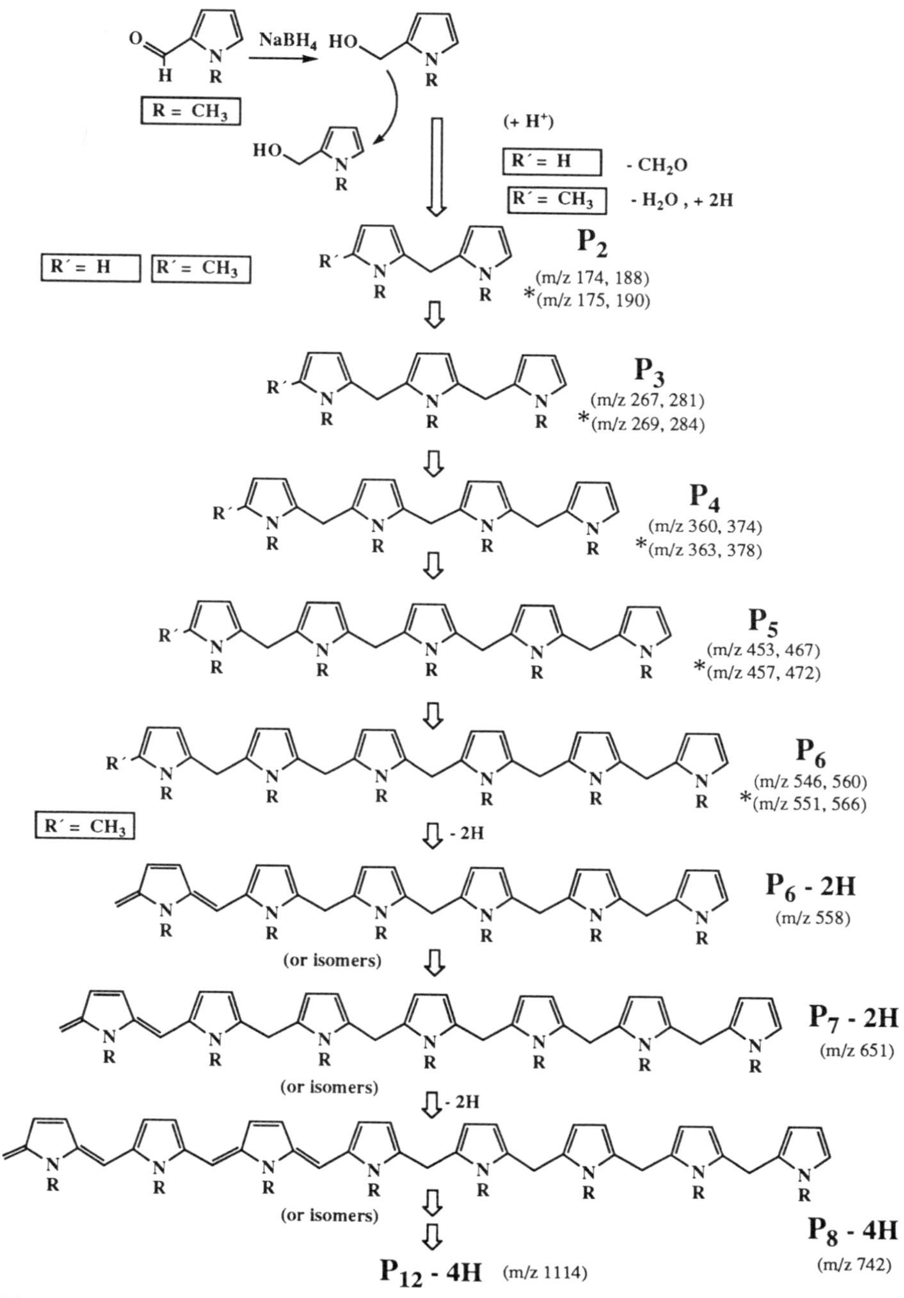

Figure 3. Formation of linear melanoidin-like polymers of type I from 2-hydroxymethyl- N-methyl-pyrrole. * indicates m/z values of the polymer generated from 2-[^{13}C]hydroxymethyl-N-methyl- pyrrole

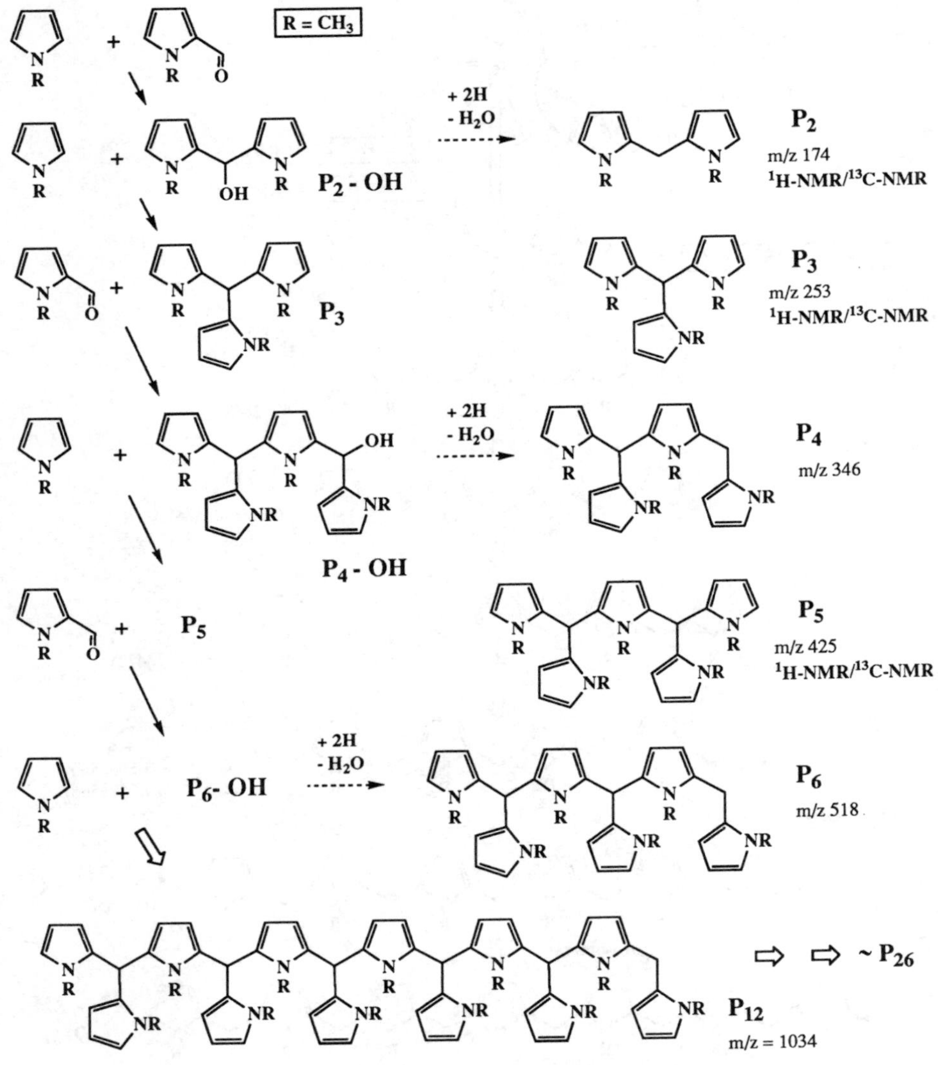

Figure 4. *Formation of melanoidin-like polymers of type II from N-methylpyrrole and 2-formyl-N-methylpyrrole (or furfural)*

Acknowledgements

This work was financially supported by the EU-program FAIR CT96-1080: Optimization of the Maillard reaction: a way to improve quality and safety of thermally processed foods.

References

1. L.-C. Maillard, Action des acides amines sur les sucres; Formation des melanoidins par voie methodique, *C.R. Hebd. Seances Acad. Sci.*, 1912, **154**, 66-68.
2. A. Arnoldi, E.A. Corain, L. Scaglioni and J. M. Ames, New colored compounds from the Maillard reaction between xylose and lysine, *J. Agric. Food Chem.*, 1997, **45**, 650-655.
3. F. Ledl and Th. Severin, Browning reactions of pentoses with amines. Studies on Maillard reaction, XIII, *Z. Lebensm.-Unters. Forsch.* 1978, **167**, 410-413.
4. R. Tressl, E. Kersten and D. Rewicki, Formation of 4-aminobutyric acid specific Maillard products from [1-^{13}C]-D-glucose, [1-^{13}C]-D-arabinose, and [1-^{13}C]-D-fructose, *J. Agric. Food Chem.* 1993, **41**, 2278-2285.
5. R. Tressl, Ch. Nittka, E. Kersten and D. Rewicki, Formation of isoleucine-specific Maillard products from [1-^{13}C]-D-glucose and [1-^{13}C]-D-fructose, *J. Agric. Food Chem.*, 1995, **43**, 1163-1169.
6. R. Tressl, G.T. Wondrak, R.-P. Krüger and D. Rewicki, New melanoidin-like Maillard-polymers from 2-deoxypentoses, *J. Agric. Food Chem.*, 1997, in press.
7. T. Nyhammer, K. Olsson and P.-A. Pernemalm, On the formation of 2-acylpyrroles and 3-pyridinols in the Maillard reaction through Strecker degradation, *Acta Chem. Scand.*, Ser. B, 1983, **37**, 879-889.
8. G.T. Wondrak, R. Tressl and D. Rewicki, Maillard reaction of free and nucleic acid-bound 2-deoxy-D-ribose and D-ribose with ω–amino acids, *J. Agric. Food Chem.*, 1997, **45**, 321-327.
9. R.Tressl, E. Kersten, G.T. Wondrak and D. Rewicki, Mechanistic studies on the Maillard reaction of [^{13}C]-labelled pentoses, in preparation.
10. R.Tressl, E. Kersten, G.T. Wondrak and D. Rewicki, Mechanistic studies on the Maillard reaction of [^{13}C]-labelled disaccharides, in preparation.
11. F.J.Hidalgo and R. Zamora, Fluorescent pyrrole products from carbonyl-amine reactions, *J. Biol. Chem.*, 1993, **268**, 16190-16197.

Recent Advances in the Analysis of Coloured Maillard Reaction Products

Jennifer M. Ames, Richard G. Bailey and John Mann[1]

DEPARTMENT OF FOOD SCIENCE AND TECHNOLOGY, THE UNIVERSITY OF READING, WHITEKNIGHTS, READING RG6 6AP, UK; [1]DEPARTMENT OF CHEMISTRY, THE UNIVERSITY OF READING, WHITEKNIGHTS, READING RG6 6AD, UK

Summary
The objective of the work was to separate and identify low molecular weight heterocyclic reaction products from xylose-glycine and glucose-glycine model systems refluxed for 2 h with the pH maintained at 5. Diode-array spectra and HPLC retention times were obtained for the resolved peaks of these model systems and for a series of standard compounds. Two peaks (one of which was coloured) were isolated from the xylose system and five peaks (two of which were coloured) were isolated from the glucose system. They were subjected to ^{1}H- and ^{13}C-NMR and mass spectral analyses. The colourless peaks were identified as the known structures 4-hydroxy-5-methyl-3(2H)-furanone (from xylose), 4-hydroxy-2-(hydroxymethyl)-5-methyl-3(2H)-furanone (from glucose), 2,3-dihydro-3,5-dihydroxy-6-methyl-4H-pyran-4-one (from glucose) and 5-hydroxymethylfurfural (from glucose). The coloured compounds were the novel structure, 2-acetyl-5-hydroxymethyl-4,5-dihydropyridin-4-one (from xylose) and two isomers of a related acetyldi(hydroxymethyl)dihydropyridinone (from glucose).

Introduction

The coloured products of the Maillard reaction are of two types: the high molecular weight macromolecular materials, commonly referred to as the melanoidins, and the low molecular weight coloured compounds. Very little is known about the nature of the melanoidins, and their analysis, leading to the unravelling of their structures, represents an enormous challenge to the food chemist.[1]

A range of low molecular weight coloured compounds, with two or three heterocyclic rings, has been isolated from Maillard model systems where a sugar and an amino acid have been used as the pair of reactants. The first such compound, 2-furfurylidene-4-hydroxy-5-methyl-3(2H)-furanone, was isolated from xylose heated with either glycine or lysine by Severin and Krönig in 1972.[2] Subsequently, other coloured compounds, with two or more heterocyclic rings, were isolated from xylose and glycine[3-5] and xylose and lysine.[6-8] Recently, the characterization of the first coloured compound with three rings from an aqueous xylose-lysine system was achieved.[8]

Progress in the structural elucidation of the low molecular weight coloured compounds from sugar-amino acid systems has been slow. The reaction product mixture varies in complexity according to the reactants.[9] In addition, individual compounds are usually present in small amounts and are themselves often sufficiently reactive to be transformed by usual isolation procedures, such as TLC. This gives two problems. First, it is often unclear which compounds should be targeted for isolation and structural analysis. The second difficulty is that mild but efficient procedures must be used to isolate the reaction products. Due to the enormity of the task, the efficient allocation of resources is important. The strategy we have adopted is to isolate compounds that appear to be present in model systems prepared from different starting materials or heated under different conditions, in order to attempt to establish reaction pathways that could be common to several systems.

Reaction Mechanisms

Where different combinations of reactants are used, compounds that appear to be structurally similar are targeted for analysis.

The aim of the work described here was to apply this strategy to aqueous xylose-glycine and glucose-glycine model systems.

Materials and Methods

Aqueous solutions (one molal with respect to xylose or glucose and glycine) were refluxed for 120 min with the pH controlled at 5 by the automatic addition of 3M sodium hydroxide. After cooling, the reaction mixture was analysed by HPLC with diode-array detection using an ODS2 reversed phase column and a water-methanol gradient. A selection of standard furans, pyrroles, pyridines and pyrazines was also analysed using the same conditions. Experimental details were similar to those used by Bailey et al.[9]

The cooled reaction mixture was extracted using ethyl acetate. The solvent was removed by rotary evaporation at 40°C and the residue was dissolved in water prior to semi-preparative reversed-phase HPLC. The resolved components observed by HPLC of the mixture were isolated from the ethyl acetate extracts. Experimental details have been described previously.[10]

NMR and mass spectra were obtained as described previously.[10]

Results and Discussion

The xylose-glycine and the glucose-glycine model systems were both dark brown in colour after refluxing for 2 hours. Both model systems gave simple chromatograms by reversed-phase HPLC. The dominant feature was material that eluted in the void volume of the HPLC column. This material was coloured and gave relatively featureless UV-visible spectra which tailed into the visible region. It is presumably polar and/or of high molecular weight. In addition, only two resolved peaks (one of which was coloured) were observed for the xylose system, while only five resolved peaks (two of which were coloured) were seen for the glucose system. In contrast, the chromatograms for the analogous systems, prepared using lysine as the amino acid, were very complex showing a greater range of resolved peaks and different types of broad band in addition to material that was not retained by the HPLC column.[9]

Characteristic diode-array spectra were obtained for standard furans, furanones, pyrroles and pyrazines, but the spectra of pyridine standards were variable and no characteristic pyridine spectrum was observed. A library of spectra for the standard compounds was compiled and searched with spectra acquired from the two model systems. The spectra from the three coloured peaks overlapped almost exactly, suggesting the presence in the systems, of three structurally related compounds. Thus, the resolved peaks from the two model systems, labelled I-VII, could be placed into one of three groups, furan-like, furanone-like and coloured compounds of unknown structure (see Table 1).

The MS and ^{1}H and ^{13}C NMR data for **I** support its identification as 4-hydroxy-5-methyl-3(2H)-furanone (Figure 1). Furthermore, the HPLC retention time and diode-array spectrum of the compound were identical to those of an authentic sample. This well known compound is formed via 2,3-enolisation of the Amadori rearrangement product (ARP) when the starting sugar is a pentose.[11] Compound **V** was identified as 5-hydroxymethylfurfural, by

Table 1 *Analytical data for the peaks resolved by reversed phase HPLC of the xylose-glycine and glucose-glycine model systems*

Peak No.	Group	MW daltons	λ_{max} (nm)	^{1}H-NMR data Coupling constants (ppm) Tetramethylsilane standard	^{13}C-NMR data Coupling constants (ppm) Tetramethylsilane standard
Xylose-glycine					
I	Furanone-like	114	287	(CDCl$_3$) 2.3 (s, 3), 4.5 (s, 2), 6.78 (s, 1)	-
II	Coloured	169	371	CD$_3$OD/CDCl$_3$ (1:9) 2.40 (m, 2), 2.42 (s, 3), 3.62 (m, 1), 3.78 (m, 2), 5.6 (s, 1)	DMSO 25.480 (CH$_3$), 37.834 (CH$_2$), 53.948 (CH), 63.223 (CH$_2$), 100.577 (CH), 152.733 (C=C), 195.698 (C=C), 196.840 (C=C)
Glucose-glycine					
III	Furanone-like	144	289	DMSO 2.12 (s, 3), 3.62 (m, 1), 3.76 (m, 1), 4.45 (m, 1), 5.10 (m, 1) 8.20 (s, 1)	DMSO 12.734 (CH$_3$), 60.198 (CH$_2$), 83.736 (CH), 134.100 (C=C), 172.494 (C=C), 194.632 (C=C)
IV	Furanone-like	144	297	DMSO 2.05 (s, 3), 4.09 (s, 2), 4.32 (s, 1), 5.90 (s, 1) 7.60 (s, 1)	DMSO 15.305 (CH$_3$), 67.389 (CH), 71.321 (CH$_2$), 131.306 (C=C), 157.732 (C=C), 187.358 (C=O)
V	Furan-like	126	283, 229	-	-
VI	Coloured	199	373	-	-
VII	Coloured	199	371	-	-

'-', analysis not performed.

comparison of the retention time and diode-array spectral data with those for the standard compound. It is established as a major product formed via 1,2-enolization of the ARP based on hexose.[11]

The identification of **III** and **IV** proved more complex because the two compounds gave very similar mass spectra and the ^{1}H-NMR data were unusually sensitive to the NMR solvent.[10] However, spiking experiments showed **IV** to be identical (with respect to HPLC retention time and diode array spectra) to an authentic sample of 4-hydroxy-(2-hydroxymethyl)-5-methyl-3(2*H*)-furanone, run under the same conditions. Furthermore, the

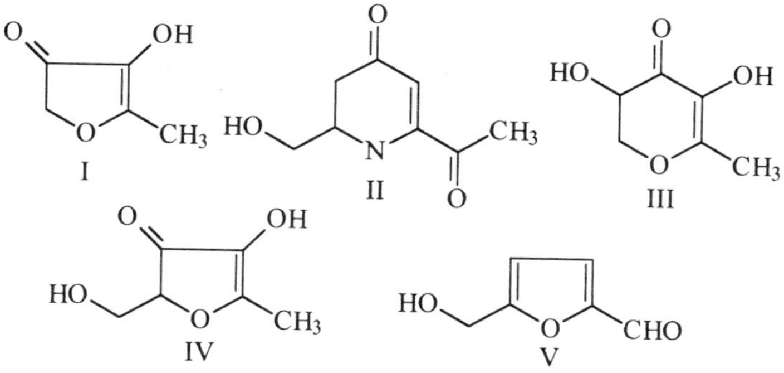

Figure 1. *Structures of compounds I - V.*

^{1}H-NMR spectra, in DMSO, of **IV** and the authentic compound were the same. Compound **IV** is formed from hexoses and is analogous to **I**.

The ^{13}C spectrum of **III** in DMSO showed signals for six carbon atoms, one of which was a carbonyl carbon. The Distortionless enhancement by polarization transfer (DEPT) spectrum showed the presence of one methyl carbon, one methylene carbon and one methine carbon, the two remaining carbon atoms, therefore, being quarternary. In addition, the ^{1}H-NMR spectrum of **III** in DMSO was non-first-order. However, it showed five one-proton signals, as required for the pyranone structure. The combined evidence suggested the identity of **III** to be 2,3-dihydro-3,5-dihydroxy-6-methyl-4*H*-pyran-4-one, a structure supported by the mass spectral data. This γ-pyranone also forms via 2,3-enolization of the ARP formed from hexoses.

The ^{1}H-NMR spectrum of the coloured compound, **II**, was also complex and non-first-order, but the ^{13}C-NMR spectrum gave enough information to allow structures to be proposed. It showed eight carbon atoms, two of which were carbonyl carbons. The DEPT spectrum showed one methyl carbon, one methylene carbon, and two carbons attached to a single proton. The electron impact (EI) and chemical ionization (CI) mass spectra showed, respectively, a molecular ion at *m/z* 169 and a prominent ion at *m/z* 170. The high-resolution EI mass spectrum revealed an accurate molecular weight of 169.0738 amu, with an error of 0.3 ppm for a formula of $C_8H_{11}NO_3$,. The combined spectral evidence suggested that **II** was the novel compound, 2-acetyl-5-hydroxymethyl-4,5-dihydropyridin-4-one. It is the first one-ringed compound to be identified from a sugar-amino acid model system that is reported to be coloured.

Other pyridinones, such as 5,6-dihydro-3-hydroxypyridinone, have been isolated from model systems based on disaccharides, such as maltose, but not from monosaccharides, such as glucose.[12,13] As well as playing a role in colour of foods, such compounds may also possess pharmacological properties. For example, 2-pyridinones are being investigated as potential non-nucleoside reverse-transcriptase inhibitors, for use as anti-HIV agents[14] while both 2- and 4-pyridinones are being investigated as possible antimicrobial agents.[15,16] The activity, if any, of **II** remains to be investigated.

Low-resolution EI mass spectra of **VI** and **VII** each showed very similar fragmentation patterns to each other and to the EI mass spectrum of **II** (see Table 2). The molecular ion at *m/z* 199 for **VI** and **VII** was confirmed by CI for both compounds. By high resolution EI

Table 2. *Low resolution EI MS data for* **II**, **VI** *and* **VII**.

	Abundance				Abundance		
m/z	II	VI	VII	m/z	II	VI	VII
199		10 (M)	10 (M)	96	15		
169	18 (M)			84			10
168		5	5	69		5	18
138	100	100	78	68	16	5	
128		15	63	55			10
111		8	16	54	16	16	
110	5			43	25	25	25
109		8	18	41			21
97		25	100	28	5	5	20

MS, an accurate mass was obtained for the molecular ion of both compounds of 199.0842 amu, with an error of 1.1 ppm for the best fit formula of $C_9H_{13}NO_4$,. Thus, **II** differs from **VI** and **VII** by CH_2O, or an hydroxymethyl group, minus a proton. Preliminary NMR data for these compounds support this.

Although **I**, **III** and **IV** were classified by their diode-array spectra as furanone-like, only **I** and **IV** were actually furanones. However, all three compounds possess an α,β-unsaturated ketone chromophore, and this was the basis of the original classification. Similarly, the fact that the diode-array spectra of **II**, **VI** and **VII** overlapped so closely, is evidence for a common chromophore in these compounds, and provides further justification for the assignment of dihydropyridinone structures to all of them.

Two further compounds were expected in the xylose system. Furfural is well established as a reaction product of this system[11] but it was not present in detectable amounts. Although the second compound, 2-furfurylidene-4-hydroxy-5-methyl-3(2H)-furanone, did not elute during the solvent gradient, a small peak was observed during solvent re-equilibration. It was isolated and analysed by ^{1}H-NMR. The spectrum matched that of 2-furfurylidene-4-hydroxy-5-methyl-3(2H)-furanone[4] which was first reported from xylose and glycine by Severin and Krönig.[2] It is formed from furfural and 4-hydroxy-5-methyl-3(2H)-furanone and its presence proves the formation of furfural in our xylose system.

In conclusion, the reaction products of aqueous xylose-glycine and glucose-glycine model systems have been analysed by HPLC and the resolved components have been chemically classified based on their diode-array spectra. The two resolved components from the xylose system were the known compound, 4-hydroxy-5-methyl-3(2H)-furanone, and a novel coloured compound, 2-acetyl-5-hydroxymethyl-4,5-dihydropyridin-4-one. Chemically related compounds were isolated from the glucose system, i.e., the known compounds, 4-hydroxy-2-(hydroxymethyl)-5-methyl-3(2H)-furanone, 2,3-dihydro-3,5-dihydroxy-6-methyl-4H-pyran-4-one and 5-hydroxymethylfurfural, and two isomers of a coloured acetyldi(hydroxymethyl)dihydropyridinone.

Acknowledgements

Mr P. Heath (Department of Chemistry, The University of Reading) is thanked for running the NMR spectra. Professor D. Cardin (Department of Chemistry, The University of Reading) is thanked for organising the mass spectral analyses. We are grateful to Professor A. Arnoldi (University of Milan, Italy) for supplying an authentic sample of 4-hydroxy-(2-hydroxymethyl)-5-methyl-3(2H)-furanone and to Mr J. Knights (Tastemaker) for supplying an authentic sample of 4-hydroxy-5-methyl-3(2H)-furanone. This work was financed by the Biotechnology and Biological Sciences Research Council (UK).

References

1. G. Rizzi, Chemical structure of colored Maillard reaction products, *Food Rev. Int.*, 1997, **13**, 1-28.
2. T. Severin and U. Krönig, Studien zur Maillard-reaktion. Struktur eines farbigen produktes aus pentosen, *Chem. Mikrobiol. Technol. Lebensm.*, 1972, **1**, 156-157.
3. F. Ledl and T. Severin, Browning reactions of pentoses with amines, *Z. Lebesm.-Unters.-Forsch.*, 1978, **167**, 410-413.
4. H. E. Nursten and R. O'Reilly, Coloured compounds formed by the interaction of glycine and xylose, in 'The Maillard Reaction in Foods and Nutrition', G. R. Waller and M. S. Feather (eds), ACS Symposium Series 215; American Chemical Society, Washington DC, 1983, pp. 103-121.
5. H. E. Nursten and R. O'Reilly, Coloured compounds formed by the interaction of glycine and xylose, *Food Chem.*, 1986, **20**, 45-60.
6. S. B. Banks, J.M. Ames and H.E. Nursten, Isolation and characterisation of 4-hydroxy-2-hydroxymethyl-3-(2'-pyrrolyl)-2-cyclopenten-1-one from a xylose/lysine reaction mixture, *Chem. Ind.*, 1988, 433-434.
7. J. M. Ames, A. Apriyantono and A. Arnoldi, Low molecular weight coloured compounds formed in xylose-lysine model systems, *Food Chem.*, 1993, **46**, 121-127.
8. A. Arnoldi, E. Corain, L. Scaglioni and J.M. Ames, New colored compounds from the Maillard reaction between xylose and lysine, *J. Agric. Food Chem.*, 1997, **45**, 650-655.
9. R.G. Bailey, J.M. Ames and S.M. Monti, An analysis of the non-volatile reaction products of aqueous Maillard model systems at pH 5, using reversed-phase HPLC with diode-array detection, *J. Sci. Food Agric.*, 1996, **72**, 97-103.
10. J. M. Ames, R. G. Bailey and J. Mann, Identification of furanone, pyranone and dihydropyridinone compounds from sugar-glycine model Maillard systems, *J. Agric. Food Chem.*
11. F. Ledl and E. Schleicher, New aspects of the Maillard reaction in foods and in the human body, *Angew. Chem. Int. Ed. Engl.*, 1990, **29**, 565-594.
12. C. Schoetter, M. Pischetsrieder, M. Lerche and T. Severin, Formation of aminoreductones from maltose, *Tetrahedron Lett.*, 1994, **35**, 7369-7370.
13. M. Pischetsrieder and T. Severin, Advanced Maillard products of disaccharides: Analysis and relation to reaction conditions, in 'Chemical Markers for Processed and Stored Foods', T.-C. Lee and H.-J. Kim, (eds) ACS Symposium Series 631; American Chemical Society, Washington DC, 1996, pp. 14-23.
14. T. J. Tucker, W. C. Lumma and J. C. Culberson, Development of nonnucleoside HIV reverse transcriptase inhibitors, *Methods in Enzymology*, 1996, **275**, 440-472.
15. G. M. Eliopoulos, C. B. Wennersten, G. Cole, D. Chu, D. Pizzuti and R. C. Moellering, In-vitro activity of a-86719.1, a novel 2-pyridinone antimicrobial agent, *Antimicrobial Agents and Chemotherapy*, 1995, **39**, 850-853.
16. D. D. Erol and N. Yulug, Synthesis and antimicrobial investigation of thiazolinoalkyl-4(1H)-pyridones, *Eur. J. of Med. Chem.*, 1994, **29**, 893-897.

Determination of the Chemical Structure of Novel Coloured Compounds Generated during Maillard-type Reactions

Thomas Hofmann

DEUTSCHE FORSCHUNGSANSTALT FÜR LEBENSMITTELCHEMIE, LICHTENBERGSTR. 4, D-85748 GARCHING, GERMANY

Summary
The formation of coloured compounds by Maillard reactions of 2-furaldehyde with primary amino acids, secondary amino acids and proteins was investigated. When 2-furaldehyde was heated with L-proline in aqueous solution, the novel intense yellow coloured 5-(S)-(2-carboxy-1-pyrrolidinyl)-2-hydroxy-(E,E)-2,4-pentadienal-(S)-(2-carboxypyrrolidine) imine (**1**) is formed as the main coloured reaction product. Prolonged heating led to the conversion of **1** into (E)-4,5-bis-[(S)-2-carboxy-1-pyrrolidinyl]-2-cyclopenten-1-one (**2**), which, to our knowledge, has as yet not been reported in the literature. On the other hand, reacting 2-furaldehyde with L-alanine, resulted in the formation of another novel, but red coloured compound, namely (S)-N-(1-carboxyethyl)-2-(E)-(2-furyl-methylene)- (**3a**) and (S)-N-(1-carboxyethyl)-2-(Z)-(2-furyl-methylene)-4-(E)-(1-formyl-2-furyl-1-ethenyl)-5-(2-furyl)-3(2H)-pyrrolinone (**3b**). The analogous 3(2H)-pyrrolinones **4a** and **4b** were isolated after substitution of L-alanine with n-propylamine. An aqueous solution of casein, when heated in the presence of 2-furaldehyde, also led to rapid colour formation. After enzymatic digestion of the isolated melanoidin-type colourants, the previously unknown amino acids (S)-N-(5-amino-5-carboxy-1-pentyl)-2-(E)-(2-furyl-methylene)- (**5a**) and (S)-N-(5-amino-5-carboxy-1-pentyl)-2-(Z)-(2-furyl-methylene)-4-(E)-(1-formyl-2-furyl-1-ethenyl)-5-(2-furyl)-3(2H)-pyrrolinone (**5b**) were identified, demonstrating a lysine side chain of a protein linked to a chromophoric structure. This is the first time in the literature that a protein-bound intense coloured Maillard reaction product has been characterized as a substructure of melanoidin. Site-specific [^{13}C] labelling of the precursor 2-furaldehyde and monitoring of the [^{13}C] enriched positions in **3a/b** by ^{13}C NMR spectroscopy led to an unequivocal elucidation of the reaction pathways leading to coloured 3(2H)-pyrrolinones.

Introduction

The Maillard reaction between reducing sugars and amino acids is known to be mainly responsible for the development of desirable colours and flavours that occurs, e.g. during thermal processing of foods, such as baking of bread or roasting of coffee. Whereas a lot of work has focused on the odorants produced during this reaction, surprisingly little is known about the chromophores responsible for the typical brown colour of the foods affected. Several model studies have been conducted in order to provide information on the structures of coloured compounds. However, most reactions have been carried out in organic solvents rather than in aqueous solution and using amines instead of amino acids. The results of such studies can, therefore, not easily be extrapolated to the situation in food systems. Recently, a yellow coloured compound was isolated from a heated xylose-lysine solution. However, the authors were not able to reveal which carbohydrate-derived intermediates were involved in the formation of the colourant.[1]

To gain more detailed insight into formation pathways, a more promising approach might be, therefore, to identify the chromophoric structures formed by reacting certain carbohydrate degradation products with one another under conditions relevant to food. For example, 2-furaldehyde, one of the main reaction products formed from pentoses, was found as a colour precursor in the presence of methylene-active carbohydrate degradation products, such as norfuraneol.[2] Although, as yet, not investigated, it might be predicted that the reaction of 2-furaldehyde with amino acids contributes to colour formation.

Knowledge of such chromophoric structures would also provide important information on the structures of melanoidin-type compounds, because it is likely that such coloured compounds represent some of the substructures incorporated into melanoidins, e.g. via cross-links between chromophores and non-coloured proteins.

The objectives of the present paper were, therefore: to identify coloured compounds formed from 2-furaldehyde in the presence of primary and secondary amino acids; and to characterize cross-links between chromophores and proteins in order to get more detailed insights into reaction pathways leading to coloured compounds.

Materials and Methods

Preparation of reaction mixtures and isolation of colourants generated
2-Furaldehyde/L-prolin: 2-furaldehyde and L-proline was heated in neutral aqueous solution for 15 min at 50°C. For the isolation of the main yellow component, the reaction mixture was freeze-dried, the residue was fractionated by column chromatography on silica gel, followed by flash chromatography on RP 18 material and preparative RP-HPLC.

2-Furaldehyde/L-alanine: 2-furaldehyde was reacted with L-alanine in aqueous solution at pH 7.0. For the isolation of the two main red coloured reaction products, the reaction mixture was extracted with ethyl acetate, the organic layer was distilled under high vacuum and the residue was fractionated by column chromatography and thin layer chromatography, followed by anion-exchange chromatography. Final purification was achieved by RP-HPLC.

2-Furaldehyde/casein: 1.5 g casein was reacted with 5 mMol 2-furaldehyde in phosphate buffer (pH 7.0) for 1 h at 70°C. For isolation of the melanoidin-type colourants generated, the high molecular fraction (MW>10000 Da) was separated by ultrafiltration.

The isolation and spectroscopical structure characterization of the following compounds is described in more detail in the literature: 2-(2-furyl-methylene)-5-(2-furyl)-3(2H)-pyrrolinone[3], **4a/b** (Ref. 3), **1** (Ref. 4), **2** (Ref. 4), **3a/b** (Ref. 4), **5a/b** (Ref. 6) and [^{13}C]-2-furaldehyde.[7]

Results and Discussion

Thermal treatment of neutral aqueous solutions of 2-furaldehyde in the presence of L-proline, L-alanine or casein led to rapid colouration of the reaction mixtures.

Reaction of 2-furaldehyde with the secondary amino acid L-proline
Heating of 2-furaldehyde and L-proline in neutral aqueous solution gave a deep yellow solution after heating for 15 min at 50°C. An intense yellow-coloured reaction product with an absorption maximum at 446 nm could be detected by RP-HPLC/DAD. After isolation, this compound was identified as 5-(S)-(2-carboxy-1-pyrrolidinyl)-2-hydroxy-(E,E)-2,4-pentadienal-(S)-(2-carboxypyrrolidine) imine (**1**; Figure 1) by using several 1D and 2D NMR techniques and, in addition by MS, UV and IR spectroscopy.[4] The novel, intense coloured N-cyanine **1** ($\varepsilon = 5.55 \times 10^4$ l mol^{-1} cm^{-1}) was shown to exist as two isomers (Figure 1).

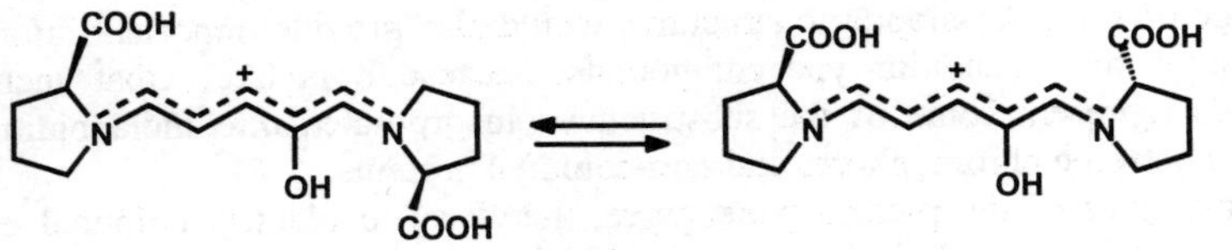

Figure 1. *Structure of the intense yellow colourant 1 identified as 5-(S)-(2-carboxy-1-pyrrolidinyl)-2-hydroxy-(E,E)-2,4-pentadienal-(S)-(2-carboxypyrrolidine) imine*

A reaction mechanism for the formation of compound **1** is proposed in Figure 2. The reaction of L-proline with 2-furaldehyde results in a furfuryliminium ion, facilitating the nucleophilc attack of a second molecule of L-proline. An intramolecular elimination reaction of the vinyloguous hemiaminal then gives, upon ring opening, rise to N-cyanine **1**.

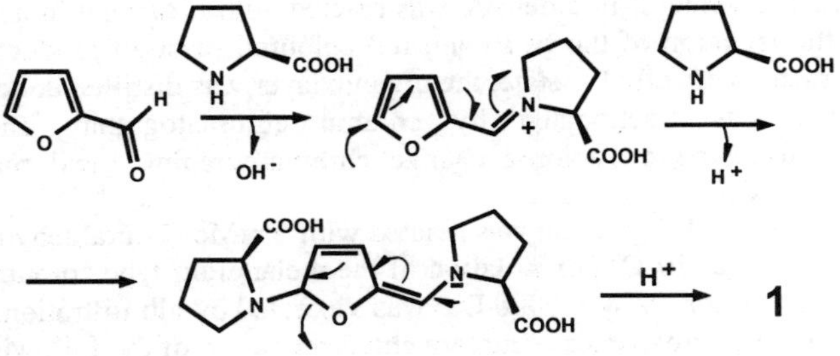

Figure 2. *Reaction pathway leading to 1 from 2-furaldehyde and L-proline*

Further thermal treatment of the colourant **1** generated another previously unknown Maillard reaction product, the (E)-4,5-bis-[(S)-2-carboxy-1-pyrrolidinyl]-2-cyclopenten-1-one (**2**; Figure 3).[4] The (E,E)-configured N-cyanine **1** might tautomerize, upon thermal treatment, into the (Z,E)-isomer, which, upon intramolecular 1,5-cyclization, then results in the cyclopentenone derivative **2**.

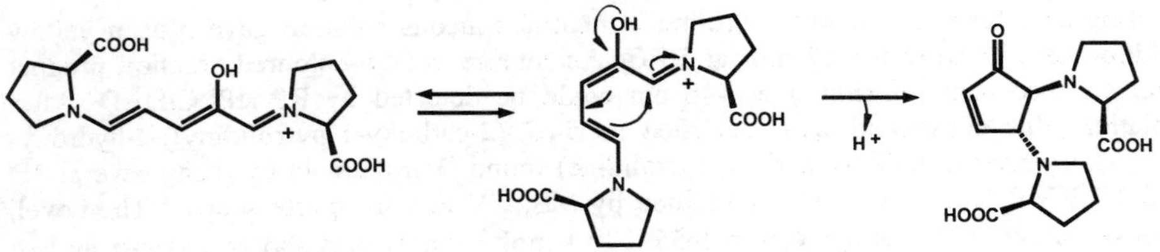

Figure 3. *Formation of (E)-4,5-bis-[(S)-2-carboxy-1-pyrrolidinyl]-2-cyclopenten-1-one (2) from N-cyanine 1*

Reaction of 2-furaldehyde with the primary amino acid L-alanine

When 2-furaldehyde was reacted with the primary amino acid L-alanine, two red coloured reaction products were detected by HPLC exhibiting similar UV/VIS (λ_{max}=330, 414, 480 nm) and identical MS spectra. After isolation of these compounds, their structures were identified as (S)-N-(1-carboxyethyl)-2-(E)-(2-furyl-methylene)- (**3a**, Figure 4) and (S)-N-(1-carboxyethyl)-2-(Z)-(2-furyl-methylene)-4-(E)-(1-formyl-2-furyl-1-ethenyl)-5-(2-furyl)-3(2H)-pyrrolinone (**3b**, Figure 4) by several spectroscopic techniques (1D-, 2D-NMR, MS, UV, IR) as well as synthetic experiments.[3,4] These intense coloured Maillard reaction products ($\varepsilon = 2.9 \times 10^4$ l mol^{-1} cm^{-1}) are reported for the first time in the literature. The (E,E)-configuration of the major isomer **3a** was unequivocally deduced from Nuclear Overhauser effects observed by 2D-NOESY and 2D-ROESY NMR spectroscopy.[3]

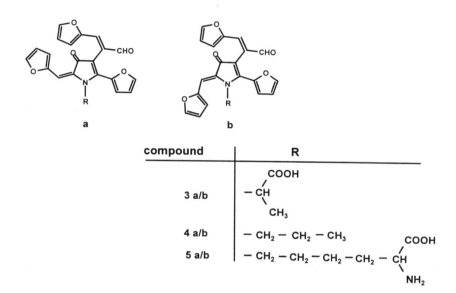

Figure 4. *Structures of red coloured N-substituted 2-(E)-(2-furyl-methylene)- and 2-(Z)-(2-furyl-methylene)-4-(E)-(1-formyl-2-furyl-1-ethenyl)-5-(2-furyl)-3(2H)-pyrrolinones*

Although **3a** and **3b** have only one chiral center, both compounds surprisingly showed a diastereomeric splitting of the NMR spectrum. Substitution of the chiral L-alanine with the achiral n-propylamine afforded N-(n-propyl)-2-(E)-(2-furyl-methylene)- (**4a**, Figure 4) and N-(n-propyl)-2-(Z)-(2-furyl-methylene)-4-(E)-(1-formyl-2-furyl-1-ethenyl)-5-(2-furyl)-3(2H)-pyrrolinone (**4b**, Figure 4), which did not show splitting of their NMR spectra.[3] Conformational analysis performed by 2D-NOESY and 2D-ROESY NMR spectroscopy indicated that the observed presence of two diastereomers in **3a/b** is the result of the existence of two atropisomers.[3] This means, that the β-furyl acroleine branch in the molecule is not coplanar with the planar 2-(E)-(2-furyl-methylene)-5-(2-furyl)-3(2H)-pyrrolinone system, but is positioned either above or below the plane of the three-ring system.

This non-planarity implies that the furyl acrolein moiety is not in conjugation with the three-ring system and should, therefore, play only a minor role in extending the chromophoric system. This suggestion could be confirmed by comparison of the UV maximum of **3a** (λ_{max}=414 nm) with that of synthetic 2-(2-furyl-methylene)-5-(2-furyl)-3(2H)-pyrrolinone, which showed a similar maximum at 406 nm thereby confirming the 2-(2-furyl-methylene)-5-(2-furyl)-3(2H)-pyrrolinone structure as the main chromophore in **3a** (Ref. 3). The UV maximum of (E)-β-furyl acrolein (λ_{max}= 324 nm) was in the range of the second UV maximum of **3a** (λ=330 nm) indicating the isolated furyl acolein moiety. The less intense UV maximum of **3a** at λ=480 nm might be due to the short-time elongation of the chromophoric system at the moment of coplanarity of the molecule during the flipping of the furyl acrolein branch.[3]

Reaction of 2-furaldehyde with casein
When caseine was heated in aqueous solution in the presence of 2-furaldehyde, the colour of the reaction mixture turned rapidly brown-orange. Melanoidin-type colourants (MW > 10000 Da) were isolated by ultracentrifugation and, after enzymatic digestion[5] of the protein backbone, the hydrolysate was analysed by HPLC.[6] A peak, corresponding to a red coloured compound with a molecular mass of m/z 476 was detected showing absorption maxima at 330, 414 and 480 nm. A minor compound, which showed similar UV/VIS and identical MS spectrum, was monitored in a ratio of 1:7 compared with the major component. The comparison of retention times, UV/VIS and MS spectra with those of the synthetic reference compounds, led to the unequivocal identification of the novel amino acids **5a** and **5b** as (S)-N-(5-amino-5-carboxy-1-pentyl)-2-(E)-(2-furyl-methylene)- and (S)-N-(5-amino-5-carboxy-1-pentyl)-2-(Z)-(2-furyl-methylene)-4-(E)-(1-formyl-2-furyl-1-ethenyl)-5-(2-furyl)-3(2H)-pyrrolinone (Figure 4).[6] This is the first time that the identification of a cross-linked amino acid connected with a bulging chromophoric structure (ε = 2.9×10^4 l mol^{-1} cm^{-1}) has been reported.

These results give an idea of how melanoidin-type colourants might be formed by a cross-linking and chromophore-generating reaction between carbohydrate-derived intermediates, such as 2-furaldehyde, and a lysine residue of a non-coloured protein backbone (Figure 5). These findings confirm the idea that low molecular weight coloured compounds might represent some of the substructures incorporated into melanoidins.

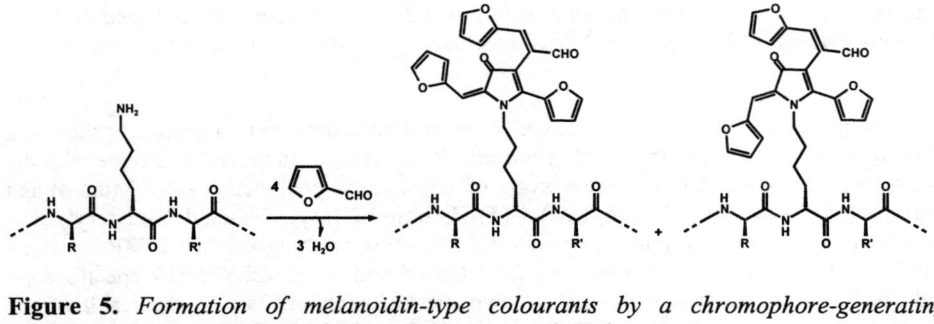

Figure 5. *Formation of melanoidin-type colourants by a chromophore-generating Maillard reaction of 2-furaldehyde and a lysine residue of a protein*

[^{13}C] Labelling studies on the formation of coloured 3(2H)-pyrrolinones

To clarify the formation pathway of the 3(2*H*)-pyrrolinones, a labelling experiment using [^{13}C] site-specific enriched 2-furaldehyde was performed. The positions of labelling, exemplified by **3a/b**, were then monitored by ^{13}C-NMR spectroscopy.[7] The four carbon sites marked in structures **3a/b** in Figure 6 were found to be enriched in [^{13}C] implying the outlined reaction mechanisms as the formation pathway of these 3(2*H*)-pyrrolinones.[7]

It is obvious that, in analogy to the 2-furaldehyde/L-proline reaction, in a first step, the backbone of the colourant is formed by ring opening of 2-furaldehyde. In the following steps, three methylene-active functions of this backbone are condensed with 2-furaldehyde, giving rise to the 3(2*H*)-pyrrolinones. Since these methylene groups might also be saturated by other carbohydrate-derived carbonyls, e.g. glyceraldehyde, a wide range of coloured 3(2*H*)-pyrrolinones may be formed via this reaction pathway. In addition, the amino function can be substituted by a variety of different amino acid moieties and can, furthermore, represent the cross-link between the chromophoric structure and proteins leading to melanoidin-type colourants.

Based on the presented findings, it has to be accepted that the elucidated formation pathway represents a principle reaction route leading to coloured 3(2*H*)-pyrrolinones in Maillard model systems as well as in thermally processed foods.

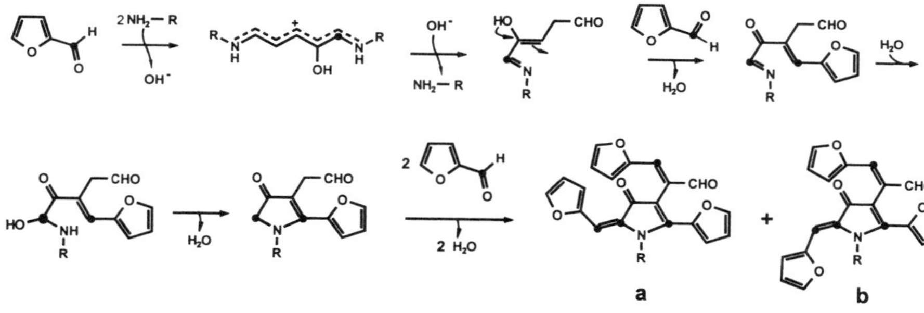

compound	R
3 a/b	−CH(COOH)(CH$_3$)
4 a/b	−CH$_2$−CH$_2$−CH$_3$
5 a/b	−CH$_2$−CH$_2$−CH$_2$−CH$_2$−CH(COOH)(NH$_2$)

Figure 6. *Formation pathway leading to 3(2H)-pyrrolinones from 2-furaldehyde and primary amino compounds (●: [^{13}C] labelled carbon site)*

References

1. A. Arnoldi, E.A. Corain, L. Scaglioni, J.M. Ames, New colored compounds from the Maillard reaction between xylose and lysine, *J. Agric. Food Chem.*, 1997, **45**, 650-655.
2. Th. Severin, U. Krönig, Structure of a colored compound from pentoses (in German), *Chem. Mikrobiol. Technol. Lebensm.*, 1972, **1**, 156-157.

3. T. Hofmann, Determination of the chemical structure of novel coloured 3(2H)-pyrrolinone derivatives formed by Maillard-type reactions, *Helv. Chim. Acta.*, 1997, **80**, 1843-1856.
4. T. Hofmann, Characterization of the chemical structure of novel colored Maillard reaction products from furan-2-aldehyde and amino acids, *J. Agric. Food Chem.*, 1997, submitted.
5. T. Henle, H. Klostermeyer, Evaluation of the extent of the early Maillard reaction in milk proteins by direct measurement of the Amadori product lactuloselysine, *Z. Lebensm. Unters. Forsch..*, 1991, **193**, 119-122.
6. T. Hofmann, Studies on melanoidin-type colourants generated from the Maillard reaction of protein-bound lysine and furan-2-carboxaldehyde - Chemical characterization of a red coloured domaine, *Z. Lebensm. Unters. Forsch.*, 1997, submitted.
7. T. Hofmann, Application of site-specific ^{13}C enrichment and ^{13}C NMR spectroscopy for the elucidation of the formation route leading to a red colored 3(2*H*)-pyrrolinone during Maillard reaction of furan-2-aldehyde and L-alanine, *J. Agric. Food Chem.*, 1997, submitted.

Formation and Structure of Melanoidins in Food and Model Systems

Sonia M. Rogacheva, Margarita J. Kuntcheva, Tzvetan D. Obretenov and Gaston Vernin[1]

DEPARTMENT OF ORGANIC CHEMISTRY, THE HIGHER INSTITUTE OF FOOD AND FLAVOUR INDUSTRIES, 26 MARITZA BLVD, B-4000 PLOVDIV, BULGARIA; [1]LABORATOIRE DE CHIMIE DES AROMES-OENOLOGIE, FACULTE DES SCIENCES ET TECHNIQUES DE ST-JEROME, Av. ESCADRILLE NORMANDIE-NIEMEN, F-13397 MARSEILLE CEDEX 20, FRANCE

Summary
Melanoidins, the final products of the Maillard reaction, are of particular interest in food chemistry. This is due not only to their importance as food components but also to their biological activity. The present study was undertaken to find the connection between the low molecular weight products of thermal degradation of melanoidins, and the structures of the isolated polymers.

The main volatile compounds formed by thermal degradation of food melanoidins and those formed in related model systems were identified by GC-MS. Heterocyclic compounds (mainly furans), alkyl phenols and other compounds were found.

An attempt was made to elucidate the mechanism of melanoidin formation in L-ascorbic acid—α-amino acid model systems. The primary intermediates were isolated and identified. Heterocyclic compounds and some products of the retro-aldol degradation of the α-dicarbonyl compounds play a significant role. A certain amount of the same low molecular compounds was found both among the degradation products of model systems and food melanoidin (or their intermediates). An hypothesis about their participation in melanoidin structures is proposed.

Introduction

As a result of numerous transformations (mainly aldolization reactions), the extremely intricate structures of melanoidins cannot be easily clarified. The formation of melanoidins *in vivo* and their isolation in order to determine their biological activity is increasingly relevant.[1] For 25 years, the goal of our work in this field has been to obtain a valuable set of data about food melanoidins and related model systems. This paper shows some similarities between different melanoidins in spite of great differences in their chemical structures.

Materials and Methods

Melanoidins were obtained from model systems,[2-4] including the interaction of reducing sugars: D-glucose, D-fructose, and L-ascorbic acid, with the α-amino acids glycine, L-lysine, L-glutamic acid and cysteine. Food melanoidins[5-7] were isolated from beer, bread, coffee, malt, cooked fish and meat (beef), and apple juice. They were analysed by spectroscopic methods and gel-permeation chromatography. Volatile extracts obtained by thermal degradation were subjected to GC-MS analysis and identification was achieved using numerous existing mass-spectra libraries and data banks.

Results and Discussion

All melanoidin samples used in this study were obtained under similar experimental conditions.

Elemental analysis
The higher hydrogen content in the melanoidins derived from model systems containing lysine (7%) compared with other amino acids (5%) suggests a higher degree of saturation in such melanoidin molecules. A similar result was observed for food melanoidins obtained from coffee, bread crust and beer. Fish melanoidins are characterized by a high nitrogen content (16%). This high N-content is also a feature of melanoidins obtained from the ascorbic acid - lysine model systems (8%).

UV spectra
All melanoidins absorb strongly in the 260-290 nm region. Half of them show no clear maxima while others absorb as a plateau. A second maximum was observed in the 314-340 nm region for the ascorbic acid-lysine, fructose-glycine, and the glucose-α-amino acid model systems, and for coffee. The second absorption peak could be due to the presence of low molecular weight products.

IR spectra
Melanoidins possess characteristic absorption bands at 1000-1200, 1400-1450, 1530-1680, 1700-1750 and 3000-3900 cm^{-1}, corresponding to the vibrations C-O-C sym., O-H, C-OH; C-H, C-O; C=C, C=N-; C=O and NH, hydrogen bonds, respectively.

Gel-permeation chromatography
Water-soluble melanoidins were fractionated on Sephadex G-50. They were eluted from 70 to 280 mL (Vo=70 mL). Food melanoidins yielded main fractions between 180-200 mL. Melanoidins from model systems were eluted within the same range but they were highly dependent on synthesis factors.

Mass spectra
Direct MS analysis of melanoidins shows the presence of a series of similar fragments: at m/z 39, 41-45, 55, 56, 60, 61, 69, 73, 85, 95-99, 126, 132 and 135. They probably arise either from fragmentation of the melanoidin molecule upon EI or from thermal degradation in the ionization chamber. Such an approach has been already reported.[8-10]

Thermal degradation of melanoidins
Several volatiles, resulting from thermal degradation of melanoidins obtained from the ascorbic acid-glycine, lysine, or glutamic acid model systems were identified (Table 1). Furans, N-containing heterocyclic compounds, O-containing aliphatics, hydrocarbons, O-containing carbocyclic compounds and phenols were also found among the intermediates in the model systems. Degradation products of ascorbic acid were the most abundant.[11] Some of these compounds are also produced during thermal degradation of different food melanoidins. From the foregoing, the following conclusions concerning the structure of melanoidins can be

drawn:
- the melanoidins contain different amounts of nitrogen depending on reaction conditions, and the nature of the model system, or food products;
- the UV absorption spectrum of melanoidins frequently shows a plateau with no clearly defined maxima due to the presence of a large number of chromophores;
- according to their IR spectra, melanoidins are formed from a finite number of closely related structural monomers;
- melanoidins possess different molecular masses whose absolute values still remain unknown.

Table 1. *Number of components from thermal degradation of melanoidins from model systems and apple juice*

Compounds	Model system*			Apple juice
	AA-Gly	AA-Lys	AA-Glu	
Aliphatic hydrocarbons				15
Aromatic hydrocarbons	4	20	3	11
Aliphatic O-containing compounds	7	1	3	4
Aliphatic amines	8	13	4	
Carbocyclic O-containing compounds	3	2	2	
Furans	13	3	10	11
Benzofurans				11
N-containing heterocyclic compounds	31	76	12	
Phenols				14
Unidentified	7	10	32	18

* AA - L-ascorbic acid, Gly - glycine, Lys - lysine, Glu - glutamic acid

Melanoidins from food and model systems give some EI fragments of identical mass. Consequently the presence of identical or closely related structural monomers is probable. This observation deserves a detailed discussion.

The mass spectrum of melanoidin obtained from cooked beef shows about 30 fragments. Among them, fragments at m/z 519, 517 and 515 were observed. Their intensities account for about 1% of the base peak at m/z 44. Additional fragments at m/z 39, 43, 45, 61, 135 and 147 are of low intensity (10%). A careful investigation of this mass spectrum has allowed us to predict likely fragmentation patterns.

Fragments at m/z 519, 517 and 515 result in others at m/z 386, 384 and 383 resulting from the loss of 133, 133 and 132 mass units, respectively. The fragment at m/z 519 also gives a fragment at m/z 279. The latter fragment and that at m/z 135 are thought to be key structures in the melanoidin molecule as they can give all the other ones of lower molecular mass. Thus, the molecular organization of melanoidin could be suggested from the combination of five types of fragment as follows: [m/z 116 + m/z 133 + 2 (m/z 135)]n1; [m/z 133 + m/z 386]n2; [m/z 69 + m/z 133 + m/z 135 + m/z 182]n3; [m/z 133 + m/z 135 + m/z 251]n4; and [m/z 519]n.

Some compounds identified appear to be common to several systems despite the large number of low molecular mass melanoidin precursors and the number of thermal degradation products. The most important are listed in Table 2. Therefore, the process of melanoidin formation in food products and in model systems, which differ in composition or are heated under different conditions, is likely to include parallel and successive reactions that are closely related. The presence of intermediates of similar structure explains the presence of related structural units in the polymer chain. This is not to understate the great number of specific differences in melanoidins.

Table 2. *Identity of melanoidin intermediates and thermal degradation compounds of melanoidins from food and model systems*

Identity	L-Ascorbic acid-amino acid models	AA Models	Thermal degradation of melanoidins from*						
			Glc-Cys	Beer	Barley	Rye	Coffee	Beef	Apple Juice
Aliphatic hydrocarbons	+							+	+
Alkyl benzenes	+	+		+					+
Aliphatic O-containing compounds	+	+		+		+	+		+
2-Hydroxycyclopent-2-en-1-one	+				+	+			
2-Methylcyclohex-2-en-1-one	+					+			
2-Hydroxy-3-methyl-cyclohex-2-en-1-one			+		+	+	+		
C_2H_5 – Furans	+	+	+			+			
2-Hydroxymethylfuran	+		+	+	+	+	+		
2-Acetylfuran	+	+	+			+			
Furfural	+	+	+	+	+				+
5-Methylfurfural			+	+	+	+	+		
5-HMF	+		+	+	+		+		+
Pyrrole	+	+				+			
Pyridine	+	+				+			
CH_3-Pyridines	+	+				+			
Phenols				+	+	+		+	+
Maltol				+		+			
Thiophenes			+			+		+	

* AA - L-ascorbic acid, Glc - glucose, Cys - cysteine

The most interesting differences were those obtained in heated apple juice. Nitrogen-containing compounds were not identified among the 66 volatile compounds obtained by thermal degradation; hydrocarbons and alkylphenols dominated (Table 1). This is probably due to the presence of terpenes and polyphenols in the apple juice. This suggests that, not only low molecular weight products formed from amino-carbonyl interactions, but also available food components could take part in melanoidin formation. It may involve the addition of nitrogen-free components of the reaction mixture to the growing polymer.

The strategy of thermal degradation we have adopted suggests possible regeneration of compounds that appear to be structurally similar to the low molecular weight intermediates of the Maillard reaction. Thus, upon high-temperature treatment, melanoidins influence not only the colour but the aroma of foodstuffs as well.

Acknowledgements

This work was financed by the National Science Fund of the Ministry of Education, Science and Technologies of Republic Bulgaria and the TEMPUS JEP 3529-92/1. Thanks are also expressed to the Aroma Chemistry and Oenology Laboratory at Aix-Marseilles III University, France.

References

1. R. Ikan (ed.), 'The Maillard Reaction. Consequences for the Chemical and Life Sciences', J.Wiley & Sons, Chichester, 1996.
2. T.D. Obretenov, M.J. Kuntcheva and I.N. Panchev, Influence of the reaction conditions on the formation of nondialyzable melanoidines from D-glucose and L-amino acids, *J. Food Process. Preserv.*, 1990, **14**, 309-324.
3. M.J. Kuntcheva, I.N. Panchev and T.D. Obretenov, Influence of reaction conditions on the formation of nondialyzable melanoidines from D-fructose and glycine, *J. Food Process. Pres.*, 1994, **18**, 9-21.
4. S.M. Rogacheva, M.J. Kuntcheva, I.N. Panchev and T.D. Obretenov, L-Ascorbic acid in nonenzymatic reactions. Reaction with glycine, *Z. Lebensm. Unters. Forsch.*, 1995, **200**, 52-58.
5. M.J. Kuntcheva and T.D. Obretenov, Isolation and characterization of melanoidins from beer, *Z. Lebensm. Unters. Forsch.*, 1996, **202**, 238-243.
6. M. Kuntcheva, S. Ivanova and T. Obretenov, Methode zur quantitativen Bestemmung von Fleischmelanoidinen, *Nahrung*, 1983, **27**(10), 897-900.
7. T.D. Obretenov, M.J. Kuntcheva, S.C. Mantchev and G.D Valkova, Isolation and characterization of melanoidines from malt and malt roots, *J. Food Biochem.*, 1991, **15**, 279-294.
8. J.J. Boon, J.W. Leeuw, Y. Rubinsztain, Z. Aizenshtat, P. Ioselis and R. Ikan, Thermal evaluation of some model melanoidins by Curie-point pyrolysis-mass spectrometry and gas chromatography-mass spectrometry, *Org. Geochem.*, 1984, **6**, 805-811.
9. H. Kato and H. Tsuchida, Estimation of melanoidin structiure by pyrolysis and oxidation, *Prog. Food Nutr. Sci.*, 1981, **5**, 147-156.
10. Y. Rubinsztain, P. Ioselis, R. Ikan and Z. Aizenshtat, Investigations on the structural units of melanoidins, *Org. Geochem.*, 1984, **6**, 791-804.
11. G. Vernin, S. Chakib, S.M. Rogacheva, T.D. Obretenov and C. Parkanyi, Thermal decomposition of ascorbic acid, *Carbohydrate Research*, in press.

Formation of N^ε-Carboxymethyllysine from Different Sugars and Glyoxal

Axel Ruttkat and Helmut F. Erbersdobler

INSTITUT FÜR HUMANERNÄHRUNG UND LEBENSMITTELKUNDE, UNIVERSITY OF KIEL, DÜSTERNBROOKER WEG 17, D-24105 K I E L, GERMANY

Summary
Equimolar (0.5 M each) aqueous solutions either of lysine monohydrochloride and selected reducing sugars or glyoxal, and a solid mixture containing casein, 10% reducing sugars and 10% distilled water received heat treatment at 98°C; the glyoxal system also received heat treatment at 40°C. The content of N^ε-carboxymethyllysine (CML) was measured by reversed-phase HPLC analysis. The results show that CML may be formed by the reaction of lysine with aldoses or ketoses like fructose or sorbose. Therefore, CML unlike furosine, is a suitable marker of heat damage involving reactions of ketoses. CML was formed very rapidly after heating a lysine-glyoxal solution at 98°C, reaching a plateau very early. The results suggest that CML is formed from the reaction of lysine with glyoxal formed via sugar oxidation or from the Schiff's bases in the initial stage of the Maillard reaction. However, it appears that under the conditions usually applied in food technology, CML is also formed by oxidative cleavage of the Amadori and possibly the Heyns products.

Introduction

During heat treatment, the ε-amino group of lysine in proteins reacts with reducing sugars forming, in the initial stage of the Maillard reaction, either Amadori products (ketoselysines) in the case of reaction with aldoses or Heyns products (aldoselysines) if reacting with ketoses. The Amadori compounds are more common because glucose is more abundant in processed food systems, either as free glucose or as a constituent of lactose or maltose. They can easily be detected by the furosine method.[1,2] A secondary product of the Maillard reaction is N^ε-carboxymethyllysine (CML) which is formed especially under oxidative conditions by splitting the Amadori compound ε-fructoselysine to CML and erythronic acid.[3] Wells-Knecht and co-workers[4] have recently proposed an alternative pathway for the formation of CML in which the sugar moiety undergoes oxidation to glyoxal, which then reacts with lysine to form CML. CML is more stable than the Amadori product and can be used as an indicator even in severely processed foods. Moreover, it is obviously formed during heating of lysine with ketoses[5] and in this way allows insight into the lysine damage caused by this type of Maillard reaction.

Materials and Methods

CML was synthesized by Lüdemann[6] and was compared with standard material obtained from Nestlé (Lausanne, Switzerland). Other reagents were of the highest quality obtainable from Merck (Darmstadt, Germany), Sigma-Chemie (Deisenhofen, Germany) and Serva (Heidelberg, Germany).

A total of 5 mL of equimolar aqueous solutions (0.5 M each) of lysine monohydrochloride and the tested sugars were heated for 3-24 h in sealed 20mL Pyrex tubes in an oven at 98°C. Under the same conditions, 2.5 mL aqueous solutions of 1 M lysine and 0.1 M glyoxal were heated at 40° and 98°C for 0.5-24 h. After heating, the tubes were removed and cooled immediately in an ice bath. For the preparation of the casein-sugar mixtures, 160 g acid-precipitated casein, edible quality, 87 % protein (Nx6.25), was mixed with 20 g of the sugars and 20 mL of distilled water. Samples of 45 g were filled into tin cans (Siladur, Germany) with a volume of 0.5 L and 0.1 m I.D. Afterwards, the cans were sealed and heated in an oven for 1-16 hours at 98°C. For each heating point, two replicates were prepared in the lysine-sugar model and four replicates in all other experiments.

Sample preparation and reversed-phase HPLC analysis was performed as described elsewhere.[7] The lysine-sugar (-glyoxal) samples were analysed without prior hydrolysis while the casein-sugar samples were hydrolysed with 6.0 M hydrochloric acid after reduction with sodium borohydride. This procedure reduces fructoselysine to hexitol-lysine and, thus prevents secondary CML production during hydrolysis. The identification of CML in the lysine-ketose models was confirmed by the addition of standard CML and by performing gas-liquid chromatography on the heptafluorobutyryl isobutylester of CML as described before.[8] Glyoxal was analysed as the Girard-T derivative by a somewhat modified reversed-phase HPLC method as described by Wells-Knecht and coworkers.[4]

Results and Discussion

As shown in Figure 1, CML formation is approximately linear in relation to time of heating, which is consistent with previous results.[6,9] The reaction with fructose produced slightly less CML than heating with sorbose or glucose but significantly less than the reaction with

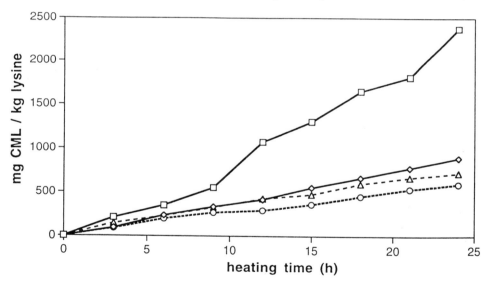

Figure 1. *Formation of N^ε-carboxymethyllysine (CML) in solutions (0.5M) with lysine monohydrochloride and various reducing sugars after heating at 98°C; galactose(—□—); - glucose (—◇—); fructose (---o---); sorbose (--- △ ---)*

galactose. The much higher CML values formed by the reaction of lysine with galactose can be explained in part by a rapid and extensive formation of the Schiff's base as described by Bunn and Higgins[10] who found a 4.6-fold increase of Schiff's base in incubations of hemoglobin with galactose, compared with glucose. Henle and coworkers[11] found higher furosine and pyridosine values in model systems containing tagatoselysine, the Amadori product of the lysine-galactose reaction. Possibly the tagatoselysine molecule forms higher amounts of CML in the same way as it contributes to the formation of furosine and pyridosine.

Figure 2. shows the results of heating casein with sugars `in the dry state´. The water content was much lower in the casein-sugar models than in the lysine-sugar mixtures. The values for CML are given in relation to protein and not lysine as in Figure 1. This means that the absolute CML values in the casein-sugar models are more than 10-fold higher than in the lysine-sugar mixtures. It is known that the Maillard reaction is enhanced in the semi-dry state. However, a higher oxidative influence under the dry conditions may contribute too. In contrast to the results of the lysine-sugar model systems the difference between the sugars is not so distinct in the casein models. Yet, here galactose tended to be more active too. Generally, there was an earlier plateau in CML values, which may be caused by the early occupation of the accessible ε-amino groups of lysine in the caseins. The comparatively lower CML values in the casein-glucose samples heated for 8 and 16 hours may be associated with the earlier and more intensive browning of these samples. It has previously been suggested by Lüdemann[6] that the formation of CML is in competition to browning. This would also be explained by the fact that the casein-fructose samples, which were less browned, showed a higher slope after 4 hours than the other mixtures.

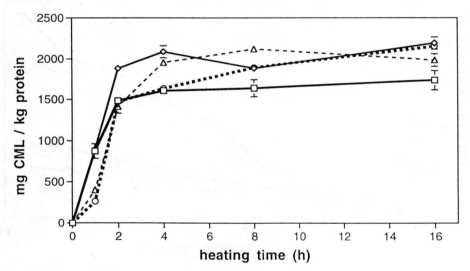

Figure 2. *Formation of N^ε-carboxymethyllysine in a model system with casein, selected reducing sugars and water after heating at 98°C; galactose*(——□——); *-glucose* (——◇——); *fructose* (---o---); *sorbose* (--- △ ---)

As shown in Table 1, CML was formed rapidly after heating a lysine-glyoxal solution at 98°C reaching a plateau very early. On the other hand, the heating at 40°C produced only small amounts of CML.

Glyoxal is formed by heat treatment of sugars. Table 2 shows results from experiments in which 0.2 M solutions of different sugars were heated for up to 8 hours. Some solutions heated for longer periods were extremely browned and were not suitable for quantitative analysis. The pH values of the solutions before heating were between 5.76 in the galactose model and 6.38 in the sorbose solution. During heating, the pH decreased to 3.17 in the sorbose solution and 3.96 in the galactose solution. The addition of iron as Fe(III)chloride increased the formation of glyoxal significantly, in all systems except glucose, which was hardly changed. The addition of nitrite had no remarkable influence (results not shown).

The mechanisms by which CML is formed from the aldoseamines (Heyns products) are not completely clear at the moment. Theoretically, it is possible that the aldoseamines are split to CML by oxidation of the dehydro sugar compound similar to the Amadori products but at a different site of the sugar molecule. However, the mechanism involving glyoxal[4] is conceivable as well. It is obvious from Figure 1 that glucose produces more CML than the ketoses, especially in the longer-heated samples. Wells-Knecht and coworkers[4] stated that the production of CML from the Amadori compounds takes longer than the direct reaction of lysine with glyoxal to form CML. It is possible that, initially, similar amounts of CML are formed from the aldoses and ketoses via the glyoxal route, while, after longer heating, additional CML is formed from the Amadori compounds. On the other hand, in the casein-sugar models, the differences between the sugars were not as distinct as in the lysine-sugar models. The mixture containing fructose produced even more CML than the model with glucose. Possibly, in the almost-dry state of the casein models there was a higher oxidative influence, decomposing either more of the Heyns product or generating more glyoxal, which then could react with lysine.

Table 1. *Formation of N^ε-carboxymethyllysine in a model system with lysine and glyoxal (molar ratio 1:0.1) after heating at 40° or 98°C*

Temperature	Heating time (h)			
	0.5	1	6	24
	mg CML / kg lysine			
40°C	22	78	136	315
98°C	2095±276	4421±5	5683±144	5898±229

Table 2. *Formation of Glyoxal in model systems with sugar solutions (0.2 M) with or without 40 (M Fe(III) after heating at 98°C*

	Heating time (h)			
	2	2+ Fe(III)	4	8
	mg Glyoxal / kg of sugar			
Glucose	1.59±0.20	1.35±0.12	1.32±0.05	2.07±0.13
Galactose	1.12±0.07	30.8±4.1	1.34±0.22	2.75±0.48
Fructose	1.42±0.05	70.6±4.3	1.18±0.09	1.42±0.07
Sorbose	2.84±0.11	44.5±3.8	3.51±0.12	3.32±0.21

As mentioned above, additional CML is formed from Amadori products during acid hydrolysis without the prior reduction step. In the fructose-lysine mixtures, CML was not formed during hydrolysis in contrast to the lysine-glucose and lysine-sorbose mixtures. This may lead to the suggestion that the Heyns product, glucose-lysine, is not split to CML at least under the conditions of acid hydrolysis.

In summary, the results indicate that CML is formed via the glyoxal route[4] from all the sugars studied. However, formation via the Heyns products, if this pathway exists at all, appears to be substantially lower than from the Amadori products, of which the tagatoselysine shows the highest rate of production. Figure 3 summarizes the presently known facts.

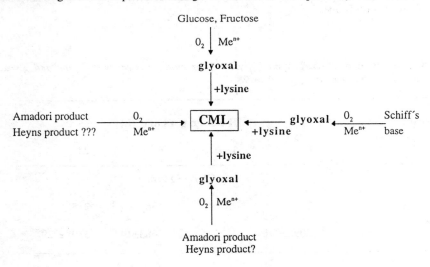

Figure 3. *Proposed pathways for the formation of CML from carbohydrates*

Acknowledgement
We gratefully acknowledge the financial support of the "Institut Danone für Ernährung", Germany

References

1. H.F. Erbersdobler and H. Zucker, Untersuchungen zum Gehalt an Lysin und verfügba-rem Lysin in Trockenmagermilch, Milchwissenschaft, 1966, **21**, 564-568.
2. H.F. Erbersdobler and A. Hupe, Determination of lysine damage and calculation of lysine bioavailability in several processed foods, *Z. Ernährungswiss.*, 1991, **30**, 46-49.
3. M. U. Ahmed, S. R. Thorpe and J. W. Baynes, Identification of N^ε-carboxymethyl-lysine as a degradation product of fructoselysine in glycated protein, *J. Biol. Chem.*, 1986, **261**, 4889-4894.
4. K.J. Wells-Knecht, D.V. Zyzak, J.E. Litchfield, S.R. Thorpe and J. W. Baynes, Mechanisms of autoxidative glycosylation: Identification of glyoxal and arabinose as intermediates in the autoxidative modification of proteins by glucose, *Biochemistry*, 1995, **34**, 3702-3709.
5. A. Ruttkat and H. F. Erbersdobler, N^ε-carboxymethyllysine is formed during heating of lysine with ketoses, *J. Sci. Food Agric.*, 1995, **68**, 261-263.
6. G. Lüdemann, Untersuchungen über Entstehen und Vorkommen von N^ε-Carboxymethyllysin in Lebensmitteln, in 'Agr. Sci. Dissertation Series', H.F. Erbersdobler (ed.), 1989, Vol. 6, Institut für Humanernährung und Lebensmittelkunde, Kiel Germany.
7. J. Hartkopf, C. Pahlke, G. Lüdemann and H.F. Erbersdobler, Determination of N^ε-carboxymethyllysine by a reversed-phase high-performance liquid chromatography method, *J. Chromatogr.*, 1994, **672**, 242-246.
8. W. Büser, H.F. Erbersdobler and R. Liardon, Identification and determination of N^ε-carboxymethyllysine by gas-liquid chromatography, *J. Chromatogr.*, 1987, **387**, 515-519.
9. J. Hartkopf and H.F. Erbersdobler, Modelluntersuchungen zu Bedingungen der Bildung von N^ε-Carboxymethyllysin in *Lebensmitteln, Z. Lebensm. Unters. und -Forschg.*, 1993, **198**, 15-19.
10. H.F. Bunn and P.J. Higgins, Reactions of monosaccharides with proteins: Possible evolutionary significance, *Science*, 1981, *213*, 222-224.
11. T. Henle, V. Herdegen, J. Kurschewitz and H. Klostermeyer, Bildung von Pyridosin und Furosin während der Säurehydrolyse proteingebundener Amadori-Verbindungen, In: Yearly report, research center for milk and foods, Weihenstephan, Germany, 1992, p. 141.

Maillard Compounds as Crosslinks in Heated β-Casein-Glucose Systems

Luisa Pellegrino, Martinus A. J. S. van Boekel, Harry Gruppen, Pierpaolo Resmini*

WAGENINGEN AGRICULTURAL UNIVERSITY, DEPARTMENT OF FOOD SCIENCE, CENTRE FOR PROTEIN TECHNOLOGY P.O. BOX 8129, 6700 EV WAGENINGEN, The Netherlands. *STATE UNIVERSITY OF MILAN, DEPARTMENT OF FOOD SCIENCE AND TECHNOLOGY, VIA G. CELORIA 2, 20133 MILAN, Italy

Summary

Heat-induced covalent aggregation of β-casein reported to be sugar-dependent has been studied with respect to both the early and advanced stages of the Maillard reaction. The aggregation of β-casein via reaction with glucose evaluated by gel-permeation chromatography in model solutions heated under a wide range of concentrations appeared to occur free of interference only in very dilute solutions (1 mg protein/mL), and was not related to the amount of fructoselysine and lysyl pyrraline. It was clearly related to the formation of pentosidine which, however, never exceeded 2 mmol/mol β-casein. The heat-induced incorporation of [U-^{14}C]-labelled Lys or Arg into β-casein in the presence of glucose simulated intermolecular crosslinking and suggested that only few reactive residues can act as acceptors. The radioactive patterns obtained by Ion-exchange chromatography and RP-HPLC, after either acid or enzymatic hydrolysis of the labelled β-casein, showed that highly basic and hydrophobic molecules involving mutual interactions of Arg and Lys are responsible for β-casein aggregation.

Introduction

It is recognized that the Maillard reaction (MR) at the advanced stage leads to progressive protein polymerization. Damage to amino acid residues, development of fluorescence and reduction of enzymatic digestibility are modifications that parallel the sugar-mediated polymerization of proteins. Covalent non-reducible polymerization of casein has been reported to occur during storage of UHT milk.[1,2] In a recent work,[3] we studied the covalent aggregation of β-casein (BC) in aqueous solutions heated in the presence or absence of glucose (GLU). Because this protein has an extended random coil conformation without cyst(e)ine residues, interference due to heat-induced molecular unfolding and S-S bridge rearrangements can be excluded. In the absence of GLU, irreversible polymerization of BC could not be detected using different analytical techniques, whereas the highest levels of lysinoalanine (LAL) were present. This suggests that the formation of intermolecular LAL was negligible indicating that mainly MR compounds were involved. Because of the presence of 5 Arg residues, including the N-terminal, and 11 Lys residues, an high susceptibility of BC to crosslinking reactions due to Maillard compounds is expected. Effect of the MR and role of Arg and Lys in covalent aggregation of BC are studied here.

Materials and Methods

Chromatographically pure BC was prepared from raw bulk milk according to Hipp *et al.*[4] including purification on DEAE cellulose. L-lysine, L-arginine and D-glucose were obtained from Merck. [U-^{14}C]-labelled Lys and Arg (300 nCi/nmol) were from DuPont-NEN.
    Solutions containing the specified concentrations of BC and GLU in 0.1 M sodium phosphate buffer (pH 7.0) were aseptically filtered and heated in sealed vials under the

specified conditions. Both before and after heating, a 12-h incubation at 2°C was done. Gel--permeation chromatography (GPC) of incubated solutions previously diluted with urea up to 5 M and kept at 4°C overnight was performed at 4°C on a Superose 6 HR10/30 column (Pharmacia) with 0.05 M sodium phosphate and 0.1 M sodium chloride solution as eluting solvent and detection at 280 nm. Previously reported methods were adopted for measuring furosine,[5] which was converted to fructoselysine (FL) equivalents by the factor 3.12 (mol to mol), lysyl pyrraline (LPA)[6] and pentosidine (PTD).[3] The incubated solutions were directly subjected to acid hydrolysis for FL and PTD analysis, while for LPA, precipitation of the protein in the presence of 12% TCA (v/v) followed by washing with water was conducted before enzymatic hydrolysis.

Labelled Arg or Lys previously diluted with the corresponding unlabelled molecules (final specific activity from 10 to 50 nCi/nmol) were incorporated into BC following reaction with GLU by heating the solutions at the specified concentrations for 88 h at 73°C. The corresponding blank samples were kept for 88 h at 2°C. The protein was precipitated in the presence of 12% TCA, filtered on a medium-speed paper filter and exhaustively washed with 12% TCA solution. The filter was directly subjected to liquid scintillation counting (LSC) and the activity value of the related blank subtracted from that of the sample. Alternatively, the washed protein was solubilized from the filter with 0.1 N NaOH, neutralized, dried under vacuum and submitted to either acid[3] or enzymatic[7] hydrolysis. The hydrolysates were analyzed for amino acids (AA) with a Biochrom 20 (Pharmacia) and the eluted radioactivity was determined by LSC on 1-min fractions.

Results and Discussion

Covalent aggregation of protein via reaction with GLU takes place after prolonged incubation periods and is dependent on the formation of degradation products arising from the MR.[8] The progress of BC aggregation could be followed sensitively by GPC, but the degree of polymerization could not be precisely estimated because of the unstructured conformation of the molecule. The Amadori compound FL on one hand, and the crosslinks LPA (a putative Lys-Lys crosslink[6,9]) and PTD (identified as a Lys-Arg crosslink in human tissues[10] and food[11]) on the other hand were used as indicators of the extent of glycation and the advanced MR, respectively.

The rate of PTD formation was studied in parallel with the rate of glycation under a wide range of heating conditions and in a highly dilute model solution (Figure 1) in order to prevent self-aggregation of BC.[12] With the level of FL being always at least 30-fold that of PTD, no direct correlation was found between the levels of the two molecules. No more than 2 mmol PTD/mol BC accumulated at the highest temperatures before degradation took place, while heating at 60°C for 2048 h gave 1.45 mmol only (results not shown). Despite these low levels, also found in the literature,[2] formation of PTD was a sensitive indicator of BC aggregation, which occurred when 0.1-0.2 mmol PTD/mol BC accumulated, irrespective of heating conditions. Crosslink formation started even at a very low level of BC glycation (7 mmol FL/mol) provided heating intensity was sufficient (Figure 2). However, the progress of glycation was not directly related to that of aggregation. Systems with comparable low levels of FL showed different extents of aggregation (Figure 2, set A) and comparable levels of aggregation could be observed at very different levels of glycation (Figure 2, AIII, BII, CI). Formation of LPA was related to the progress of the MR only and not to the aggregation extent. In contrast, PTD accumulation mainly followed the behaviour

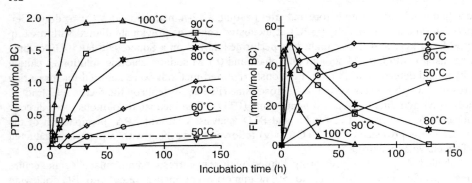

Figure 1. *Pentosidine (PTD) and fructoselysine (FL) formation in model systems of β--casein (47 nmol/mL) and glucose (7.5 μmol/mL) heated at different time/temperature conditions.(---, minimum level of PTD at which aggregation was detected by GPC)*

of protein aggregation at the conditions tested, irrespective of the extent of the MR. In a second experiment, the concentration of both BC and GLU in model systems heated at 80°C for different times was increased by a factor of up to 20, at a constant molar ratio (1:120). At the lowest concentrations (Figure 3, set D), the amount of PTD increased by a factor 10 when heating time increased, while the amount of FL increased from 33 to 46 mmol/mol BC only. When the concentration of the reactants was increased, promoting a

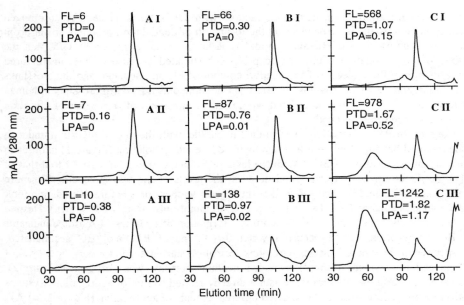

Figure 2. *β-casein (BC) aggregation evaluated by GPC in model systems containing 47 nmol BC/mL and 2 (set A), 20 (set B) or 200 (set C) μmol glucose/mL, heated at 70°C for 12h (set I), 24h (set II) or 48h (set III). Fructoselysine (FL), pentosidine (PTD) and lysyl pyrraline (LPA) concentrations are as mmol/mol BC*

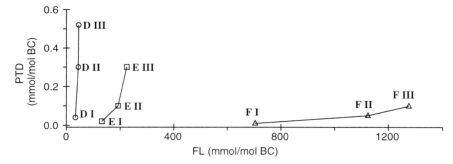

Figure 3. *Pentosidine (PTD) and fructoselysine (FL) in model systems containing 47 and 7.5 (set D), 250 and 30 (set E) and 1250 and 150 (set F) nmol β-casein/mL and μmol glucose/mL, respectively, and heated at 80°C for 4h (set I), 8 h (set II) and 16 h (set III)*

linear increase of FL (from set D to F), PTD formation was progressively depressed. The same behaviour was found for BC aggregation (Figure 4).

These results are consistent with BC aggregation not being directly dependent on the extent of glycation, in the absence of which, however, aggregation could not occur. BC aggregation was related to heating conditions more than to an intensive MR, which can reduce the number of the reactive amino groups acting as acceptors in crosslinks. However, the inhibiting effect of the excess of sugar on formation of advanced glycation products can not be excluded.[13] The protective effect of the increased concentration BC can be also related to BC micellization, which can depress the mobility of the molecules and shield the reactive groups involved in crosslinking. The behaviour of BC aggregation is better evaluated in highly diluted model solutions because, at high concentrations, mutual interactions of protein and rearrangements of the sugar derivatives can interfere with the covalent polymerization. In all our experiments, BC aggregation correlated well with PTD formation only; both may be regulated by a common mechanism but the low amounts of

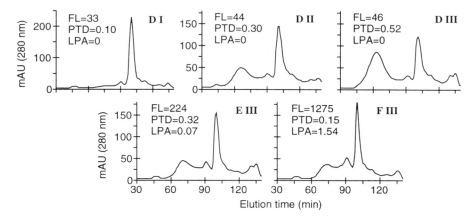

Figure 4. *β-casein aggregation evaluated by GPC in some of the model systems described in Figure 3. Fructoselysine (FL), pentosidine (PTD) and lysyl pyrraline (LPA) concentrations are in mmol/mol BC*

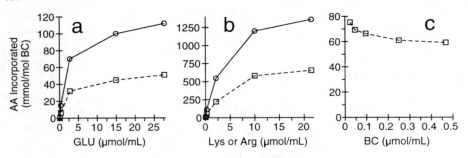

Figure 5. *Incorporation of Lys (——) or Arg (----) into β-casein (BC) in model systems heated at 73°C for 88h containing: (a) 47 nmol BC/mL, 235 nmol Lys or Arg/mL and different amounts of glucose (GLU); (b) 47 nmol BC/mL, 27.7 μmol GLU/mL and different amounts of Lys or Arg; (c) 160 μmol GLU/mL and different amounts of BC and Arg at a constant molar ratio of 1:5*

PTD cannot explain quantitatively the extent of aggregation found. MR molecules other than the crosslinks tested in this study must be involved, including a possible role of 3--deoxyglucosone (3-DG).[14]

In order to elucidate the interactions involved in crosslinking, either [U-^{14}C]-labelled Arg or Lys were incorporated by heating into BC via reaction with GLU. This approach has been used to obtain labelled intermolecular crosslinks.[15] The incorporation rate was high at low concentrations of GLU (Figure 5a) but progressively decreased with increasing sugar concentration, and levelled off at between 55 and 110 mmol/mol BC (less than 0.1% of free AA) depending on the AA. When the concentration of the AA was progressively increased by up to 100-fold, the incorporation was up to 10-fold higher but followed the same behaviour (Figure 5b). A slight decrease in incorporation was observed when both BC and AA concentrations were increased (Figure 5c). The incorporation of Arg was not improved by a preceding 48-h glycation of BC (Table 1, exp. 1 and 2), but it was significantly improved by a 48-h preceding reaction of Arg with glucose (exp. 3).

Although the covalent aggregation of BC during heating can compete with AA

Table 1. *Incorporation of Arg into β-casein (BC) in model systems either containing the three reactants (exp. 1 and 4) or with preceding glycation of BC (exp. 2) or Arg (exp. 3)*

Experiment	Heating time at 73°C (h)	Reactants (nmol/mL) BC (47)	GLU (27700)	Arg (235)	Incorporated Arg (mmol/mol BC)
1	24	x	x	x	23.8
2	48 + 24	x	x	x	21.1
3	48 + 24	x	x	x	49.5
4	72	x	x	x	50.0

incorporation, a parallel behaviour of the two phenomena is evident. AA incorporation seems to be limited by excess GLU and lowered by increasing the concentration of BC.

In both cases, extensive glycation may be the reason. Furthermore, the maximum amount of incorporated Arg and Lys was very low, compared with the number of amino groups theoretically acting as acceptors. Possibly, the same Arg and Lys residues are involved in both aggregation and AA incorporation. It seems that Arg or Lys incorporation is stimulated by their reaction with GLU before linking to BC, but the prior glycation of BC has no effect. It is possible that during the preceding glycation of BC, the few GLU moities resulting in crosslinks react quickly with the reactive amino groups acting as acceptors. In contrast, the Amadori compounds of free Arg cannot easily cross-react in the absence of BC and remain available for further incorporation. The reactive residues not easily glycated themselves but having strong affinity for the GLU moiety of Amadori compounds can act as acceptors, and the guanido group of Arg has such a characteristic.[16] The Arg N-terminal residue is easily glycated because of the conformation of BC and can have a specific role as a donor of the GLU moiety. PTD formation is reported to follow a comparable pathway.[17]

Acid or enzymatic hydrolysates of BC with incorporated labelled Arg or Lys were subjected to IEC and the eluted radioactivity was detected. The chromatograms obtained for acid and enzymatic hydrolysates were almost comparable as well as those for Arg and Lys incorporation. A separation example is reported in Figure 6. Presence of the labelled AA in the related pattern can be due to a partial decomposition of crosslink molecules during hydrolysis.[18] A radioactive peak of unretained material, observed in all the patterns and higher in enzymatic than in acid hydrolysates, is probably due to acid peptides of low-MW.[7] The radioactive components eluted in the area of ammonia have been identified as Arg-derivatives formed by the reaction with either GLU or 3DG,[18] but comparable radioactive peaks are also shown by chromatogram of BC containing labelled Lys. The radioactivity of the PTD peak, eluted just before the LiOH front, was less than 1% of the injected radioactivity, confirming the low quantitative significance of this molecule in crosslinks. A major radioactive peak representing up to 40% of the bound radioactivity was found in all IEC patterns, corresponding to the front that contained the most basic and strongly retained compounds. This material when subjected to RP-HPLC gave radioactive peaks that were strongly retained and eluted after PTD. Some of these peaks were common for both BC labelled with Lys or Arg. The presence of the labels in such basic and hydrophobic compounds suggests that the crosslinks of BC, like PTD, derive mainly from mutual interactions of Arg and Lys via reaction with GLU and include a cyclic moiety. Recently, another compound of this type acting as a Lys-Lys crosslink and containing an imidazole group has been identified in tissues.[19] These results indicate that the mechanism of formation of such molecules could elucidate that of covalent aggregation of BC.

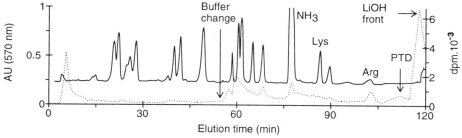

Figure 6. *Ion exchange chromatography and eluted radioactivity (----) of the enzymatic hydrolysate of β-casein with incorporated [U-^{14}C]-Arg*

References

1. M. Zin El-Din and T. Aoki, Polymerization of casein on heating milk, *Int. Dairy J.*, 1993, 3, 581-588.
2. T. Henle, U. Schwarzenbolz and H. Klostermayer, Irreversible crosslinking of casein during storage of UHT-treated skim milk, *IDF Bulletin n° 9602*, 290-298.
3. L. Pellegrino, M.A.J.S. van Boekel, H. Gruppen and P. Resmini, Heat induced aggregation and covalent linkages in β-casein model systems, 'Proceedings of Hannah Symposium 1997', in press.
4. N.J.Hipp, M.L. Groves, J.H. Custers and T.L. McMeekin, Separation of α-, β- and γ-casein, *J.Dairy Sci*, 1952, 35, 272-281.
5. P. Resmini, L. Pellegrino and G. Battelli, Accurate quantification of furosine in milk and dairy products by a direct HPLC method, *Ital. J. Food Sci.*, 1990, 2, 173-183.
6. P. Resmini and L. Pellegrino, Occurrence of protein-bound lysylpyrrolaldehyde in dried pasta, *Cereal Chem.*, 1994, 71, 254-262.
7. De Noni, L. Pellegrino, P. Resmini, and P. Ferranti, About presence of free phosphoserine in ripened cheese and in enzymatic hydrolysate of casein, *Nahrung-Food*, 1997, in press.
8. Shaw and M.J.C. Crabbe, Monitoring the progress of non-enzymatic glycation *in vitro*, *Int. J. Peptide Protein Res.*, 1994, 44, 594-602.
9. R.H. Nagaraj, M. Portero-Otin and V.M.Monnier, Pyrraline ether crosslinks as a basis for protein crosslinking by the advanced Maillard reaction in aging and diabetes, *Arch. Biochem. Biophys.*, 1996, 325, 152-158.
10. D.R. Sell and V.M. Monnier, Structure elucidation of a senescence crosslink from human extracellular matrix, *J. Biol. Chem.*, 1989, 264, 21597-21602.
11. T. Henle, U. Schwarzenbolz and H. Klostermeyer, Detection and quantification of pentosidine in foods, *Z. Lebensm. Unters Forsch.*, 1997, 204, 95-98.
12. W. Buchheim and D.G. Schmidt, On the size of monomers and polymers of β-casein, *J. Dairy Res.*, 1976, 46, 277-280.
13. R.G. Khalifah, P. Todd, A. Booth, S.X. Yang, J.D. Mott and B.G. Hudson, Kinetics of Nonenzymatic glycation of ribonuclease A leading to advanced glycation end products, *Biochemistry*, 1996, 35, 4645-4654.
14. H. Kato, D.B. Shin and F. Hayase, 3-Deoxyglucosone crosslinks proteins under physiological conditions, *Agric. Biol. Chem.*, 1987, 51, 2009-2011.
15. M. Prabhakaram and B.J. Ortwerth, Determination of glycation crosslinking by the sugar-dependent incorporation of [^{14}C]Lysine into protein, *Anal. Biochem.*, 1994, 216, 305-312.
16. C.L. Borders, J.A. Broadwater, P.A. Bekeny, J.E. Salmon, A.S. Lee, A.M. Eldrige and V.P.Pett, A structural role for arginine in proteins: multiple hydrogen bonds to backbone carbonyl oxygens, *Protein Sci.*, 1994, 3, 541-548.
17. S.K. Grandhee and V.M. Monnier, Mechanim of formation of the Maillard protein cross-link pentosidine, *J. Biol. Chem.*, 1991, 266, 11649-11653.
18. Y. Konishi, F. Hayase and H. Kato, Novel imidazole compound formed by the advanced Maillard reaction of 3-deoxyglucosone and arginine residues in proteins, *Biosci. Biotech. Biochem.*, 1994, 58, 1953-1955.
19. R.H. Nagaraj, I. Shipanova and F. M. Faust, Protein cross-linking by the Maillard reaction, *J. Biol. Chem.*, 1996, 271, 19338-19345.

The Maillard Reaction of Ascorbic Acid with Amino Acids and Proteins - Identification of Products

Monika Pischetsrieder, Bernd Larisch, and Theodor Severin

INSTITUT FÜR LEBENSMITTELCHEMIE, LUDWIG-MAXIMILIANS-UNIVERSITÄT MÜNCHEN, SOPHIENSTR. 10, 80333 MUNICH, GERMANY

Summary
The covalent binding of L-ascorbic acid (AA) or its degradation products to proteins (protein ascorbylation) is of importance for food chemistry and medical science. The main reaction product of mixtures of AA and alkylamines, such as propylamine or N^α-acetyllysine, was 3-deoxy-3-(alkylamino)ascorbic acid (3-DAA). When L-dehydroascorbic acid (DHA), the primary oxidation product of AA, was reacted, several products could be detected using HPLC-DAD. The four major compounds were identified as 2-deoxy-2-(alkylamino)ascorbic acid (2-DAA), 3-DAA, oxalic acid monoalkylamide (OMA) and oxalic acid dialkylamide (ODA). In the presence of arginine derivatives and oxygen, the main degradation product of AA was identified as N^δ-(4-(1,2-dihydroxy-3-propyliden)-3-imidazolin-5-on-2-yl)-L-ornithine (DPI).
Since OMA was a major product when DHA was reacted with lysine derivatives, it was speculated that OMA may also be formed as an ascorbylation product of proteins. Thus, OMA-modified protein was synthesized and a polyclonal antibody was produced that was highly specific for OMA. In a competitive ELISA, ascorbylated protein inhibited antibody binding, indicating that OMA is formed as an ascorbylation product of proteins. Oxygen is necessary for the generation of OMA. The antibody did not show cross-reactivity with several proteins that had been glycosylated with other carbohydrates, including AGE-protein.

Introduction

During the Maillard reaction, free amino acids or side chains of proteins react with reducing sugars, resulting in the formation of a variety of products. These compounds are formed during processing and storage of foodstuffs and *in vivo*. They have a major influence on the quality of food and on many processes *in vivo*. Recently, it was discovered that in addition to sugars, L-ascorbic acid (AA) can undergo a Maillard-type reaction.[1] In particular, protein ascorbylation, the covalent binding of AA or its degradation products to proteins, deserves attention. This reaction leads to various changes of the physical and physiological properties of proteins, such as browning, formation of fluoresent compounds,[2] protein cross linking[1] and protein precipitation.[3] It has been suggested that the Maillard reaction of AA has anti-nutritional effects,[4] and causes discolouration[5] and off-flavour formation[6] during food processing. Furthermore, there is strong evidence that under oxidative conditions, protein ascorbylation can also occur *in vivo*,[7] resulting in undesirable physiological effects, such as cataract formation.[2]

However, little is known about reaction mechanisms that lead to protein ascorbylation and products have not been identified. Thus, the objective of this study was to elucidate the structure of ascorbylation products. First, alkylamines and amino acid derivatives were reacted with AA under various conditions and the major products were identified. In a second step, immunological methods were used to show that oxalic acid monoamide (OMA), one of the main products of the reaction of AA with alkylamine, is also formed during protein ascorbylation.

Materials and Methods

3-Deoxy-3-(alkylamino)ascorbic acid (3-DAA)
3-DAA was isolated from a reaction mixture of AA and propylamine or N^α-acetyllysine by preparative HPLC and the structure was determined by spectroscopic data.[8]

2- Deoxy-2-(alkylamino)ascorbic acid (2-DAA), Oxalic acid monoalkylamide (OMA), and Oxalic acid dialkylamide (ODA)
The products were isolated as described before.[9] L-Dehydroascorbic acid (DHA) was heated with propylamine or N^α-acetyllysine and the compounds were isolated by preparative HPLC. Identification of the products was achieved by interpretation of spectroscopic data and comparison of chromatographic and spectroscopic properties with those of synthesized reference compounds.

N^α-acetyl-N^δ-(4-(1,2-dihydroxy-3-propyliden)-3-imidazolin-5-on-2-yl)-L-ornithine (DPI)
DPI was prepared by the reaction of DHA with N^α-acetylarginine and purified using preparative HPLC.[10] Identification was achieved with the help of spectroscopic data.

Analysis of OMA-protein by ELISA
Immunological assays were performed as described before.[11] OMA-protein was synthesized by the reaction of oxalic acid bis(N-hydroxysuccinimide) ester with proteins and polyclonal anti-OMA antibody was obtained by injecting OMA-protein into two rabbits. The antibody was characterized by noncompetitive ELISA, and cross-reactivity with various compounds was determined in a competitive assay.

Results and Discussion

Although the Maillard reaction of AA is of great importance in food chemistry and medicine, little is known about reaction mechanisms and products. Amino acid analyses of ascorbylated proteins have revealed that AA attacks mainly lysine chains and, to a minor extent, arginine and histidine.[12] For the present study, free amino acids were heated with AA under various conditions and the reactions were monitored by HPLC. The major Maillard products of AA were isolated and their structures identified.

Reaction of AA with lysine derivatives under nonoxidative conditions
When AA was heated with N^α-acetyllysine, a major product, which had a UV maximum at 278 nm, could be detected by HPLC. The compound was isolated and its structure determined by spectroscopic data and chemical methods. It was found that the hydroxyl group in position 3 of AA had been substituted by the ε-aminogroup of lysine, resulting in the formation of 3-deoxy-3-(N^ϵ-(N^α-acetyllysin)-yl)-ascorbic acid (3-DAA) (Figure 1). 3-DAA is readily formed during heating of a mixture of AA and N^α-acetyllysine under reflux, but also during incubation for several days at 37 °C.

Reaction of AA with lysine derivatives under oxidative conditions
When a mixture of AA and N^α-acetyllysine was reacted under oxidative conditions or when L-dehydroascorbic acid (DHA), the primary oxidation product of AA was used as a reactant, several products were detected by HPLC. Identification of the five main products was achieved by the use of propylamine as a model compound for N^α-acetyllysine. Later it was shown that analogous products are also formed as derivatives of N^α-acetyllysine.

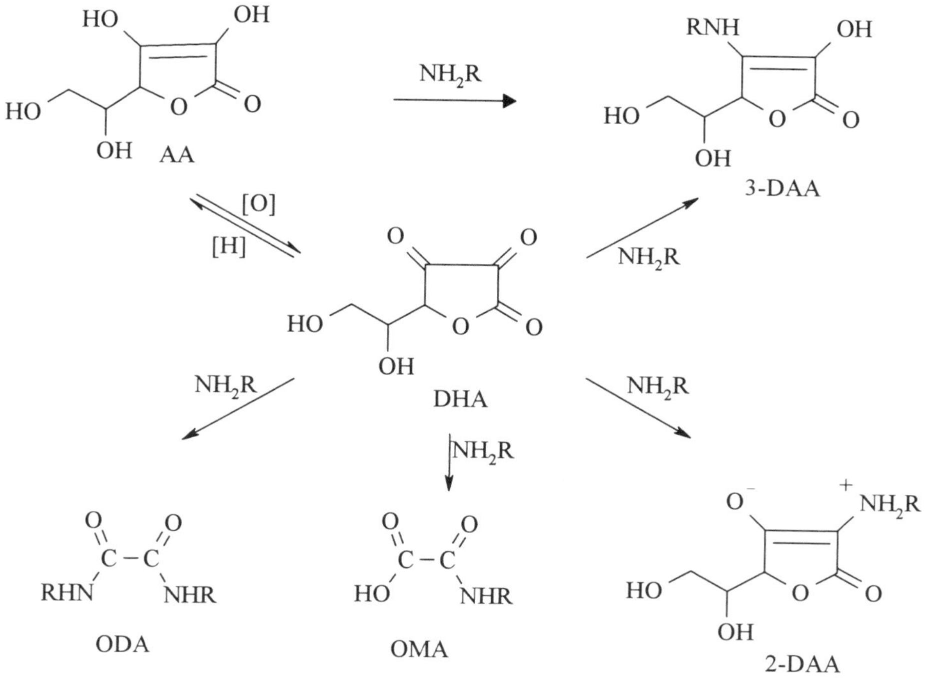

Figure 1. Reactions of AA with alkylamines and N^α-acetyllysine

NH_2R = e.g. propylamine, N^α-acetyllysine

Heating of DHA with propylamine leads to the formation of two minor products which were identified as AA and 3-DAA. It can be assumed that DHA is first reduced by reactive intermediates to give AA which reacts further to produce 3-DAA as described above. The three main products were isolated and their structures elucidated: 2-deoxy-2-propylamino-ascorbic acid (2-DAA), oxalic acid monopropylamide (OMA), and oxalic acid dipropyl-amide (ODA) (Figure 1). It is likely that 2-DAA and 3-DAA have strong anti-oxidative and - in the presence of metal ions - pro-oxidative properties, which even exceed those of AA.[13] Thus, their formation may be of importance in food processing and *in vivo*. Since ODA has two molecules of amine incorporated, it can be assumed that the ODA-protein adduct contributes to AA-induced protein cross linking.

Reaction of AA with arginine derivatives
Incubation of AA with various arginine derivatives, such as N^α-acetylarginine leads to the formation of a major product with characteristic UV maxima at 238 and 283 nm. The compound was isolated and identified as N^δ-(4-(1,2-dihydroxy-3-propyliden)-3-imidazolin-5-on-2-yl)-L-ornithine (DPI, Figure 2).[10] The presence of oxygen was essential for the formation of DPI. Since DPI can also be formed from DHA and L-xylosone, it was concluded that AA is first oxidized and decarboxylated to give L-xylosone. Subsequently, L-xylosone condenses with the guanidinium group of arginine, and DPI is formed following dehydration.

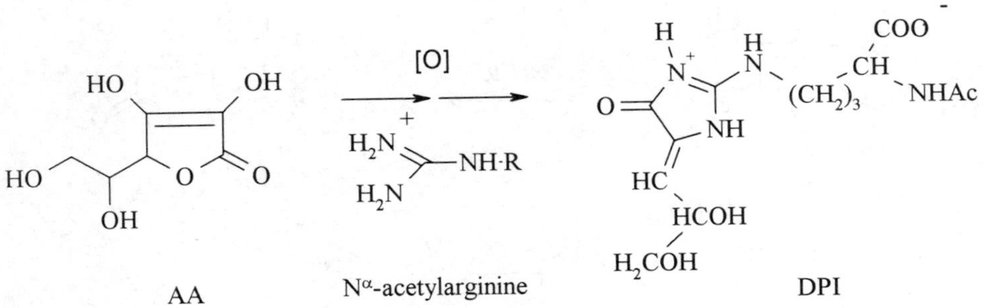

HO—CH₂—CH(OH)—[ring: HO—C=C—OH, O, C=O]   +   H₂N—C(=NH)—NH-R (H₂N⁺)   →[O]   [imidazolone structure with HC(OH)—HCOH—H₂COH side chain and NH—CH(COO⁻)—(CH₂)₃—NHAc]

AA   Nᵅ-acetylarginine   DPI

Figure 2. Reaction of AA with Nᵅ-acetylarginine

With the formation of 2-DAA, 3-DAA, OMA, ODA, and DPI, 5 Maillard products of AA have been isolated, which can be formed as derivatives of free lysine and arginine, or of the side chains of proteins. Assuming their generation during food processing or *in vivo*, such products can contribute to various effects of protein ascorbylation.

Immunochemical detection of OMA as an ascorbylation product of proteins
Since most of the above mentioned Maillard products are not stable under the conditions of acidic or alkaline hydrolysis, it is difficult to prove that they are also formed during protein ascorbylation. Consequently, a polyclonal antibody was raised against OMA-protein, because OMA is one of the main products of the Maillard reaction of AA with low molecular weight amines. The antibody was used in competitive and noncompetitive ELISAs. In a competitive ELISA, total binding inhibition was achieved by the addition of purified oxalic acid mono(Nᵋ-(Nᵅ-acetyllysin)-yl)amide. This result indicated that the antibody was highly specific against OMA. Furthermore, products similar to OMA-protein, such as acetylated and CML-protein, did not show cross-reactivity, confirming the specificity of the antibody (Figure 3).

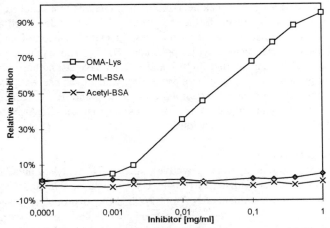

Figure 3. ELISA competition curves to characterize the specificity of the anti-OMA antibody

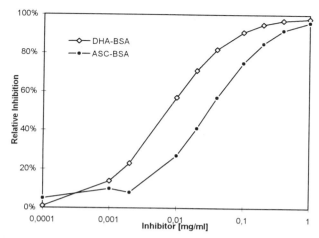

Figure 4. ELISA competition curves with ascorbylated proteins as inhibitors

Subsequently, ascorbylated protein was subjected to competitive ELISA, and dose-dependent binding inhibition of up to 100 % was observed (Figure 4). This result shows that OMA represents an ascorbylation product of proteins. Formation of OMA from AA and proteins was highly dependent on the presence of oxygen. Protein which was incubated in the presence of DHA showed higher reactivity than protein which was treated with AA in the presence of oxygen. In the absence of oxygen, AA did not generate OMA. Glycated or AGE-protein (advanced glycation end-product) did not show cross-reactivity. In addition, OMA could not be detected on proteins, which had been glycosylated by other carbohydrates, such as fructose or lactose. Only ribosylated protein displayed significant cross-reactivity. Thus, it can be concluded that the antibody can be used to distinguish ascorbylated from glycated protein.

References

1. B. Ortwerth and P. Olesen, Ascorbic acid-induced crosslinking of lens proteins: evidence supporting a Maillard reaction, *Biochim. Biophys. Acta*, 1988, **956**, 10-22.
2. K. Bensch, J. Fleming and W. Lohmann, The role of ascorbic acid in senile cataract, *Proc. Natl. Acad. Sci. USA*, 1985, **82**, 7193 - 7196.
3. B. Ortwerth, M. Feather and P. Olesen, The precipitation and cross-linking of lens crystallins by ascorbic acid, *Exp. Eye Res.*, 1988, **47**, 155 - 168.
4. I. Ziderman, K. Gregorski, S. Lopez and M. Friedman, Thermal interaction of ascorbic acid and sodium ascorbate with proteins in relation to nonenzymatic browning and Maillard reactions in foods, *J. Agric. Food Chem.*, 1989, **37**, 1480 - 1486.
5. S. Rogacheva, M. Kuntcheva, I. Panchev and T. Obretenov, L-Ascorbic acid in nonenzymatic reactions I. Reaction with glycine, *Z. Lebensm. Unters. Forsch.*, 1995, **200**, 52 - 58.
6. H. Sakurai, K. Ishii, H. Nguyen, Z. Reblova, H. Valentova and J. Pokorny, Condensation reactions of dehydroascorbic acid with aspartame, in 'Chemical Reactions in Foods III - Proceedings', J. Velisek and J. Davidek (eds), Czech. Chemical Society, Prague, 1996, p. 112.

7. J. Hunt, Ascorbic acid in diabetes mellitus, in 'Subcellular Biochemistry, Vol. 25: Ascorbic acid: Biochemistry and Biomedical Cell Biology', J.R. Harris (ed.), Plenum Press, New York, 1996, p. 369.
8. M. Pischetsrieder, B. Larisch and Th. Severin, Reaction of ascorbic acid with aliphatic amines, *J. Agric. Food Chem.*, 1995, **43**, 3004 - 3006.
9. B. Larisch, M. Pischetsrieder and Th. Severin, Reactions of dehydrascorbic acid with primary aliphatic amines including N^{α}-acetyllysine, *J. Agric. Food Chem.*, 1996, **44**, 1630 - 1634.
10. M. Pischetsrieder, Reaction of L-ascorbic acid with L-arginine derivatives, *J. Agric. Food Chem.*, 1996, **44**, 2081 - 2085.
11. M. Pischetsrieder, B. Larisch and W. Seidel, Immunochemical detection of oxalic acid monoamides which are formed during the oxidative reaction of L-ascorbic acid and proteins, *J. Agric. Food Chem.*, 1997, **45**, 2070 - 2075.
12. S. Slight, M. Feather and B. Ortwerth, Glycation of lens proteins by the oxidation products of ascorbic acid, *Biochim. Biophys. Acta*, 1990, **1038**, 367 - 374.
13. M. Pischetsrieder and Th. Severin, New aspects on the Maillard reaction - formation of aminoreductones from sugars and L-ascorbic acid, *Recent Res. Devel. in Agricultural & Food Chem.*, 1997, **1**, 29 - 37.

Chemiluminescent Products of the Maillard Reaction: Studies on Model Systems

Nobutaka Suzuki,[1] Hideo Hatate,[1] Iwao Mizumoto,[2] and Mitsuo Namiki[3]

[1]NATIONAL UNIVERSITY OF FISHERIES, DEPARTMENT OF FOOD SCIENCE, SHIMONOSEKI 759-65, JAPAN. [2]TOYAMA NATIONAL COLLEGE OF MARITIME, TOYAMA 933-02, JAPAN. [3]TOKYO UNIVERSITY OF AGRICULTURE, SETAGAYA, TOKYO 156, JAPAN

Summary
Chemiluminescent intermediates were prepared from a model Maillard reaction between phenylglyoxal and benzylamine under an oxygen atmosphere as a solid material that gives light emission in the visible region. Such chemiluminescence is similar to that first detected by us from representative Maillard reactions, such as those from glucose and lysine or methylglyoxal and methylamine, both in the presence or absence of oxygen. This suggests that the material in the present studies contains peroxide compound(s). The model Maillard reaction did not give near-infrared luminescence at 1270 nm, which is characteristic of singlet oxygen formation. Structural studies were attempted on the solid material by means of chemical, chromatographic, and spectral methods.

Introduction

Concerning the mechanism of the Maillard reaction, the scheme proposed by Hodge,[1] involving Amadori rearrangement as a key step, has been accepted widely as being the most appropriate. However, the development of novel free-radical products at an early stage of the reaction has been found by Namiki et al.,[2] and a new mechanism has been demonstrated that involves fragmentation of the Schiff's base product followed by the formation of a pyrazinium free-radical product as well as browning products.[3,4] Glomb and Monnier[5] established that half of the N^ε-(carboxymethyl)lysine formed in a glucose-lysine system originates from oxidation of the Amadori product, and the other half originates from a pre-Amadori stage, largely independent of glucose autoxidation. In addition, the generation of weak chemi- luminescence (CL) during the amino-carbonyl reaction has been reported by Bordalen[6] and Namiki et al.,[7] and the effects of various reaction conditions on CL formation have been examined by Kurosaki et al.[8]

This paper discusses our recent work on the development of novel chemiluminescent products in a Maillard reaction, and describes some properties of CL in the model Maillard reaction between phenylglyoxal as a carbonyl compound and benzylamine as an amino compound in relation to free-radical formation and browning.

Materials and Methods

Chemicals
Phenylglyoxal and *Cypridina* luciferin analogue (CLA-Phenyl)[9] were purchased from Tokyo Chemical Co., and benzylamine and NaN_3 from Wako Pure Chemical Co., Osaka, Japan.

Instruments
FT-NMR (JEOL JNM-Ex90, 90 MHz), UV-Vis absorbance photometer (JASCO V-520), IR (JASCO IRA-1), and an ultra-sensitive multi-spectrophotometer (Otsuka Electronics,

IMUC-7000) were employed for the spectral elucidation. A Lumicounter 1000 (Nichion) was used for the quantitative detection of luminescence (Photomultiplier: Hamamatsu Photonics, R878, max. sens. at 410 nm). A near-infrared spectrophotometer[10] was constructed using a Ge-PIN detector (Applied Detector, Model 403L) cooled by liquid nitrogen and a lock-in amplifier (Princeton Applied Research: Model 124A).

Preparation of solid itermediates
Benzylamine solution (879 mg, 8.2 mmol) in water (30 mL) was added dropwise into phenylglyoxal solution (500 mg, 3.7 mmol) in MeOH (30 mL) at 35 °C with stirring under oxygen streams for 10 min. Pale yellow solid precipitates were collected on a glass filter *in vacuo* and washed with cold water three times, resulting in a yield of 295 mg (Sample A).
A second precipitate was collected and washed resulting yellow solid with a yields of 374 mg (Sample B). These two materials were seen to be identical as assessed by TLC (silica gel, CH_2CL_2-n-hexane 1:1) and continied three components initially, but decomposed to at least five components. No starting compounds were found in the products. However, these products were unstable at room temperature and gave light emission in the visible region in the presence or absence of O_2 (λmax at ~ 550 nm: Figures 1 & 2). The changes were monitored by HPLC (a JASCO 807-IT integrator with a 880-PU Intelligent HPLC pump; analyzed on a JASCO FINPAK Silicone C_{18}-5 column, detected by a JASCO 875-UV/Vis detector, at 245 nm, eluted with MeOH)(Figure 3). Its LC-MS was measured (Figure 3).

Chemiluminescence measurement
CL in the visible region was measured using the Lumicounter 1000 CL detector system (sensitivity, 10-14 W; wave length region, 300-650 nm). The counts per second shown on the vertical axis in Figure 1 is CL intensity. It is not a direct count of the number of photons but a unit by which one count corresponds to ~ 10 photons.
[Experiment A] A solution of the amino compound in phosphate buffer or Merczen buffer and a solution of the carbonyl compound in phosphate buffer were mixed in a sample cell of the Lumicounter 1000 and heated to 35 °C in the sample chamber. CL was recorded as a function of time. Effects of additives were also observed. The CL of Samples A and B was observed in MeOH or DMSO solutions under the same conditions.
[Experiment B] A mixture of the amino and carbonyl compounds was heated in a flask. Aliquots taken at time intervals during heating were used for the determination of browning by absorption at 420 nm and CL was determined using the Lumicounter 1000 system.
[Experiment C] For the measurement of CL in reaction system under oxygen-free conditions, frozen solutions of amino and carbonyl compounds were deaerated and exchanged with pure argon gas by degassification *in vacuo* followed by saturation with pure argon gas after thawing. These reactant solutions were mixed in a quartz cuvette under an argon gas atmosphere and CL was measured instantaneously on heating using the Lumicounter 1000 apparatus (Figure 1). All materials used were of reagent grade.

Results and Discussion

Among the carbonyl compounds examined, methylglyoxal was the most reactive, and the development of CL was observed in the reaction with most amino acids and proteins. The effectiveness of systems in the formation of CL are closely realted to their behaviour in promoting free-radical formation and browning.[3] For methylglyoxal which is classified as an inactive compound in free-radical formation,[11] we found a weak ESR signal only in the initial stage of its Maillard reaction with methylamine.[7] This short-life free-radical formation seems to be correlated with high activity in the development of CL. In the present study, we employed phenylglyoxal and benzylamine as substrates to improve the stability of the products.

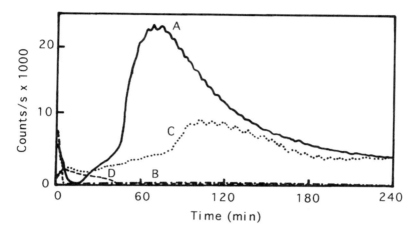

Figure 1. CL-time curves in the visible region. CL of $PhCOCHO$-$PhCH_2NH_2$ in the presence (A) or absence (B) of O_2 and CL of Sample A in the presence (C) or absence (D) of O_2

Examination of the relationship between CL and browning showed that CL was observed at an early stage of the reaction (intensity maximum at ~ 30-60 min), prior to browning. These features in the development of CL are similar to those observed in free-radical formation in the Maillard reaction.[3] Integrated CL curves corresponded well to the browning curves at the same pH, and indicated that the CL-forming reaction product(s) may directly participate in further reactions giving rise to melanoidin.[7]

CL is generated from excited molecules such as singlet oxygen ($^1O_2$) and carbonyl compounds and oxygen play an important role in the formation of such excited molecules.[12] In fact, nitrogen gas bubbling eliminated almost all of CL aerated in the Maillard reaction.[7,8] However, the ESR signal of Maillard reaction systems is detectable in an open test tube and disappears rapidly upon bubbling of air, indicating that the free radical product is fairly stable in the reaction mixture;[13] browning proceeds even in the absence of oxygen, although the effect of oxygen on the reaction remains to be elucidated.[4]

Thus, we examined the effect of oxygen on the CL formed during the Maillard reaction by preparing a strictly anaerobic reaction system using Experiment C. As shown in Figure 1, the reaction of phenylglyoxal-benzylamine as well as methylglyoxal-methylamine in this system showed only a little CL at an early stage, and it increased dramatically with the introduction of air.[7] This indicates that most of the CL is caused by the reaction of oxygen with some reactive product(s), such as a free-radical compound and/or a highly oxidizable compound.

A visible spectrum of the CL developed in the present Maillard reaction system at 35 °C was measured as well as the CL of methylglyoxal-methylamine in aqueous and ethanolic solutions at 50 °C by the diode-array spectrometer.[7] As shown in Figure 2, all of the spectra gave maxima at around 550-580 nm. These were the first reported spectra of CL generated in the Maillard reaction, but it could not be confirmed from these spectra that $^1O_2$ in involved in the reaction.

The fluorescence spectrum of the reaction mixture of methylglyoxal-methylamine, excited at 330 nm, showed an emission maximum at 430 nm at an initial stage then shifted to 580 nm. The reaction mixture did not emit light at 50 ns after excitation, indicating no occurrence of phosphorescence and no contribution of a triplet excited state in the CL development.[7]

We also tried to detect the near-infrared luminescence at ~1270 nm in the CL of the Maillard model systems as direct evidence for the presence of singlet oxygen. However, it was undetectable for the present reaction as well as for the other Maillard reactions examined.

Upon using various radical scavenging and/or reducing agents in conjunction with Experiment A, CL was decreased instantaneously and markedly by the addition of ascorbic acid or cysteine. The mechanism of such activity is not clear. However, it proved the existence of free-radical products and/or production of some active oxygen species.[14,15]

However, no clear evidence of formation of $^1O_2$, a probable factor in the generation of CL, in such reaction systems has yet been demonstrated. The fact that the addition of ascorbic acid caused a decrease in CL suggests $^1O_2$ does not contribute to the generation of CL from ascorbic acid; it was also shown that no detectable CL was measured in solutions of ascorbic acid with or without the presence of Fe or Cu ions.

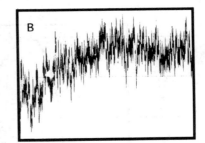
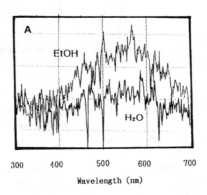

Figure 2. *CL spectra in the visible region*
A: CL spectrum of MeCOCHO-MeNH$_2$ at 50 °C in H$_2$O and EtOH.
B: CL spectrum of PhCOCHO-PhCH$_2$NH$_2$ at 35 °C in MeOH.

These facts suggest that no appreciable amount of $^1O_2$ (leading to CL) is formed by the reaction of a reductone compound with oxygen even in the presence of a metal ion. We also conducted a preliminary experiment to measure CL development during heating of solutions of Amadori product, alone or with an amino acid, using N-(1-deoxy-D-fructosyl)-L-leucine.[7] In this experiment, CL formation was small in the case of the Amadori compound alone compared with the reaction with amino acid. Thus, we concluded[7] that there is no appreciable contribution of the Amadori product to the CL formed in the Maillard reaction.[8]

Addition of CLA-Phenyl[9] enhanced the CL of the present reaction system and subsequent addition of NaN$_3$ gave no further enhancement, indicating the presence of superoxide instead of $^1O_2$. No significant effect on CL was observed upon addition of NaN$_3$ and triethylene diamine (DABCO), suggesting that $^1O_2$ is not involved in the CL development.

Sensitizers of lowest singlet excited state (S$_1$) (9,10-diphenylanthracene, perylene, and rubrene) and sensitizers of S_1 and lowest triplet excited state (T_1) (9,10-dibromoanthracene and dibromoanthracenesulfonic acid) enhanced CL in dilute solutions, while they quenched CL in concentrated solutions, indicating the presence of some singlet excited state molecule(s).

9-Acridone-2-sulfonic acid (a sensitizer of S_1 and O_2^-)[16] enhanced CL, suggesting the presence of S_1 and O_2^-, but no effect on CL was observed upon the addition of superoxide dismutase. CL development was inhibited by 2,4,6-tri-*tert*-butylphenol, and higher concentrations of 2,2'-azobisisobutyronitrile (AIBN) and benzoylperoxide. The fact that it was also enhanced by dilute solutions of AIBN and benzoylperoxide suggests the contribution of some radical intermediate(s) in CL development.

Samples A and B contained three components, which decomposed on dissolution in MeOH or DMSO to give a complex mixture. Some information on their structures was obtained. LC-MS suggested that they have molecular weights larger than the 1:2-adduct of phenylglyoxal and benzylamine.

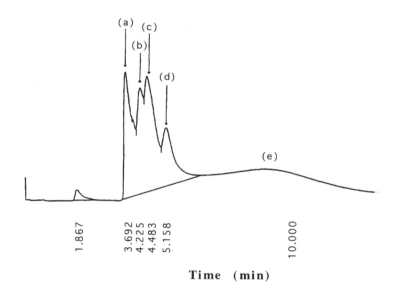

Figure 3. *LC-Mass spectra of Sample A Peaks (a)-(c) exist from the initial stage and peaks (d) and (e) increase gradually. (a): $M^+ = 447$; (b): $M^+ = 449$; (c): $M^+ = 483$; and (d): $M^+ = 539$.*

Acknowledgements

We thank the International Technology Exchange Society, Brunswick, Ohio, USA, for generously providing us with a near-infrared luminescence spectrophotometer. We also acknowledge JEOL Co., Tokyo, Japan, for the LC-MS and International Reagents Co., Kobe, Japan, for kindly supplying facilities.

References

1. J. E. Hodge, Chemistry of browning reaction in model systems, *J. Agric. Food Chem.*, 1953, **1**, 928-943.
2. M. Namiki, T. Hayashi and S. Kawakishi, Free radicals developed in the amino- carbonyl reaction of sugars with amino acids, *Agric. Biol. Chem.*, 1973, **37**, 2935- 2937.
3. M. Namiki and T. Hayashi, A new mechanism of the Maillard reaction involving sugar fragmentation and free radical formation, *ACS Symp. Ser.*, 1983, 255, 21-46.
4. M. Namiki, Recent studies on the browning reaction mechanism and the development of antioxidants and mutagens, *Adv. Food Res.*, 1988, **32**, 115-184.
5. M. A. Glomb and V. M. Monnier, Mechanism of protein modification by glyoxal and glycoaldehyde, reactive intermediates of the Maillard reaction, *J . Biol. Chem.*, 1995, **270**, 10017-10026.
6. B. E. Bordalen, 'Analytical Application of Bioluminescence and Chemiluminescence', Academic Press, New York, 1984, pp. 577-579.
7. M. Namiki, M. Oka, M. Otsuka, T. Miyazawa, K. Fujimoto, K. Namiki, N. Kanamori and N. Suzuki, Chemiluminescence developed at an early stage of the Mailard reaction, in 'Maillard Reactions in Chemistry, Food, and Health', T. P. Labuza, G. A. Reineccius, V. M. Monnier, J. O'Brien, and J. W. Baynes, Editors., Royal Society of Chemistry, Cambridge, 1994. pp. 88-94.
8. Y. Kurosaki, H. Sato and M. Mizugaki, Extra-weak chemiluminescence of drugs. VI. Extra-weak chemiluminescence arising from the amino-carbonyl reaction, *J. Biolum. Chemilum.*, 1989, **3**, 13-19.
9. M. Nakano, Determination of superoxide radical and singlet oxygen based on chemiluminescence of luciferin analogues, *Methods Enzymol.*, 1990, **186**, 585-591.
10. N. Suzuki, I. Mizumoto, Y. Toya, T. Nomoto, S. Mashiko and H. Inaba, Steady-state near-infrared detection of $^1O_2$: A Stern-Volmer quenching experiment with luminol, SOD, and *Cypridina* luciferin analogues, *Agric. Biol. Chem.*, 1990, **54**, 2783-2787.
11. T. Hayashi, Y. Ohta and M. Namiki, Electron resonance spectral study on the structure of the novel free radical products formed by the reactions of sugars with amino acids and amines, *Agric. Food Chem.*, 1977, **25**, 1282-1287.
12. A. K. Campbell, 'Principle and Application in Biology and Medicine', Ellis Horwood, Chichester, 1988, pp. 15-125.
13. M. Namiki and T. Hayashi, Development of novel free radicals during the aminocarbonyl reaction of sugars with amino acids, *J. Agric. Food Chem.*, 1975, **23**, 487-491.
14. I. Fridovich, Oxygen radicals, hydrogen peroxide and oxygen toxicity, in 'Free Radical in Biology', W. A. Pryor, ed., Academic Press, New York, 1975, pp. 239-277.
15. P. A. Seib and B. M. Tolbert, 'Advances in Chemistry', ACS, Washington, DC, 1982, **200**, p. 163.
16. N. Suzuki, B. Yoda, T. Nomoto, I. Mizumoto, H. Inaba and T. Goto, Studies on the chemiluminescent detection of active oxygen species: 9-Acridone- 2-sulfonic Acid, a specific probe for superoxide, *Agric. Biol. Chem.*, 1991, **55**, 1561- 1564.

Food Technology

The Use of Capillary Electrophoresis to Investigate the Effect of High Hydrostatic Pressure on the Maillard Reaction

Vanessa M. Hill, Jennifer M. Ames, Dave A. Ledward and Louise Royle

DEPARTMENT OF FOOD SCIENCE AND TECHNOLOGY, THE UNIVERSITY OF READING, WHITEKNIGHTS, PO BOX 226, READING, RG6 6AP, UK

Summary
High pressure processing using pressures in the range 100 - 1000 MPa is now a commercially viable preservation technique for selected foods. Such foods, typically jams and fruit juices, are believed to posses superior colour and flavour, retaining many of their natural qualities. Very little is known about the effect of pressure in this region on the Maillard reaction. Aqueous glucose - lysine solutions (initial pH 10.1) were reacted at 60°C at either atmospheric pressure or 600 MPa. The total reaction mixtures were analysed by capillary electrophoresis with diode-array detection. The data suggested that the atmospheric and high-pressure systems were similar qualitatively but that some large quantitative differences existed. Larger amounts of a "pyridine-like" compound were seen in the high-pressure system, whilst other compounds were present at either lower or higher amounts in the high-pressure system.

Introduction

The use of high pressure as a means of processing and preserving foods has been recognized since the turn of this century. Progress has, however, been slow and high-pressure processing was not a commercially viable technique until recently. Large-scale high-pressure equipment is now available to subject foods to the pressures necessary to either preserve or modify their eating quality.[1] High-pressure-processed foods, such as jams and juices, are currently on sale in Japan, whilst pressure-treated orange juice has recently become available in France, and pressurised avocado paste has gone on sale in America.

It is reported that pressures in the range of 100 - 1000 MPa can have similar effects to thermal processing for the preservation of foods. Most microorganisms are inactivated, a likely cause being altered permeability of cell membranes.[2] Unfortunately, spores (e.g., those of *Clostridium botulinum*) are resistant to pressure,[3] but the use of moderate temperatures, such as 40 - 60°C, coupled with pressurization, may be an efficient means of destroying them.[4]

In contrast to heat treatment, high pressure does not affect the vitamin content, flavour, or colour of foods under most circumstances, because covalent bond rupture will not occur under pressure. Thus, pressure-treated foods, such as fruit juices and jams, should have improved eating qualities.[2,5] However, the use of moderate temperatures, in combination with pressurization, may influence chemical reactions, such as lipid oxidation and Maillard browning.[6,7] In fact, all chemical reactions occurring with a decrease in total molar volume will be enhanced by pressure.[8]

The Maillard reaction is known to be affected by many factors, including temperature and pH,[9] but little is known about the influence of high pressure on it. Tamaoka *et al.*[7] measured the velocities of the condensation reaction between various amino and carbonyl compounds, and of the subsequent development of browning. They stated that the development of brown colour was greatly suppressed when pressures of 200 - 400 MPa were applied to a xylose - lysine system (prepared in 100 mM NaHCO$_3$, pH 8.2), at 50°C, compared with the same system incubated at atmospheric pressure. Hill et al.[10] incubated glucose - lysine systems at 50°C over the pH range 5.1 - 10.1. When the initial pH was above ~ 7, the rate of colour development at 600 MPa was enhanced, compared with that at atmospheric pressure, whilst the reverse was seen at lower pH values, apparently disagreeing with the results of Tamaoka et al.[7] They also repeated the xylose - lysine reaction of Tamaoka et al.[7] The solution of xylose - lysine was prepared in pH 8.2 buffer, as described by Tamaoka et al.,[7] but this gave a final solution of only pH 6.3 before incubation. Results were in agreement with Tamaoka et al.[7], browning being suppressed under pressure. Therefore, both studies show that browning is suppressed in Maillard systems with pH values less than ~ 7.

Hill et al.[11] also studied the influence of high pressure on the formation of volatile reaction products in an aqueous glucose - lysine solution incubated at 60°C, initial pH 10.1. For samples with the same degree of colour development (as assessed by absorbance at 420 nm), those subjected to high pressure had far lower levels of volatile compounds than those reacted at atmospheric pressure.

The aim of the study reported here was to compare the total reaction products of an aqueous glucose - lysine model system, reacted at either atmospheric pressure or 600 MPa, using capillary electrophoresis (CE).

Materials and Methods

The aqueous model system, initial pH 10.1, comprised glucose and lysine (one molal with respect to each reagent); 10 mL aliquots were incubated in sealed polyethylene bags, either at atmospheric pressure and 60°C ± 0.1°C, or in a high pressure rig at 600 MPa and 60°C ± 2°C. Samples were incubated for either 8 h at atmospheric pressure or for 5 h at 600 MPa. Further details are reported in Ref. 10. Absorbance values at 420 nm were measured using a Perkin-Elmer (Beaconsfield, UK) 552 spectrophotometer. Six replicate solutions were prepared and aliquots were incubated at both atmospheric pressure and 600 MPa. A blank was prepared from water (pH 9.2) and incubated at both atmospheric pressure and 600 MPa. This pH was chosen as it was between the initial and final pH values of the sample solution. After incubation, solutions were cooled, diluted ten-fold with water and passed through a 0.2 μm filter, prior to CE analysis.

Ultrafiltration of the samples incubated at atmospheric pressure was performed using an Amicon (Stonehouse, UK) ultrafiltration cell fitted with a 3,000 or a 1,000 Da nominal molecular weight cut-off ultrafiltration membrane. Further details are given in Ref. 12. Ultrafiltered samples were analysed by CE.

CE was performed using a Hewlett Packard[3D] (Bracknell, UK) system equipped with a diode-array detector. Data were analysed using HP Chemstation software. An uncoated, fused-silica capillary (50 μm internal diameter, 48.5 cm total length, 40 cm to the detector) was used. Samples were applied to the anionic end of the capillary using pressure injection (5 s at 50 mbar). The capillary was maintained at 25 °C and the maximum threshold value for the current was set at 100 μA. Separation was achieved by application of a voltage of 25 kV for 15 min using a running buffer of 50 mM tetraborate at pH 9.3. Electrode buffer was renewed every 3 runs. Further details are reported in Ref. 12. Differences in peak areas at atmospheric pressure and 600 MPa were assessed using Student's t-test.

Results and Discussion

When the glucose-lysine model system was incubated at 60°C at either atmospheric pressure or 600 MPa, a brown colour and a sweet 'digestive biscuit-like' aroma developed with time. After incubation, the pH fell from an initial value of 10.1 to ~ 8.9, in both the atmospheric and high-pressure systems. The high initial pH chosen is representative of pH values used in the production of Class III (ammonia) caramels.[13] Whether with or without high pressure, plots of absorbance at 420 nm versus incubation time showed an induction period before the absorbance increased rapidly, and apparently linearly, with time.[10] However, the rate of increase in absorbance was greater at 600 MPa. Model systems incubated at 600 MPa required an incubation of only 5 h, compared with the 8 h required at atmospheric pressure, to reach the same degree of colour development (0.44 - 0.48 absorbance units for a 250-fold dilution, measured at 420 nm).

A previous study that compared CE and reverse phase HPLC, for their abilities to separate reaction products from an aqueous xylose - glycine model system, showed that CE gave superior separation,[12] Therefore, CE was chosen as a suitable separation technique for this study. However, although CE enables simultaneous analysis of all charged solutes, in the simplest mode of operation; capillary zone electrophoresis, which was used in this study, any neutral compounds will co-migrate and hence, may not be resolved.

Figure 1 shows electropherograms of the model system, incubated with or without pressure treatment, monitored at 200 nm. It is clear that many reaction products can be resolved by this technique. The electropherograms obtained for both the atmospheric pressure and high-pressure systems, are highly complex (no significant peaks were seen in the blank runs). They show many well-resolved peaks, eluting superimposed on a broad band of poorly resolved reaction products. This broad band contains coloured material and it has been previously demonstrated that it can be attributed to high molecular weight Maillard reaction products (i.e., melanoidins).[12] CE of the ultrafiltered fraction, possessing a nominal molecular weight below 1000 Da, from the system incubated at atmospheric pressure, showed a large reduction in the size of the broad band. Royle et al.[12] suggested that the broadness of the band may be due to components with a wide range of molecular weights and closely related charge-to-mass ratios. They proposed that the components constituting the broad band comprised similar sub-units, accounting for little variation in charge-to-mass ratio with size.

A comparison of the electropherograms for the atmospheric and high-pressure treated systems shows both similarities and differences in the profiles (Figure 1). The atmospheric system gave thirty-five resolved peaks, while the pressure-treated system gave forty-two. The areas of the resolved peaks and the broad band were integrated separately and there was no significant difference between the sum of the resolved peak areas for the two systems. Both orders of migration and spectral characteristics were used to compare peaks from the two systems. Nine peaks were clearly common to both, while several appeared to be present in only one system (Figure 1). However, some components may be present, but below the limit of detection, in the second system.

It can be seen that, of the nine common peaks, six have significantly different areas, while only three of the areas were not significantly different between the two systems (Table 1). Peak A is due to residual lysine and its area was similar in the two systems, indicating that lysine was used to a similar extent at both pressures. A large difference

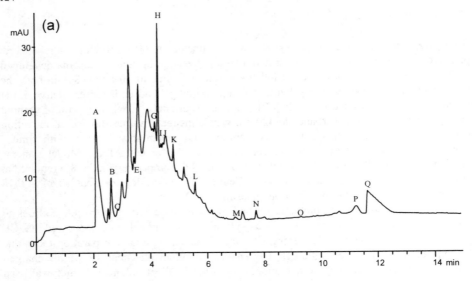

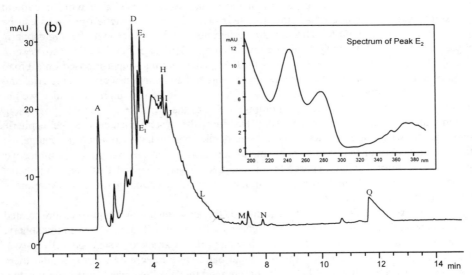

Figure 1. *Electropherograms of 1 molal glucose - lysine model systems, initial pH 10.1, incubated at (a) atmospheric pressure (8 h) or (b) 600 MPa (5 h) at 60 °C. Solutions run on CE at 1:10 dilution.*

between the two systems can be seen for the area of peak E_1, which was dramatically reduced in size at atmospheric pressure. Two other peaks, D and E_2, were detected only at 600 MPa. All three of these peaks possess similar diode-array spectra, E_1 and E_2 being almost identical (the spectrum of peak E_2 is shown in the inset to Figure 1b). This suggests that compounds responsible for these peaks posses similar structures.

Table 1. *Peak areas and levels of significant difference of peaks detected in both the atmospheric and high-pressure systems.*

Peak Label	Peak Area^a		P-Value	Significance^b
	Atmos. pressure	600 MPa		
A	119.0	123.4	0.241	ns
E_1	3.5	17.2	<0.000	***
H	32.5	11.4	<0.000	***
I	2.1	6.1	<0.000	***
J	20.6	5.5	<0.000	***
L	3.9	0.9	<0.000	***
M	1.8	1.7	0.306	ns
N	4.2	2.9	0.001	**
Q	81.0	85.9	0.116	ns

^{a} Areas are an average of 6 replicate runs (average coefficient of variation < 28%).
^{b} Level of significance indicated by * ($p \leq 0.05$) ; ** ($p \leq 0.01$) ; *** ($p \leq 0.001$); ns = not significant.

Previous work by Hill et al.[11] investigated the volatile compounds generated in the same model system, incubated under identical conditions. The application of pressure caused a substantial decrease in the yields of the most abundant reaction products, the pyrazines, which were dominant at both atmospheric and high pressure. 2,5- and/or 2,6-Dimethylpyrazine was the most abundant reaction product at both atmospheric pressure and 600 MPa. Representatives of the volatile compounds identified were analysed by CE and a library of diode-array spectra was compiled. As pyrazines were dominant in the volatile analysis, a wide range was analysed. All of the pyrazines showed similar spectra and eluted around 2.5 to 3.0 min. However, no spectra typical of pyrazines were obtained for either the atmospheric or high-pressure samples. The spectra of peaks D, E_1, and E_2 (Figure 1) were similar to that obtained for 3-pyridine carboxaldehyde, with some additional absorbance in the visible region. The area of peak H was about three times in the atmospheric pressure system that of the corresponding peak from the system treated at 600 MPa, and it possessed a diode-array spectrum similar to that of 4-hydroxy-2,5-dimethyl-3-(2H)-furanone. Both furanones and pyridines were identified among the volatile reaction products of this model system.[11]

In conclusion, CE of a glucose-lysine model Maillard system has proved to be a useful method for separating the total reaction products. A comparison of the data obtained for this system, incubated with or without elevated pressure treatment, has shown some significant quantitative differences in peak areas. For example, three peaks with diode-array spectra similar to standard pyridines, possessed significantly different peak areas depending on the pressure. Pyrazines, which dominated the volatile fraction of this system, were not detected by CE and hence, may not be major reaction products.

Acknowledgement
This work was funded by The Ministry of Agriculture, Fisheries and Food (MAFF), UK, via a studentship awarded to Vanessa M. Hill.

References

1. D. A. Ledward, 'High Pressure Processing of Foods'. Eds. Ledward, D. A., Earnshaw, R. G., Johnson, D. E. and Hasting, A. P. M. Nottingham University Press, Loughborough, 1995, p. 1-6.
2. R. J. Swientek, High hydrostatic pressure for food preservation, *Food Processing*, 1992, **53,** 90-91.
3. G. W. Gould, 'High Pressure Processing of Foods'. Eds. Ledward, D. A., Earnshaw, R. G., Johnson, D. E. and Hasting, A. P. M. Nottingham University Press, Loughborough, 1995, p. 27-36.
4. D. G. Hoover, Pressure effects on biological systems. *Food Technology*, 1993, **47,** 150-155.
5. D. Knorr, 'High Pressure Processing of Foods'. Eds. Ledward, D. A., Earnshaw, R. G., Johnson, D. E. and Hasting, A. P. M. Nottingham University Press, Loughborough, 1995, p. 81-98.
6. P. B. Cheah and D. A. Ledward, High pressure effects on lipid oxidation. *J. Am. Oil Chem Soc.*, 1995, **72,** 1059-1063.
7. T. Tamoaka, N. Itoh and R. Hayashi, High pressure effect on Maillard reaction, *Agric. Biol. Chem.*, 1991, **55,** 2071-2074.
8. N. S. Isaacs, 'Liquid Phase High Pressure Chemistry', J. Wiley and Sons, Chichester, 1981, p. 182-319.
9. J. M. Ames, Control of the Maillard reaction in food systems. *Trends in Food Sci Technol.*, 1990, **1,** 150-154.
10. V. M. Hill, J. M. Ames and D. A. Ledward, Influence of high hydrostatic pressure and pH on the rate of Maillard browning in a glucose-lysine system, *J. Agric. Food Chem.*, 1996, **44,** 594-598.
11. V. M. Hill, D. A. Ledward and J. M. Ames, 'Flavour Science ; Recent Developments'. Eds. Taylor, A. J. and Mottram, D.S. Royal Society of Chemistry, Cambridge, 1996, p. 235-238.
12. L. Royle, R. G. Bailey and J. M. Ames, Separation of Maillard reaction products from xylose - glycine and glucose - glycine model systems by capillary electrophoresis and comparison with reverse phase HPLC, *Food Chem.*, submitted.
13. J. Coulson, 'Development in Food Colours - 1'. Ed. Walford, J. Applied Science Publishers Ltd, London, 1980, p. 189-218.

Dehydroascorbic Acid Mediated Crosslinkage of Proteins Using Maillard Chemistry - Relevance to Food Processing

Juliet A. Gerrard, Siân E. Fayle,[1] Kevin H. Sutton and Andy J. Pratt.[1]

NEW ZEALAND INSTITUTE OF CROP & FOOD RESEARCH LTD, PRIVATE BAG 4704, CHRISTCHURCH, NEW ZEALAND. [1]CHEMISTRY DEPARTMENT, UNIVERSITY OF CANTERBURY, CHRISTCHURCH, NEW ZEALAND.

Summary
Dehydroascorbic acid (DHA) is the first stable oxidation product of ascorbic acid and can participate in redox reactions such as the oxidative crosslinking of proteins through disulfide bonds. Emphasis on the redox chemistry of ascorbic acid and DHA has led to its Maillard type chemistry being somewhat over-shadowed. Model studies of DHA with both amino acids and proteins have demonstrated that dehydroascorbic acid is particularly susceptible to Maillard reactions and can crosslink proteins by mechanisms that do not involve disulfide bonding. This may partly explain the improver action of ascorbic acid in flour-based products.

Introduction

Ascorbic acid (vitamin C) is a common food additive and is one of the few 'chemicals' that consumers are happy to ingest in large quantities. It is generally added to food either to improve the nutritive qualities, or to function as an anti-oxidant. In addition, it has been used as a flour improver for many years; addition of ascorbic acid to a dough improves its elasticity and gas-retaining properties and results in a larger loaf with improved texture.[1] The action of ascorbic acid as a flour improver has generally been attributed to the first stable oxidation product of ascorbic acid, dehydroascorbic acid (DHA). The two compounds exist in a redox equilibrium (Figure 1), and have equivalent biological utility.[1]

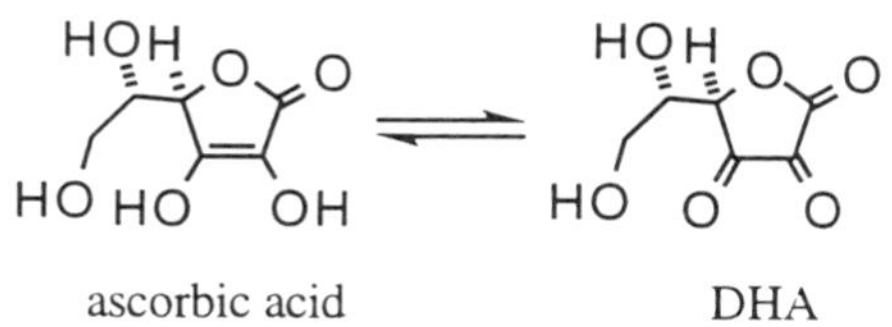

Figure 1. *Ascorbic acid and DHA*

It has been assumed that DHA acts as an oxidizing agent and forms disulfide crosslinks in the gluten matrix, regenerating ascorbic acid.[2] That DHA can form disulfide bonds from thiols nonenzymatically has been demonstrated unequivocally[3] and DHA-mediated crosslinkage of hemoglobin has been shown to be due to disulfide bonding.[4] However, workers in other areas have suggested that other mechanisms are responsible for protein aggregation (e.g. a hydrophobic interaction).[5] Ascorbic acid is known to react with proteins in Maillard-type reactions,[6] presumably via oxidation to DHA and reaction of the central carbonyl group. Work on eye lens proteins has suggested that Maillard chemistry is responsible for dehydroascorbic acid mediated crosslinking.[7] We postulate that this chemistry may also occur during food processing, and that this may be at least partly responsible for the flour improving qualities of ascorbic acid.

Materials and Methods

Materials
Unless otherwise stated, all materials were obtained from Sigma and were of analytical grade. Solvents were purchased from BDH and were of analytical grade. Methanol for HPLC was purchased from BDH. Water for HPLC was produced with a Milli-Q Water Purification System (Millipore). Dehydroascorbic acid (DHA) was prepared by the method of Ohmiri et al.[8] Purity of the resulting product was assessed by ^{1}H- and ^{13}C- NMR (recorded on a Varian Unity 300 at 300 and 75 MHz, respectively) and UV spectroscopy (recorded on a Hewlett Packard 2452A Diode Array Spectrophotometer interfaced with a personal computer running HP 89532A UV-visible operating software). High molecular weight glutenins were prepared by the method of Sutton.[9] Polyacrylamide gel electrophoresis (PAGE) was carried out according to the method of Dunn[10] with a 3% stacking gel and a 12.5% separating gel.

HPLC Analyses
Method 1: analyses were conducted using a Waters 626 solvent delivery/control system with a Waters WISP717 automatic sample injector and a Waters 996 diode array detector. The column was an Applied Biosystems Brownlee Spheri-5 RP18 (220 x 4.6 mm) fitted with an Applied Biosystems Brownlee guard column (15 x 3.2 mm). 5 µL of sample was injected. Instruments were controlled and data were recorded using the Waters Millenium 2010 software operating on a personal computer. Spectral data were recorded over the wavelength range 190-600 nm and 'extracted' at four wavelengths (254 nm, 280 nm, 360 nm and 460 nm). The column was maintained at 30 °C using a Waters column heater; a flow rate of 1 mL min^{-1} and a data acquisition time of 66 minutes were used; the solvent gradient was varied from 5% MeOH:95% water to 15% MeOH:85% water over 60 minutes.

Method 2: analyses were conducted using a Philips PU4100 Liquid Chromatograph System equipped with a Philips PU4120 Diode Array Detector interfaced to a personal computer running Philips PU6003 Diode Array Detector System Software (V3.0) and a colour plotter. The column was an Alltech Analytical Econosphere C_8 (250 x 4.6 mm; 5 µm particle size) fitted with an Alltech C_8 guard column. A sample volume of 20 µL of sample was injected. A flow rate of 5 mL min^{-1} and a data acquisition time of 20 minutes were used. Best results were obtained using isocratic elution with 75% MeOH:25% water:0.05% trifluoroacetic acid.

Preparation of the amino acid model systems
L-Lysine monohydrochloride (18.26 g, 0.1 mol) and DHA (17.6 g, 0.1 mol) or D-(+)- xylose (15.0 g, 0.1 mol) were dissolved in distilled water (100 mL) and the solution was refluxed for the required time. For UV analysis, a 30 µL aliquot was removed every 10 minutes, for 150 minutes, made up to 2.5 mL, and the absorbance measured at 460 nm.

Preparation of the protein model systems
Ribonuclease (Type XII-A: from bovine pancreas) (15 mg) was dissolved in water or buffer (600 µl) and 200 µl was removed to act as a control; the remaining solution was vigorously shaken with 10 mg of DHA and divided into aliquots. Each aliquot and the control were incubated for a set time at a set temperature (typically 37°C over several days); samples were then removed, prepared for PAGE,[10] and stored frozen at -20°C prior to electrophoresis. Glutenin samples were treated in the same fashion except that, for full solubility, a Tris.HCl, pH 6.8 buffer, containing 0.1% dithiothreitol and 8M urea was required.

Food Technology 129

Results and Discussion

In order to gain as much insight as possible as to the nature of the interaction of DHA with proteins, we studied two types of model systems: the reaction of DHA with amino acids, and the reaction of DHA with model proteins.

The reaction of DHA with amino acids

Previous work on the reaction of DHA with amino acids has been rather limited.[11] We began by adopting methodology recently developed for amino acid-sugar systems,[12] and comparing the reactivity of DHA and lysine with that of xylose and lysine, the latter being well-characterized.[12] Figure 2 shows a comparison of the absorbance at 460 nm for the two systems. DHA is seen to be a much more reactive Maillard reagent than xylose under these conditions.

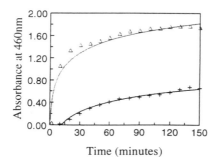

Figure 2. *Comparison of the reaction between DHA and lysine ($\triangle$) and xylose and lysine (+) as judged by absorbance at 460 nm*

Separation of the products using the method of Bailey *et al.*[12] (HPLC method 1) led to very different results for each reaction. The xylose-lysine system contained unretained peaks, unresolved material running as a convex broad band, unresolved material running as a tailing broad band, and resolved peaks, as has been noted previously.[12] In contrast, the DHA-lysine system contained mainly unresolved peaks, suggesting that a different separation procedure was required. Representative results are shown in Figure 3; for the DHA-lysine system, similar traces were found at all wavelengths.

A new HPLC method (method 2) was developed to try and effect better separation of the DHA-lysine model system. Figure 4 shows the improved resolution obtained by this method. Resolution was sufficient to allow identification of two products, ascorbic acid and a red pigment (Figure 5) which was identified by comparison of its UV trace to that reported in the literature for this compound.[13] We are currently trying to improve this resolution further and perform preparative experiments. We also hope to develop capillary electrophoresis methods to separate these products, as has recently been reported for the separation of products from related reactions.[14]

The presence of this red pigment is clear evidence that DHA forms Schiff's bases with lysine residues. It was also detected in the reaction of DHA with αt-BOC-lysine (data not shown).

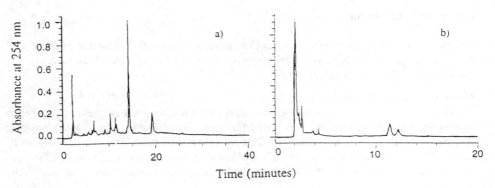

Figure 3. *Typical comparison of the reaction products from the reaction between DHA and lysine(△), and xylose and lysine (✦), after 90 minutes, as judged by HPLC analysis*

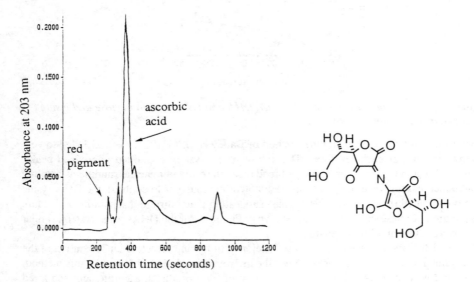

Figure 4. *Improved HPLC separation of products from the DHA and lysine model system*

Figure 5. *The structure of the red pigment isolated by method 2*

The reaction of DHA with proteins

DHA was reacted with two model proteins, ribonuclease A and high molecular weight glutenins, under a variety of reaction conditions. Following the reaction at 460 nm gave a comparable increase in the absorbance to the amino acid systems (data not shown). Ribonuclease A was found to undergo covalent crosslinking (Figure 6). These crosslinks were not broken by treatment with reducing agents such as dithiothreitol. The amount of crosslinking increases with time and temperature, ultimately producing large aggregates that do not enter a polyacrylamide gel; we are currently investigating reactions at higher temperatures over shorter times. Since the oligomerization of the protein is DHA dependent, and the covalent crosslinks are not broken by treatment with reducing agents, we propose that they are formed by Maillard chemistry, beginning with Schiff's base formation by DHA and a lysine residue.

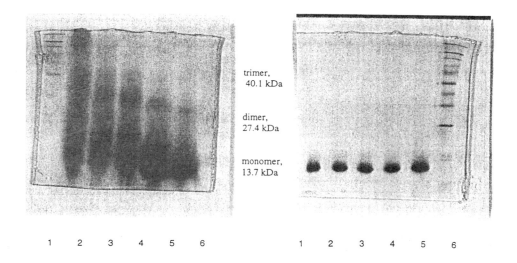

Figure 6. SDS-PAGE gel of ribonuclease A incubated in the presence of DHA (left hand side) and in the absence of DHA (right hand side). Lanes 1-6 (left hand side) correspond to: molecular weight markers, 8 days, 6 days, 4 days, 2 days and 0 days incubation at 37°C, respectively. Lanes 1-6 (right hand side) correspond to: 8 days, 6 days, 4 days, 2 days and 0 days incubation at 37°C, and molecular weight markers, respectively.

Many of the reactions undertaken, including those of the high molecular weight glutenins, rapidly produced an insoluble aggregated product. This is consistent with the observations of previous work on ovalbumin.[5] We are therefore developing other methods to study these reactions and gather evidence that DHA forms Maillard products with proteins, specifically gluten. Visual inspection of the proteins during incubation shows a gradual increase in coloured products. Ribonuclease A that has been treated with DHA, but has not yet formed oligomers, has been analysed by size-exclusion HPLC with diode array detection (data not shown). A clear shift in the UV-visible absorption spectrum of the protein is seen, consistent with Schiff's base formation. We are also investigating the relative rates of formation of disulfide bonds and crosslinks induced by the Maillard reaction, in an effort to establish which phenomenon has the greater significance during food processing.

Acknowledgements

This work was funded by the New Zealand Foundation for Research, Science and Technology. We thank Drs J.M. Ames and R.G. Bailey for valuable discussions.

References

1. M.B. Davies, B.E. Conway and David A. Partridge, 'Vitamin C: Its Chemistry and Biochemistry', Royal Society of Chemistry, Cambridge, 1991, pp.4-5.
2. P.A. Voysey and J.C. Hammond, Reduced-additive breadmaking technology, in 'Technology of Reduced-Additive Foods', J. Smith (ed.), Blackie Academic and Professional, London, 1993, pp. 80-94.
3. B.S. Winkler, Unequivocal evidence in support of the nonenzymatic redox coupling between glutathione/glutathione disulfide and ascorbic acid/dehydroascorbic acid, *Biochim. Biophys. Acta*, 1992, **1117**, 287-290.
4. S. Deb, S. Som, S. Basu and I.B. Chatterjee, Interaction of mammalian hemoglobins with DHA, *Experientia*, 1981, **37**, 940-941.
5. K. Nishimura, M. Ohtsuru and K. Nigota, Effect of DHA on ovalbumin, *J. Agric. Food Chem.*, 1989, **37**, 1539-1543.
6. M. Friedman, Food browning and its prevention: an overview, *J. Agric. Food Chem.*, 1996, **44**, 631-653.
7. B.J. Ortwerth, M. Linetsky and P. R. Oleson, Ascorbic acid glycation of lens proteins produces UVA sensitizers similar to those in human lens, *Photochem. and Photobiol.*, 1995, **62**, 454-462.
8. M. Ohmiri, H. Higashioka and M. Takagi, Pure DHA prepared by oxygen oxidation of L-ascorbic acid with active charcoal as catalyst, *Agric. Biol. Chem.*, **47(3)**, 607-608.
9. K.H. Sutton, Qualitative and quantitative variation among high molecular weight subunits of glutenin detected by RP-HPLC, *J. Cereal Sci.*, 1991, **14**, 25-34.
10. M.J. Dunn, Initial Planning, in 'Protein Purification Methods', E.L.V. Harris and S.Angal (ed.s), IRL Press at OUP, Oxford, 1989, p. 21-25.
11. B. Larisch, M. Pischetsrieder and T. Severin, Reactions of DHA with primary aliphatic amines including N^{α}-acetyllysine, *J. Agric. Food Chem.*, 1996, **44**, 1630-1634.
12. R.G. Bailey, J.M. Ames and S.M. Monti, An analysis of the non-volatile reaction products of aqueous maillard model systems at pH 5, using RP- HPLC with diode-array detection, *J. Sci. Food Agric*, 1996, **72**, 97-103.
13. T. Kurata, M. Fujimaki and Y. Sakurai, Red pigment produced by the reactionof DHA with α-amino acid, *Agric. Biol. Chem.*, 1973, **37**, 1471-1477.
14. A.J. Tomlinson, J. Mlotkiewicz and I.A.S. Lewis, Application of capillary electrophoresis to the separation of coloured products of Maillard reactions, *Food Chem.*, 1994, 219-223.

Linking Proteins Using the Maillard Reaction and the Implications for Food Processors

Sandra Hill and Azar Mat Easa

UNIVERSITY OF NOTTINGHAM, DEPARTMENT OF APPLIED BIOCHEMISTRY AND FOOD SCIENCE, SUTTON BONINGTON, LOUGHBOROUGH, LE12 5RD, UK

Summary
Heating proteins and reducing sugars together reduces the solubility of the proteins. This can be seen as gel formation and the inability of the gel fragments or powders to go into solution in solvents that contain agents that should break disulfide bonds and non covalent associations. It would appear that the proteins become covalently linked one to another in an organized manner in dilute solutions. The creation of crosslinks could explain many of the characteristics of the sugar-protein gels, such as their high breakstrengths and clarity. Under conditions of high temperature and pressure that occur within the barrel of an extruder, protein crosslinking occurs if a reducing sugar is present. In extreme conditions, the reaction prevents a textured product being formed. Control of the induced crosslinking could enable better control of processes used extensively by food manufacturers.

Introduction

It has been shown that the heating of solutions containing proteins and carbonyl groups causes the formation of gels that have higher breakstrengths than comparable samples not containing carbonyl groups.[1] The brown-coloured gels formed in this way have been termed Maillard gels. Sugars that are considered to form the greatest amount of colouration when heated with a protein cause gels to be formed in the shortest time. Glyoxal, glyceraldehyde, ribose, xylose, glucose, fructose and lactose have all been shown to promote gelation. Increasing the concentration of sugar decreases the time required for gels to form. For example, it took 160 minutes to gel a 3% bovine serum albumin (BSA) solution in the presence of 2% xylose, while in the presence of 4% sugar, the gelation time was reduced to 90 minutes. It was predicted that only 1.7% of xylose or 0.7% of ribose was required to promote gelation. Work carried out by Mat Easa[2] indicated that the minimum amount of BSA required to form a gel was 7%, but this could be reduced to 1.2% or 0.5% if 3% xylose or 3% ribose, respectively, were incorporated into the system. Temperature is also a very important factor in the time required to initiate the gel. The gelation rate is increased 4.7 times for each 10°C rise in temperature.[2]

Several globular proteins that show an ability to set thermally have been used to study Maillard gelation. Egg white, lupin isolates, soya isolate and BSA have all demonstrated changed gelling properties in the presence of reducing sugars.[1] The last two proteins have been studied in most detail and the work has indicated that the reaction between the sugars and proteins produces several features relevant to the gelation process. Increase in the negative charge on the protein, decrease in the isoelectric point of the protein and the decrease in pH of the system as the Maillard reaction progresses all go some way to explain the increase in gel strength and low syneresis levels associated with the Maillard gels. However, the most important change that incorporation of a sugar induces is a crosslink between the protein molecules.[3] The number of crosslinks is dependent on the concentration of carbonyl groups, duration of heating, temperature and type of sugar. Promotion of this crosslink and its relevance to texture is described in this paper.

Methods

Dilute solutions
Solutions of 3% (w/v) BSA and 3% BSA + 2% xylose were prepared in distilled water and heated at 95 °C for between 10 and 80 minutes. The heated samples were diluted to 3mg protein/mL solvent, with the final solution containing 1% sodium dodecyl sulfate (SDS) and 1% β-mercaptoethanol (ME). Dynamic light-scattering data were obtained on this sample using a Malvern Instruments System 4700C photometer and obtaining the translational diffusion coefficient. Sedimentation velocity estimates were made using a XLA (Beckman, USA) analytical ultracentrifuge equipped with absorption optics (280nm). Experiments were carried out at 20 °C using rotor speeds of 15000-40000 rpm. The methodology used for these two techniques and subsequent analysis were those of Errington et al.[4]

Gel preparation
Gels were prepared by adding the protein, BSA (3-12%) or soya protein isolate (SPI) (10-20%), to water either containing or not containing carbonyl compounds. Gelation was initiated by heating the samples in a water bath, in a rheometer, or in a steam atmosphere (121°C) in a laboratory autoclave.

Extruded products
Soya grits (Foodmaker Ltd, Corby Northants) were dry mixed with either 5 or 10 % xylose, glucose or sucrose for 12 min and then left to stand for 24 hours before extrusion. The mixtures were extruded in a Clextral BC21 twin-screw extruder at a screw speed of 300 rpm and a water feed rate of 1.1 kg/h. A single die of 3mm diameter was used and the temperature at the die end was varied from 85 to 190 °C.

Estimation of protein solubility
Solids were homogenized and passed through a sieve of linear aperture 211μm. Particulates (0.5g) were mixed with 10mL of solutions containing 1% SDS and 1% ME. After an extraction time of 14 hours, any undissolved material was removed and the protein concentration estimated[5] in the supernatant after suitable dilution.

Results and Discussion

Dilute solution study
The dilute solution study indicated that the pH of the system containing xylose decreased throughout the heating period. A final pH of 6.0 was reached after heating for 80 minutes at 95°C. BSA heated alone showed an initial decrease in the diffusion coefficient and a rapid increase in the sedimentation coefficient during the first ten minutes of heating; the values then remained constant. The pattern observed for 3% BSA heated in the presence of 2% xylose was approximately the same as seen for the BSA alone for the initial ten minutes. On further heating, the sedimentation values continued to be approximately the same, both the BSA and BSA with xylose having a sedimentation coefficient ($S_{20,w}$) of between 70 and 90s during 20 to 80 minutes of heating. However, the fall in the diffusion coefficient ($D_{20,w}$) for the system containing the sugar continued as heating proceeded. After 80 minutes of heating, the diffusion coefficient for the BSA was 0.32×10^6 cm^2 s^{-1} while that of the BSA and xylose was 0.12×10^6 cm^2 s^{-1}. To obtain information about the bonds maintaining the

association of the molecules, the heated samples were diluted in SDS and ME. In the case of the xylose system the results remained identical to those found when the diffusion coefficient was estimated in buffer. These results suggest that the aggregates formed in the xylose system involve non disulfide covalent linkages resulting from the Maillard reaction.

Apparent molar masses can be obtained by combining the sedimentation and diffusion coefficients (Figure 1). The values for the BSA alone would indicate that initially a dimer of native BSA was formed and that association continued until 30 molecules were aggregated. In the presence of xylose, the molecule weights continued to increase and, therefore, it can be assumed that the molecules aggregated until finally the system gelled. Some information about the shape of the aggregates can be obtained from double logarithmic plots of the molar mass of the samples against the sedimentation and diffusion coefficients.[6] Based on the slopes of the plots, it was suggested that in the xylose systems the aggregates become larger by protein molecules associating in a linear (i.e. end-to-end) fashion.[3] This is consistent with the so-called "string of beads" model originally proposed to explain the structure of protein gels.[7] It could be imagined that this organization would lead to fine-stranded gels.

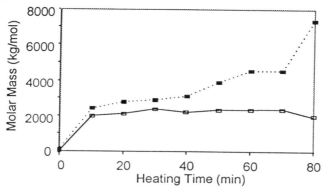

Figure 1. *Variation in molar mass as calculated from the sedimentation and diffusion coefficients of 3% bovine serum albumin (–□–) and 3% bovine serum albumin + 2% xylose (--■--) as a function of heating time at 95°C*

Gels

Initial studies on gel formation were carried out at 121°C and indicated that brown gels of high breakstrength, low levels of syneresis and some clarity were formed by the inclusion of a carbonyl group.[8] Figure 2 indicates that gels formed from BSA or SPI were totally soluble in the mixed SDS / ME solvent. If the gels were made in the presence of 3% xylose the solubility of the gels decreased. In the presence of the BSA no appreciable quantity of soluble protein was estimated, but with the soya ~ 40% of the protein remained in a soluble form. It has been postulated that this might reflect that the 7s soya globulin might not have participated in the Maillard crosslinking and hence retained its solubility.[9]

It was found that gels could be formed from BSA at 90 °C if heating was in excess of 60 minutes. If carbonyl groups were present, the duration of heating to form a gel depended

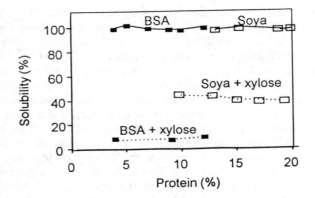

Figure 2. *Solubility of gels made from heating (60 min at 121°C) solutions of BSA (■) and soya isolate (□), with (----) and without (——) 3% xylose. Solvent: 1% ME and 1% SDS*

on the type and concentration of the carbonyl group. Solubility of the proteins also depended on these factors (Figure 3a). Unlike gels formed in the absence of carbonyl groups, where the gels were fully soluble in the solvents, solubility of Maillard gels decreased dramatically after a certain heating period (Figure 3b). It was noted that gelation and changes in solubility would occur if the non-gelled samples containing carbonyl groups were stored at ambient temperature after the initial heating process. It is therefore clear that heating proteins in the presence of a carbonyl group reduces their solubility in solvents that would be expected to break non-covalent bonds and disulphide bridges. The formation of covalent links, or agents that can later go on to form covalent links, during the heating process would seem to occur in these systems.

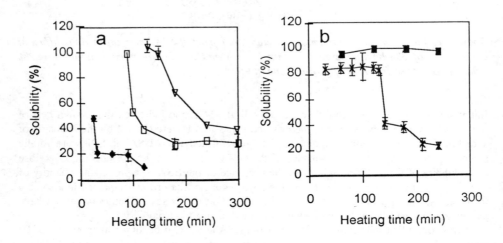

Figure 3. *The solubility of gels in 1% SDS + 1% ME as a function of heating time at 90°C. Graph a represents BSA samples at 3% heated with 0.25% glyoxal (♦), 6% xylose (□) and 3% xylose (∇). Graph b represents 9% BSA heated in the presence of 3% xylose(X) or heated by itself (■). The error bars represent the standard deviation.*

Extrusion

The formation of homogenous products often with an expanded texture can be achieved by the use of twin- or single-screw extruders. Mitchell and Areas[10] have predicted that formation of covalent crosslinks, which would not breakdown within an extruder, would reduce the formation of a homogenous melt within the barrel. This would be expected to reduce the stability of the extrudate matrix and possibly stop the texturization of the product as it leaves the barrel. The results of studies of heating protein solutions with reducing sugars indicate that Maillard crosslinking could represent the type of chemical modification that renders some products difficult to extrude. To investigate this, soya grits were extruded with or without sugars at a range of temperatures.

Under the conditions used, soya extruded without sugars or with 5% sucrose always formed a textured product. The residence time within the barrel was typically less than one minute. If 10% glucose or 5% xylose were added to the feed and the temperature held at less than 100°C, then a light-coloured textured product was formed. If the temperature was increased the product became darker in colour. At a certain temperature the product failed to form a continuous extrudate. This occurred at a die-end temperature of >110°C when xylose was incorporated at 5%. Glucose incorporation at 10% caused failure in textured product formation at 130 °C.

Extruded samples were assessed for their solubility in 1% SDS + 1% ME solutions. The results (Figure 4) indicate that samples incorporating the reducing sugars show a marked decrease in solubility at the higher die-end temperatures. The marked fall in solubility actually occurs after the product has failed to form a stable extrudate. If samples of extrudate containing sugar, that show only a limited loss of solubility, are further heated, then a further loss of solubility is seen.

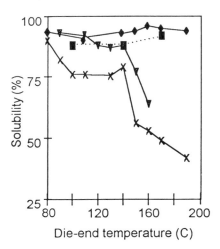

Figure 4. *Solubility in 1% SDS + 1% ME of soya extrudates without sugar (♦), with 5% sucrose (■), 10% glucose (▼) or 5% xylose (X) as a function of die- end temperature*

It would appear that Maillard crosslinking can be introduced into systems by extrusion. If the process is not too harsh, then further processing can encourage more crosslinks to occur. It would appear that excessive crosslinking within the barrel of an extruder stops the product forming an homogenous textured product. This might explain the narrow optimum range of conditions quoted for the practical extrusion of some products. It has been suggested that the Maillard reaction may have a negative effect on extrusion behaviour of proteins, even in the absence of added carbohydrate.[11] However, the action, if understood, may have positive features. For example, the incorporation of 5% reducing sugars, in conjunction with heating and extrusion, render a soya-based animal feedstuff more resistant to microbial attack within the ruminant gut.[12] We suggest that Maillard crosslinking might be occurring in a range of food products, whether it is required or not. Heating of dilute solutions or high concentrations of materials does seem to encourage the proteins to link one with another and hence alters the behaviour of the protein.

References

1. S.E. Hill, J.R. Mitchell and H.J. Armstrong, The production of heat stable gels at low protein concentrations by the use of the Maillard reaction, in 'Gums and Stabilisers in the Food Industry', G.O. Phillips, D.J. Wedlock and P.A. Williams (eds.), Elsevier, London, 1992, Vol. **6**, pp. 471-478.
2. A. Mat Easa, 'Factors Affecting Maillard Induced Gelation of Protein-Sugar Systems', PhD Thesis, University of Nottingham, 1996.
3. A. Mat Easa, H.J. Armstrong, J.R. Mitchell, S.E. Hill, S.E. Harding and A.J. Taylor, Maillard induced complexes of bovine serum albumin - a dilute solution study, *Int. J. Bio. Macromol.*, 1996, **18**, 297-301.
4. N. Errington, S.E. Harding, K.M. Varum and L. Illum, Hydrodynamic characterization of chitosans varying in degree of acetylation, *Int. J. Biol. Macromol.*, 1993, **15**, 113-117.
5. O.H. Lowry, N.J. Rosebrough, A. Farr and R.J. Randell, Protein measurement with the Folin phenol reagent. *J. Biol. Chem.*, 1951, **193**, 265-275.
6. S.E. Harding, K.M. Varum, B.T. Stoke and O. Smidsrod, Molecular weight determination of polysaccharides, in 'Advances in Carbohydrate Analysis', C.A. White (ed.), JAI Press, Connecticut, 1991, Vol. **1**, pp. 63-144.
7. M.P. Tombs, Alterations to proteins during processing and the formation of structures, in 'Proteins as Human Food', R.A. Lawrie, (ed.), Butterworths, London, 1970, pp. 126-138.
8. H.J. Armstrong, 'Enhanced Protein Gelation Using the Maillard Reaction and Elevated Temperatures', PhD Thesis, University of Nottingham, 1994.
9. O. Cabodevila, S.E. Hill, H.J. Armstrong, I. De Sousa and J.R. Mitchell, Gelation enhancement of soy protein isolate using the Maillard reaction and high temperatures. *J. Food Sci.*, 1994, **59**, 872-875.
10. J.R. Mitchell and J.A.G. Areas, Structural changes in biopolymers during extrusion, in 'Extrusion Cooking Science and Technology', J.L. Kokini, C.T. Ho and M.W. Karwe (eds.), Marcel Dekker, New York, 1992, pp. 349-360.
11. D.A. Ledward and J.R. Mitchell, Protein extrusion - more questions than answers, in 'Food Structure - Its Creation and Evaluation', J.R. Mitchell and J.M.V. Blanshard (eds.), Butterworths, London, 1988, pp. 219-230.
12. A. Egurrola Leyzagoyen, Oil seed flour with added sugar used as animal feed - is made by mixing, heating and extruding, Patent: ES 2 074 023, 1995.

Kinetics and Analytical Aspects

Modelling of the Maillard Reaction: Rate Constants for Individual Steps in the Reaction

B.L. Wedzicha and Lai Peng Leong

PROCTER DEPARTMENT OF FOOD SCIENCE, UNIVERSITY OF LEEDS, LEEDS LS2 9JT, UK

Summary

This paper illustrates the way in which the reactivity of sulphite species (S^{IV}) in aldose-amino acid-S^{IV} mixtures provides kinetic data on the individual steps in the aldose-amino acid browning reaction, not previously accessible. Examples show the effects of reaction conditions on browning, the cocktail effect of mixtures of amino acids and, particularly, their synergistic behaviour, and the use of S^{IV} to probe the importance of specific reaction intermediates such as Amadori compounds in Maillard browning. The approach may also be used to obtain similar kinetic data on the browning of oligosaccharides and fructose; the latter shows interesting contrast with the browning of glucose. Overall, the approach has the potential of an effective technique for the *predictive* modelling of Maillard browning in complex systems, including those where migration of reactants takes place during food processing.

Introduction

The Maillard reaction is a cascade of consecutive and parallel reaction steps, whose complexity is well illustrated by the thousands of known reaction products and intermediates. The kinetic approach tends to present a much simpler view of this mechanism, because it is based only on the rate-determining steps in the reaction. It is powerful because rate-determining steps provide control-points. The important requirement is to be able to identify these steps correctly and, when such knowledge is based on the formation or loss of characterized intermediates, then the kinetic approach becomes fundamental.

It is well known that sulphite species (S^{IV}) inhibit all the consequences of the Maillard reaction and the substance acts, therefore, on a key intermediate before the flavour- and colour-forming pathways diverge. The mechanism of the inhibition of browning has been demonstrated rigorously; at pH 4-6, its kinetics involve the formation of 3-deoxyhexosulose (DH) in the first slow step, which is then converted to a reactive species in a second slow step.[1-4] Rate constants for both these steps can be measured unambiguously by measuring the rate of reaction of S^{IV} in aldose-amino acid-S^{IV} systems. Overall, DH is converted to 3,4-dideoxy-4-sulphohexosulose (DSH) which is a relatively stable product with respect to browning.

There is general agreement that the kinetics of browning in the Maillard reaction can be represented by a series of 3 consecutive rate-determining steps[5,6] although the extreme complexity of the process of modelling this reaction is well appreciated.[7] A new approach to deriving parameters for kinetic models of the Maillard reaction was first introduced at the 5th International Symposium on the Maillard Reaction in Minneapolis[8] and is based on recognizing that, by measuring the rate of reaction of S^{IV} in aldose-amino acid-S^{IV} systems, we can obtain the rate of conversion (in mol L^{-1} h^{-1}) of reducing sugar to key precursors of melanoidins. Success in this approach stems from the fact that *independently* measured rate-constants for the first two steps apply to complete Maillard systems and represent appropriate control-points in the reaction. Our approach to model the Maillard reaction of glucose+glycine has been validated further through the successful calculation of the rate of browning from first principles, using kinetic data on the glucose-glycine-S^{IV} reaction, and the browning of glucose-glycine and DH-glycine mixtures.[9]

The overall kinetic model of the aldose-amino acid reaction, showing the point at which S^{IV} interrupts the pathway but omitting the amino acid for clarity, is illustrated below,

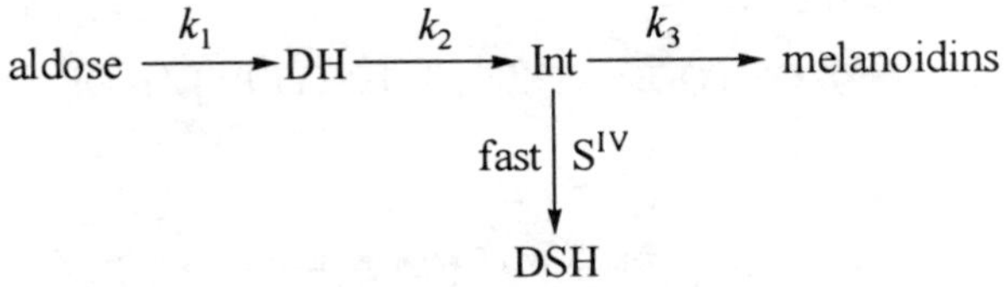

$$\text{aldose} \xrightarrow{k_1} \text{DH} \xrightarrow{k_2} \text{Int} \xrightarrow{k_3} \text{melanoidins}$$
$$\text{fast} \downarrow S^{IV}$$
$$\text{DSH}$$

where k_1, k_2 and k_3 are rate constants, and Int is an unspecified, but nitrogen-free, intermediate. The first two steps are kinetically dependent on amino acid concentration. The essential features of this mechanism are, therefore, two steps common to systems with or without S^{IV}, and a third step exclusively to the pathway of colour formation. For small degrees of browning, the integrated rate equation for the formation of colour (measured as absorbance at 470 nm) has been derived by Davies et al.,[9] as follows:

$$A_{470} = \frac{k_1 k_2 k_3}{6} t^3$$

where
$k_1 = k_1'[\text{aldose}][\text{amino acid}]$
$k_2 = k_2'(1+k[\text{amino acid}])$
t = time

The formal involvement of amino acids as catalysts in the first two steps of browning is established and, while none has yet been identified in the final step in the formation of melanoidins, amino acids are known to be incorporated into the structures of these products. Therefore, we propose that, since the mechanisms of these three steps are likely to be different, the activities of different amino acids in promoting these reactions are also expected to be different. This paper reports for the first time, the values of the three rate constants for a range of amino acids with different tendencies towards browning. The aim is to demonstrate the cocktail behaviour of amino acid mixtures and speculate on why mixtures of amino acids show synergistic behaviour.

Materials and Methods

The procedures used to measure the rates of browning and of the aldose-amino acid-S^{IV} reaction are the same as reported previously.[9] The value of k_1 was obtained by fitting a line to the linear portion of [S^{IV}]-time graphs.[1] Values of k_2 and k_3 were obtained by non-linear regression (Microcal ORIGIN 4.1) of [S^{IV}]-time and A_{470}-time data to the integrated rate equation.[9]

Results and Discussion

Type of amino acid

All experiments reported here were carried out at constant pH. The reactivities towards browning of the chosen range of amino acids are illustrated in Figure 1, representing a 65-

fold range of values of $k_1k_2k_3$ as the amino acid is changed from lysine to alanine. The way that this difference is generated by the variation in the values of the individual rate constants is shown in Table 1, where the results have been arranged in order of decreasing k_1. It is evident that different amino acids are, indeed, able to promote the steps in the reactions to different extents and the reason why one amino acid browns faster than another is far from straightforward.

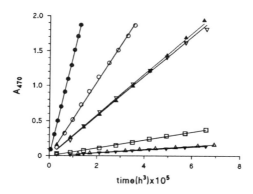

Figure 1. *Graphs illustrating the dependence of absorbance (A_{470}) on $(time)^3$ for mixtures of glucose and lysine ●, arginine ○, glycine ▽, serine ▲, valine ◻, glutamic acid △, alanine ▼. The function is plotted as $(A_{470}-a)=bt^3$ where a is an intercept ($a<0.12$) and b is the slope. The values of slope $\times 10^5$ h^{-3} were, respectively, 1.44, 0.529, 0.293, 0.285, 0.058, 0.024, 0.022. Reaction conditions: [glucose] = 1.0 mol L^{-1}, [amino acid] = 0.5 mol L^{-1} except valine which was 0.25 mol L^{-1}, pH 5.5 (0.2 M acetate buffer), 55 °C*

Table 1. *Values of rate constants k_1, k_2 and k_3 as defined in the kinetic model. The reaction conditions are as those given in the legend to Figure 1. The values for valine were obtained at 0.25 M; the values of k_1 and k_2 should be doubled for direct comparison with the other amino acids*

amino acid	$k_1(mol\ L^{-1}\ h^{-1}) \times 10^{-4}$	$k_2(h^{-1}) \times 10^{-2}$	$k_3(mol^{-1}\ L\ h^{-1})$
lysine	3.06	1.31	21.6
arginine	2.90	4.43	2.47
glycine	1.62	0.58	18.2
serine	1.29	1.15	11.9
valine	1.21	1.62	1.77
glutamic acid	1.11	1.81	0.71
aspartic acid	0.92	1.10	1.58
alanine	0.74	1.13	1.60

Behaviour of glucose-arginine-glycine mixtures

Preliminary experiments have shown that the browning of mixtures of amino acids is often synergistic. A good example we have identified is the glucose-arginine-glycine reaction. To consider the source of this effect, we need to assume that, in the mixed amino acid system, the kinetic model applies to arginine and glycine individually, and that the rate constants are as given in Table 1. Thus, the two reactions run in parallel as follows,

$$\text{glucose} \xrightarrow{\text{gly}} \text{DH} \xrightarrow{\text{gly}} \text{Int} \xrightarrow{\text{gly}} \text{M1}$$

$$\text{glucose} \xrightarrow{\text{arg}} \text{DH} \xrightarrow{\text{arg}} \text{Int} \xrightarrow{\text{arg}} \text{M2}$$

where M1 and M2 represent the different chromophores formed from glycine and arginine, respectively, and the boldness of the arrows illustrates the predominant reactions. Qualitatively, the synergistic behaviour is explained by the common intermediates DH and Int. Numerical modelling of the kinetic behaviour of the 3 steps in this mixed amino acid system confirms the additive behaviour of the rate constants in each of the first two steps; apparent values of k_1 and k_2 in the glucose-arginine-glycine reaction were 3.8×10^{-4} mol L^{-1} h^{-1} and 4.6×10^{-2} h^{-1}, respectively. On the other hand, the rate of development of colour in the final step does not realize the full potential of the glycine-dependent reaction; indeed the overall value of k_3 (14 mol^{-1} L h^{-1}) tends towards the average of the values for the individual reactions. It is likely that a model in which different chromophore formation is imagined to occur independently for the two amino acids is probably too simple. Currently, we are investigating the stoichiometries of incorporation of the individual amino acids into the melanoidin product, in relation to the relative values of k_3 for the individual reactions, to provide evidence of the formation of 'mixed' polymers and, possibly, 'mixed' chromophores.

Other systems

Any intermediate in the main reaction pathway should react at least as quickly as the principal reactant (*i.e.*, glucose, DH or Int) in the relevant rate-determining step of the kinetic model. Thus, the classical model of the Maillard reaction at pH 5.5 places the Amadori rearrangement product on the path from glucose to DH. It is found,[10] however, that at 40 mmol L^{-1}, the maximum concentration measured in glucose-glycine-S^{IV} mixtures, monofructose glycine reacts with S^{IV} at a negligible rate compared with the rate of reaction in a corresponding glucose-glycine-S^{IV} reaction. This observation adds further weight to the belief[11] that, in slightly acidic solution corresponding to the pH of many foods, the Amadori rearrangement product is not an important reaction intermediate in the formation of colour.

In principle, this approach allows the effect of any reaction variables on the values of the rate constants to be explored. To provide an example, Bellion and Wedzicha[12] reported the effect, on the value of k_1, of adding glycerol to reaction mixtures. This showed that the expected increase in rate of browning as water activity is decreased is reflected in the increase in the value of this rate constant. Similarly, Wedzicha and Kedward[13] demonstrated that the reaction of oligosaccharides and maltose with glycine+S^{IV} follows the same kinetics as the glucose-glycine-S^{IV} reaction. The value of k_1 derived from the established model is proportional to the concentration of reducing groups as illustrated in Figure 2. This suggests that the reactivity of a wide range of oligosaccharides is independent of their degree of polymerization. Furthermore, it was found that, with the exception of maltose, k_1 is a good indicator of the relative rates of browning of glucose and the oligosaccharides of different dextrose equivalents.

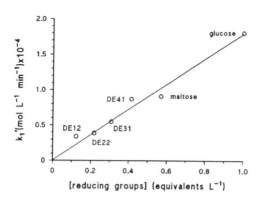

Figure 2. *Relationship between k_1 and the reducing group concentration for glucose, maltose and a range of oligosaccharides with their dextrose equivalents shown. The concentration of each sugar was 180 g L^{-1}. Based on illustration in reference 13*

The kinetics of the fructose-amino acid-S^{IV} reaction differ fundamentally from those of reactions involving aldoses. While the reaction of fructose also gives rise to DSH as a major product,[14] it shows kinetic behaviour consistent with only one rate-determining step, which is relatively insensitive to the presence of amino acids, as demonstrated in Table 2.

Table 2. *Effect of amino acids (0.5 mol L^{-1}) on the value of k_1 for the fructose-amino acid-S^{IV} reaction. From reference 14*

amino acid	$k_1(\mu mol\ L^{-1}\ h^{-1})$
none	53
glycine	53
glutamic acid	55
arginine	50
proline	71
lysine	76

These results now offer the potential for the modelling of glucose-fructose-amino acid mixtures which occur widely in foods undergoing Maillard browning.

Conclusions

The approach to the modelling of the kinetics of Maillard browning described in this paper has been tested rigorously and is now providing new insight into specific rate-determining processes within the reaction. As with any kinetic study, the validity of any conclusions is critically dependent on the model used. Its success can be judged from the excellent correlations which are being obtained.

Acknowledgements
The authors are grateful to BBSRC for support of the formative research over a period of several years. Most of the results are from research which has been carried out with financial support from the Commission of the European Communities, Agriculture and Fisheries (FAIR) specific RTD programme, CT96-1080, "Optimization of the Maillard Reaction. A Way to Improve the Quality and Safety of Thermally Processed Foods". It does not necessarily reflect its views and in no way anticipates the Commission's future policy in this area.

References

1. B.L. Wedzicha, A kinetic model for the sulphite-inhibited Maillard reaction, *Food Chem.*, 1984, **14**, 173-184.
2. B.L. Wedzicha and D.N. Garner, Formation and reactivity of osuloses in the sulphite-inhibited Maillard reaction of glucose and glycine, *Food Chem.*, 1991, **39**, 73-86.
3. B.L. Wedzicha and J. Kaban, Kinetics of the reaction between 3-deoxyhexosulose and sulphur(IV) oxospecies in the presence of glycine, *Food Chem.*, 1986, **22**, 209-223.
4. B.L. Wedzicha and N. Vakalis, Kinetics of the sulphite-inhibited Maillard reaction: the effect of sulphite ion, *Food Chem.*, 1988, **27**, 259-271.
5. G. Haugaard, L. Tumerman and H. Silvestri, A study of the reaction of aldoses with amino acids, *J. Am. Chem. Soc.*, 1951, **73**, 4594-4600.
6. H. Kato, M. Yamamoto and M. Fujimaki, Mechanisms of browning degradation of D-fructose in special comparison with D-glucose-glycine reaction, *Agric. Biol. Chem.*, 1969, **33**, 939-948.
7. T.P. Labuza, Interpreting the complexity of the kinetics of the Maillard reaction, in 'Maillard Reaction in Chemistry, Food and Health', T.P. Labuza, G. Reineccius, V. Monnier, J. O'Brien and J. Baynes (eds), Royal Society of Chemistry, Cambridge, UK, 1994, pp. 176-181.
8. B.L. Wedzicha, I.R. Bellion, and G. German, New insight into the mechanism of the Maillard reaction from studies of the kinetics of its inhibition by sulfite, in 'Maillard Reaction in Chemistry, Food and Health', T.P. Labuza, G. Reineccius, V. Monnier, J. O'Brien and J. Baynes (eds), Royal Society of Chemistry, Cambridge, UK, 1994, pp. 82-87.
9. C.G.A. Davies, B.L. Wedzicha and C. Gillard, A kinetic model of the glucose-glycine reaction, *Food Chem.*, 1997, **60**, 323-329.
10. M.D. Molero-Vilchez and B.L. Wedzicha, A new approach to study the significance of Amadori compounds in the Maillard reaction, *Food Chem.*, 1997, **58**, 249-254.
11. R. Tressl, C. Nittka and E. Kersten, Formation of isoleucine-specific Maillard products from [1-^{13}C]-D-glucose and [1-^{13}C]-D-fructose, *J. Agric. Food Chem.*, 1995, **43**, 1163-1169.
12. I.R. Bellion and B.L. Wedzicha, Effect of glycerol on the kinetics of the glucose-glycine-sulphite reaction, *Food Chem.*, 1993, **47**, 285-288.
13. B.L. Wedzicha and C. Kedward, Kinetics of the oligosaccharide-glycine-sulphite reaction: relationship to the browning of oligosaccharide mixtures, *Food Chem.*, 1995, **54**, 397-402.
14. S.E. Swales and B.L. Wedzicha, Kinetics of the sulphite-inhibited browning of fructose, *Food Add. Cont.*, 1992, **9**, 479-483.

Molecular Modelling Study of the Mechanism of Acid-Catalysed Disaccharide Hydrolysis: Implications for Nonenzymatic Browning Reactions

John O'Brien and Peter Bladon[1]

THE UNIVERSITY OF SURREY, SCHOOL OF BIOLOGICAL SCIENCES, GUILDFORD GU2 5XH, UK; [1]THE UNIVERSITY OF STRATHCLYDE, DEPARTMENT OF PURE AND APPLIED CHEMISTRY, THOMAS GRAHAM BUILDING, 111 CATHEDRAL ST, GLASGOW G1 1XL, UK

Summary
The rates of hydrolysis of several disaccharides, including sucrose and trehalose in solution (pH 2.5), were compared. Molecular parameters were calculated *in vacuo* using the MOPAC package and the semiempirical AM1 model. Calculations were conducted on the intact disaccharides and possible carbenium ions. Although steric acceleration may influence the rate of hydrolysis of sucrose, it is likely that electronic effects play the major role. Interestingly, the charge on the putative carbenium ion derived from fructose appears to be delocalized compared with that on the putative carbenium ion derived from the glucosyl moiety. Such charge delocalization would stabilize the fructose carbenium ion improving its activity as a leaving group in the hydrolysis reaction. It is suggested that carbenium ions formed in such hydrolysis reactions may participate directly as reactants in Maillard reactions.

Introduction

Disaccharide hydrolysis frequently precedes or accompanies nonenzymatic browning (NEB) reactions. In the case of NEB of systems containing nonreducing disaccharides, such as sucrose or trehalose, disaccharide hydrolysis is a prerequisite for browning to occur and the rate and extent of hydrolysis appears to be a rate-limiting step in the development of NEB reactions. Hydrolysis of reducing disaccharides may significantly influence the kinetics of NEB reactions by increasing the molar concentration of sugar reactants. As such, the rate of disaccharide hydrolysis may be an important determinant of the rate of nonenzymatic browning reactions under some conditions (e.g. dehydrated systems) and, therefore, the stability of stored food and biological systems.[1] Prevention of hydrolysis is a feature of sucrose refining where pH control is critical. By comparison, acid hydrolysis is actually desirable in the first stage in the dehydration of prunes; the products of hydrolysis subsequently participate in Maillard and caramelization reactions.[2]

Sucrose appears to be particularly unstable among the disaccharides. O'Brien[3] reported recently that the rate constants for brown colour formation (A_{420}) were of the order of ~200 - 2000-fold greater for a dehydrated sucrose system than for a system containing the non-reducing disaccharide trehalose. Significantly, browning rates for sucrose were similar to those of a glucose system under some conditions suggesting rapid hydrolysis of the glycosidic linkage.

Mechanistically, acid-catalysed disaccharide hydrolysis is an A1 unimolecular reaction involving protonation of the glycosidic oxygen followed by heterolysis of the resultant oxonium ion, to yield a monosaccharide molecule and a carbenium ion. The developing carbenium ion will be stabilized relative to the substrate by substituents that donate electrons. The ion subsequently reacts with water to produce a second molecule of monosaccharide. This acid-catalysed mechanism of sucrose hydrolysis appears to apply under most conditions in food systems – including mild alkaline conditions (up to ~pH 8.3). Cyclic carbenium ions have been suggested as a possible mechanism for mutarotation of monosaccharides, but experiments using $H_2^{18}O$ have demonstrated the pathway, although plausible, not to be significant.[4]

Although sucrose hydrolysis was the first reaction to be studied kinetically (by Wilhelmy in 1850) and was used by Arrhenius to develop his equation describing the effect of temperature on reaction rate, little is known of the molecular features that influence the rates of acid-catalysed disaccharide hydrolysis. This study was conducted to compare the rates of hydrolysis of selected disaccharides at pH 2.5 and to examine, using molecular models, the molecular features that may influence reaction rates. Molecular models were prepared for each disaccharide oxonium ion and the carbenium ions corresponding to the constituent monosaccharides.

Materials and Methods

Model systems
All chemicals were obtained from Sigma Chemical Co., Poole, UK. Model systems for the examination of hydrolysis rates contained 0.1M sugar plus 0.1M citric acid adjusted to pH 2.5 using concentrated NaOH. Aliquots (2mL) were heated in duplicate in 7mL Bijou bottles. Hydrolysis rates were monitored by measuring specific monosaccharides – glucose in the case of cellobiose, lactose, maltose sucrose and trehalose, and galactose in the case of lactulose. Glucose was measured using a coupled glucose oxidase assay (kit no. 510-DA, Sigma Chemical Co.).

Crystal data
Data for crystal structures for the following were obtained from the Cambridge Crystallographic Database: α–D-galactose;[5] fructose;[6] α–maltose (α–D-glucopyranosyl-(1→4)-glucopyranose);[7] β–lactose (β–D-galactopyranosyl-(1→4)-glucopyranose);[8] lactulose (β–D-galactopyranosyl-(1→4)-fructopyranose);[9] β–cellobiose (β–D-glucopyranosyl-(1→4)-glucopyranose);[10] α–α–D-trehalose;[11] and sucrose.[12]

Calculations
Calculations were conducted on a Silicon Graphics Inc Indy Workstation fitted with a MIPS R44100 processor. Calculations were done *in vacuo* using the MOPAC package and the semiempirical AM1 model.[13] The graphics interface package used was INTERCHEM, the code for which was written by one of us (P.B.). Disaccharides and disaccharides protonated on the glycosidic oxygen were minimized and the following calculated: partial atomic charges; glycosidic bond lengths; glycosidic bond angles; glycosidic torsion angles; and steric energy maps. Atom numbering of the glycosidic bond atoms is shown in Figure 1. Carbenium ions were modelled by removing the anomeric –OH group, adding a charge of +1 to the carbon, and re-minimizing the resultant structure. The C-O bond then takes on a double-bond character (Figure 1). The partial atomic charges and heats of formation were determined for the resulting structures. The ions are in the half-chair conformation for the pyranoses and are near planar for the furanoses. It was assumed that carbenium ions could only form between the anomeric carbon and the ring oxygen as other positions (e.g. C4) would not confer the necessary stability.

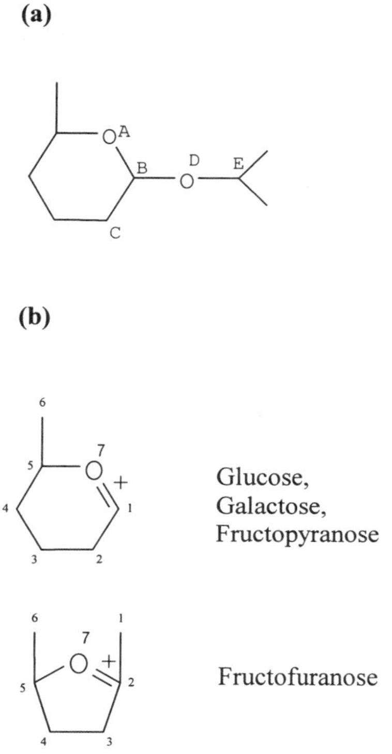

Kinetics and Analytical Aspects

(a)

[structure showing pyranose ring with OA, B, C labels and O-D-E linkage]

(b)

[structure of 6-membered ring carbenium ion with numbering 1-7] Glucose, Galactose, Fructopyranose

[structure of 5-membered ring carbenium ion with numbering 1-7] Fructofuranose

Figure 1. *Atom numbering of the glycosidic linkage (a) and structures and atom numbering of carbenium ions (b)*

Results and Discussion

Disaccharide hydrolysis rates in model systems are represented in Figure 1. Clearly, sucrose is uniquely unstable in acid, compared with the other disaccharides studied. The acid stability of disaccharides in the present assay was as follows: trehalose > cellobiose > lactose = lactulose = maltose > sucrose. The order of stability in acid is consistent with literature values for different studies.[14] Previous studies have suggested the hydrolysis rate of lactose to be greater than that of maltose, but there was no difference under the conditions employed in the present study.

In the case of all the reducing disaccharides studied, the optimized bond lengths indicate that the glycosidic O–C1 bond stretches more than the glycosidic O–C4 bond. This observation suggests that the former is more likely to break following protonation resulting in the formation of the carbenium ions examined below (Table 1). Relief of steric crowding appears to be a significant driving force in the acid hydrolysis of many glycosides.[15] Such steric acceleration may play a significant role in the hydrolysis of sucrose as evidenced by examination of steric energy maps (data not shown). Protonation of the glycosidic oxygen of sucrose and, to a certain extent, maltose exacerbates the steric crowding. This is illustrated by the large change in glycosidic bond angle observed on protonation of sucrose and maltose compared with other sugars (Table 1). It is noteworthy that the most stable disaccharides in the present study, trehalose and cellobiose, exhibit the fewest steric barriers to rotation about the glycosidic bond.

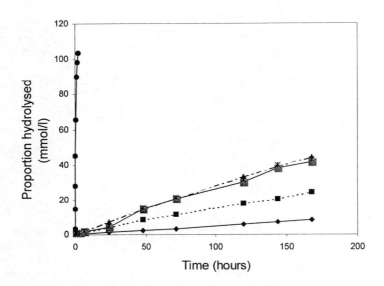

Figure 2. *Hydrolysis profiles of selected disaccharides in model systems.*
Cellobiose (··■··); Lactose (─■─); Lactulose (─▲─);
Maltose (─✱─); Sucrose (─●─); Trehalose (─◆─)

The anomeric carbon of the glucopyranose carbenium ion is clearly more electropositive than that of fructofuranose (Table 2), which appears to be due to charge delocalization in the case of the fructofuranose ion. This should make the fructofuranose carbenium ion a better leaving group in the heterolysis and, therefore, may explain the instability of sucrose. The partial atomic charge on C1 of the galactose carbenium ion was intermediate between those of glucose and fructofuranose. However, C2 of the fructopyranose ion bore a strong positive charge, presumably making it a poor leaving group. As there appears to be a significant proportion of lactulose present in the fructopyranose configuration in solution (see O'Brien[16]), the above differences in carbenium ion electronic structure may explain the observed greater stability of lactulose compared with sucrose.

Table 1. Effect of protonation on glycosidic bond lengths and bond angle

Sugar	Bond B-D (Å)	Bond D-E (Å)	Angle B-D-E (°)
Cellobiose	1.409	1.440	114.477
Cellobiose –H$^+$	1.492	1.518	113.293
Δ	0.083	0.078	-1.184
Lactose	1.412	1.435	115.207
Lactose –H$^+$	1.509	1.514	113.880
Δ	0.097	0.079	-1.327
Lactulose	1.414	1.423	112.912
Lactulose –H$^+$	1.505	1.493	112.439
Δ	0.091	0.070	-0.473
Maltose	1.416	1.433	117.114
Maltose –H$^+$	1.547	1.500	120.117
Δ	0.131	0.067	3.003
Sucrose	1.425	1.428	114.230
Sucrose –H$^+$	1.523	1.507	119.60
Δ	0.098	0.079	5.37
Trehalose	1.417	1.417	114.704
Trehalose –H$^+$	1.492	1.529	114.355
Δ	0.075	0.112	-0.349

Δ = difference between protonated and unprotonated forms

Table 2. Calculated partial atomic charges on selected carbenium ions

Ion	Atom number/ partial charge						
	1	2	3	4	5	6	7
Glucose	**0.2532**	-0.0687	-0.0060	0.0024	-0.0277	-0.0277	-0.0547
Galactose	**0.2478**	-0.0893	-0.0213	0.0011	0.0125	-0.0241	-0.0747
Fructo-pyranose	-0.0363	**0.2748**	-0.0379	-0.0156	-0.0171	-0.0363	-0.0677
Fructo-furanose	-0.0207	**0.2110**	-0.0480	-0.0304	-0.0454	-0.0218	-0.0598

Anomeric carbons are presented in bold

Calculated heats of formation (ΔH) represent a measure of thermodynamic stability and therefore are directly related to reactivity. As shown in Table 3, the calculated heats of formation suggest the following order of reactivity for the carbenium ions studied: fructofuranose > galactopyranose > glucopyranose > fructopyranose. This implies that while the fructofuranose carbenium ion benefits from stabilization due to charge delocalization, it remains the most thermodynamically reactive of the carbenium ions studied.

Table 3. *Calculated heats of formation of carbenium ions*

Carbenium ion	ΔH (kcal/mol)
Fructofuranose	-38.15
Galactopyranose	-41.99
Glucopyranose	-42.39
Fructopyranose	-47.47

The present observations on the mechanistic reasons for the lability of sucrose are consistent with previous publications. For example, labelling experiments have established that the fructofuranosyl bonds in sucrose are three times as labile as the glucopyranosyl bonds.[17] Despite such observations, fructose is generally observed at lower concentrations than glucose during degradation of sucrose which is at least partially due to the greater reactivity of fructose.[18] However, such observations alone fail to explain why sucrose should be ~1000-fold more labile to acid than trehalose.

Similarly, hydrolysis rates of sucrose fail to account fully for the high rates of non-enzymatic browning of such systems in the presence of amino acids at the same pH. This may be due to amino acid catalysis of the hydrolysis reaction. Alternatively, direct participation of carbenium ions in the Maillard reaction might also account for the high rates of browning observed in the presence of amino acids. Mechanistically, there appears to no reason why amination of carbenium ions by amino acids may not be possible. We therefore propose that carbenium ion intermediates may represent an important substrate for Maillard reactions under some circumstances. Further work is being conducted to test this hypothesis. In summary, the present work suggests that both electronic effects (carbenium ion stability and bond stretching on the oxonium ion) and steric acceleration/entropy effects play a role in determining the hydrolysis rates of the sugars studied. The present study has not considered the effect of solvent on hydrolysis and NEB reactions. There is evidence that solvent plays a significant role in such reactions modifying both steric and electronic effects[19] and such interactions should be considered in future work.

References

1. M.Karel and T.P. Labuza, Nonenzymatic browning in model systems containing sucrose, *J. Agric. Food Chem.*, 1968, **61**, 717-719.
2. L.G. Wilford, H. Sabarez and W.E. Price, Kinetics of carbohydrate change during dehydration of d'Agen prunes, *Food Chem.*, 1997, **59**, 149-155.
3. J. O'Brien, Stability of trehalose, sucrose and glucose to nonenzymatic browning in model systems, *J. Food Sci.* 1996, 61, 679-682.
4. L. Anderson and J.C. Garver, Computer modeling of the kinetics of tautomerization (mutarotation) of aldoses: implications for the mechanism of the process, in 'Carbohydrates in Solution', H.S. Isbell (ed.), American Chemical Society, 1973, pp.20-38.
5. J. Ohanessian & H. Gillier-Pandraud Structure cristalline de l'α–D-galactose, *Acta Crystallogr.*, Sect. B, 1976, **32**, 2810-2813.
6. J.A. Kanters, G. Roelofsen, B.P. Alblas & I. Meinders, The crystal and molecular structure of β–D-fructose, with emphasis on anomeric effect and hydrogen bond interactions, *Acta Crystallogr.*, Sect. B, 1977, **33**, 665-672.
7. F. Takusagawa & R.A. Jacobson, Crystal and molecular structure of α–maltose, *Acta Crystallogr.*, Sect. B, 1978, **34**, 213-218.
8. K. Hirotsu & A. Shimada, Crystal and molecular structure of β–lactose, *Bull. Chem. Soc. Japan*, 1974, **47**, 1872-1879.
9. G.A. Jeffrey, R.A. Wood, P.E. Pfeffer & K.B. Hicks, Crystal structure and solid-state NMR analysis of lactulose, *J. Am. Chem. Soc.*, 1983, **105**, 2128-2133.
10. S.S.C. Chu & G.A. Jeffrey, Refinement of the crystal structures of β–D-glucose and cellobiose, *Acta Crystallogr.*, Sect. B, 1968, **24**, 830-838.
11. G.A. Jeffrey & R. Nanni, The crystal structure of anhydrous α,α–trehalose at -150°C, *Carbohydr. Res.*, 1985, **137**, 21-30.
12. R.C. Hynes & Y. Le Page, Sucrose, a convenient test crystal for absolute structures, *J. Appl. Chrystallogr.*, 1991, **24**, 352-354.
13. M.J.S. Dewar, E.G. Zoebisch, E.F. Healy and J.J.P. Stewart, AM1: a new general purpose quantum mechanical molecular model, *J. Am. Chem. Soc.*, 1985, **107**, 3902-3909.
14. J.N. Be Miller, Acid-catalysed hydrolysis of glycosides, *Adv. Carbohydr. Chem.*, 1967, **22**, 25-108.
15. D. Cocker and M.L. Sinnott, Steric acceleration in the acid-catalysed hydrolysis of 1-adamantyl β–D-glucopyranoside. The origin of the high rates of hydrolysis of tertiary glycosides, *J. Chem. Soc. Chem. Comm.*, 1972, 414-415.
16. J. O'Brien, Heat-induced changes in lactose: isomerization, degradation, Maillard browning, in 'Heat-Induced Changes in Milk', P.F. Fox (ed.), International Dairy Federation, 1995, pp. 134-170.
17. F. Eisenberg, Relative stability of the interglycoside bonds of sucrose to acid hydrolysis in $H_2^{18}O$, *Carbohydr. Res.*, 1969, **11**, 521-530.
18. G.N. Richards, Chemistry and processing of sugar beet and sugar cane, in 'Chemistry and Processing of Sugarbeet and Sugarcane', M.A. Clarke and M.A. Godshall (eds), Elsevier, Amsterdam, 1988, pp.253-264
19. G. Eggleston, J.R. Vercellotti, L.A. Edye and M.A. Clarke, Effects of salts on the thermal degradation of concentrated aqueous solutions of sucrose, *J. Carbohydr. Res.*, 1996, **15**, 81-94.

Heating of Sugar-Casein Solutions: Isomerization and Maillard Reactions

M.A.J.S. van Boekel and C.M.J. Brands

WAGENINGEN AGRICULTURAL UNIVERSITY, DEPARTMENT OF FOOD SCIENCE, PO BOX 8129, 6700 EV WAGENINGEN, THE NETHERLANDS

Summary
The objective was to study the heat-induced reaction of each of the monosaccharides glucose, fructose and galactose with casein at 120 °C and to build a kinetic model for the reaction. In addition to participating in the Maillard reaction, glucose isomerized into fructose, fructose into glucose and galactose into tagatose, while no mannose, psicose or talose were detected. Furthermore, the sugars degraded into formic acid and other, unidentified, fragments. The reaction products of the Maillard reaction were the corresponding Amadori products of glucose or galactose, but in the case of fructose no Amadori or Heyns product could be detected. The Amadori product was further degraded into formic acid and acetic acid (plus unidentified fragments). Despite the absence of an Amadori or Heyns product in fructose-casein samples, lysine loss was comparable to that of glucose-casein and galactose-casein. Lysine loss was much greater than the concentration of Amadori product, which means that the Amadori product was rather quickly degraded again. Based on these results, a kinetic model was built that could quantitatively predict the behaviour of monosaccharides in the Maillard reaction as well as in the simultaneously occurring sugar isomerization/degradation.

Introduction

Reducing sugars react with lysine residues of casein in the Maillard reaction during heating. However, sugars are at the same time subject to isomerization and degradation reactions, catalysed by protein.[1,2] In heated milk, isomerization/degradation reactions were reported to be much more important from a quantitative point of view than the Maillard reaction.[1,2] Because these sugar reactions occur simultaneously with the Maillard reaction, the Maillard reaction becomes even more intricate. Kinetic modelling opens the possibility of unravelling the various reactions.[2] In the present study, various sugars were compared in their reaction behaviour towards casein; reaction products and reaction pathways were identified and kinetic parameters for the reaction network were established by kinetic modelling.

Materials and Methods

A reducing sugar (150 mM of glucose, fructose or galactose) and sodium caseinate (3 % w/w) were dissolved in a phosphate buffer at pH 6.8. These model systems were heated in closed glass tubes in an oil bath at 120 °C ranging from 0 to 40 minutes.

After heating, sugars and organic acids were separated from the protein via Sephadex G25 disposable columns (NAP-25, Pharmacia) and determined by HPLC using an ion-exchange column (ION-300, Interaction Chromatography Inc.). Sugars were detected by monitoring the refractive index; organic acids by their UV absorbance at 210 nm. Available lysine residues were determined fluorimetrically after derivatization with *ortho*-phthaldialdehyde.[3] The Amadori compound was determined by measuring furosine content, using HPLC[4]; a factor of 3.1 was used to convert furosine concentration to that of the Amadori compound.[5] The amount of Heyns compound was estimated via periodate oxidation and carboxymethyllysine (CML) determination,[6] using HPLC after derivatization with dabsyl.[7]

Kinetic modelling was done by setting up differential equations for the various reaction steps, numerically solving them, and fitting them to the data by non-linear regression using the determinant criterion, as described before.[2]

Results and Discussion

Sugar isomers were quantitatively the most important reaction products detected (Figure 1). Isomerization occurs via the Lobry de Bruyn-Alberda van Ekenstein transformation, a base-catalysed enolization of an aldose or ketose to the enediol, followed by isomerization,[8] also occurring in neutral or acidic environment in the presence of a protein.[1] Glucose, fructose and mannose are in equilibrium with the same 1,2-enediol; galactose, tagatose and talose are in equilibrium with another 1,2-enediol. Fructose is in equilibrium with psicose via the 2,3-enediol.[8] However, in the present study mannose, talose and psicose were not found in detectable amounts.

Formic acid and acetic acid were detected as sugar degradation products (Figure 2). Saccharinic acids were also detected but in much lower amounts. Formic acid was produced at an earlier stage than acetic acid but, eventually, the amount of acetic acid was higher. Formic acid is supposedly formed via the 1,2-enediol by C1-C2 cleavage, and acetic acid via the 2,3-enediol by C2-C3 cleavage.[8] At the same time, of course, C5 and C4 fragments are formed, which were not identified in this study. Possible C5 fragments are furfuryl alcohol, 3-deoxypentulose, 2-deoxyribose, 2-deoxyxylose, and possible C4 fragments are erythrose and threose. Other possible C1 and C2 fragments are formaldehyde and acetaldehyde, but these compounds were not found in detectable amounts. The formation of organic acids caused a pH decrease, typically from pH 6.7 to pH 6.4-6.3 after 40 min heating at 120 °C (considerable buffering by casein occurred). The formation of acid, as determined by titration, was in agreement with the amounts of acetic and formic acid found by HPLC.

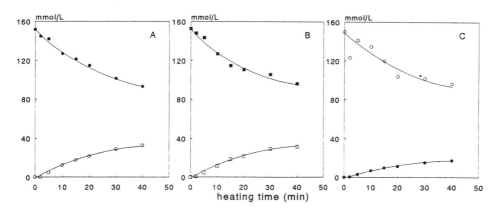

Figure 1. Sugar-casein solutions heated at 120°C. A. Glucose-casein, B. Galactose-casein, C. Fructose-casein. Glucose (●), fructose (O), galactose (■), tagatose (□)

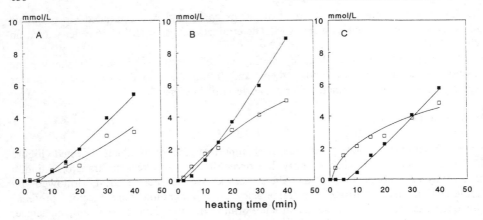

Figure 2. Formation of organic acids in sugar-casein solutions heated at 120°C. A. Glucose-casein, B. Galactose-casein, C. Fructose-casein. Formic acid (□), acetic acid (■)

Protein-bound fructosyllysine (glucose-casein) and tagatosyllysine (galactose-casein) were detected as Amadori products. Loss of available lysine was also monitored (Figure 3). The formation of Amadori compound did not follow the loss of lysine (except in the very beginning). Apparently, the Amadori compound was subject to substantial breakdown, and the increasing loss of lysine must have been due to the formation of advanced Maillard reaction products in which lysine residues were incorporated. The Heyns compound glucosyllysine was not detected in the heated fructose-casein system via the carboxymethyllysine (CML) method.[6] This may mean that it was not formed at all, or that it was quickly degraded again after formation (the CML method should be suitable to detect Heyns compounds[9]). Formation of the Heyns compound has been reported for proteins incubated with fructose at physiological temperature,[10] but it is not known what happens at higher temperatures. A negligible amount of Amadori compound was formed in the fructose-casein system, probably arising from reaction of the isomer glucose.

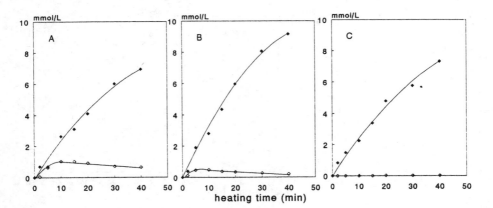

Figure 3. Loss of lysine (♦) and formation of Amadori product (◊) in sugar-casein solutions. A. Glucose-casein, B. Galactose-casein, C. Fructose-casein

Heating of glucose or fructose solutions (i.e. without protein) established that sugars were indeed precursors of organic acids, formic acid being the predominant one in both cases, but the level of acetic acid was somewhat higher in the case of fructose than in the case of glucose. In sugar-casein solutions, it was not clear whether organic acids were exclusively formed from sugar isomerization/degradation or also from the Amadori/Heyns product. Therefore, formation of the protein-bound Amadori product was induced by incubating 150 mM glucose with 3% casein at 65 °C for 15 h. After cooling to room temperature, the glycated protein was separated from sugars and reaction products as described in the Material and Methods section; the resulting protein-bound Amadori compound was present at 0.43 mmol/l. The solution was heated at 120 °C and acetic and formic acid were both formed, but acetic acid in about 1.5-fold higher amounts. Remarkably, the breakdown of the Amadori product was almost quantitatively characterized by formation of formic and acetic acid (and, as yet, unidentified sugar fragments). There is hardly any literature on formation of organic acids from Amadori compounds, apart from the fact that C1 and C2 compounds are sometimes mentioned as being formed.[11] We tried to isolate the protein-bound Heyns compound by the same procedure (incubating fructose with casein at 65 °C), but subsequent heating of the incubated protein did not result in formation of organic acids. This confirmed our observation mentioned above that the protein-bound Heyns product could not be detected.

Kinetic modelling
An attempt was made to build a kinetic model based on the results obtained (Figure 4). Formic acid was assumed to be mainly formed from the sugar via 1,2-enolization, and acetic acid mainly from the Amadori product; in the case of a ketose, acetic acid could be formed via 2,3-enolization of the ketose. Loss of lysine was explained from its reaction with intermediate and advanced Maillard reaction products arising from degradation of the Amadori product. As mentioned above, acid formation via sugars occurs via the enediols, but on the assumption of steady-state behaviour of the enediols, it can be shown that acid formation is directly proportional to the sugar concentration.[8] Consequently, the reaction rate constants k are pseudo reaction rate constants, describing the intermediate steps.

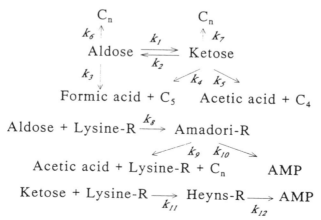

Figure 4. Simplified kinetic model for sugar-casein reactions. C_n =sugar fragments ($n \leq 5$), AMP=advanced Maillard products, k= reaction rate constant, R=protein chain

Table 1. Rate constants (± standard deviation) as found by kinetic modelling for the model displayed in Figure 4

rate constant	glucose-casein	galactose-casein	fructose-casein
k_1 (s^{-1} × 10^4)	1.52 ± 0.02	1.90 ± 0.10	1.52 ± 0.02
k_2 (s^{-1} × 10^4)	0.80 ± 0.01	4.67 ± 0.5	0.80 ± 0.01
k_3 (s^{-1} × 10^4)	0.09 ± 0.004	0.18 ± 0.003	0
k_4 (s^{-1} × 10^4)	0	0	0.18 ± 0.007
k_5 (s^{-1} × 10^4)	0	0	0.15 ± 0.007
k_6 (s^{-1} × 10^4)	0.32 ± 0.07	0.19 ± 0.08	0
k_7 (s^{-1} × 10^4)	0	0	0.85 ± 0.14
k_8 (L.mmol^{-1}.s^{-1} ×10^4)	0.04 ± 0.001	0.06 ± 0.003	0
k_9 (s^{-1} × 10^4)	24.33 ± 1.55	80.33 ± 9.28	0
k_{10} (s^{-1} × 10^4)	33.83 ± 2.37	101.17 ±9.93	0
k_{11} (L.mmol^{-1}.s^{-1} ×10^4)	0	0	0.03 ± 0.001
k_{12} (s^{-1} × 10^4)	0	0	710 ± 51.67

Further simplification of the model is possible. In the case of an aldose, formation of formic and acetic acid via the ketose can be neglected because of the relatively low concentration of the ketose (hence, k_4, k_5, k_7, k_{11}, k_{12} = 0). Likewise, in the case of a ketose, k_3, k_6, k_8, k_9 and k_{10} = 0. Preliminary simulations showed that this simplification did not change the fits. Table 1 shows rate constants obtained from the best fits. Figure 5 shows the fit for glucose and fructose at 120 °C, in which case the values for k_1 and k_2 were the same. The fit was excellent for glucose, but somewhat less for fructose as far as the organic acids were concerned. The very rapid degradation of the Heyns compound (if formed at all) in the fructose system is reflected in the very high value of k_{12} (compare k_9/k_{10} for glucose with k_{12}). The reaction routes in the fructose-casein system clearly need further investigation, especially with respect to organic acid formation.

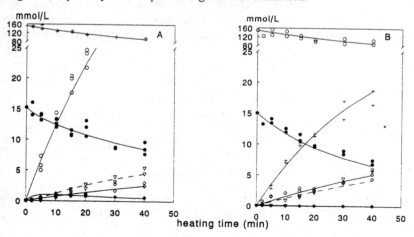

Figure 5. Simulations (drawn lines) based on the kinetic model of Figure 4 for glucose-casein (A) and fructose-casein (B) heated at 120°C. Glucose (+), fructose (O), lysine (●), acetic acid (---) (∇), formic acid (◊), Amadori/Heyns compound (♦)

The difference between the glucose-casein and galactose-casein systems is notable in the rate constants k_8, k_9 and k_{10}, which reflects the observation that the Amadori compound tagatosyllysine was less stable than fructosyllysine. Also, formic acid formation via the sugar was faster as reflected in k_3. (The higher value for k_2 in the galactose system should be confirmed in experiments with tagatose to establish this rate constant more precisely.) The more rapid reaction of galactose compared with glucose is in line with the literature.[12]

Preliminary results reported here indicate that the kinetic model is able to predict the behaviour for glucose-casein systems heated in the range 80-140 °C. Further research is underway to test the model more extensively and this will allow calculation of the temperature dependence of each reaction step in the model.

Acknowledgements
This study was financially supported by the EU-program FAIR CT96-1080: Optimization of the Maillard reaction: a way to improve quality and safety of thermally processed foods.

References

1. H.E. Berg and M.A.J.S. van Boekel, Degradation of lactose during heating of milk. 1. Reaction pathways, *Neth. Milk Dairy J.*, 1994, **48**, 157-175.
2. M.A.J.S. van Boekel, Kinetic modelling of sugar reactions in heated milk-like systems, *Neth. Milk Dairy J.*, 1996, **50**, 245-266.
3. F.J. Morales, C. Romero and S. Jiménez-Pérez, New methodologies for kinetic study of 5-(hydroxymethyl)-furfural formation and reactive lysine blockage in heat-treated milk and model systems, *J. Food Prot.*, 1995, **58**, 310-315.
4. P. Resmini, L. Pellegrino and G. Batelli, Accurate quantification of furosine in milk and dairy products by a direct HPLC method, *Ital. J. Food Sci.*, 1990, **3**, 173-183.
5. P.A. Finot, R. Deutsch and E. Bujard, The extent of the Maillard reaction during the processing of milk, *Prog. Food Nutr. Sci.* 1981, **5**, 345-355.
6. R. Badoud, and L. Fay, Mass spectrometric analysis of N-carboxymethylamino acids as periodate oxidation derivatives of Amadori compounds. Application to glycosylated haemoglobin. *Amino Acids*, 1993, **5**, 367-375
7. J.K. Lin and J.Y. Chang, Chromophoric labeling of amino acids with 4-dimethylaminoazobenzene-4'-sulfonyl chloride, *Anal. Chem.*, 1975, **47**, 1634-1638.
8. J.M. de Bruijn, 'Monosaccharides in Alkaline Medium: Isomerization, Degradation, Oligomerization.', PhD Thesis, University of Technology, Delft, 1986.
9. W.L. Dills, Protein fructosylation: fructose and the Maillard reaction, *Am. J. Clin. Nutr.*, 1993, **58**, 779S-787S.
10. A. Ruttkat, 'Untersuchungen neuer Reaktionswege zur Bildung von N^ε-Carboxymethyllysin und dessen Eignung zur Qualitätsbeurteilung hitzegeschädigter Lebensmittel', PhD Thesis, Christian-Albrechts-Universität, Kiel, 1996.
11. R. Tressl, C. Nittka, E. Kersten, and D. Rewicki, Formation of Isoleucine-specific Maillard products from [1-^{13}C]-D-glucose and [1-^{13}C]-D-fructose, *J. Agric. Food Chem.* 1995, **43**, 1163-1169
12. Y. Kato, T. Matsuda, N. Kato, K. Watanabe and R. Nakamura. Browning and insolubilization of ovalbumin by the Maillard reaction with some aldohexoses. *J. Agric. Food Chem.*, 1986, **34**, 351-355

Relationship between Ultraviolet and Visible Spectra in Maillard Reactions and CIELAB Colour Space and Visual Appearance

Douglas B. MacDougall and Mirjana Granov

DEPARTMENT OF FOOD SCIENCE AND TECHNOLOGY, THE UNIVERSITY OF READING, WHITEKNIGHTS, READING RG6 6AP, UK

Summary

The coloured products of the Maillard reaction are usually measured by selected wavelength absorption but their colour is seldom reported in colorimetric terms. Solutions of glucose-lysine, glucose-glycine, xylose-lysine and xylose-glycine were heated under mildly alkaline conditions to produce a selection of coloured reaction products. Their spectra were transformed into CIELAB colour space and compared. The solutions darkened at different rates to produced the yellow to brown colours. The relationship of hue to chroma was similar for each solution except for xylose-glycine which became green before darkening to brown then red.

Introduction

The early stages of the Maillard reaction result in colourless compounds that subsequently polymerise to produce the brown colours that are distinctive of the reaction.[1-3] The compounds formed early in the reaction absorb in the ultraviolet (UV) (e.g., 5-hydroxymethylfurfural, HMF),[1] but do not absorb in the visible region of the spectrum. As the reaction progresses, the spectrum changes and colour develops. This is typified by very strong absorption in the UV region, the tail of which at the long-wave end of the absorption curve progressively protrudes into the blue region of the visible spectrum and causes the yellow, orange and brown colours to appear. Differences in hue are caused by both the concentration of the pigments formed and wavelength distribution of the spectrum.

A recent review on the use of spectrophotometery and colorimetry in the measurement of browning in foods[4] reported that most researchers use only selected wavelengths (e.g., 420-460 nm) although some have used CIE tristimulus values. However, such measurements do not define the colours of the yellow to brown pigments or how they are related to concentrations of reactants and products.

For any reducing sugar-amino acid combination at defined pH and temperature, the UV and visible spectrum will have a distinctive, but changing, absorption profile. The characteristics of the colours can thus be defined by absorption (and reflectance) colorimetry based on their spectra. A previous study on caramel showed that, although 'brown' colours may be similar, serial dilution of their tinting strength gave different curvilinear loci in CIELAB colour space.[5] Similar L*a*b* maps could be produced for Maillard colours in solution and possibly in light-scattering suspensions.

This paper reports similarities and differences in the colours, the spectra and calculated locations of the colours in CIELAB space produced by a selection of sugar-amino acid mixtures heated under mildly alkaline conditions.

Materials and Methods

Aqueous sugar-amino acid mixtures (30 ml) were heated in 1.5 cm diameter test tubes in a covered water bath at 95 (±0.5) °C. Contents of the tubes reached thermal equilibrium within 5 minutes. The mixtures were glucose and lysine, glucose and glycine, xylose and lysine and xylose and glycine. The chemicals were obtained from Aldrich Chemical Co.

(Gillingham, Dorset, UK) and were D-glucose (ACS reagent), D-xylose (99%), L-lysine hydrochloride (99%), glycine (99%) and glycine hydrochloride (98%). Three concentrations of sugars were used for each mixture, 1%, 3% and 5%, with the amino acid concentration at 1% throughout. The solutions were all adjusted to pH 8.0 (±0.2) using 3M NaOH prior to heating to maximise the rate of browning. During heating, pH was maintained at between 7.7 and 8.2 by addition of NaOH. Two test tubes (60 mL) were removed from the water bath after 15, 30 and 45 minutes, and after 1, 2, 3, 4 and 6 hours. They were cooled in melting ice. Upon reaching a temperature of 25°C, pH was determined and cooling was completed to <2°C. On the basis of such pH measurements, solutions remaining in the water bath were adjusted by dropwise addition of NaOH to maintain pH close to 8.0. The cold solutions were transferred to brown glass bottles and held at +2°C until measured, viewed and photographed.

Colour was measured in CIELAB space[6] by transmission spectrophotometry (%T) in a 1 cm cell in a Hunter Colorquest Diode Array Spectrophotometer, set to illuminant D_{65} and the 10° standard observer. Data are reported as L*, uniform lightness and the chromaticness coordinates a* (+red to -green) and b* (+yellow to -blue) and the psychologically interpretable values of hue, $h^* = \tan^{-1}(b^*/a^*)$, and chroma, $C^* = (a^{*2} + b^{*2})^{0.5}$. Absorption and transmission spectra were measured in a 1 cm quartz cell in a Perkin-Elmer Lambda-2 Spectrophotometer from 200 to 1000 nm. In each case, the reference was the appropriate sugar solution at the same concentration but without the amino acid. Samples in 1.5 cm diameter clear-glass straight sided bottles were observed in a viewing cabinet under Artificial Daylight Fluorescent Illumination (General Electric) and photographed as transparancies on AGFA film.

Results and Discussion

Changes in the Spectra
During the reaction of the sugars with the amino acids, the development of the absorption / transmission spectra (Figure 1) progressed from colourless to deep 'brown'. At the start of the reaction of the glucose-lysine system, there was strong absorption in the far-UV, in the region of 220 nm, from the amino acid. Early during heating, prior to any absorption in the visible region of the spectrum, the colourless stages in the Maillard reaction sequence produced intense absorption in the near-UV in the region of 350 nm. Absorption increased rapidly to an A of >4 (<0.01%T) and, at the same time, the absorption band became much less sharply defined, while the width of the band increased both towards the far-UV and into the blue end of the visible spectrum, producing the first pale yellow colours. Visually this occurs by reduction in blue (B) cone response relative to green (G) and red (R) response in the tristimulus mechanism of colour vision. It means that high G plus high R results in the sensation of yellow or colours near to it. Similarly, low G plus R in the absense of B results in brown. In the Maillard sequence, the perceived colours would usually be described, in increasing order as the spectral profile moves redwards, as yellow, to pale orange brown, to brown, to reddish brown, to deep or dark reddish brown. The brilliance or distinctiveness of the hue associated with the 'brown' colour depends on the slope of the transmission curve and the adapting light. Steep profiles result in brilliant colours of defined hue, whereas shallow profiles result in dull colours. The glucose-lysine spectra in Figure 1 were essentially similar in character to those of glucose-glycine and xylose-lysine, but different from xylose-glycine, in which early absorption spectra had a peak at 625nm and a %T band with λ_{max} of 530 nm.

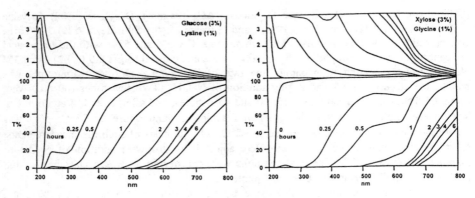

Figure 1. *Absorption (A) and transmission (%T) spectra of glucose (3%)-lysine (1%) and xylose (3%)-glycine (1%) mixtures heated at 95 °C, pH 8. The timing of the sequence of absorptiion and transmission spectra are the same.*

Colour development

CIELAB lightness (L*) and chroma (a*,b*) plots of the progress of colour development are given in Figure 2 for the glucose-amino acid mixtures and in Figure 3 for the xylose-amino acid mixtures. The effect of sugar concentration on the rate of loss of lightness (L*) is clearly illustrated, the 1% being much slower than the 5% with the 3% between, but closer to the 5%. Both lysine and glycine produced similar patterns of lightness loss in the presence of glucose, but the rates for xylose, as expected, were considerably faster.[3] In addition, the pattern of lightness loss for the xylose-glycine mixtures was different from both glucose mixtures and the xylose-lysine mixtures.

The development of the hue and chroma of the colours are shown in the a*b* diagrams. The centre of the diagram (a*=0, b*=0) indicates complete lack of chromatic colour, but, as coloured products begin to form, with the exception of the xylose-glycine mixtures, the colour locus moves towards yellow, virtually along the b*+ axis, with a hue angle of h*=93-95° and continues in the form of a loop with hue changing from yellow to orange (orange brown) to the point of maximum chroma or brilliance (h*=67°). The locus then decreases in b*+, though still increasing in a*+ towards reddish brown and finally both b*+ and a*+ turn towards the origin. These later colours with low chroma are extremely dark. The most important colorimetric finding from both glucose-amino acid mixtures is the location of the individual colours on the locus, irrespective of which sugar concentration generated them. For each amino acid, they lie on an identical locus, which indicates that the quality of the colour is not altered by sugar concentration at the concentration of amino acid used in the experiment. The chroma diagrams for the two amino acids with glucose were similar, but not identical, indicating that the sequence of Maillard pigments formed during their respective reactions were not identical. The xylose-lysine locus was identical to that of glucose-lysine, although the locus extended further because of the very dark colours (h*= approximately 0°). The absorption / transmission spectra of the two reactions were the same over their common range. This raises the question as to whether it is the amino acid rather than the type of sugar that has the controlling role in the character of the pigments formed.

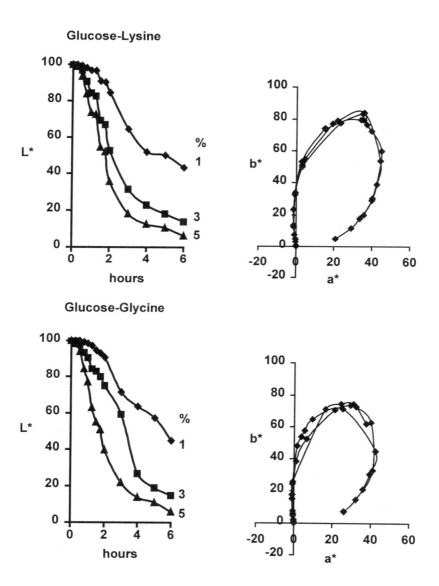

Figure 2. *CIELAB lightness (L*) and chroma (a*b*) plots of colour development of glucose (1,3,5%)-lysine (1%) and glucose (1,3,5%)-glycine (1%) mixtures heated at 95 °C, pH 8*

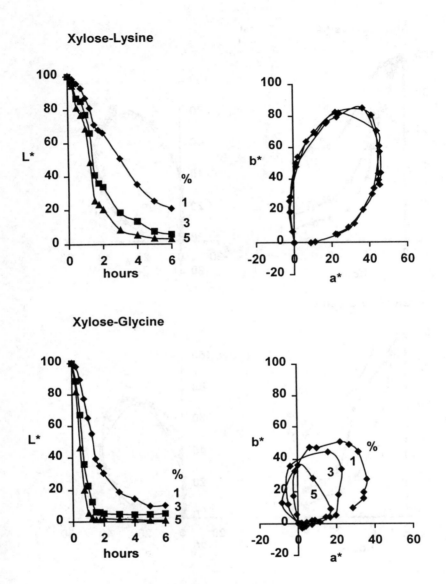

Figure 3. *CIELAB lightness (L*) and chroma (a*b*) plots of colour development of xylose (1,3,5%)-lysine (1%) and xylose (1,3,5%)-glycine (1%) mixtures heated at 95 °C, pH 8*

The xylose-glycine reaction was distinctive in that it produced unexpected colours. The reactions for the 3% and 5% xylose solutions were exceedingly fast, the colours darkening completely within an hour, while the 1% reacted as rapidly as any of the other 5% mixtures examined. The locus, or in this case loci, of colour changes were different for each sugar concentration and produced separate loops on the chroma diagram. It may be that 3% and 5% xylose-glycine reactions were so fast that their chroma maxima were missed, but sufficient samples of the 1% were examined to confirm maximum chroma hue as $h^*=67°$, although its chroma of $C^*=54$ was considerably less brilliant than that of the other mixtures ($C^*=81$-93). However, it is not the lower maximum chroma that is so noteworthy, but that the initial colour was unambiguously green with $h^*=113°$. Its greenness is explained by the development of the transmission band in the green region of the spectrum (Figure 1). The initial green progressed to a more typical brown and, ultimately, to a deep 'claret' or 'blood' red with the lowest hue recorded, $h^*=308°$. The possibility that such differences from the other mixtures might be due to impurities was tested by using glycine or glycine hydrochloride from a variety of sources, all of which developed the green and deep red hues. Since green did not occur in the case of glucose-glycine or with xylose-lysine, it is concluded that the green compound is a genuine product of the alkaline Maillard reaction, although probably transient in nature at high temperature.

This paper has shown that the compounds produced in the Maillard reaction under alkaline conditions exhibit distinct characteristics in both the UV and visible spectral regions with different absorption values during their development. The commonly used technique of measuring single-wavelength absorbance between 420 to 460nm for estimating Maillard colour may give a reasonable indication of reaction rates and / or the extent of the reaction, but cannot provide sufficient information to define the quality of the colours, which should not simply be called 'brown'.

Acknowledgement

D.B.MacD. wishes to thank Professor H. E. Nursten for our interesting discussions and his helpful comments and advice.

References

1. R.G. Bailey, J.M. Ames and S.M. Monti, An analysis of the non-volatile reaction products of aqueous Maillard model systems at pH 5, using reversed-phase HPLC with diode-array detection, *J. Sci. Food Agric.*, 1996, **72**, 97-103.
2. J.M. Ames, A comparison of the colour properties of six heated sugar/amino acid model systems, *Chem. Ind.*, 1987, September, 659-661.
3. V.M Hill, D.A. Ledward and J.M. Ames, Influence of high hydrostatic pressure and pH on the rate of Maillard browning in a glucose-lysine system, *J. Agric. Food Chem.*, **44**, 1996, 594-598.
4 J.B. Hutchings, 'Food Colour and Appearance', Blackie Academic & Professional, London, 1994, pp. 293-296.
5. D.B. MacDougall, Measurement of food and beverage colour appearance, in 'Distilled Beverage Flavour', J.R. Piggott, and A. Paterson (eds), Ellis Horwood Ltd, Chichester, UK, 1989, pp. 85-96.
6. R.G.W. Hunt, 'Measuring Colour', Ellis Horwood, London, (2nd ed.), 1995.

Evaluation of the Acyclic State and the Effect of Solvent Type on Mutarotation Kinetics and on Maillard Browning Rate of Glucose and Fructose

C.G.A. Davies,[1] A. Kaanane[2], T.P Labuza[1], A.J. Moscowitz[3] and F. Guillaume[3]

[1]DEPARTMENT OF FOOD SCIENCE AND NUTRITION, UNIVERSITY OF MINNESOTA, ST PAUL, MN 55108 USA; [2]DEPARTMENT DE CHIMIE-BIOCHIMIE ALIMENTAIRE, INSTITUT AGRONOMIQUE ET VETERINAIRE, HASSAN II, RABAT, MOROCCO; [3]DEPARTMENT. OF CHEMISTRY, UNIVERSITY OF MINNESOTA, MINNEAPOLIS, USA.

Summary

The concentration of acyclic sugar in D_2O solutions of glucose or fructose using FTIR spectroscopy was determined. This showed that for glucose, since no IR peak was found, the acyclic concentration is less than the detection limit of 0.001% at all pHs studied (5-10), while there was a just detectable peak for fructose, so that the acyclic concentration for fructose is around 0.001% in the pH range 6 to 10. These values are less than those reported by other methods (polarography and circular dichroism).

The browning rate was less in D_2O than in H_2O for both sugars, but did not follow the expected mass ratio of solvent. Also the rate of browning did not follow the same ratios as the mutarotation rates in the two solvents studied, i.e. fructose browned about 2.5 times faster than glucose, but had a mutarotation rate that was 10 times faster than glucose. The addition of electrolytes has a significant effect on the browning rate in water, while the effect on the mutarotation rates of glucose and fructose in water is insignificant at a level of $p = 0.05$.

Introduction

Burton & McWeeny[1] were the first to suggest that the rate of Maillard browning increases as the concentration of acyclic sugar increases. An equilibrium mixture of a monosaccharide reducing sugar contains, in varying concentrations, two pyranose forms (α and β), two furanose forms (α and β), the acyclic form, and solvated and ionic modifications. Methods for determining the acyclic concentration include a polarography technique[2,3] and circular dichroism,[4] but seem to be unsatisfactory.

It is well known that the carbonyl group absorbs in the infra-red region at about 1700-1750 cm^{-1}. Fourier transform infra-red (FTIR) spectroscopy is about 2000 times more sensitive than the classical infra-red grating spectroscopy. In general, FTIR spectra can only be recorded if the species being studied is dissolved in D_2O, since water interferes with measurements. However, changing the solvent from H_2O to D_2O changes the degree of ionization and solvation of the solute and, thus, may cause spectral shift. The ion product of D_2O is 1.95×10^{-15} M at 25 °C whereas that of H_2O is 1.0×10^{-14} M (Ref. 5). Thus, reactions that involve protons such as the rate of formation of acyclic sugar will be affected by solvent type since it involves the transfer of three electrons and three protons from the solvent. Reactions involving proton transfer (i.e. acid catalysis) are faster in D_2O than in H_2O with an isotope effect that ranges from 0.39 to 0.69.[6]

There is little work on the relationship between the concentration of acyclic form of different sugars and the rate of browning. Burton *et al.*[7] found that the browning rate was galactose > mannose > glucose, following the acyclic concentration pattern of Cantor and Peniston.[3] Burton and McWeeny[1] concluded that the aldose-glycine reaction appeared to depend initially on the conformational stability of the aldose molecules with pentoses more reactive than hexoses. Thus, browning is faster because of the higher the concentration of the acyclic form. Kaanane and Labuza[8] using data from Bunn and Higgins[9] concluded that there was no clear relationship between glycosylation rate and concentration of acyclic form when aldoses and ketoses are considered together.

The objectives of the present work were to verify the amount of acyclic form of glucose and fructose by using an FTIR technique and to study the effects of solvent type on mutarotation kinetics and on the Maillard reaction of these sugars with glycine.

Materials and Methods

Measurement of the rate mutarotation
Optical rotation of solutions containing α-D-glucose (0.5 M) or β-D-fructose (0.5 M) in D_2O or distilled water was measured at room temperature with and without added $FeCl_2$ or $CuCl_2$ (20 ppm). Readings were taken every minute for 20 min and then every 10 min until equilibrium was reached using a PolyScience SK 6 polarimeter (Fischer Scientific). The first order rate constants for mutarotation for each solution were calculated by linear regression of the data using equation 1:

$$Ln(a-a_{eq})/(a_o-a_{eq}) = -(k_1+k_2) t \qquad (1)$$

where a = rotation angle at time t; a_{eq} = rotation angle at equilibrium; a_o = rotation angle at t = 0; k_1+k_2 = overall rate constant (equal to the slope of $ln(a-a_{eq})$ vs. t); t = time (min)

Measurement of concentration of acyclic sugar
Infrared measurements were made at 4 cm^{-1} resolution using a Sirius 100 FTIR spectrometer (Mattson Instruments Inc., Madison, Wisconsin). The spectrometer was first purged with N_2 for 30 min. Samples were placed in a cell with CaF_2 windows and Teflon (0.1 mm) spacers. Each spectrum is an average of 100 interferograms.

For quantitative analysis and to determine the minimum concentration of carbonyl group that can be detected by FTIR, butyraldehyde was used as a standard. A series of solutions of butyraldehyde in D_2O containing 1.6×10^{-5} - 2.2×10^{-4} M were prepared and analyzed by FTIR. A plot of peak area versus concentration gave a linear fit ($r^2 = 0.99$). The minimum concentration of carbonyl group that can be detected by FTIR technique appears to be around 1.6×10^{-5} M or less than 0.002%.

Solutions containing fructose (1.5 M) or glucose (1.5 M) were prepared in D_2O at different pHs (5, 6, 7 and 10) using phosphate buffers and left at room temperature until equilibrium was reached (~ 8 h).

Measurement of browning
Solutions containing glucose (0.25 M) or fructose (0.25 M) with glycine (1.25 M) were prepared using either D_2O or HPLC-grade H_2O as solvent and incubated at 60°C. The pH of the solutions (Orion model 811 pH meter, Orion Research Inc., Boston, MA) was in the range 5.5-6.9. Every hour, samples were removed from the incubator and the absorbance at 420 nm was measured using a Beckman DB-G spectrophotometer. Kinetic analysis was done by the method of Baisier and Labuza.[10]

Results and Discussion

FTIR

Figure 1 shows differential infra-red absorption spectra (1500 - 2000 cm^{-1}) of fructose (1.5 M) in D_2O. These spectra show an absorption band at ~1725 cm^{-1}. In order to confirm that this peak was due to carbonyl absorption, a fructose solution (1.5 M) was spiked with butyraldehyde (0.0015% w/w) and analyzed (Figure 1, spectrum E). The addition of butyraldehyde increased the absorption at 1725 cm^{-1}, confirming that this peak is due to the carbonyl groups.

Table 1 shows the calculated percentage acyclic form of fructose and glucose over the pH range 5.5 - 10.4 compared with previously published results.[2,4] As seen, the acyclic concentration measured using FTIR was lower than that reported in previous studies. At pH > 6.11, the concentration of acyclic fructose was calculated to be less than 0.001%. At pH 6.11 - 10.4, the concentration is ~0.001% which is still less than reported values. Glucose (1.5 M), pH 5 - 10, gave no carbonyl peak. Thus, the predicted acyclic concentration is less than 1.6×10^{-5} M.

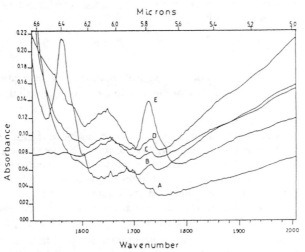

Figure 1. *FTIR Spectra of fructose (1.5 M) at different pHs (A, pH 5.50; B, pH 6.11; C, pH 7.21; D, pH 10.39). E: Fructose with butyraldehyde (0.0015 %)*

Mutarotation

The mutarotation first order rate constants as affected by solvent type and added metal ions at 24°C are shown in Table 2. As can be seen, the rate of mutarotation for either sugar is significantly slower in D_2O than in H_2O. Table 2 also shows the solvent isotope effect (SIE) value for both glucose and fructose. The SIE is defined as the ratio of a rate constant measured (k_{H_2O}) in H_2O to the corresponding value (k_{D_2O}) in D_2O. Langhton and Robertson[11] showed that theoretical limits for the SIE of glucose mutarotation are from 3 to 6. The value obtained, 4.4, is in the expected range. One possible mechanism for the mutarotation of glucose, which has five exchangeable electrons, involves pre-equilibrium proton transfer, with a different proton being removed in the transition state.

Table 1. *Concentration of acyclic forms of fructose and glucose measured using different methods*

Method	pH	% Acyclic form	
		Glucose	Fructose
Circular Dichroism[4]	5.2-7.0	0.002	0.700
Polargraphy[2]	6.5	0.012	-
	7.0	0.022	-
	7.5	0.04	-
FTIR*	5.5	<0.001	<0.001
	6-10	<0.001	0.001

*this study (standard = butyraldehyde)

Thus, mutarotation reactions are catalysed by both hydrogen and hydroxyl ions with the rate expressed as in equation of the type.[12]

$$k = A + B[H^+] + C[OH^-] \quad (2)$$

where A, B and C are constants.
For glucose, in H_2O at 20°C the equation is:

$$k_{H_2O} = 0.006 + 0.18[H^+] + 16000[OH^+] \quad (3)$$

In D_2O, the mutarotation rate will take the same form as equation (3) with the hydrogen and hydroxyl ions replaced by D^+ and OD^- ions. Thus,

$$k_{D_2O} = A_1 + B_1[D^+] + C_1[DO^-] \quad (4)$$

Hence, according to equations 2 and 4, the mutarotation rate of sugars depends on the ion product of the solvent which at 25 °C is 1.95×10^{-15} M for D_2O, while that for water is 1.0×10^{-14} M.[6] Therefore, the ratio of the ion product of H_2O to D_2O is 5.1, in agreement with the ratio that was observed in this study for the mutarotation rates of glucose and fructose seen in Table 2. Addition of ~20 ppm Cu (as $CuCl_2$) or Fe (as $FeCl_2$) had no significant effect on the mutarotation rate in H_2O.

Rate of Browning
Table 3 shows the extent of browning at 60°C for both glucose and fructose at the same concentration in the presence of glycine at a 1:3 molar ratio in water or D_2O. Table 3 summarizes the rate constants assuming a pseudo zero order reaction.[10] As expected, fructose browns significantly faster than glucose in both solvents. The calculated browning ratio (k_{H2O}/k_{D2O}) for fructose is 1.33 and 3.42 for glucose. The browning rate for fructose is 2.8 times that of glucose in water but 7.3 times faster in deuterium oxide.

Table 2. *Effect of solvent type on first order mutarotation rate constants for glucose and fructose at 24 ± 1°C*

Sugar	k_{H_2O} (min^{-1})	k_{D_2O} (min^{-1})	Solvent Isotope Effect k_{H_2O}/k_{D_2O}
Glucose	0.0299	0.0068	4.40
+ Fe	0.0320		
+ Cu	0.0240		
Fructose	0.280	0.060	4.70
+ Fe	0.216		
+ Cu	0.236		

Table 3. *Effect of Solvent on the zero-order rate constants for browning of glucose-glycine or fructose-glycine mixtures at 60°C*

Zero order rate constants	Glucose	Fructose	Fructose/Glucose
H_2O	3.9	11.0	2.8
D_2O	1.14	8.3	7.28
H_2O/D_2O	3.42	1.33	0.39

The addition of Cu and Fe ions to water-based sugar-amino acid solutions had a significant effect on the browning rate, increasing the rate 5.6 times for glucose and 3.6 times for that of fructose (data not shown).

Conclusions

FTIR measurements showed that the acyclic concentration of glucose and fructose in D_2O was lower than previously measured for sugars in H_2O. The solvent isotope effect on glucose and fructose shows that glucose is more strongly influenced by the solvent than fructose. The mutarotation rate was more strongly affected by the solvent isotope effect than the browning reaction. This suggests that the formation of or concentration of acyclic sugars is not the rate limiting step for the early Maillard reaction.

The importance of the lower acyclic sugar concentration observed in D_2O cannot be over stressed. Many studies of food materials use D_2O as a substitute for water when H_2O cannot be used as the solvent. For example, Richardson et al.[13] used D_2O to study the mobility of H_2O in corn starch suspensions by NMR and Fisher et al.[14] used FTIR spectroscopy to study the kinetics and the mechanisms of the hydrolysis of ß-lactam antibiotics by ß-lactamase enzyme. Such studies rarely discuss the SIE, while we have shown that the reaction rate is significantly altered by changing the solvent from H_2O to D_2O. Thus, it is important to consider the potential errors that may occur when kinetic data obtained for reactions measured in D_2O are extrapolated to aqueous food systems.

As noted earlier, the literature[1,7] notes that the Maillard reactivity of sugars generally increases as the acyclic concentration increases. However, as we show, the concentrations of the acyclic form for fructose (which is the more reactive) and glucose are very small and the values obtained by FTIR cannot be used to explain the order of reactivity observed for different reducing sugars. Thus, we still cannot be sure if the concentration of the acyclic form is the controlling in the Maillard reaction since the levels are too low to quantify.

Acknowledgements
This project was supported in part by the USAID- Institut Agronomique et Vétérinaire Hassan II, MOROCCO project and in part by the National Science Foundation (Grant Number; NSF/CBT - 8512914).

References

1. H.S. Burton and D. J. McWeeny, Nonenzymatic browning reactions: Considerations for sugar stability *Nature,* 1963, **197**, 266-268.
2. S.J. Angyal, The compositions of reducing sugars in solution, *Adv. Carbohydr. Chem. Biochem.,* 1984, **42,** 15-68.
3. S.M. Cantor and Q.P. Peniston, The reduction of aldose at the dropping mercury cathode: Estimation of the aldehydo structure in aqueous solutions, *J. Assoc. Off. Anal. Chem.,* 1940 **62,** 2113-2121.
4. L.D. Hayward and S.J. Angyal, Asymmetry rule for the circular dichroism of reducing sugars and the proportion of carbonyl forms in aqueous solutions thereof, *Carbohydr. Res.,* 1977, **53,** 13-20.
5. W.F.K. Wynne-Jones, The electrolytic dissociation of heavy water, *Trans. FaradaySoc.,* 1936, **32,** 1397-1402.
6. H.S. Isbell and W. Pigman, Mutarotation of sugars in solutions: Part II, in: 'Advances in Carbohydrate Chemistry & Biochemistry', Vol. 24, M.L. Wolfrom & Tipsom (eds.), Academic Press, New York, 1969, pp.13-65.
7. H.S. Burton, D. J. McWeeny and D.O. Biltchiffe, Sulphites and aldose-amino reactions, *Chem. Ind.*, 1962, **219**, 693-695.
8. A. Kaanane & T.P. Labuza, The Maillard Reaction in Foods, in: 'The Maillard Reaction in Aging, Diabetes and Nutrition', J. Baynes, (ed.) A.R. Liss Press, Inc., New York, 1989, pp. 301-327.
9. H.F. Bunn and P.J. Higgins, Reaction of monosaccharides with proteins: possible evolutionary significance, *Science,* 1981, **213,** 222-224.
10. W.M. Baisier and T.P. Labuza, Maillard Browning kinetics in liquid model systems, *J. Agric. Food Chem.,* 1992, **40,** 707-713.
11. P.M. Langhton and Robertson, Solvent isotope effects for equilibria and reactions, in 'Solute-Solvent Interaction', Coetzee and Ritchie, (eds.), Marcel Dekker, Inc., New York, (1969) 396-537.
12. W. Pigman and H.S. Isbell, Mutarotation of sugars in solution: Part I: in 'Advances in Carbohydrate Chemistry', Vol. 23,. Wolfrom, M.L. & Tipson, R.S.(eds.), Academic Press, New York., 1968, pp. 11-57.
13. S.J. Richardson, I.C. Baianu and M.P. Steinberg, Mobility of water in corn starch suspensions determined by NMR, *Starch/Stärke,* 1987, **39,**79-83.
14. J. Fisher, J.G. Belasco, S. Khosla and J.R. Knowles, β-Lactamase proceeds via an acyl-enzyme intermediate. Interaction of *Escherchia coli* RTEM enzyme with cefoxitin *Biochemistry,* 1980, **19,** 2895-2901.

Influence of D-Glucose Configuration on the Kinetics of the Nonenzymatic Browning Reaction

Jörg Häseler, Beate Beyerlein and Lothar W. Kroh

TECHNISCHE UNIVERSITÄT BERLIN - INSTITUT FÜR LEBENSMITTELCHEMIE
GUSTAV-MEYER-ALLEE 25, D-13355 BERLIN, GERMANY

Summary
The kinetics of the nonenzymatic browning reaction, especially caramelization, was influenced by the configuration of D-glucose. Spectrophotometric analysis suggests a higher reactivity of β-D-glucose in comparison with α-D-glucose during caramelization. More 5-hydroxymethyl-2-furaldehyde (HMF) and 2-furaldehyde (FF) were obtained from β-D-glucose than from α-D-glucose. In addition to the 3-deoxyglucosone pathway, an alternative mechanism via two C3-fragments was proposed.

The formation of 1,6-anhydro-β-D-glucose and glucobioses was independent of the configuration of glucose because it passes through a glycosyl cation. Due to steric preference, the 1→6-linked glucobioses isomaltose and gentiobiose are favoured.

Under Maillard reaction conditions, the formation of HMF was also influenced by the configuration. At the start of the reaction the Amadori compound was formed slightly easier from α-D-glucose.

Introduction

The constitution and stereochemistry of carbohydrates influences their reactivity in thermolysis/hydrothermolysis reactions fundamentally, for example, the configuration of D-glucose affects the kinetics of nonenzymatic browning reactions.[1] Under given conditions the β-anomer shows higher reactivity.

D-glucose is the main product of the thermolysis of α-glucans.[2] The individual steps in such reactions are important. Either glucose reacts via transglycosylation to branched maltodextrins or it is an intermediate in the formation of 1,6-anhydro-β-D-glucose (AHG). A glycosyl cation intermediate could exert a considerable influence on the reaction.[4] The preferred pathway is possibly influenced by stereochemical effects.

It is known that enolization is a much slower reaction than mutarotation, nevertheless at temperatures above the melting point mutarotation takes place.[3] While at 20 °C mutarotation is a matter of hours, at higher temperatures reaction will proceed magnitudes faster. If mutarotation dominates in the initial stage of the browning reaction no difference in the behaviour of the two anomers should be observable.

Materials and Methods

Reaction mixtures (100 mg) containing glucose or mixtures of glucose-DL-alanine (equimolar) were heated in a closed system at 160 °C or 180 °C for caramelization and in an open system for the Maillard reaction at 135 °C. After heating the samples were dissolved in 2 mL distilled water (filtered if necessary). Spectrophotometric measurements at 280 nm and 420 nm were used for the determination of browning intensity using a UVIKON 930 spectrophotometer (Pharmacia).

Reaction products 5-hydroxymethyl-2-furaldehyde (HMF) and 2-furaldehyde (FF) were detected by HPLC: pump, Pharmacia LKB 2248; column, 2 x RP-C18, Spherisorb ODS2, 250 mm x 4.6 mm, 5 µm; eluent, 0.01 N hexanesulfonic acid-methanol (77:23), flow rate, 0.7 mL/min; detector, UV/VIS (VWM 2141), λ=280 nm; oven temperature, 55 °C; injection volume, 20 µL.

Carbohydrates and Amadori compounds were analyzed by RI-HPLC: pump, Shimadzu LC-9A; column, IEC Aminex HPX 87C; eluent, 0.01 N $Ca(NO_3)_2$; flow rate, 0.5 mL/min; detector, Shimadzu RID-6A; oven temperature, 80 °C; injection volume, 20 µL.

The non-volatile polar fraction formed upon D-glucose thermolysis was characterized by GC/EI-MS after reductive cleavage[5,6], GC, HP 5890 Series II gas-liquid chromatograph; column, SE-54 (0.25 mm x 25 m); temperature, 100 °C for 3 min, 2 °C/min to 250 °C, hold for 10 min; MS, HP 5989B; scan, 40-550 amu.

Results and Discussion

The chosen compounds allow different reaction pathways and mechanisms during the nonenzymatic browning reaction to be studied. At 280 nm intermediates such as enediols, dicarbonyls and aldehydes were measured while at 420 nm the higher molecular fraction as well as coloured compounds and caramel were detected. The typical products of carbohydrate degradation during caramelization and Maillard reaction are represented by the furan derivatives 5-hydroxymethyl-2-furaldehyde (HMF) and 2-furaldehyde (FF).

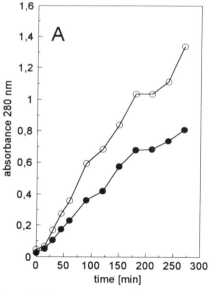 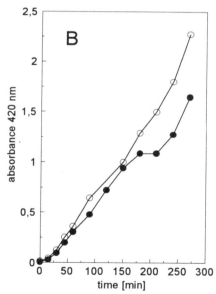

Figure 1. Browning intensity of α- and β-D-glucose (160 °C) at 280 nm (A; diluted 1:200) and 420 nm (B; diluted 1:50). α-D-glucose, •; β-D-glucose, o.

The potential influence of different configurations of carbohydrate in the early phase of the Maillard reaction may be highest in the case of glycosylamine and Amadori compounds as typical intermediates. Reactions of the Amadori compound are influenced more by the degradation of the carbohydrate than by carbohydrate configuration, therefore differences due to the configuration are rather small in Maillard reaction.

Finally, the kinetics of the formation of AHG and the glucobioses were studied in caramelization, since the formation of these products is an important step toward further reactions. Similar amounts of AHG were formed, therefore it might be concluded that the mechanism yielding AHG is not influenced by the position of the anomeric hydroxyl group.

Caramelization

At 160 °C, the β-D-configuration produces higher browning intensity (Figure 1). This can be explained by the initiating reactions of nonenzymatic browning. Ring opening, enolization, further reactions like abstraction of water, oxidative processes and retro-aldolreactions were faster in β-D-glucose. During the observed time (4 ½ h) the β-D-glucose shows always a higher absorbance at 420 nm (Figure 1B). It also results in higher formation of furan derivatives (Figure 2). The amount of HMF formed from α-D-glucose was approximately half as much as formed from β-D-glucose. Furthermore it was interesting that the amount of HMF increases markedly from the start of the reaction, while for the formation of FF an induction phase was observed. These results were consistent with the spectrophotometric data. The higher reactivity of the β-configuration could be explained by the following considerations. In the early phase, configuration and conformation may be decisive. The velocity of ring opening might differ between the α- and β-anomers or, if the velocity of ring opening is equal, the difference might be explained by favoured fragmentation of the ring structure of the two anomers; for example, in the C3-C4 position.

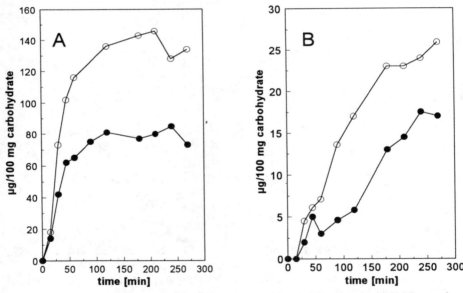

Figure 2. Formation of HMF (A) and FF (B) during heating of α- and β-D-glucose at 160 °C. α-D-glucose, •; β-D-glucose, o

The formation of AHG appears not to be influenced by the configuration of the starting material. Both graphs show that neither configuration was preferred (Figure 3B). This suggests that the glucosyl cation might be here an important intermediate.

In contrast, the amount of degraded glucose depended on the configuration of the starting material; the loss of β-glucose was greater than that of α-glucose (Figure 3A).

Figure 4 shows the chromatograms of carbohydrate derivatives after reductive cleavage. The proportion of single acetylated derivatives is presented. Therefore the formation of different linked glucobioses upon transglycosilation, was not influenced by the configuration of the starting material.

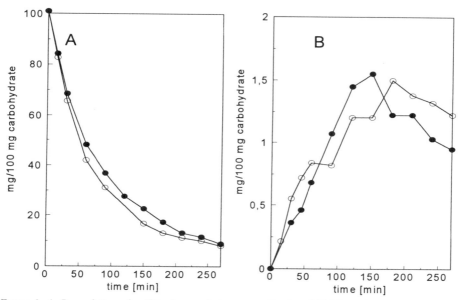

Figure 3. A, Degradation of α-/β-D-glucose during caramelization (160 °C) and B, Formation of AHG during caramelization (160 °C). α-D-glucose, •; β-D-glucose, o

Maillard reaction

The absorbance at 280 nm and 420 nm and the formation of Amadori compound and HMF appear to be slightly higher in the case of α-D-glucose (Figures 5 and 6). This could be caused by an easier formation of the β-glucosylamine from α-D-glucose as a precursor for the Amadori compound.

The β-anomer, solved in n-octanol, shows a higher browning activity in comparison with α-glucose.[7] Without a solvent, however, during the observed time α-D-glucose browns slightly more compared to the β-configurated carbohydrate. Also HMF was formed more from the α-anomer, whereas the Amadori rearrangement compound shows this behaviour only in the increasing phase (Figure 6).

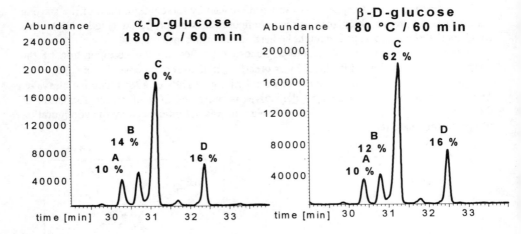

Figure 4: Formation of different linked glucobioses during caramelization of D-glucose (160 °C);
A = 4-O-acetyl-1,5-anhydro-2,3,6-tri-O-methyl-D-glucitol (indicating maltose and cellobiose);
B = 2-O-acetyl-1,5-anhydro-3,4,6-tri-O-methyl-D-glucitol (indicating kojibiose and sophorose);
C = 6-O-acetyl-1,5-anhydro-2,3,4-tri-O-methyl-D-glucitol (indicating AHG, gentiobiose and isomaltose);
D = 3-O-acetyl-1,5-anhydro-2,4,6-tri-O-methyl-D-glucitol (indicating nigerose and laminaribiose)

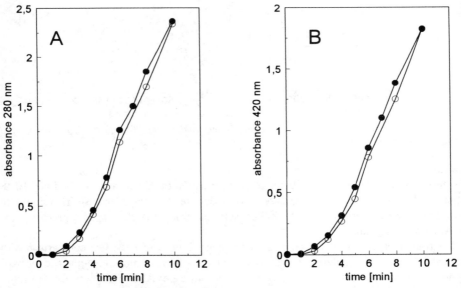

Figure 5. Browning intensity of α- and β-D-glucose with alanine (135 °C) at 280 nm (A, 1:200 dilution) and 420 nm (B, 1:50 dilution). α-D-glucose, •; β-D-glucose, o

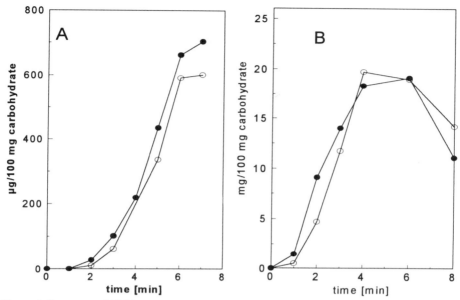

Figure 6. Formation of HMF (A) and fructose alanine (B) during the Maillard reaction at 135 °C. α-D-glucose, •; β-D-glucose, o

Acknowledgements

The authors sincerely thank the Deutsche Forschungsgemeinschaft Grand No. Kr 1452/2-1 for financial support, further we would like to thank Renate Brandenburger for technical assistance and Anke Hollnagel for scientific discussion.

References

1. L. W. Kroh and G. Westphal, Die Maillard-Reaktion in Lebensmitteln, *Mitteilungsblatt der Chemischen Gesellschaft der DDR*, 1986, **35**, 73-80.
2. L. W. Kroh, W. Jalyschko and J. Häseler, Non-volatile reaction products by heat-induced degradation of α-glucans. Part I: Analysis of oligomeric maltodextrins and anhydro-sugars, *Starch/Stärke*, 1996, **48**, 426-433.
3. R. Schallenberger and G. G. Birch, Sugar Chemistry, The Avi Publishing Company. Inc., Westport, Connecticut, 1975.
4. T. L. Lowary and G. N. Richards, Mechanism in pyrolysis of polysaccharides: Cellobiitol as a model for cellulose. *Carbohydr. Res.*, 1990, **198**, 79-89.
5. I. Ciucanu and F. Kerek, A simple and rapid method for the permethylation of carbohydrates, *Carbohydr. Res.*, 1984, **131**, 209-217.
6. J.G. Jun and G.R. Gray, A new catalyst for reductive cleavage of methylated glycans, *Carbohydr. Res.*, 1987, **163**, 247-261.
7. G. Westphal and L. W. Kroh, *Nahrung*, 1985, **29**, 757-764.

Protein-Bound Maillard Compounds in Foods: Analytical and Technological Aspects

Thomas Henle, Uwe Schwarzenbolz, Axel W. Walter and Henning Klostermeyer

LEHRSTUHL FÜR MILCHWISSSENSCHAFT, TECHNISCHE UNIVERSITÄT MÜNCHEN, VÖTTINGER STR. 45, D-85350 FREISING, GERMANY

Summary
By means of suitable analytical techniques, the formation of protein-bound Maillard compounds in foods was studied. The lysine derivative pyrraline could be detected in enzymatic hydrolysates of milk and bakery products at levels up to 3700 mg/kg protein by ion-exchange chromatography (IEC) with photodiode-array measurement and by ion-pair RP-HPLC. During storage of freeze-dried milk, the rate of formation of pyrraline correlated directly with the water content of the samples. In acid hydrolysates, sensitive determination of the fluorescent arginine-lysine crosslink pentosidine was achieved using IEC with direct fluorescence detection. Levels of pentosidine in various foods ranged between 'not detectable' (< 50 µg/kg protein) and 35 mg/kg protein, indicating that pentosidine does not play a major part in crosslinking of food proteins. A previously unknown protein-bound arginine derivative, N^δ-(5-methyl-4-oxo-5-hydroimidazol-2-yl)-L-ornithine, was isolated from acid hydrolysates of bakery products. The ornithinoimidazolinone is formed by direct condensation of the guanido group of arginine and the sugar degradation product methylglyoxal, thus representing a new post-translational modification of food proteins. For certain foods (baking products, roasted coffee), between 20 and 50 % of the arginyl residues might react with methylglyoxal during food processing.

Introduction

Functional and nutritional properties of foods are significantly influenced by chemical reactions occurring during heating or storage. In this context, post-translational modifications of proteins by carbohydrates are of particular importance. However, compared to the knowledge we have concerning the degradation of carbohydrates and the formation of aroma compounds during the Maillard reaction,[1] only limited information is available about the formation of protein-bound amino acid derivatives during the various stages of nonenzymatic browning reactions. For years, the Amadori products lactuloselysine and fructoselysine, first isolated from heated milk in 1977, and their oxidative degradation product N-ε-carboxymethyllysine (CML), detected in small amounts in hydrolysates of milk proteins, represented the only known protein-bound Maillard compounds formed during food processing.[2,3] Consequently, research in our laboratory was focused on the development of suitable analytical techniques for the detection and quantification of amino acid derivatives and studies were carried out concerning the formation of individual compounds during heating and storage of various foods.

Materials and Methods

Sample preparation
Food samples were obtained from local retail stores. For storage experiments, skim milk was freeze-dried and stored in a climatic oven at 70 °C under varying humidity (20 - 60 % r.h.). For amino acid analysis, samples were hydrolyzed enzymatically using four enzymes (pepsin, pronase, aminopeptidase, prolidase), or by 6 N hydrochloric acid as described previously.[4]

Chromatography

Amino acid analysis was perfomed on an Alpha Plus amino acid analyzer (LKB Biochrom, Freiburg, Germany), using a stainless steel column (150 x 4 mm) filled with ion-exchange resin, DC4A-spec, sodium form (Benson, Reno, USA). Elution buffers, ninhydrin reagent as well as operating conditions are described elsewhere.[4-6] Amino acids were detected initially using either a photodiode array detector PDA 996 (Waters, Eschborn, Germany) or a fluorescence detector SFM 25 (Kontron, Eching, Germany), connected between the outlet of the column and the ninhydrin reaction coil.[4,5] Following subsequent ninhydrin derivatization, amino acids were detected at 570 nm. Quantification of pyrraline via isocratic ion-pair reversed-phase HPLC with UV detection was performed as described previously.[7] Reference material, synthesized as described in previous reports, was used for calibration.[5,6,8]

Results and Discussion

It is well known from several studies, that α-dicarbonyl compounds, which are formed during the advanced Maillard reaction and carbohydrate degradation in acidic and alkaline media, are highly reactive towards proteins, reacting with lysine and arginine residues in particular. An acid-labile pyrrole derivative resulting from the reaction of the ε-amino group of lysine and 3-deoxyglucosulose, namely 2-amino-6-(2-formyl-5-hydroxymethyl-1-pyrrolyl)-hexanoic acid ('pyrraline'), was first identified in heated reaction mixtures of lysine and glucose.[9]

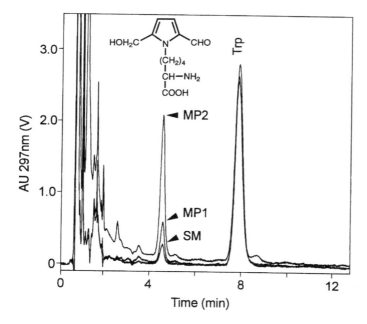

Figure 1. *Ion-pair RP-HPLC with direct UV-detection at 297 nm: Enzymatic hydrolysates of sterilized milk (SM) and commercial milk powder samples (MP1, MP2). Inset: Chemical structure of pyrraline*

Following complete enzymatic hydrolysis, we were able to detect and quantify pyrraline along with all other amino acids including the Amadori products lactuloselysine and fructoselysine using ion-exchange chromatography with photodiode-array measurement and subsequent ninhydrin derivatization.[4] Alternatively, fast and sensitive quantification of pyrraline within a run time of 10 minutes at levels lower than 10 mg/kg of protein can also be achieved using isocratic ion-pair RP-HPLC and direct UV-detection at 297 nm (Figure 1). Values for pyrraline in milk products ranged from 'not detectable' (raw, pasteurized and UHT-treated milk) up to ~ 150 mg/kg of protein (sterilized milk, evaporated milk, most milk powder samples). During storage of freeze dried milk under varying relative humidity, a linear increase of the amount of pyrraline could be observed, with the rate of the formation correlating directly with the water content of the samples (Figure 2). Increased values found for certain samples of milk powder or whey powder, respectively (up to 3100 mg/kg of protein), thus can be explained by 'overprocessing' and/or storage of the products under adverse conditions. Remarkably high amounts of pyrraline were measured in bakery products (200 to 3700 mg/kg of protein), indicating that up to 15 % of the lysine residues may react to the pyrrole derivative during baking - an observation rising questions concerning possible nutritional and physiological consequences.[7]

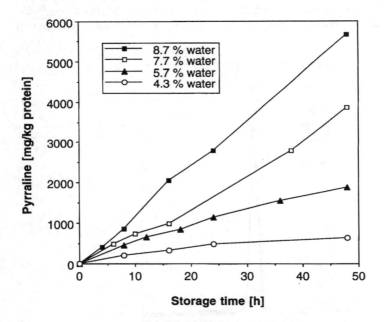

Figure 2. *Formation of pyrraline during storage of freeze dried milk at 70 °C. Inset: Water content after equilibration of samples. Equilibration was complete after ~2 h storage time*

The direct participation of the Maillard reaction in the crosslinking of proteins during aging *in vivo* was demonstrated with the identification of pentosidine, an imidazopyridinium compound, in which one lysine and one arginine residue are linked together by a ribose moiety.[10]

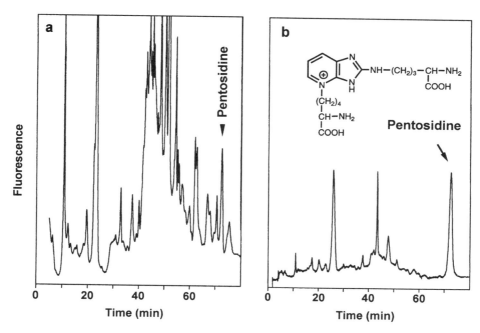

Figure 3. *Ion-exchange chromatography with direct fluorescence detection at λ_{ex} = 335 nm and λ_{em} = 385 nm. (a) Acid hydrolysate of roasted coffee (37 pmol pentosidine/mg of protein) (b) Acid hydrolysate of human plasma (7.5 pmol pentosidine/mg of protein). Inset: Chemical structure of pentosidine*

Following the synthesis of pentosidine reference material, we were able to detect very small amounts of the crosslink in addition to common amino acids in a variety of foods for the first time using ion-exchange chromatography with direct fluorescence detection (Figure 3 a). The detection limit was lower than 50 µg/kg protein. Levels of pentosidine in food ranged between 'not detectable' and 2 - 5 mg/kg protein for certain milk products (sterilized milk, evaporated milk). Up to 35 mg/kg protein could be quantified for some bakery products and coffee samples, which is in a range of concentration comparable to the amounts of pentosidine usually found in physiological samples like urine or plasma (Figure 3 b). During storage of UHT-treated skim milk at 37 °C for up to 3 months, a significant increase in the amount of pentosidine could be observed, which correlated with the irreversible (non-reducible) polymerization of the casein fraction.[11] However, compared with the amounts of the crosslinking amino acids lysinoalanine and histidinoalanine (up to 3000 mg/kg of protein), the measured values for pentosidine were more than three orders of magnitudes lower (up to 5 mg/kg of protein), indicating that pentosidine does not play a major part in the crosslinking of food proteins.

The observation that heating or storage of food proteins in the presence of carbohydrates leads to a significant modification of arginine has been well known for years,[12] although, to date, no individual arginine derivatives have been isolated from foods. Based on model experiments, one possible pathway, the formation of heterocyclic imidazolones via a reaction of the guanido group of arginine with α-dicarbonyls like 3-deoxyglucosulose, 3-deoxypentosulose or methylglyoxal, was established.[13-15]

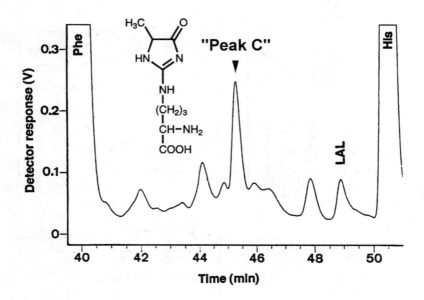

Figure 4. *Ion-exchange chromatography of the acid hydrolysate of an alkali treated pretzel with ninhydrin derivatization and detection at 570 nm. Inset: Chemical structure of N^δ-(5-methyl-4-oxo-5-hydroimidazol-2-yl)-L-ornithine*

During routine amino acid analysis, we detected a previously unknown ninhydrin-positive compound ("peak C") in acid hydrolysates of alkali-treated bakery products (pretzels, snack bars; Fig. 4). The amount of peak C correlated with the decrease in arginine. Following preparative isolation from a food sample, the compound was unequivocally identified by FAB-MS, ^{1}H- and ^{13}C-NMR as the protein-bound ornithinoimidazolinone, N^δ-(5-methyl-4-oxo-5-hydroimidazol-2-yl)-L-ornithine.[6] Independent synthesis proved that the amino acid derivative is a condensation product of the arginine side chain and the sugar degradation product methylglyoxal, thus representing a new post-translational modification of food proteins. For certain bakery products and roasted coffee beans, the amounts of the ornithinoimidazolinone ranged between 400 and 1300 mg per 100 g protein, indicating that 20 to 50 % of the arginyl residues might have reacted with methylglyoxal during baking or roasting. As methylgloyxal and other α-dicarbonyls, in particular glyoxal and 3-deoxyglucosulose, are observed as sugar degradation products under physiological conditions, the presence of such arginine derivatives and similiar compounds in biological systems seems likely.[15]

By the use of suitable analytical techniques, various protein-bound amino acid derivatives originating from the reaction of proteins with carbohydrates or their degradation products can be quantified in foods. Individual compounds can serve as suitable indicators for monitoring the advanced stages of the Maillard reaction and may be used to charaterize the influence of processing and storage on food quality. Due to the remarkably high concentrations of pyrraline and the previously unknown ornithinoimidazolinone in certain food samples, more information is needed about the physiological fate of such modified proteins during digestion and excretion. Corresponding studies are currently in progress.[16]

References

1. J.M. Ames, The Maillard reaction, in 'Biochemistry of Food Proteins', B.J.F. Hutton (ed.), Elsevier Applied Science, London, 1992, pp. 99-153.
2. A.B. Möller, A.T. Andrews and G.C. Cheeseman, Chemical changes in ultra-heat-treated milk during storage. II. Lactuloselysine and fructoselysine formation by the Maillard reaction, *J. Dairy Res.*, 1977, **44**, 267-275.
3. W. Büser and H.F. Erbersdobler, Carboxymethyllysine, a new compound of heat damage in milk products, *Milchwissenschaft*, 1986, **41**, 780-785.
4. T. Henle and H. Klostermeyer, Determination of protein-bound 2-amino-6-(2-formyl-5-hydroxymethyl-1-pyrrolyl)-hexanoic acid ("pyrraline") by ion-exchange chromatography and photodiode array detection, *Z. Lebensm. Unters. Forsch.*, 1993, **196**, 1-4.
5. T. Henle, U. Schwarzenbolz and H. Klostermeyer, Detection and quantification of pentosidine in foods, *Z. Lebensm. Unters. Forsch.*, 1997, **204**, 95-98.
6. T. Henle, A.W. Walter, R. Haessner and H. Klostermeyer, Detection and identification of a protein-bound imidazolone resulting from the reaction of arginine residues and methylglyoxal, *Z. Lebensm. Unters. Forsch.*, 1994, **199**, 55-58.
7. T. Henle and A. Bachmann, Studies on the formation of pyrraline in milk products, *Z. Lebensm. Unters. Forsch.*, in press.
8. T. Henle and A. Bachmann, Synthesis of pyrraline reference material, *Z. Lebensm. Unters. Forsch.*, 1996, **202**, 72-75.
9. T. Nakayama, F. Hayase and H. Kato, Formation of ε-(2-formyl-5-hydroxymethyl-pyrrol-1-yl)-L-norleucine in the Maillard reaction between D-glucose and L-lysine, *Agric. Biol. Chem.*, 1980, **44**, 1201-1202.
10. D.R. Sell and V.M. Monnier, Structure elucidation of a senescence cross-link from human extracellular matrix, *J. Biol. Chem.*, 1989, **264**, 21597-21602.
11. T. Henle, U. Schwarzenbolz and H. Klostermeyer, Irreversible crosslinking of casein during storage of UHT-treated skim milk, *Int. Dairy Fed. Special Issue*, 1996, **No. 9602**, 290-298.
12. A. Mohammad, H. Fraenkel-Conrad and H.S. Olcott, The "browning" reaction of proteins with glucose, *Arch. Biochem.*, 1949, **24**, 157-178.
13. Y. Konishi, F. Hayase and H. Kato, Novel imidazolone compound formed by the advanced Maillard reaction of 3-deoxyglucosone and arginine residues in proteins, *Biosci. Biotech. Biochem.*, 1994, **58**, 1953-1955.
14. R. Sopio and M. Lederer, Reaction of 3-deoxypentosulose with N-methyl and N,N-dimethylguanidine as model reagents for protein-bound arginine and creatine, *Z. Lebensm. Unters. Forsch.*, 1995, **201**, 381-386.
15. P.J. Thornalley, Methylglyoxal, glyoxalases and the development of diabetic complications, *Amino Acids*, 1994, **6**, 15-23.
16. T. Henle, R. Deppisch and E. Ritz, The Maillard reaction - from food chemistry to uremia research, *Nephrol. Dialysis Transplant.*, 1996, **11**, 1719-1722.

Flavour Chemistry

Formation of Maillard Aromas During Extrusion Cooking

Chi-Tang Ho and William E. Riha III

DEPARTMENT OF FOOD SCIENCE, RUTGERS UNIVERSITY, NEW BRUNSWICK, NJ 08903, USA

Summary
Extrusion cooking is a powerful food processing technique that utilizes high temperature and high shear to produce food products with unique physical and chemical characteristics. The conditions employed during extrusion cooking are generally ideal for the production of aroma volatiles via the Maillard reaction.
2% levels of the amino acids glutamine, glutamic acid, and arginine were separately added to corn flour with 5% glucose and 18% moisture. The volatiles produced were dictated by the amino acid added. Glutamine produced the highest levels of volatiles, specifically pyrazines, and the addition of glutamic acid produced higher levels of furans.

Introduction

Extrusion cooking is a food processing technique that involves the application of high shear and high heat to a material such as wheat flour. When inside the extruder, turning screws mix and transport the material to a small opening at one end of the extruder, the die. Inside the extruder, before the material is pushed through the die, pressure builds up and as the material leaves the extruder the high temperature and sudden pressure drop causes water to instantly vaporize. The steam produced creates the familiar and characteristic puffed product, but, also strips away important flavour volatiles which were produced during the extrusion[1].

This steam distillation of volatiles is a problem processors have had to deal with, usually by adding flavour back to a product post-extrusion. This results in an unevenly flavoured product and the use of expensive flavourings. Processors have also tried adding flavours which are less volatile and will not be stripped as easily at the die[2,3].

Another solution may be the addition of reactive precursors to the flour being extruded prior to extrusion[4]. While this technique will not prevent volatiles being lost at the die, the added precursors will result in higher levels of volatiles being produced. The addition of the reactive precursors may also be used to alter the flavour of the extrudate. Theoretically, this is an excellent way to improve the flavour of extruded products. The conditions used during extrusion are ideal for the production of volatiles via the Maillard reaction. This reaction which occurs at relatively low moisture contents and high temperatures involves the reaction of an amine and a carbonyl[5]. There are many steps in this reaction and therefore many different types of volatiles can be produced[6].

For example, pyrazines, which are nitrogen-containing heterocyclics are important for giving foods roasted, toasted, or nutty aromas[6,7]. These compounds would obviously be very important to most of the products made by extrusion cooking, specifically breakfast cereals and snack foods. Pyrazines can be formed through the Strecker degradation. In this reaction, two or three carbon sugar fragments formed in the earlier stages of the Maillard reaction can react with α-amino groups of amino acids or ammonia which can be released by the deamidation reaction of glutamine or asparagine[8].

The addition of the amino acid cysteine to wheat flour during extrusion has recently been reported[9]. Due to the presence of both a thiol group and an amino group on the amino acid cysteine, one would except a large number of different volatiles to be produced. These researchers added 0, 0.25, 0.5, 0.75, or 1.0% cysteine to high gluten wheat flour and extruded the mixture on a twin-screw extruder at 185°C, 500 rpm, and 16% moisture. The concentration of most classes of volatiles identified increased up to 0.5% cysteine added; above this concentration there was very little change in the volatile concentration. This may be due to the fact that cysteine has a great effect on the physical properties of the extrudate, such as expansion ratio and oil/water binding capacity, and perhaps the extrudate, with its limited ability to retain volatiles, becomes saturated with flavour compounds. Although the majority of compounds identified were sulfur containing, the addition of cysteine did increase the production of pyrazines.

This paper reports the generation of Maillard aromas by the addition of arginine glutamine, or glutamic acid as amino source and glucose as carbonyl source during single-screw extrusion of corn flour.

Materials and Methods

Extrusion of corn meal with glutamine, glutamic acid, and arginine
The amino acids glutamine, glutamic acid, and arginine were separately added at a 2% level to three batches of corn meal. The three batches all had 5% glucose added. Distilled water was added dropwise during mixing to raise the moisture content from 12% to 18%. A control batch with no amino acid or glucose added was also adjusted to 18% moisture content. The production of extrusion samples was accomplished using a C. W. Brabender type 2003 single-screw extruder with a 1.9 cm barrel diameter and an L/D ratio of 20:1. The die diameter was 6.5 cm. The extruder was brought to equilibrium conditions using untreated corn meal. The screw speed was set to 100 rpm, the two heating zones were set to 100°C and 130°C, and the die temperature was measured at 110°C. Once the extruder reached operating conditions, the corn meal with precursors was fed into the extruder and samples were collected ten minutes after the change in feed was made. Samples were ground in the presence of dry ice in a Waring blender and stored in mason jars in a freezer.

Volatile isolation
Eight grams of each of the four ground extrudates was weighed into a glass cylinder. The samples were then spiked with internal standard (1 ml of 1 mg/mL d-8-toluene in methanol) and connected to a Thermal Desorption Sample Collecting System (SIS, Ringos, NJ). The volatiles were then purged and trapped into thermal desorption tubes containing two equal beds of the resins Tenax TA and Carbotrap. The trapping conditions were as follows: oven block temperature, 80°C; nitrogen flow, 40 ml/min; and trapping time, one hour.

Volatile analysis
The trapped volatiles were desorbed directly into the GC injection port (220° C, 5 min.) using a Model TD-1 Short Path Thermal Desorption Apparatus (SIS, Ringos, NJ). Separation of the volatiles was accomplished using a Varian 3400 gas chromatograph equipped with a nonpolar fused silica capillary column (60m x 0.32 (i.d.), 1 mm film thickness, DB-1; J&W Scientific). The GC was operated with an injector temperature of 270° C, a helium carrier gas flow rate of

1 mL/min., a split ratio of 10:1, and a detector temperature of 300° C. The column temperature program was from -20° C to 280° C at 2° C /min. Quantification was accomplished by the use of an internal standard. The concentration of each component was converted to part per billion of the ground extrudate.

GC/MS analysis
Tentative identification of the volatiles was accomplished by GC/MS using a Varian 3400 gas chromatograph coupled to a Finnigan MAT 8230 high-resolution mass spectrometer, using the same GC program as the separation. Mass spectra were obtained by electron ionization at 70 eV and a source temperature of 250° C. The filament emission current was 1 mA and spectra were recorded on a Finnigan MAT SS 300 data system.

Results and Discussion

Table 1 lists all of the volatiles identified from the extruded samples. The sample with glutamine added produced the highest concentration of volatiles, followed by glutamic acid and finally arginine. Pyrazines were the most abundant class of volatiles produced in the glutamine, arginine, and control samples. Of the pyrazines, methylpyrazine was produced at the highest concentration in all samples.

The high concentrations of unsubstituted pyrazine and methylpyrazine indicates that there may have been a great deal of sugar fragmentation occurring. The early stages of the Maillard reaction results in the formation of a Schiff's base and, finally, an Amadori product. In the present study, the glucose added was reacting with either the amino group of the amino acids or free ammonia which was being released to form the Amadori product. The Amadori product can then either produce 1 or 3 deoxyglucosones which through retroaldol can lead to the formation of pyruvaldehyde, glyceraldehyde, and glycolaldehyde among others. Glycolaldehyde is a two carbon dicarbonyl which can react with the α-amino group of an amino acid to form an α-aminoketone. Two of these compounds can condense to form an unsubstituted pyrazine. Pyruvaldehyde is a three-carbon dicarbonyl which can similarly react with the α-amino group of an amino acid to form a three carbon α-aminoketone. Upon condensation with a two carbon aminoketone, methylpyrazine is formed.

Ammonia, unlike the α-amino groups will not react with the dicarbonyls; it will react with α-hydroxyketones[10]. In this case, ammonia can react with glyceraldehyde, a three carbon α-hydroxyaldehyde, to form an aminoketone which can condense with a two carbon aminoketone to form methylpyrazine. Due to the very high concentration of methylpyrazine observed, it is very likely that high levels of these two and three carbon sugar fragments were formed.

The second most abundant type of pyrazine identified was the dimethylpyrazines, such as 2,5-dimethylpyrazine and 2,6-dimethylpyrazine. Dimethylpyrazines are easily formed from the condensation of two three-carbon aminoketones as just described. However, 2,3-dimethylpyrazine requires the condensation of a four carbon fragment, such as 2,3-butanedione (diacetyl), and a two carbon fragment. Trimethylpyrazine requires the participation of a four carbon fragment and a three carbon fragment.

Table 1. *Volatiles tentatively identified from the extrusion of corn meal with various amino acids and glucose added*

Compounds	Glutamine	Arginine	Glutamic acid	Control
Pyrazines				
pyrazine	2227.0	1664.2	801.8	218.1
methylpyrazine	15221.0	3424.2	1595.0	200.3
ethylpyrazine	417.1		1145.1	246.6
2,5-dimethylpyrazine	1001.7	57.4	317.9	69.2
2,6-dimethylpyrazine	2036.0	933.3	383.4	
2,3-dimethylpyrazine	294.5		104.4	
vinylpyrazine	859.0	62.9	74.3	
2-ethyl-3-methylpyrazine	153.3	56.5	31.2	17.7
trimethylpyrazine	58.1	56.9	14.8	
2-methyl-5-vinylpyrazine	311.2	29.5		54.2
2-methyl-6-vinylpyrazine	12.2			
isopropenylpyrazine	17.9		17.3	
2-methyl-5-*trans*-propenyl-pyrazine	20.4		6.9	
2-acetyl-5-methylpyrazine	59.5	25.8	41.8	
2-acetyl-6-methylpyrazine	89.4	10.4	19.9	
6,7-dihydro-5-methyl-cyclopentapyrazine	6.1	8.9	15.5	
Furans				
2-methylfuran	8.5	14.6	14.3	
2-furfural	5750.0	22.2	8723.5	452.0
2-furfuryl alcohol	242.5		115.3	8.3
5-methyl-2-furfural	246.2			
Others				
2,3-butanedione	36.8	9.1	83.8	
3-methylbutanal	33.5	39.5	97.1	
acetic acid		137.3		
1-hydroxy-2-propanone	68.7	31.7	157.7	
2-acetyl-1,4,5,6-tetrahydro-pyridine	63.2			

Of the furans produced, 2-furfural reached the highest concentration in all the samples. 2-Furfural is produced from sugar degradation which may be catalyzed by amino groups via the Maillard reaction. Ammonia may also catalyze this breakdown in sugars which will result in formation of furans. However, as illustrated by the higher concentration of furans from the glutamic acid sample, ammonia was not key in the formation of furans. More than likely, when ammonia was present in the glutamine sample, the sugar fragments that were produced preferentially participated in the formation of pyrazines.

Of the other volatiles, 2-acetyl-1,4,5,6-tetrahydropyridine, which was found only in the glutamine samples, is important because is has a unique cracker-like aroma[6]. Many of the other compounds, such as 2,3-butanedione and 3-methylbutanal are formed from sugar degradation.

From this experiment, it is reasonable to assume that ammonia is released from amino acids and is incorporated into the Maillard reaction. Past evidence from Hwang et al.[8] indicates that deamidation of glutamine can occur under the conditions employed and participate in pyrazine formation. Sohn and Ho[11] showed that the deamidation of glutamine did occur at high levels between the temperature range of 110 to 180°C. This study was of course a model system utilized a reaction bomb for two hours. When the reaction time of such a model system is compared to the very short time the material remains in the extruder (about 30 seconds), one may wonder whether the two systems can be compared. However, one must realize that the time for the material in the reaction bomb to reach reaction temperature must be very long compared with the extruder, which will almost instantly heat the material due to the excellent contact with the heating surface. In addition, extrusion adds mechanical energy into the equation in the form of the turning screws. Such shear forces may play a role in facilitating the reactions. More importantly, the mixing that takes place during extrusion will undoubtedly bring reactants into contact with each other.

Conclusions

It has been shown that the generation of pyrazines at very high levels, well over the threshold levels can be produced by adding reactive precursors. When glutamine and glucose was added to corn meal and extruded on a single-screw extruder. The levels of pyrazines produced were much higher than those found in the corn meal control. Glutamine is an interesting amino acid because it contains an amide side chain. This side chain has the ability to release ammonia, which can react in the degradation of sugar and can also react with some of the sugar fragments formed, the hydroxyketones. Of course, glutamine also has an α-amino group which, can also react with sugars, such as glucose, and can react with the dicarbonyl fragments formed. Glutamine therefore would be expected to be an excellent choice as a precursor if one were interested in producing higher levels of pyrazines.

Another class of compounds formed by the Maillard reaction are the furans. Furans are oxygen-containing, five-membered heterocyclics which provide sweet or caramel-like aromas to foods. These volatiles are also very important to the flavour of extruded products, especially corn based products. Furans are formed in the earlier stages of the Maillard reaction, before the retroaldolization of the Amadori compounds which yields the small dicarbonyls necessary for pyrazine formation. Once again, by the addition of precursors we were able to preferentially produce furans. This was accomplished using the amino acid glutamic acid. Glutamic acid is the acidic form of glutamine. The two amino acids are structurally very similar; however, there was a large difference between the volatiles produced from the addition of glutamine versus glutamic acid. Furans were the most abundant form of volatiles produced from the extrusion with glutamic acid. Obviously, the acidic side chain is very important to the formation of furans. The large discrepancy between glutamine and glutamic acid also reiterates the importance of deamidation and ammonia on flavour generation.

Another indication that deamidation is important to flavour production during extrusion is that the addition of arginine to corn meal did not produce many volatiles at all. Arginine has a guanidino side chain which contains three nitrogen atoms in addition to the α-amino group. Despite the abundance in nitrogen very few pyrazines were produced.

References

1. J.M. Harper, Comparative analysis of single- and twin-screw extruder, in 'Food Extrusion and Technology', J.L. Kokini, C.-T. Ho and M.V. Karwe (eds), Marcel Dekker Inc., New York, 1992, pp.139-148.
2. J.A. Maga, Flavor formation and retention during extrusion, in 'Extrusion Cooking', C. Mercier, P. Linko and J.M. Harper (eds), American Association of Cereal Chemists, St. Paul, 1989, pp. 387-398.
3. M. Mariani, A. Scotti and E. Colombo, Flavours for foodstuffs obtained by extrusion-cooking: analytical and application aspects, in 'Progress in Flavour Research 1984', J. Adda (ed), Elsevier, Amsterdam, 1985, pp. 549-562.
4. W.E. Riha III and C.-T. Ho, Formation of flavors during extrusion cooking. *Food Rev. Int.*, 1996, **12**, 351-373.
5. H.I. Hwang, T.G. Hartman, M.V. Karwe, H.V. Izzo and C.-T. Ho, Aroma generation in extruded and heated wheat flour, in 'Lipids in Food Flavors', C.-T. Ho and T.G. Harman (eds), American Chemical Society, Washington, DC, 1994, pp. 144-157.
6. C.-T. Ho, Thermal generation of Maillard aromas, in 'The Maillard Reaction: Consequences for the Chemical and Life Sciences', R. Ikan (ed.), John Wiley & Sons Ltd, Chichester, 1996, pp. 27-53.
7. J. Maga, Pyrazine update. *Food Rev. Int.*, 1992, **8**, 479-558.
8. H.I. Hwang, T.G. Hartman, R.T. Rosen and C.-T. Ho, Formation of pyrazines from the Maillard reaction of glucose and glutamine-amine-^{15}N, *J. Agric. Food Chem.* 1993, **41**, 2112-2115.
9. W.E. Riha III, C.F. Hwang, M.V. Karwe, T.G. Hartman and C.-T. Ho, Effects of cysteine addition on the volatiles of extruded wheat flour. *J. Agric. Food Chem.* 1996, **44**, 1847-1850.
10. H. Weenen, S.B. Tjan, P.J. de Valois, N. Bouter, A. Pos and H. Vonk, Mechanism of pyrazine formation, in 'Thermally Generated Flavors: Maillard, Microwave, and Extrusion Processes', T.H. Parliment, M.J. Morello and R.J. McGorrin (eds), American Chemical Society, Washington, DC, 1994, pp. 142-157.
11. M. Sohn and C.-T. Ho, Ammonia generation during thermal degradation of amino acids. *J. Agric. Food Chem.* 1995, **43**, 3001-3003.

Thermal Degradation of the Lachrymatory Precursor of Onion

Roman Kubec and Jan Velíšek

INSTITUTE OF CHEMICAL TECHNOLOGY, DEPARTMENT OF FOOD CHEMISTRY AND ANALYSIS, TECHNICKÁ 1905, 166 28 PRAGUE, CZECH REPUBLIC

Summary
The lachrymatory precursor of onion, S-(1-propenyl)-L-cysteine sulfoxide, was heated in a closed model system at different temperatures (from 80 to 200°C) in the presence of variable amounts of water (0-98%) for 1 to 60 minutes. The volatile compounds produced were isolated by extraction, analyzed, and identified by means of GC and GC-MS. The major volatile sulfur-containing compounds generated by thermal degradation of this amino acid were 2,4- and 3,4-dimethylthiophenes and alkyl substituted thiazoles. The most important non-sulfur volatile degradation products (formed only at water contents higher than 20%) were 2-methyl-2-butenal and 2-methyl-2-pentenal.

Introduction

Important naturally occurring constituents of many plants belonging to the *Liliaceae* and *Brassicaceae* families are S-alk(en)yl-L-cysteines and their sulfoxides (S-allyl-, S-propyl-, S-methyl- and S-1-propenyl-L-cysteine sulfoxide). These unique sulfur-containing non-protein amino acids are important precursors of characteristic flavors enzymatically produced on disruption of plant tissues.[1] However, culinary processing (boiling, frying, baking etc.) causes inactivation of alliinase and, thus, considerable amounts of aroma precursors remain to participate in flavor generation in thermally processed vegetables.[2-4]

The lachrymatory principal of onion has been identified as propanethial-S-oxide. This compound arises from S-(1-propenyl)-L-cysteine sulfoxide (PeCSO) by the action of C-S lyase, alliinase (EC 4.4.1.4). Propanethial-S-oxide possesses a pungent odor and is easily degraded to various volatile compounds responsible for the characteristic onion aroma. It has been shown that its sensory attributes other than lachrymatory potential, such as odor, bitter taste, and tongue-biting sensation are probably due to degradation products of PeCSO.[5] This amino acid has also been implicated in the development of pink color in onion homogenates.[6]

Along with PeCSO and other minor cysteine sulfoxides (methyl and propyl analogues), considerable amounts of the corresponding γ-glutamyl-S-alk(en)yl-L-cysteine dipeptides are also found in the intact onion bulbs. These peptides are not cleaved by alliinase, nevertheless it can be assumed that they also participate in the generation of flavor of thermally processed onion.

Materials and Methods

Synthesis of the lachrymatory precursor
S-(1-propenyl)-L-cysteine sulfoxide was synthesized by oxidation of S-(1-propenyl)-L-cysteine, which was prepared by base-catalysed isomerization of S-(2-propenyl)-L-cysteine (deoxyalliin) according to the procedure of Carson and Boggs.[7] Structure of synthesized amino acid was confirmed by ^{1}H, ^{13}C NMR, and IR spectroscopy. Purity (higher than 97%) was checked by means of HPLC after derivatization with o-phthaldialdehyde in the presence of *tert*-butyl mercaptan (OPA derivatization).

Thermal decomposition of amino acids
Amino acid (50 mg) was placed in a 5 mL glass tube; water was added and the tube was sealed. Following equilibration for 24 hours at room temperature, the tube was heated in an oven at a temperature in the range of 80°-200°C, then cooled in a freezer to -18°C and crushed under water. The resulting solution was immediately extracted with 5 mL of diethyl ether and analyzed by GLC.

Gas chromatographic analyses
A Hewlett-Packard 5890 gas chromatograph equipped with a flame ionization detector and a HP-5 or HP-INNOWax capillary column (30 m × 0.25 mm, i.d.; film thicknesses of 25 and 50 µm, respectively) was used. The operating conditions were as follows: $T_i = 220°C$, $T_d = 250°C$, nitrogen carrier gas flow rate of 2 mL min^{-1} and temperature program: from 40°C (held for 3min) to 240°C, at 4°C min^{-1}. GC-MS analysis were carried out using a Hewlett-Packard G1800A chromatograph. The operating conditions were the same as described above, with the exception of a helium carrier gas flow rate of 0.6 mL min^{-1}. Mass spectra were obtained by EI ionization at 70 eV over the range of 15-300 mass units, with an ion source temperature of 250°C.

Results and Discussion

The most important degradation pathway of S-(1-propenyl)-L-cysteine sulfoxide seems to be an internal Michael condensation to cycloalliin (3-methyl-1,4-thiazane-5-carboxylic acid-S-oxide)[8] and in smaller scale the reduction to S-(1-propenyl)-L-cysteine (Figure 1). From the latter compound 2,4- and 3,4-dimethylthiophenes are formed in the next reaction step probably *via* the condensation of 1-propenylthiyl radicals and elimination of hydrogen sulfide.[9] 2,5-Dimethylthiophene and methylthiophenes were found only in trace amounts. It has been reported that 3,4-dimethylthiophene might also be produced by cyclization of bis(1-propenyl) disulfide.[1] However, even traces of this disulfide could not be detected in the volatile products formed from thermally degraded PeCSO. Dimethylthiophenes are considered to be the typical constituents of fried onion aroma.

Total amounts of the predominant volatiles formed from PeCSO as influenced by time of heating, temperature and water content are summarized in Tables 1a-c, respectively. As can be seen, a total amount of dimethylthiophenes increases with prolonged time of heating (at a temperature of 120°C) and with temperature. Nevertheless at higher temperatures, the formation of the typical degradation products of S-containing amino acids, e.g. hydrogen sulfide, sulfur dioxide and alkyl substituted thiazoles (namely 5-propyl-, 2-ethyl-4-methyl-, 2,4-dimethyl-5-ethyl-, 2-methyl-4,5-diethyl-, 2,4-dimethyl-5-propyl-, and 2-methyl-4-ethyl-5-propylthiazoles), was evident.

The total amount of dimethylthiophenes is maximized in the presence of 10-80% water. The formation of 2-methyl-2-pentenal and 2-methyl-2-butenal dominated at higher water activity, whereas in systems with water content lower than 40% did not take place. The proposed mechanism of PeCSO degradation which is favored in the presence of higher water amounts is shown in Figure 2. It can be assumed that S-(1-propenyl)-L-cysteine sulfoxide undergoes rearrangement leading to the intermediate sulfenate (similarly to alliin[2-4]). The reduction of sulfenate would yield 1-propenol and cysteine (was detected by TLC). These compounds can form propionaldehyde and acetaldehyde, respectively, whose aldolization 2-methyl-2-butenal and 2-methyl-2-pentenal are generated.

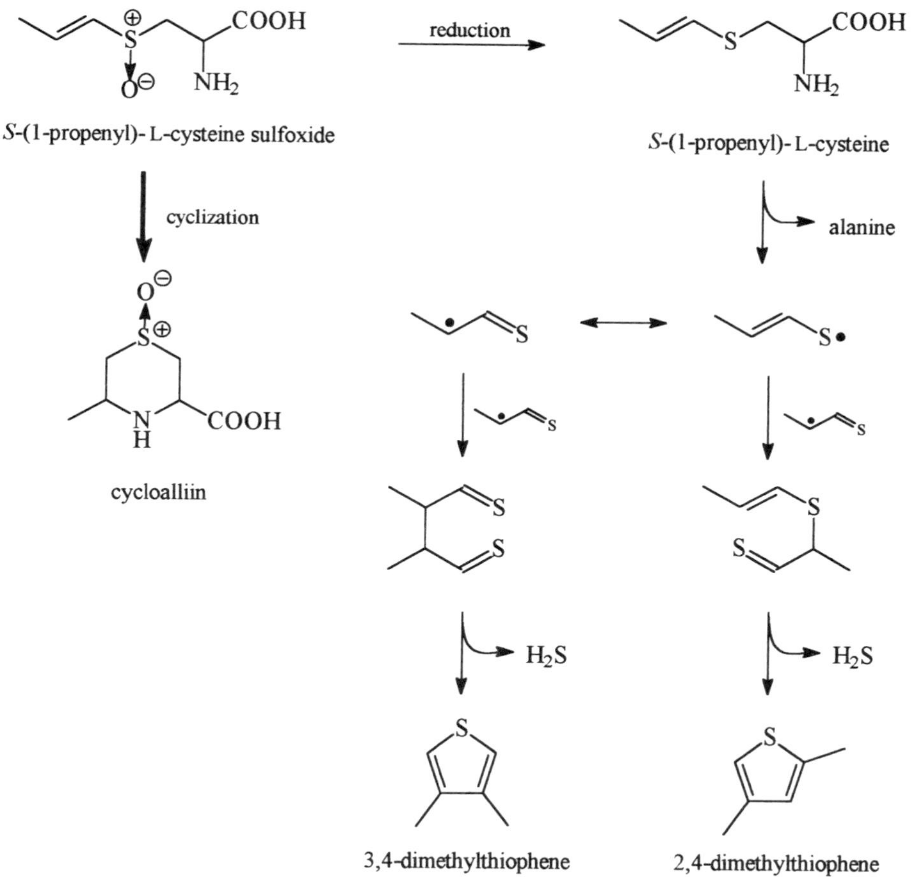

Figure 1. *Proposed mechanism of formation of dimethylthiophenes from PeCSO*

The resulting aroma of degraded PeCSO samples is strongly dependent on temperature of heating. At lower temperatures it can be described as potato- or vegetable-like; at higher temperatures it is sulfury, and a burnt, gummy flavor prevails.

In summary, there are no doubts, that the thermal decomposition of S-(1-propenyl)-L-cysteine sulfoxide (both free and bound in the γ-glutamylpeptides) can play an important role in the development of the aroma of thermally processed onion. However, in comparison with the degradation of the isomeric amino acid alliin (S-allyl-L-cysteine sulfoxide),[2-4] the amount of volatiles is much smaller, probably due to thermal stability of cycloalliin.

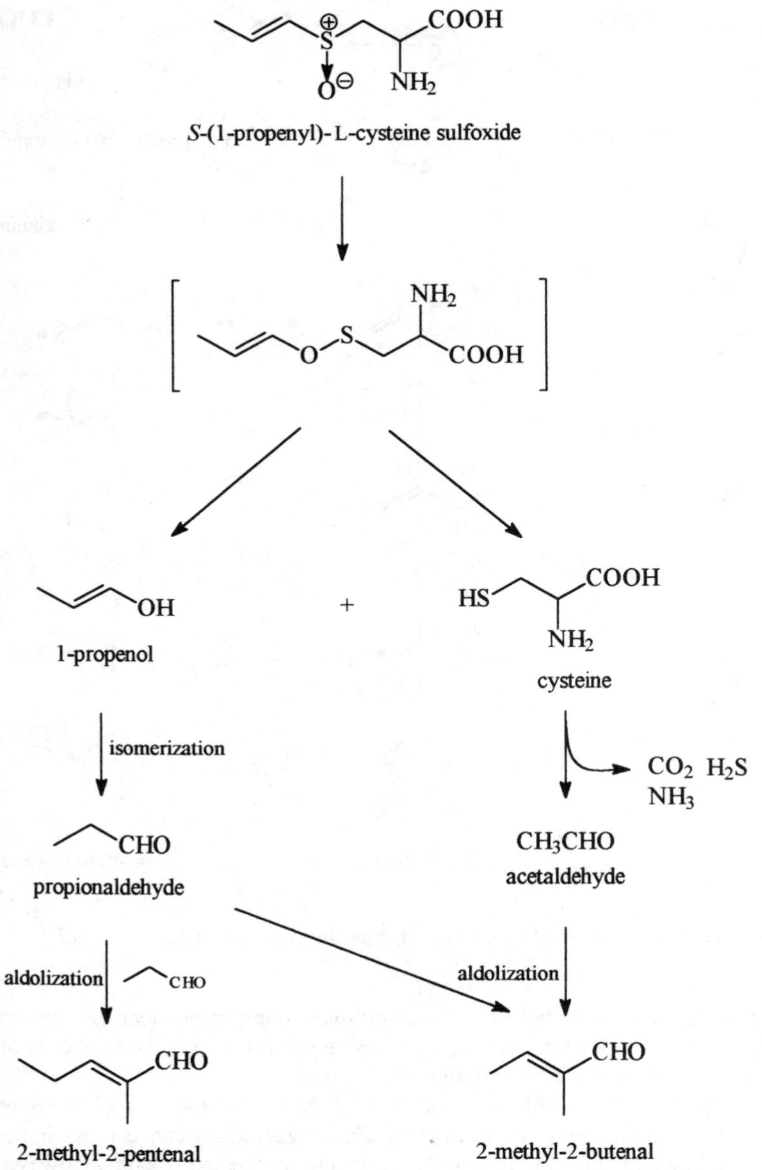

Figure 2. *Proposed mechanism of formation of 2-methyl-2-butenal and 2-methyl-2-pentenal from thermally degraded S-(1-propenyl)-L-cysteine sulfoxide*

Table 1a-c. Dependence of dimethylthiophenes and 2-methylalkenals amount on time of heating, temperature and water content (in mg g^{-1} PeCSO)

	time of heating (min)^a						
	1	2	4	8	15	30	60
2,4DMT	0.7	1.8	5.2	9.4	10.4	10.8	15.8
3,4DMT	traces	1.7	4.5	8.0	8.9	9.4	14.6

	temperature of heating (°C)^b						
	80	100	120	140	160	180	200
2,4DMT	3.4	15.2	15.8	29.3	28.0	33.7	39.5
3,4DMT	8.3	9.2	14.6	30.0	30.8	37.7	51.5

	water content (%)^c							
	0	5	10	20	40	80	90	95	98
2,4DMT	3.6	10.0	15.8	17.1	16.6	3.2	3.0	3.8	3.3
3,4DMT	9.6	12.1	14.6	13.5	44.9	34.4	18.5	17.1	15.3
2MBe	-	-	-	-	2.6	16.7	36.2	39.5	36.6
2MPe	-	-	-	1.0	5.6	10.0	16.3	20.8	21.9

^{a} At 120°C and 10% water, ^{b} for 60 minutes and 10% water, ^{c} at 120°C for 60 minutes. Abbreviations used: 2,4DMT 2,4-dimethylthiophene, 3,4DMT 3,4-dimethylthiophene, 2MBe 2-methyl-2-butenal, 2MPe 2-methyl-2-pentenal.

References

1. E. Block, The organosulfur chemistry of the genus *Allium*-Implications for the organic chemistry of sulfur, *Angew. Chem., Int. Ed. Engl.*, 1992, **31**, 1135-1178.
2. C.-T. Ho, T.-H. Yu and L.-Y. Lin, Contributions of nonvolatile flavor precursors of garlic to thermal flavor generation, in 'Food Flavors: Generation, Analysis and Process Influence', G. Charalambous (ed.), Elsevier Science B.V., Amsterdam, 1995, p. 909-918.
3. T.-H. Yu, C.-M. Wu, R.T. Rosen, T.G. Hartman and C.-T. Ho, Volatile compounds generated from thermal degradation of alliin and deoxyalliin in an aqueous solution, *J. Agric. Food Chem.*, 1994, **42**, 146-153.
4. R. Kubec, J. Velíšek, M. Doležal and V. Kubelka, Sulfur-containing volatiles arising by thermal degradation of alliin and deoxyalliin, *J. Agric. Food Chem.*, 1997, **45** (in press).
5. S. Schwimmer, Enzymatic conversion of *trans*-(+)-S-1-propenyl-L-cysteine S-oxide to the bitter and odorbearing components of onion, *Phytochemistry*, 1968, **7**, 401-404.
6. S. Shannon, M. Yamaguchi and F.D. Howard, Precursors involved in the formation of pink pigments in onion purees, *J. Agric. Food Chem.*, 1967, **15**, 423-426.
7. J.F. Carson and L.E. Boggs, The synthesis and base-catalysed cyclization of (+)- and (-)-*cis*-S-(1-propenyl)-L-cysteine sulfoxides, *J. Org. Chem.*, 1966, **31**, 2862-2864.
8. Y. Ueda, T. Tsubuku and R. Miyajima, Composition of sulfur-containing components in onion and their flavor characters, *Biosci. Biotech. Biochem.*, 1994, **58**, 108-110.
9. H. Nishimura and J. Mizutani, Photochemistry and radiation chemistry of sulfur-containing amino acids. A new reaction of the 1-propenylthiyl radicals, *J. Org. Chem.*, 1975, **40**, 1567-1575.

The Interaction of Lipid-Derived Aldehydes with the Maillard Reaction in Meat Systems

Donald S. Mottram and J. Stephen Elmore

THE UNIVERSITY OF READING, DEPARTMENT OF FOOD SCIENCE AND TECHNOLOGY, WHITEKNIGHTS, READING RG6 6AP, UK

Summary
The presence of lipids in foods, especially phospholipids, provides an extra source of reactants which can participate in the complex series of reactions that comprise the Maillard reaction. A number of volatile compounds that derive from lipid–Maillard interactions have been found in model reaction systems and in foods. In recent research a series of thiazoles and thiazolines, with long-chain alkyl substituents, have been isolated from pressure cooked beef. Studies, using model systems, demonstrated that these compounds arose from the reaction between aldehydes, produced in lipid oxidation, and simple dicarbonyls, ammonia and hydrogen sulfide, which are the intermediates in Maillard reactions involving cysteine.

Introduction

The characteristic aromas of cooked foods, including meat, are derived from thermally induced reactions occurring during heating, principally the Maillard reaction and the degradation of lipid. Both types of reaction involve complex reaction pathways leading to a wide range of products, which accounts for the large number of volatile compounds found in cooked foods.[1] Heterocyclic compounds, especially those containing sulfur, are important flavour compounds produced in the Maillard reaction, providing savoury, meaty, roast, and boiled flavours. The Maillard reaction essentially involves reactions between amino compounds, especially amino acids, and reducing sugars. In food systems the presence of other structural food components such as lipid, protein and carbohydrate could be expected to modify the nature and concentration of volatile aroma compounds derived from these reactions. Flavour formation in the Maillard reaction depends on the initial formation of reactive intermediates, such as carbonyls from sugar degradation, and simple aldehydes, ammonia and hydrogen sulfide from amino acids,[2] which undergo further reaction to give a wide range of volatile aroma compounds. The participation of lipid-derived compounds, such as aldehydes, might be expected in such reactions.

This paper reviews some of the evidence for the formation of volatiles from such interactions in model Maillard systems and in meat. Results of recent work are presented which have demonstrated the formation of novel heterocyclic compounds in cooked meat, produced by the interaction of lipid oxidation products with the Maillard reaction.

Influence of Phospholipids on Meat Flavour

The structural phospholipids have been shown to have an important role in flavour formation in cooked meat. Phospholipids are essential structural components of all cells and contain a much higher proportion of unsaturated fatty acids than the triglycerides, including significant amounts of polyunsaturated fatty acids, such as arachidonic acid (20:4), which makes them more susceptible to oxidation during heating. As well as contributing to the flavour of meat, these lipid oxidation products have been shown to modify the profile of Maillard-derived volatiles produced in cooked meat.[3,4]

Interaction of phospholipids with the Maillard reaction in meat-like model systems

A number of studies of the effect of phospholipids on the volatile products of heated aqueous solutions of amino acids and sugars have shown that Maillard reaction products are influenced by the presence of phospholipids.[5-7] The most noticeable effects were obtained when cysteine and ribose (important meat flavour precursors) were reacted in the presence of phospholipids, such as phosphatidylcholine, phosphatidylethanolamine or phospholipids extracted from meat. The reaction systems gave complex mixtures of volatiles which were dominated by sulfur-containing heterocycles, particularly thiophenes, thienothiophenes, dithiolanones, dithianones, furanthiols and thiophenethiols. In the presence of phospholipids, a reduction in the amounts of many of these volatiles was observed, compared with the system without lipid. This confirmed the observations in meat that phospholipids exert a quenching effect on the quantities of heterocyclic compounds formed in the Maillard reaction. The reaction mixtures also contained compounds derived from the interaction of lipid-derived aldehydes with Maillard reaction intermediates, the most abundant of which were 2-pentylpyridine, 2-pentylthiophene, 2-hexylthiophene and 2-pentyl-2H-thiapyran. All of these heterocyclic compounds appeared to be formed by the reaction of unsaturated aldehydes with hydrogen sulfide or ammonia derived from cysteine.[8] When a triglyceride fraction from meat was used in Maillard systems, instead of phospholipids, only trace amounts of these interaction products were found. This may be explained by the much higher proportion of polyunsaturated fatty acids occurring in the phospholipids than in the meat triglycerides. These undergo thermal oxidation much more readily than saturated or monounsaturated fatty acids.

Volatiles in Meat Formed via Lipid–Maillard Interactions

A number of volatile compounds have been found in meat and other cooked foods which could be formed from the interaction of lipid with the Maillard reaction. The occurrence of such compounds has been reviewed by Whitfield.[9] These compounds are O-, N- or S-heterocycles containing long n-alkyl substituents ($C_5 - C_{15}$). The alkyl groups usually derive from aliphatic aldehydes, obtained from lipid oxidation, while amino acids are the source of the nitrogen and sulfur.

2-Pentylpyridine is commonly found in the volatiles of cooked meat and is probably formed from 2,4-decadienal and ammonia.[8] A number of other alkylpyridines have been reported in lamb fat.[10] Related reactions between dienals and hydrogen sulfide may be responsible for the formation of 2-alkylthiophenes with $C_4 - C_8$ alkyl substituents which have been reported in pressure-cooked beef.[11,12] Other heterocyclic compounds, with long n-alkyl substituents, found in meat include butyl- and pentyl-pyrazines.[13,14] It has been suggested that these could result from the reaction of pentanal or hexanal (from lipid oxidation) with a dihydropyrazine, formed by the condensation of two aminoketone molecules.[15] The latter is a product of the Strecker degradation of amino acids with α-dicarbonyl compounds. Pentanal and hexanal also appear to be involved in the formation of 5-butyl-3-methyl-1,2,4-trithiolane and its 5-pentyl homologue, which have both been reported in fried chicken[16] and pork.[17] Trithiolanes can be formed from aldehydes and hydrogen sulfide, and the reaction of hydrogen sulfide, acetaldehyde and pentanal or hexanal has been suggested as the route to the butyl and pentyl trithiolanes.[18,19]

The aroma characteristics have only been examined for a few of these alkyl-substituted heterocyclic compounds, but those which have been reported suggest that such compounds may contribute to fatty, fried aromas.[10,19]

Thiazoles and thiazolines in cooked beef

Several thiazoles with $C_4 - C_8$ *n*-alkyl substituents in the 2-position have been reported in roast beef[20] and fried chicken.[13] Other alkylthiazoles with longer 2-alkyl substituents ($C_{13} - C_{15}$) were found in the volatiles of heated beef and chicken with the highest concentrations in beef heart muscle.[21] Aliphatic aldehydes from lipid oxidation are the likely source of the long *n*-alkyl groups in these compounds. The alkylthiazoles containing $C_{13} - C_{15}$ alkyl groups require $C_{14} - C_{16}$ aldehydes, and the most likely sources of these are the plasmalogens which contain long-chain alkenyl ether substituents, which hydrolyse to give fatty aldehydes. Heart muscle contains higher concentrations of plasmalogens which explains the higher levels of these alkylthiazoles found in heated beef heart.[21]

Recently over 40 alkyl-3-thiazolines have been isolated from cooked beef.[22] Most of the thiazolines contained C5 – C9 *n*-alkyl substituents in the 2-position. Small quantities of some thiazoles, with similar substitution to the thiazolines, were also obtained. The meat was produced in a study of the quality of beef from cattle fed diets that attempted to modify the polyunsaturated fatty acid (PUFA) composition by feeding lipid supplements. The cattle were fed on grass silage and concentrates containing different fat sources: palm oil (control), bruised linseed, fish oil, and bruised linseed plus fish oil in equal amounts. Steaks from these animals were cooked at 140 °C in an autoclave for 30 min and 40 g samples were taken for volatile analysis by headspace concentration on Tenax TA and gas chromatography–mass spectrometry.[22] Although most of the 3-thiazolines were present in meat from cattle fed control diets, their concentrations were much higher in meat from the cattle fed fish oil or linseed supplements (Table 1). Cooked meat from the animals that had been fed fish oil or linseed also had considerably higher concentrations of saturated and unsaturated aldehydes than meat from controls (Table 1).

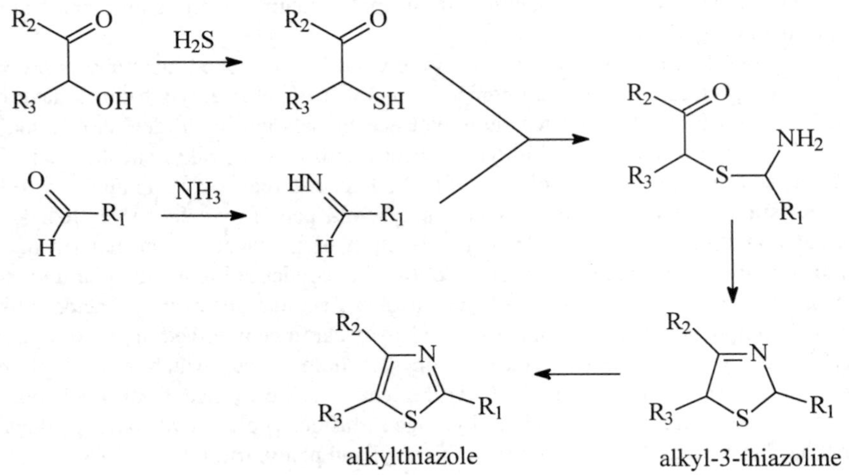

Figure 1. *Formation of alkylthiazolines and alkylthiazoles*

Table 1. Thiazolines and aliphatic aldehydes isolated from the volatiles of pressure-cooked beef from animals fed diets with different lipid supplements. Values are means of three replicate analyses.

	Amount in headspace volatiles (ng/100g)			
	control	linseed	fish oil	linseed + fish
3-Thiazolines [a]				
2-Butyl-4-methyl-	tr[b]	tr	7	12
2-Butyl-4,5-dimethyl-	4	11	25	27
2-Pentyl-4-methyl-	tr	5	63	12
2-Pentyl-4,5-dimethyl-	8	6	33	32
2-Pentyl-4-ethyl-	tr	tr	5	5
2-Pentyl-4-ethyl-5-methyl-	tr	3	13	12
2-Pentyl-5-ethyl-4-methyl-	3	3	9	7
2-Hexyl-4-methyl-	tr	8	12	31
2-Hexyl-4,5-dimethyl-	tr	17	69	73
2-Hexyl-4-ethyl-	tr	3	9	13
2-Hexyl-4-ethyl-5-methyl-	3	11	23	28
2-Hexyl-5-ethyl-4-methyl-	tr	4	14	18
2-Heptyl-4-methyl-	tr	5	5	23
2-Heptyl-4,5-dimethyl-	tr	10	46	64
2-Heptyl-4-ethyl-	tr	tr	5	8
2-Heptyl-4-ethyl-5-methyl-	tr	4	16	19
2-Heptyl-5-ethyl-4-methyl-	tr	tr	9	12
2-Octyl-4-methyl-	tr	10	11	51
2-Octyl-4,5-dimethyl-	tr	25	114	174
2-Octyl-4-ethyl-	tr	4	11	19
2-Octyl-4-ethyl-5-methyl-	3	14	40	52
2-Octyl-5-ethyl-4-methyl-	tr	6	20	28
2-Nonyl-4,5-dimethyl-	–[c]	tr	9	13
2-Nonyl-4-ethyl-5-methyl-	–	tr	tr	3
Aldehydes				
Pentanal	360	553	2528	1600
Hexanal	530	1318	3715	2040
Heptanal	273	1563	4608	3360
Octanal	210	715	1703	1195
Nonanal	320	768	1070	950
2-Hexenal	–	tr	83	50
2-Heptenal	–	tr	155	113
2-Octenal	–	60	238	115
2-Nonenal	tr	55	273	183
2,4-Heptadienal	–	tr	108	tr
2,4-Octadienal	–	–	tr	tr
2,4-Nonadienal	–	–	tr	tr

[a] 3-Thiazolines with substituents in the 5-position (methyl or ethyl) were present as pairs of geometric isomers; [b] trace (< 2 ng/kg thiazoline or < 20 ng/kg aldehyde); [c] not detected

The most likely routes to the 3-thiazolines and thiazoles are from α-hydroxyketones or α-diones, hydrogen sulfide, ammonia and aldehydes (Figure 1). Hydroxyketones and diones are sugar degradation products from the Maillard reaction, while ammonia and hydrogen sulfide may be produced from cysteine by Strecker degradation or hydrolysis. Lipid oxidation provides long-chain aldehydes. This route was confirmed by heating aqueous mixtures containing α-hydroxyketones (1-hydroxypropanone or 1-hydroxy-2-butanone or 3-hydroxy-2-butanone) or α-dicarbonyls (2,3-butanedione or 2,3-pentane-dione), alkanals (C_5 to C_{10}) and ammonium sulfide.[23] 3-Thiazolines, with C_3 to C_9 alkyl substituents in the 2-position and methyl or ethyl in positions 4 and/or 5, were major products of the reactions involving hydroxyketones (up to 35% of total volatiles formed) and small amounts of thiazoles were also obtained. Reactions involving diones also resulted in thiazolines, but they also gave thiazoles in similar quantities.

Odour-port GC assessment of the alkyl-3-thiazolines and alkylthiazoles indicated that they did not possess low odour thresholds and, therefore, may not be very important odour impact compounds. However, these compounds result from reactions that compete for the intermediates of the other flavour-forming Maillard reactions, and thereby may modify and control the generation of desirable aroma compounds.

Conclusion

The flavour of meat and other thermally processed foods derives from a complex series of reactions involving both the Maillard reaction and lipid degradation. Interactions between compounds produced by these routes leads to a number of heterocyclic compounds with long-chain alkyl substituents, such as pyridines, pyrazines, thiophenes, thiazoles and thiazolines. Some of these compounds may contribute to the aroma of heated foods. In addition, the reactions by which they are formed will compete with other flavour-forming Maillard reactions and, thus, influence the overall aroma profiles of the thermally processed foods.

References

1. D.S. Mottram, Meat, in 'Volatile Compounds in Foods and Beverages', H. Maarse (ed.), Marcel Dekker, New York, 1991, pp. 107-177.
2. H.E. Nursten, Recent developments in studies of the Maillard reaction, *Food Chem.*, 1980, **6**, 263-277.
3. D.S. Mottram and R.A. Edwards, The role of triglycerides and phospholipids in the aroma of cooked beef, *J. Sci. Food Agric.*, 1983, **34**, 517-522.
4. D.S. Mottram, The role of phospholipids in meat flavor: an overview, in 'Contribution of Low and Non-volatile Materials to the Flavor of Foods', W. Pickenhagen, C.T. Ho and A.M. Spanier (eds), Allured Publishing, Carol Stream IL, 1996, pp. 193-206.
5. L.J. Salter, D.S. Mottram and F.B. Whitfield, Volatile compounds produced in Maillard reactions involving glycine, ribose and phospholipid, *J. Sci. Food Agric.*, 1988, **46**, 227-242.
6. F.B. Whitfield, D.S. Mottram, S. Brock, D.J. Puckey and L.J. Salter, Effect of phospholipid on the formation of volatile heterocyclic compounds in heated aqueous solutions of amino acids and ribose, *J. Sci. Food Agric.*, 1988, **42**, 261-272.

7. L.J. Farmer, D.S. Mottram and F.B. Whitfield, Volatile compounds produced in Maillard reactions involving cysteine, ribose and phospholipid, *J. Sci. Food Agric.*, 1989, **49**, 347-368.
8. L.J. Farmer and D.S. Mottram, Interaction of lipid in the Maillard reaction between cysteine and ribose: effect of a triglyceride and three phospholipids on the volatile products, *J. Sci. Food Agric.*, 1990, **53**, 505-525.
9. F.B. Whitfield, Volatiles from interactions of Maillard reactions and lipids, *Crit. Rev. Food Sci. Nutr.*, 1992, **31**, 1-58.
10. R.G. Buttery, L.C. Lin, R. Teranishi and T.R. Mon, Roast lamb fat: basic volatile components, *J. Agric. Food Chem.*, 1977, **25**, 1227-1229.
11. R.A. Wilson, C.J. Mussinan, I. Katz and A. Sanderson, Isolation and identification of some sulfur chemicals present in pressure-cooked beef, *J. Agric. Food Chem.*, 1973, **21**, 873-876.
12. D.B. Min, K. Ina, R.J. Peterson and S.S. Chang, Preliminary identification of volatile flavor compounds in the neutral fraction of roast beef, *J. Food Sci.*, 1979, **44**, 639-642.
13. J. Tang, Q.Z. Jin, G.H. Shen, C.T. Ho and S.S. Chang, Isolation and identification of volatile compounds from fried chicken, *J. Agric. Food Chem.*, 1983, **31**, 1287-1292.
14. D.S. Mottram, The effect of cooking conditions on the formation of volatile heterocyclic compounds in pork, *J. Sci. Food Agric.*, 1985, **36**, 377-382.
15. E.-M. Chiu, M.-C. Kuo, L.J. Bruechert and C.-T. Ho, Substitution of pyrazines by aldehydes in model systems, *J. Agric. Food Chem.*, 1990, **38**, 58-61.
16. S.S. Hwang, J.T. Carlin, Y. Bao, G.J. Hartma and C.T. Ho, Characterisation of volatile compounds generated from the reactions of aldehydes with ammonium sulfide., *J. Agric. Food Chem.*, 1986, **34**, 538-542.
17. P. Werkhoff, J. Brüning, R. Emberger, M. Güntert and R. Hopp, Flavor chemistry of meat volatiles: New results on flavor components from beef, pork and chicken, in 'Recent Developments in Flavor and Fragrance Chemistry', R. Hopp and K. Mori (eds), VCH, Weinheim, 1993, pp. 183-213.
18. M. Boelens, L.M. van der Linde, P.J. de Valois, H.M. van Dort and H.J. Takken, Organic sulfur compounds from fatty aldehydes, hydrogen sulfide, thiols and ammonia as flavor constituents, *J. Agric. Food Chem.*, 1974, **22**, 1071-1076.
19. C.T. Ho, J.T. Carlin and T.C. Huang, Flavour development in deep-fat fried foods, in 'Flavour Science and Technology', M. Martens, G.A. Dalen and H. Russwurm (eds), Wiley, Chichester, 1987, pp. 35-42.
20. G.J. Hartman, Q.Z. Jin, G.J. Collins, K.N. Lee, C.T. Ho and S.S. Chang, Nitrogen-containing heterocyclic compounds identified in the volatile flavor constituents of roast beef, *J. Agric. Food Chem.*, 1983, **31**, 1030-1033.
21. L.J. Farmer and D.S. Mottram, Lipid-Maillard interactions in the formation of volatile aroma compounds, in 'Trends in Flavour Research', H. Maarse and D.G. van der Heij (eds), Elsevier, Amsterdam, 1994, pp. 313-326.
22. J.S. Elmore, D.S. Mottram, M.B. Enser and J.D. Wood, Novel thiazoles and 3-thiazolines in cooked beef aroma, *J. Agric. Food Chem.*, 1997, **45**, 3603-3607.
23. J.S. Elmore and D.S. Mottram, Formation of thiazoles and thiazolines in the reactions of α-hydroxyketones, aldehydes and ammonium sulfide, *J. Agric. Food Chem.*, 1997, **45**, 3595-3602.

Reaction Products of Glyoxal with Glycine

Jan Velíšek, Tomáš Davídek[1] and Jiří Davídek

INSTITUTE OF CHEMICAL TECHNOLOGY-DEPARTMENT OF FOOD CHEMISTRY AND ANALYSIS, 166 28 PRAGUE, TECHNICKÁ 1905, CZECH REPUBLIC, [1] PRESENT ADDRESS: FRISKIES CENTRE R&D NESTLÉ, AMIENS, FRANCE

Summary
Glyoxal is a very reactive compound. In aqueous solutions it undergoes the Cannizzaro reaction and autoxidation yielding glycolic and glyoxalic acid, respectively. It reacts with glycine yielding 3-carboxymethyl-1-imidazoliumethanoate as the major reaction product. Formaldehyde, aminoacetaldehyde, methylamine, ammonia and other compounds arise in the Strecker degradation of glycine. Subsequent reactions of these products lead to a variety of heterocyclic compounds such as imidazole, pyrazine, alkyl-, formyl- and acetyl pyrazines, formyl and acetyl pyrroles arising as the minor reaction products. The Cannizzaro reaction of formaldehyde yields methanol which is the precursor of formaldehyde acetals. Formaldehyde also reacts with methylamine yielding dimethylamine.

Introduction

Glyoxal (dialdehyde of oxalic acid) arises in food as a degradation product of sugars and lipids. It is believed that the major routes leading from sugars to glyoxal are the following reactions: the retroaldolization of glycosylamines [1] (*via* the corresponding imine of glycolaldehyde) and fragmentation of triose reductone.[2] Glyoxal is the simplest dialdehyde, the first member of a homologous series of α-dicarbonyl compounds to which also belong its well-known homologue methylglyoxal (pyruvaldehyde), hydroxymethylglyoxal, ethylglyoxal and various C-6 glycosuloses, such as 3-deoxy-D-*erythro*-hexosulose, 1-deoxy-D-*erythro*-2,3-hexodiulose and 4-deoxy-D-*glycero*-2,3-hexodiulose, which are well-known key intermediates in the Maillard reaction of sugars such as glucose and fructose.

Glyoxal is a very reactive compound as the presence of the α-carbonyl group in its structure increases the reactivity of the second carbonyl (-I effect) and enhances the addition reaction to the polarized carbonyl group. Therefore, it may be supposed that glyoxal (once formed from sugars) will immediately participate in the Maillard reaction.

In this paper we studied the possible role of glyoxal in the Maillard reaction. We were mainly interested in the reaction of glyoxal with glycine (the simplest amino acid) and flavour-active reaction products that may arise as the reaction products.

Material and Methods

Aqueous solutions containing either glyoxal (0.1 to 1 mol.L^{-1}) or glyoxal and glycine in molar ratios of 1:1 to 1:2 were heated in buffered solutions (of pH 3-12) at different temperaturus (25°C to 90°C). Non-volatile reaction products were separated and quantified using TLC, HPLC and capillary isotachophoretic measurements. Volatile reaction products were separated and quantified by GLC. Mass spectrometry, ^{1}H- and ^{13}C-NMR spectrometry were used for identication of the reaction products. Detailed procedures and methods have been published.[2-5]

Results and Discussion

In alkaline, neutral and even mild acid solutions, glyoxal undergoes intramolecular Cannizzaro reaction forming glycolic acid. The reaction rate is proportional to the glyoxal concentration and to the second power of hydroxyl ion concentration. By autoxidation of glyoxal, glyoxylic acid arises as a minor product (Figure 1).

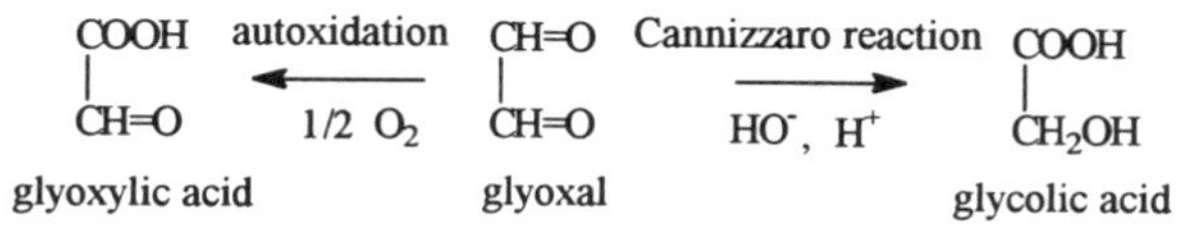

$$\underset{\text{glyoxylic acid}}{\begin{array}{c} COOH \\ | \\ CH=O \end{array}} \xleftarrow[1/2\ O_2]{\text{autoxidation}} \underset{\text{glyoxal}}{\begin{array}{c} CH=O \\ | \\ CH=O \end{array}} \xrightarrow[HO^-,\ H^+]{\text{Cannizzaro reaction}} \underset{\text{glycolic acid}}{\begin{array}{c} COOH \\ | \\ CH_2OH \end{array}}$$

Figure 1. *Cannizzaro reaction and autoxidation of glyoxal*

The reaction of glyoxal with glycine is primarily a series of oxidative decarboxylation reactions known as the Strecker degradation in which glycine is oxidized to formaldehyde while glyoxal is reduced to glycolaldehyde and its amino analogue, respectively. The initial reaction product is the corresponding Schiff's base (Figure 2). The reaction is reversible and the equilibrium constant K is ~1.3. Decarboxylation of the Schiff base gives imine A which may be partly hydrolysed to formaldehyde and the enol form of amino acetaldehyde and partly isomerized to the corresponding N-methyl imine B. Hydrolysis of the latter compound gives methylamine.

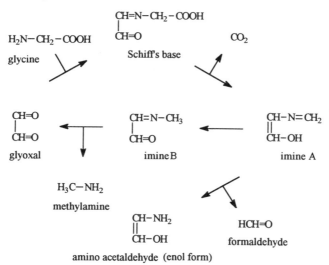

Figure 2. *Strecker degradation of glycine*

At the same time, the reaction mixture turns brown. The brown colour intensity increases with temperature and the pH value of the solution with a maximum at pH 9-10. The competitive Cannizzaro reaction also takes place. At higher pH values, it is even faster than the reaction of glyoxal with glycine. For instance, in solutions at pH 9 (at 90°C) the maximum colour intensity is achieved in 2 h and glyoxal totally disappears from the reaction mixture while only ~ 50% of the original amount of glycine enters the reaction with glyoxal. About one third of the glycine entering the reaction with glyoxal is decarboxylated. Subsequent reactions of the reactive intermediates arising by the Strecker degradation of glycine yield a variety of different secondary reaction products.

The major identified nitrogen-containing, non-volatile reaction product in acid, neutral or alkaline media is 3-carboxymethyl-1-imidazoliumethanoate, (1,3-bis(carboxymethyl)-imidazole). The greatest yield of this compound, which possesses slightly acid taste, is produced in mild acid media (up to about 40% of the theoretical yield) [5]. In much lower amounts (at the level of a few ppm), imidazole (Figure 3) and traces of 4(5)-methylimidazole are formed.

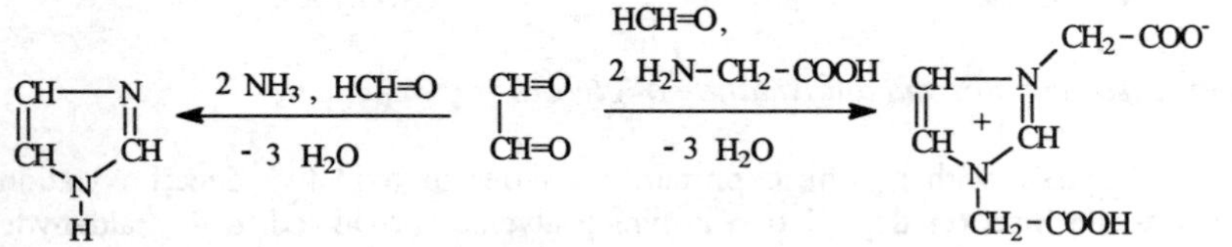

Figure 3. *Formation of imidazole and 3-carboxymethyl-1-imidazoliumethanoate*

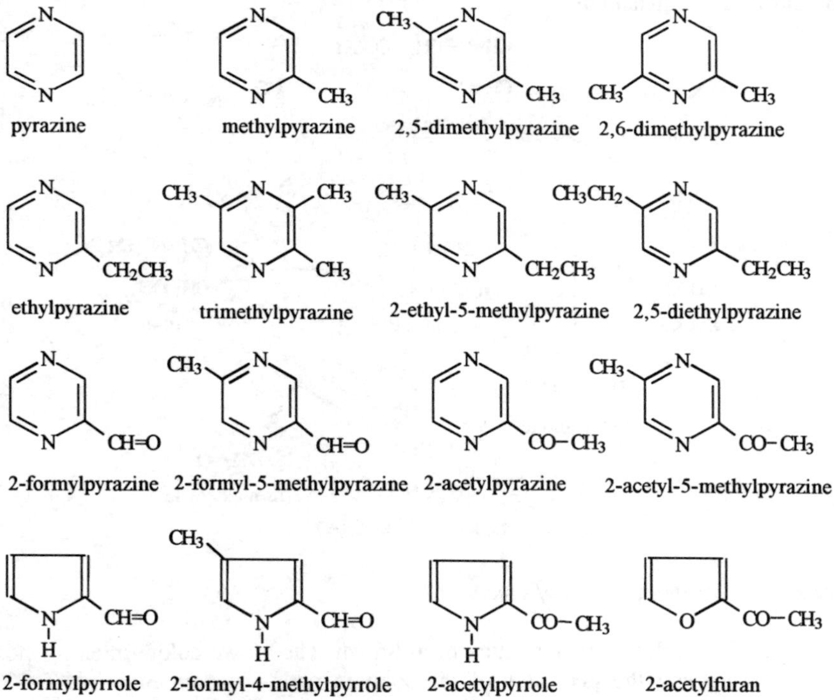

Figure 4. *Volatile heterocyclic reaction products*

Reaction of glyoxal with glycine also leads to a number of volatile products with roasted and burned flavours. In total, 16 pyrazines, 9 pyrroles and 1 furan were found as volatile reaction products. The list of the 16 identified compounds (based on mass spectral and retention time data) is given in Figure 4.

The major volatile product is unsubstituted pyrazine. Pyrazine is formed in glyoxal-glycine solutions at a level of several ppm (pH 9, 90°C). It is expected that it is produced by self-condensation of the enol form of aminoacetaldehyde, followed by autoxidation of the resulting dihydropyrazine (Figure 5).

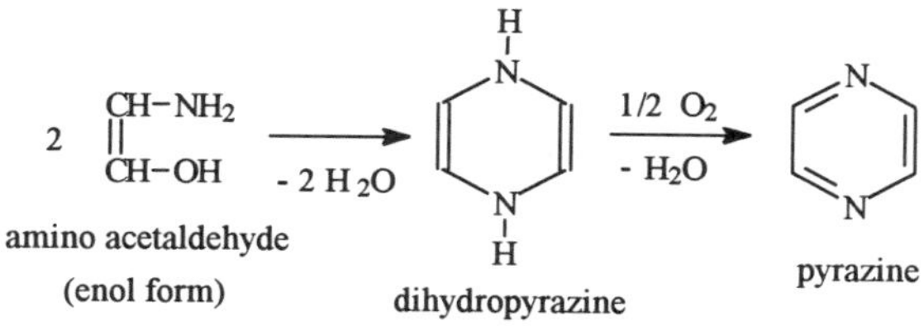

Figure 5. *Formation of pyrazine*

Aminoacetaldehyde (or its tautomers) seems to be also involved in the formation of some other reaction products. Aminoacetaldehyde can split to yield ammonia and glycolaldehyde. Aldolisation of the latter compound with formaldehyde may lead to glyceraldehyde and possibly also to higher sugars, elimination of the β-hydroxyl group of glyceraldehyde and isomerization of the product gives rise to methylglyoxal and other products (Figure 6).

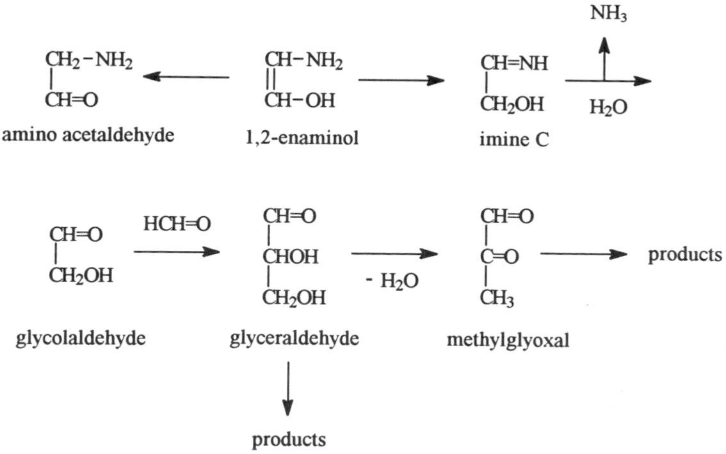

Figure 6. *Formation and subsequent reactions of glycolaldehyde*

Formaldehyde also undergoes the Cannizzaro reaction yielding methanol and formic acid. In solutions, formaldehyde occurs as the corresponding hydrate (methyleneglycol) and its polymerization gives a series of low-molecular-weight oligomeric hydrates. In acid and neutral media they react with methanol forming the corresponding acetals called oxaalkanes (Figure 7). Two oxaalkanes (2,4,6-trioxaheptane and 2,4,6,8-tetraoxanonane) were identified as the major formaldehyde acetals [4] being present in a concentration of ~ 10 nmoles / mol glycine.

Formaldehyde further acts as a methylating agent. It reacts with methylamine yielding traces of dimethylamine and possibly also trimethylamine (Figure 8).

$$HCH=O \xrightarrow{H_2O} HO-CH_2-OH \xrightarrow[-2 H_2O]{2 CH_3OH} H_3C-O-CH_2-O-CH_3$$

formaldehyde   formaldehyde hydrate   dimethoxymethane

$$(n-1) \; HO-CH_2-OH \downarrow$$

$$H(-O-CH_2-)_n^- OH \xrightarrow[-2 H_2O]{2 CH_3OH} CH_3 (-O-CH_2-)_n^- O-CH_3$$

alkoxyalkanes

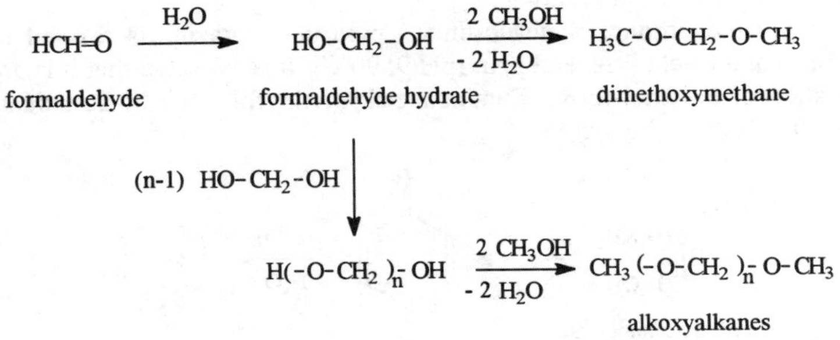

Figure 7. *Formation of formaldehyde hydrates and alkoxyalkanes*

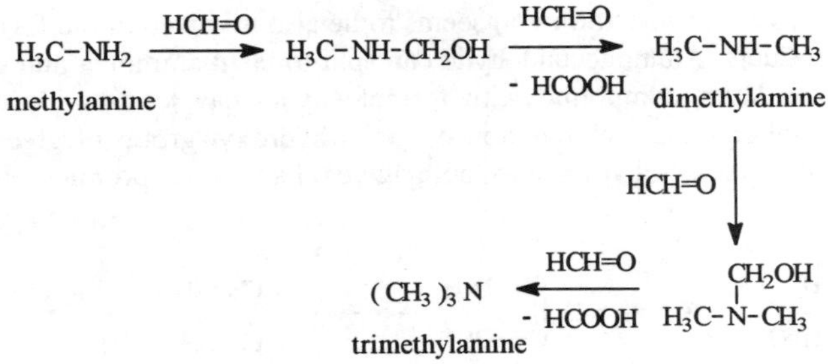

$$H_3C-NH_2 \xrightarrow{HCH=O} H_3C-NH-CH_2OH \xrightarrow[-HCOOH]{HCH=O} H_3C-NH-CH_3$$

methylamine   dimethylamine

$$HCH=O \downarrow$$

$$(CH_3)_3 N \xleftarrow[-HCOOH]{HCH=O} \underset{H_3C-N-CH_3}{CH_2OH}$$

trimethylamine

Figure 8. *Formation of aliphatic amines*

References

1. T. Hayashi and M. Namiki, in 'Amino-Carbonyl Reactions in Food and Biochemical Systems' M. Fujimaki, M. Namiki and H. Kato, eds, Kodansha Ltd, Tokyo and Elsevier Science Publishers, Amsterdam, 1986, p. 29.
2. J. Davídek, J. Velíšek and Pokorný J. eds, 'Chemical Changes during Food Processing', Elsevier Science Publisher, Amsterdam, 1990, p. 107.
3. J. Davídek, J. Velíšek and G. Janíček, Reaction products of non-enzymatic browning reaction in glyoxal-glycine system, Proc. IV Int. Congr. Food Sci. Technol., Vol. I, 1974, p. 298-305.
4. J. Velíšek, T. Davídek, J. Davídek, I. Víden and P. Trška, Some formaldehyde reaction products in non-enzymatic browning reactions. *Z. Lebensm. Unters. Forsch.*, 1989, **188**, 426-429.
5. J. Velíšek, T. Davídek, J. Davídek, P. Trška, F. Kvasnička and K. Velcová, New imidazoles in nonenzymatic browning reactions. *J. Food Sci.*, 1989, **54**, 1544-1646.

Mechanistic Studies on the Formation of the Cracker-like Aroma Compounds 2-Acetyltetrahydropyridine and 2-Acetyl-1-pyrroline by Maillard-type Reactions

Peter Schieberle and Thomas Hofmann[1]

INSTITUTE OF FOOD CHEMISTRY/TUM and [1]GERMAN RESEARCH CENTRE FOR FOOD CHEMISTRY, LICHTENBERGSTRASSE 4, 85748 GARCHING, GERMANY

Summary
1-Pyrroline, the Strecker degradation product of the amino acids proline or ornithine, was identified as the key intermediate in the formation of the intense roast-smelling food odorants 2-acetyltetrahydropyridine (ATHP) and 2-acetyl-1-pyrroline (AP). Reacting 1-pyrroline with hydroxy-2-propanone yielded high amounts of only the ATHP, whereas the reaction with 2-oxopropanal gave only AP. Synthesized 2-(1-hydroxy-2-oxo-1-propyl)-pyrrolidine was shown to be the key precursor of ATHP. The last step in AP formation was shown to be simply an air oxidation of 2-acetylpyrrolidine, which is proposed to be formed from 1-pyrroline and 2-oxopropanal via the intermediate 2-(1,2-dioxo-1-propyl) pyrrolidine.

Introduction

On the basis of high odour activities (ratio of concentration to odour threshold), the intense 'roasty'/'cracker like' flavour compounds 2-acetyltetrahydropyridine (ATHP) and 2-acetyl-1-pyrroline (AP) have been characterized as key odorants in several processed foods, such as Basmati rice,[1] wheat bread crust[2] or popcorn.[3] Both odorants have also been identified in thermally treated model systems in which the amino acids proline[4,5] and ornithine[5] were reacted with carbohydrates.

Hodge et al.[6] proposed a mechanism assuming proline and the carbohydrate fragmentation product 2-oxopropanal as precursors of ATHP (Figure 1).

Based on quantitative data, we previously showed that the thermal degradation of the amino acid ornithine, when reacted with 2-oxopropanal, is more effective in generating AP than proline.[5] A subsequent study indicated[7] that 1-pyrroline, the Strecker degradation product of proline or ornithine, is a key intermediate in the formation of AP, in which it reacts with 2-oxopropanal. However, ornithine did not yield ATHP.

Using [U-^{13}C]-glucose in the reaction with proline, we showed[8] that, in the mixture of AP isotopomers formed, predominantly the two carbons of the acetyl group were labelled. In similar experiments using [1-^{13}C]-glucose, it was recently confirmed that the labeling occurs in the acetyl group. However, 50 % of the AP was found to be unlabelled.[9]

Clearly, several details of the formation pathways leading to ATHP and AP remain to be clarified. Information on formation pathways and intermediates would be very helpful in, for example, optimizing yields during manufacture of Maillard-type process flavours. Therefore, the following investigations were conducted to gain further insights into formation mechanisms with special emphasis on identifying single-precursor intermediates.

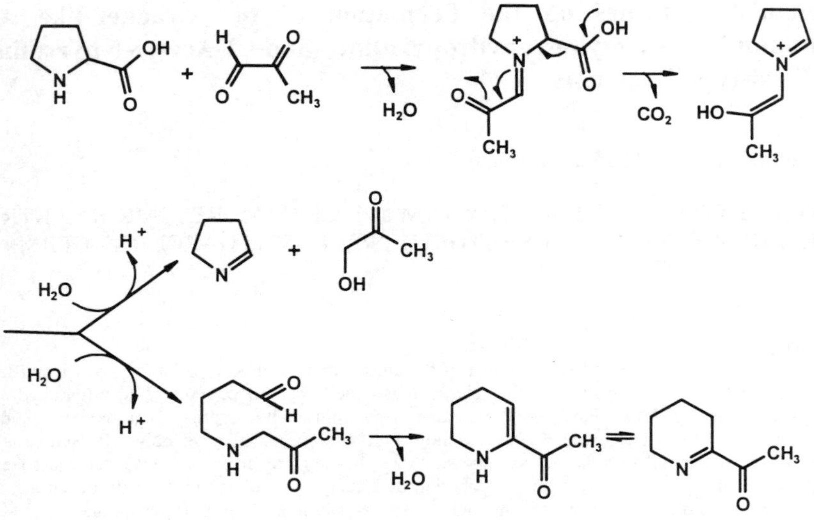

Figure 1. *Formation of 2-actyltetrahydropyridine, 1-pyrroline and hydroxy-2-propanone from proline/2-oxopropanal according to Hodge et al.*[6]

Materials and Methods

Chemicals
L-Proline, D-glucose, [U-^{13}C]-glucose, 2-methyl-1-pyrroline, hydroxy-2-propanone and 2-oxopropanal (40 % aqueous solution) were from Aldrich (Steinheim, Germany).

Syntheses
1-Pyrroline was prepared as described previously.[7] 2-Acetylpyrrolidine, 2-acetylpiperidine and 2-(1-hydroxy-2-oxo-1-propyl)-pyrrolidine were synthesized as reported recently.[10,11]

Model reactions; stable isotope dilution assays
L-proline (1 mmol) and D-glucose (2 mmol) or [U-^{13}C]-D-glucose (2 mmol) were mixed with silica gel (3 g) containing 300 µL of aqueous phosphate buffer (pH 7.0, 0.1 M) and then treated for 10 min at 160°C. The amounts of 2-acetyl-1-pyrroline and 2-acetyltetrahydropyridine formed were quantified by stable isotope dilution assays in extracts obtained by sublimation *in vacuo* and using the deuterium-labelled analogues as internal standards.[7]

Results and Discussion

In a recent study, proline and glucose were reacted for 10 min at 160°C under dry-heating conditions.[11] Extraction with diethyl ether followed by sublimation of the extract *in vacuo* afforded an intense 'roast-smelling' extract. By application of Aroma Extract Dilution Analysis (AEDA),[3] we detected six 'popcorn-like' odour regions in the gas chromatogram. The structures of the odorants corresponding to those areas are displayed in Figure 2.

Among the six odorants, the highest Flavour Dilution (FD) factor or odour activity was displayed by 2-acetyl-1-pyrroline. The experiment was repeated using [U-^{13}C]-labelled glucose instead of glucose and the relative amounts of isotopomers in the AP and ATHP formed were determined using mass spectrometry.

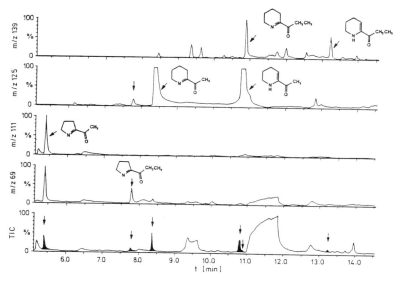

Figure 2. *Structures of odour-active, 'popcorn-like' volatiles (indicated by arrows) generated in a thermally processed proline-glucose mixture. TIC = total ion chromatogram*

The results (Table 1) indicated that, in agreement with previous data,[8] mainly two carbon atoms in the AP stem from glucose (m/z 113 vs. m/z 111). In the ATHP, however, only the isotopomer with three labelled carbons (m/z 128 vs. m/z 125) was formed, suggesting only one reaction pathway in ATHP formation.

Table 1. *Main isotopomers (represented by their molecular ions) determined in 2-acetylpyrroline and 2-acetyltetrahydropyridine generated from proline in the presence of either unlabelled or [U-^{13}C]-labelled glucose (Glc)*

Odorant	Molecular ion, m/z (%)	
	Unlabelled Glc	Labelled Glc
2-Acetyl-1-pyrroline	111 (93.9)	113 (76.8)
2-Acetyltetrahydropyridine	125 (88.3)	128 (89.2)

Formation of 2-acetyltetrahydropyridine
According to Hodge's proposal,[6] the reaction of proline with 2-oxopropanal might lead to either N-acetonyl-4-aminobutanal or, alternatively, to 1-pyrroline and hydroxy-2-propanone (Figure 1). However, recent work by de Kimpe et al.[12] questionned the importance of the N-acetonyl-derivative as the key precursor of ATHP.

As an alternative formation pathway, it might be assumed (Figure 3) that the nucleophilic hydroxy-2-propanone could react with the electrophilic 1-pyrroline yielding the intermediate 2-(1-hydroxy-2-oxo-1-propyl)-pyrrolidine (I in Figure 3). Upon ring opening it might be enolized into 4,5-dioxoheptylamine, which finally yields the ATHP upon elimination of water.

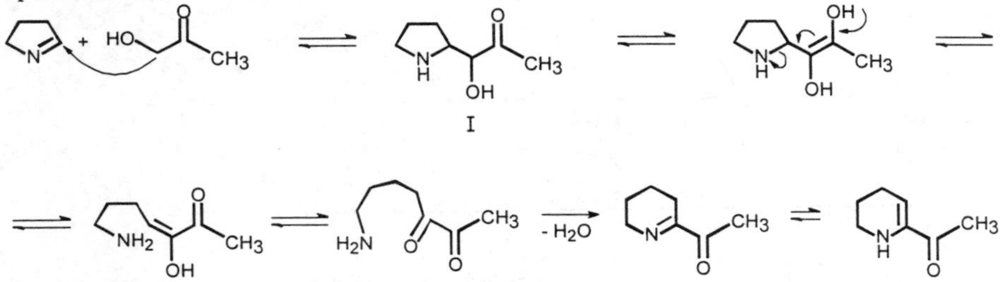

Figure 3. *Formation pathway leading from 1-pyrroline and hydroxy-2-propanone to 2-acetyltetrahydropyridine*

In an initial experiment, hydroxy-2-propanone was reacted with 1-pyrroline and the effect of pH on the amounts of ATHP formed was determined. As shown in Table 2, the precursor compounds gave significant yields of ATHP, especially at higher pHs. This is in good agreement with the first reaction step shown in Figure 3, because the reactivity of the α-hydroxyketone should be higher under basic conditions.

Table 2. *Influence of pH on the formation of 2-acetyltetrahydropyridine (ATHP) from 1-pyrroline and hydroxy-2-propanone*[a]

	ATHP	
pH	μg	mol %
3.0	<0.1	-
5.0	0.9	0.1
7.0	10.8	0.9
9.0	38.4	3.1

[a] 1-Pyrroline (10 μmol) and hydroxy-2-propanone (10 μmol) were reacted for 30 min at 100°C in phosphate buffer (5 mL; 0.5 M).

To confirm the ring enlargement proposed in Figure 3, we reacted 2-methyl-1-pyrroline with hydroxy-2-propanone. Because the ring enlargement shifts carbon-2 of the pyrroline into position 3 in the ATHP, 2-acetyl-3-methyltetrahydropyridine should be formed from the 2-methyl-1-pyrroline. This compound was indeed identified as the main reaction product. Sensory and analytical data will be reported elsewhere.[11]

As final confirmation of the reaction sequence shown in Figure 3, the key intermediate 2-(1-hydroxy-2-oxo-1-propyl)-pyrrolidine was synthesized in a four-step sequence starting from 2-hydroxyethylpyrrolidine.[11] Heating of this intermediate for 30 min at 100°C in phosphate buffer gave very high yields of ATHP (23.4 % at pH 7.0 and 35.0 % at pH 9), thereby supporting the proposed reaction mechanism.

Formation of 2-acetyl-1-pyrroline

In a previous publication,[9] a simple oxidation of 2-acetylpyrrolidine has been proposed to be the final step in the formation of 2-acetyl-1-pyrroline. In a recent publication,[13] we showed that 2-(1-hydroxymethyl)-4,5-dihydrothiazole, a tautomer of 2-acetylthiazolidine, is easily oxidized into the 2-acetyl-2-thiazoline simply by air oxygen. This result supported the idea that an oxidation of 2-acetylpyrrolidine might also yield AP very easily.

However, although sometimes mentioned in the literature,[9] a synthesis of 2-acetylpyrrolidine had not been reported at the time of the study. Using the synthetic route shown in Figure 4, we obtained the trifluoroacetate of 2-acetylpyrrolidine in a very high yield.[10] In agreement with the data for the thiazolidine,[13] the pyrrolidine also rapidly oxidized into the AP. As shown in Table 3, the oxidation was completed (nearly 100 % conversion) within 2 h upon standing in water at 25°C in the presence of oxygen. These results confirm the great susceptibility of cyclic α-aminoketones towards oxidation and, furthermore, indicate the great importance of oxidation reactions in the Maillard reaction in general. A proposal for an oxidation mechanism has recently been discussed.[13]

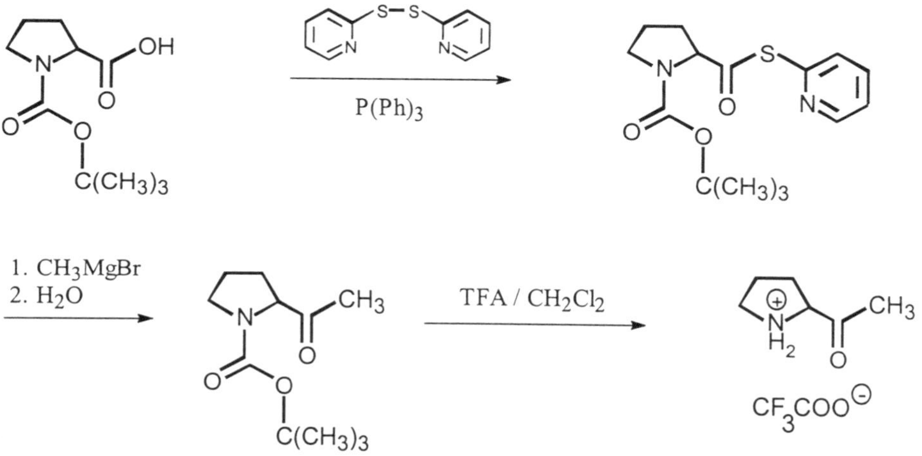

Figure 4. *Synthetic route used to prepare 2-acetylpyrrolidine (as trifluoroacetate)*

Table 3. *Time course of the formation of 2-acetyl-1-pyrroline (AP) from 2-acetylpyrrolidine (APD)*[a]

Reaction time	AP (µg)	Conversion of ADP (%)
5	12.0	26
30	32.5	72
120	44.4	>99

[a] 2-Acetylpyrrolidine trifluoroacetate (45 µg, 0.4 µmol) was dissolved in tap water (1 mL) containing dicyclohexylamine (0.4 µmol) and was allowed to stand at 25°C in a brown glass vial.

The next question was, of course: how is the 2-acetylpyrrolidine formed during the degradation of proline? Following the previously established idea[5,7] that 2-oxopropanal and 1-pyrroline are the key intermediates in AP formation, we studied whether water has an influence on the yields of the AP.

Table 4. *Influence of the presence of water on the amounts of 2-acetyl-1-pyrroline (AP) generated from 1-pyrroline and 2-oxopropanal*

Reaction system	AP (μg)
Aqueous[a]	58.5
Non-aqueous[b]	1.1

[a] 1-Pyrroline (10 μmol, 690 μg) was reacted in the presence of 2-oxopropanal (10 μmol) for 30 min at 100°C in phosphate buffer (5 mL; pH 7.0; 0.5 M).

[b] The reactants (10 μmol each) were mixed with silica gel (3 g containing 20 mg of KH_2PO_4 in 300 μL of water) and reacted for 5 min at 180°C.

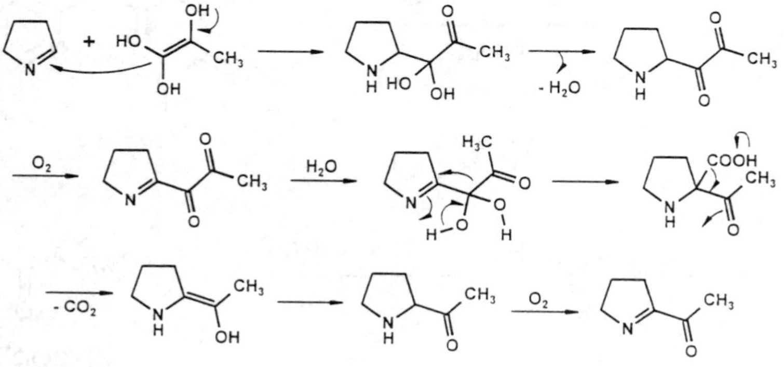

Figure 5. *Hypothetical pathway leading from 2-oxopropanal (as hydrate) and 1-pyrroline to 2-acetyl-1-pyrroline*

The results (Table 4) clearly indicated that the reaction between 1-pyrroline and 2-oxopropanal results in the formation of 50-fold more AP in an aqueous than in a non-aqueous system. Based on this result it can be assumed that the hydrated 2-oxopropanal is the reactive form and that, consequently, the first step in AP formation may take place as shown in Figure 5. In the first step, the hydrated 2-oxopropanal attacks the electrophilic 1-pyrroline yielding the 2-(1,2-dioxo-1-propyl) pyrrolidine. Its oxidation into the corresponding pyrroline derivative followed by a rearrangement into 2-acetyl-2-pyrrolidinic acid and a subsequent decarboxylation then yields 2-acetylpyrrolidine which has been shown to give the AP simply by air oxidation (see above).

References

1. R.G. Buttery, L.C. Ling and B.O. Juliano, 2-acetyl-1-pyrroline - an important component of cooked rice, *Chem. Ind.*, 1982, 958-959.
2. P.Schieberle and W.Grosch, Identification of the volatile flavour compounds of wheat bread crust - comparison with rye bread crust, *Z. Lebensm. Unters. Forsch.*, 1985, **180**, 474-478.
3. P.Schieberle, Primary odorants in popcorn, *J. Agric. Food Chem.*, 1991, **39**, 1141-1144.
4. R. Tressl, B. Helak and N. Martin, Formation of flavor compounds from L-proline, In 'Topics in Flavour Research', R.G. Berger, S. Nitz and P. Schreier (eds), Hangenham, Freising, 1985, pp.139-160.
5. P. Schieberle, The role of free amino acids in yeast as precursors of the odorants 2-acetyl-1-pyrroline and 2-acetyltetrahydropyridine in wheat bread crust, *Z. Lebensm. Unters. Forsch.*, 1990, **191**, 206-209.
6. J.E. Hodge, F.D. Mills and B.E. Fisher, Compounds derived from browned flavours, *Cereal Sci. Today*, 1972, **17**, 34-40.
7. P. Schieberle, Quantitation of important roast-smelling odorants in popcorn by stable isotope dilution assays and model studies on flavour formation during popping, *J. Agric. Food Chem.*, 1995, **43**, 2442-2448.
8. P. Schieberle, Formation of 2-acetyl-1-pyrroline and other important flavour compounds in wheat bread crust, In 'Thermal Generation of Aromas', ACS Symposium Series 409, American Chemical Society, Washington DC, 1989, pp. 268-275.
9. D. Rewicki, R. Tressl, U. Ellerbeck, E. Kersten, W. Burgert, M. Gorzinski, R.S. Hauck and B. Helak, Formation and synthesis of some Maillard generated aroma compounds. In 'Progress in Flavour Precursor Studies', P. Schreier and P. Winterhalter (eds), Allured Publishing Corp., Carol Stream, USA, 1993, pp. 301-314.
10. T. Hofmann and P. Schieberle, New and convenient syntheses of the important roasty, popcorn-like smelling food aroma compounds 2-acetyl-1-pyrroline and 2-acetyltetrahydropyridine from their corresponding cyclic α-amino acids, *J. Agric. Food Chem.*, submitted.
11. T. Hofmann and P. Schieberle, 2-Oxopropanal, hydroxy-2-propanone and 1-pyrroline - important intermediates in the generation of the roast smelling food flavor compounds 2-acetyl-1-pyrroline and 2-acetyltetrahydropyridine, *J. Agric. Food Chem.*, submitted.
12. N. de Kimpe, W.S. Dhooge, W.S. Shi, M.A. Keppens and M.M. Boelens, On the Hodge mechanism of the formation of the bread flavour component 6-acetyl-1,2,3,4-tetrahydropyridine from proline and sugars, *J. Agric. Food Chem.*, 1994, **42**, 1739-1742.
13. T. Hofmann and P. Schieberle, Studies on the intermediates generating the flavour compounds 2-methyl-3-furanthiol, 2-acetyl-2-thiazoline and Sotolon by Maillard-type reactions. In 'Flavour Science - Recent Developments', A.J. Taylor and D.S. Mottram (eds), The Royal Society of Chemistry, Cambridge, UK, 1996, pp. 182-187.

Toxicology and Antioxidants

Formation of Mutagenic Maillard Reaction Products

P. Arvidsson[1], M.A.J.S. van Boekel[2], K. Skog[1] and M. Jägerstad[1]

[1]DEPARTMENT OF APPLIED NUTRITION AND FOOD CHEMISTRY, CHEMICAL CENTRE, LUND UNIVERSITY, P.O. BOX 124, SE-221 00 LUND, SWEDEN
[2]DEPARTMENT OF FOOD SCIENCE, WAGENINGEN AGRICULTRURAL UNIVERSITY, P.O. BOX 8129, 6700 EV WAGENINGEN, THE NETHERLANDS

Summary
A model system was used to examine the kinetics of formation of polar heterocyclic amines by heating the precursors creatinine, glucose and amino acids in proportions similar to those in bovine meat but at higher concentrations. Formation of heterocyclic amines was studied between 150 and 225°C for 0.5 - 120 min, depending on temperature. Heated samples were subjected to solid phase extraction and HPLC analysis, with photodiode array detection for identification and quantification of heterocyclic amines. IQx, MeIQx, 7,8-DiMeIQx, 4,8-DiMeIQx, PhIP, harman and norharman were identified. A first-order reaction model and the Eyring equation were fitted to data for the formation of polar heterocyclic amines to obtain rate constants and to determine the effect of temperature on reaction kinetics.

Introduction

When frying and grilling meat and fish, several mutagenic Maillard reactions products are formed. The mutagenicity mainly originates from heterocyclic amines, found primarily in the crusts and pan residues. The levels of heterocyclic amines in cooked meat are highly dependent on cooking time and temperature[1] and, to some extent on type of meat.[2] Heterocyclic amines may be formed at normal cooking temperatures, e.g., 150°C - 200°C and, in general, a higher temperature or a longer cooking time increases their production.[1,2] The precursors in meat and fish are creatin(in)e, glucose, free amino acids and some dipeptides.[3,4] The IQ and IQx derivatives are assumed to arise from such precursors through the Maillard reaction and Strecker degradation upon heating.[3]

Our objectives were to examine the kinetics of formation of heterocyclic amines from precursors used in proportions similar to those in meat using a model system. After being validated in cooking experiments, it may be possible to predict the formation of heterocyclic amines using such kinetic models in order to design safer cooking equipment and procedures.

Materials and Methods

The precursors creatinine, glucose, amino acids and carnosine were heated in proportions similar to those found in bovine meat[4] but at a concentration 20 times higher (water solutions). The samples were heated in sealed quartz test tubes (100 × 4.0 mm i.d., 1.0 mm wall thickness) in an oil-bath.

Heated samples were extracted and purified using solid-phase extraction to be analysed later using reversed-phase HPLC with photodiode array detection for identification and quantification of heterocyclic amines as described by Gross *et. al.* (1992).[5] Amounts were corrected for incomplete recovery and expressed in nmol heterocyclic amines/mmol creatinine. All experiments were performed in duplicate.

Samples heated at 200°C during 2.5, 5.0, 10.0 and 15.0 min were analysed for amino acids (Biotronik LC5991), creatinine[6] and glucose (Boehringer Mannheim glucose oxidase/peroxidase method).

A first-order model was used for the formation of compounds: $C_t=C_0(1-e^{-k_1t})$, in which C_t is the concentration of a heterocyclic amine as a function of time (nmol heterocyclic amine/mmol creatinine); C_0, the concentration of the compound from which the heterocyclic amine is formed (C_0 is not known) (nmol heterocyclic amine/mmol creatinine); t, the heating time (min); and k_1, the rate constant for degradation of C_0 and, at the same time, the rate constant for formation of C_t (min^{-1}). In the case of degradation of heterocyclic amines, a first-order model was assumed, and this leads to the equation $C_t=C_0(e^{-k_2t}-e^{-k_1t})$, with k_2 as the rate constant of degradation of heterocyclic amines (min^{-1}).

The temperature dependence was analysed using the Eyring equation[7] ($k = k_bT/h \exp(\Delta S^\ne/R) \exp(-\Delta H^\ne/RT)$), in which k is the rate constant; k_b, the Boltzmann constant; h, the Planck constant; R, the molar gas constant; $\Delta S^\ne$, the activation entropy (J/mol K); $\Delta H^\ne$, the activation enthalpy (J/mol); and T, the temperature (K). All data were analysed at once in a one-step procedure[8] using the C_0-values obtained from fits described above.

Results

Average recoveries from the purification stage were 88 ± 7, 92 ± 8, 88 ± 8, 89 ± 9 and 79 ± 8 (means (%) ± standard deviations (%), n=7) for IQx, MeIQx, 7,8-DiMeIQx, 4,8-DiMeIQx and PhIP, respectively. Detection limits were ~2 ng for IQx, ~4 ng for PhIP and ~1 ng for MeIQx, 7,8-DiMeIQx and 4,8-DiMeIQx, per injection. The HPLC showed linearity in the range used (i.e., up to 2000 ng MeIQx per injection). In general, duplicate values varied by about ±5%.

Table 1. *The calculated C_0, rate constants of formation k_1 and degradation k_2, coefficient of determination R^2 and the number of data points n. (± approximate standard deviation)* [9]

	Temp.(°C)	C_0 (nmol/mmol)	k_1 (min^{-1})	k_2 (min^{-1})	R^2	n
IQx	150	0.261 ± 0.018	0.0386 ± 0.0064	0	0.90	16
	175	0.891 ± 0.020	0.361 ± 0.039	0	0.86	21
	200	1.65 ± 0.06	0.318 ± 0.042	0	0.86	26
	225	-	-	-	-	-
MeIQx	150	6.02 ± 1.15	0.00721 ± 0.00189	0	0.97	18
	175	7.40 ± 0.35	0.0643 ± 0.0063	0	0.97	22
	200	10.23 ± 0.26	0.223 ± 0.017	0	0.97	26
	225	7.02 ± 0.57	0.733 ± 0.133	0.0253 ± 0.0065	0.74	28
7,8-DiMeIQx	150	0.111 ± 0.014	0.0223 ± 0.0060	0	0.83	18
	175	0.199 ± 0.020	0.123 ± 0.030	0	0.83	18
	200	0.402 ± 0.013	0.257 ± 0.026	0	0.94	26
	225	0.409 ± 0.035	0.417 ± 0.064	0.0061 ± 0.0045	0.90	26
4,8-DiMeIQx	150	-	-	-	-	-
	175	0.490 ± 0.271	0.0161 ± 0.0113	0	0.90	16
	200	0.496 ± 0.030	0.166 ± 0.026	0	0.88	26
	225	0.617 ± 0.022	0.342 ± 0.037	0	0.91	24
PhIP	150	1.79 ± 2.23	0.00154 ± 0.00211	0	0.96	18
	175	1.50 ± 0.16	0.0250 ± 0.0037	0	0.98	18
	200	1.56 ± 0.07	0.148 ± 0.018	0	0.93	26
	225	-	-	-	-	-

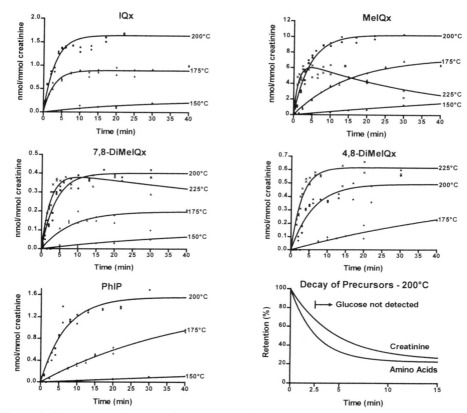

Figure 1. *Formation of heterocyclic amines at 150, 175, 200 and 225°C (note the different scales on the vertical axes) and the decay of precursors at 200°C* [9]

Heating the reagent solution resulted in formation of the polar heterocyclic amines IQx, MeIQx, 7,8-DiMeIQx and 4,8-DiMeIQx and PhIP. A first-order model appeared to fit the experimental data reasonably well (Fig. 1 and Table 1). The Eyring equation appeared to fit the experimental data reasonably well and the calculated parameters are shown in Table 2.

Table 2. *The calculated activation enthalpy $\Delta H^{\neq}$, the activation entropy $\Delta S^{\neq}$ and the number of data points n. (± approximate standard deviation)* [9]

Compound	$\Delta H^{\neq}$ (kJ/mol)	$\Delta S^{\neq}$ (kJ/(mol K))	n
IQx	68.9 ± 6.2	-110.1 ± 10.3	62
MeIQx	97.7 ± 2.3	-54.8 ± 2.3	86
7,8-DiMeIQx	61.1 ± 3.4	-131.5 ± 8.8	78
4,8-DiMeIQx	94.2 ± 5.7	-67.2 ± 5.0	75
PhIP	134.4 ± 4.2	19.8 ± 1.2	66

Consumption of initial reactants was studied at 200°C (Figure 1). After 2.5 min of heating the glucose concentration was below detection levels. The initial pH was 5.9 in the reagent solution and after heating: 5.9, 6.0, 6.5 and 7.0 following 2.5, 5.0, 10.0 and 15.0 min of heating, respectively.

Discussion

We used a complex mixture to try to simulate the frying of meat. Previous model experiments have examined one or a few amino acids at a time. The same polar heterocyclic amines, namely IQx, MeIQx, 7,8-DiMeIQx, 4,8-DiMeIQx, PhIP were found upon heating, and in about the same amounts (a few nmol heterocyclic amines/mmol creatinine) as in earlier reports and frying experiments.[10] Neither IQ nor MeIQ was detected in any of the samples and no attempts were made to analyse the apolar fraction from the purification stage.

The heat transfer through the wall of the test tubes was very efficient, which was important for calculations of reaction kinetics. It took only 30 sec for the reagent solution to reach the heating temperature.

Formation of heterocyclic amines began immediately upon heating, and within 30 sec surprisingly high amounts had been formed. After some further time, depending on the temperature, a maximum level was reached after which the concentration, except at 225°C, more or less levelled out to a plateau (Figure 1). The plateau was most probably reached as the Maillard reaction slowed down, due to rapid disappearance of glucose. A similar plateau formation was reported in fried ground beef by Knize (1994)[1] who measured the formation of MeIQx, and by Bjeldanes (1983)[11] who measured the mutagenic activity.

An interesting observation was made at 225°C. The plateau was only reached for 4,8-DiMeIQx, while the amounts of MeIQx and 7,8-DiMeIQx decreased after peaking, and no significant amount of PhIP was detected. These findings indicate that the heterocyclic amines were not only formed, but also underwent some kind of degradation. Such degradation might have occurred at all temperatures we used, but it was significant only above 200°C for the IQx derivatives and PhIP. A degradation of heterocyclic amines was reported in 1995 by Jackson and Hargraves[12], who measured the formation of MeIQx and 4,8-DiMeIQx in a model system containing threonine, glucose and creatine. This prompted us to examine the stability of the heterocyclic amines at higher temperatures. Although degradation was clearly seen upon heating mixtures of pure synthetic standards (Figure 2) at 225°C, it was not as pronounced as in the samples containing all precursors.

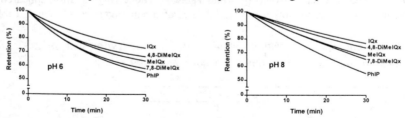

Figure 2. *Retention of heterocyclic amines in a pure synthetic standard heated at 225°C* [9]

Kinetics of formation (and subsequent degradation) could reasonably be described by the first-order models $C_t = C_0(1 - e^{-k_1 t})$ and $C_t = C_0(e^{-k_2 t} - e^{-k_1 t})$, respectively. It is, of course, clear that the actual reaction mechanism for formation of heterocyclic amines is much more complicated. However, the remarkably good fit suggests that the rate-limiting step in the cascade of reactions (Maillard and Strecker) producing heterocyclic amines is a monomolecular (first-order) or bimolecular reaction with one reactant in large excess (pseudo-first-order). The pathway for formation of heterocyclic amines is very complicated and mostly unknown. It is, however, assumed that Maillard reaction products such as pyrazines (or pyridines), Strecker aldehydes and creatine/creatinine are condensed to form the IQx (or IQ) derivatives, perhaps through free-radical reactions.[3,10,13]

The temperature dependence of the rate constants of formation was analysed using the Eyring equation ($k = k_b T/h \exp(\Delta S^\ast/R) \exp(-\Delta H^\ast/RT)$) rather than the Arrhenius equation ($k = A \exp(-E_a/RT)$), because the Eyring equation gives the activation entropy $\Delta S^\ast$, which may give an indication about the reaction mechanism. Negative values indicate a bimolecular reaction and near zero values a monomolecular reaction.[7] The calculated activation entropy (Table 2) indicates that the rate-limiting step forming all the IQx derivatives is a bimolecular reaction ($\Delta S^\ast < 0$), while the one forming PhIP indicates a monomolecular reaction ($\Delta S^\ast \approx 0$). This means that the rate-limiting step forming all the IQx-derivatives might be of the pseudo-first-order type (one of two reactants in large excess). It has been suggested[14,15] that one step in the mechanism involves a reaction between a Strecker aldehyde and creatinine, both of which are in abundance, before reacting with a pyrazine forming the IQx-derivative. Thus, the rate-limiting step might then be the reaction between creatinine-aldehyde and pyrazine. The activation enthalpy $\Delta H^\ast$ is approximately equal to the activation energy[7], and can be regarded as the amount of energy which must be supplied to the reactants in order to initiate reactions. PhIP has the highest activation energy (Table 2), and this can explain why PhIP has so seldom been reported at low temperatures.[16]

The retention of creatinine and amino acids (Figure 1) was >20%, even after 15 min of heating, while all the glucose had reacted after 2.5 min. Thus glucose was the limiting precursor, although the formation of heterocyclic amines did not generally increase with higher glucose content.[17]

The purpose of this study was to produce kinetic data for the prediction of the amount of heterocyclic amines that would be formed in fried meat when cooking time and cooking temperature are known. The time dependence of the formation of heterocyclic amines seemed to follow the first-order model $C_t = C_0(1-e^{-k_1 t})$, or $C_t = C_0(e^{-k_2 t} - e^{-k_1 t})$ at very high temperatures, such as 225°C, while the temperature dependence seemed to follow the Eyring equation. The parameters in the time models, at 150, 175, 200 and 225°C, and the Eyring equation could be estimated from our data.

Acknowledgments

This study was supported by the Swedish Cancer Foundation (1824-B95-14XAB) and has also been carried out with financial support from the Commision of the European Communities, Agriculture and Fisheries (FAIR) specific RTD programme, CT96-1080, "Optimisation of the Maillard reaction. A way to improve quality and safety of thermally processed foods." It does not necessarily reflect its views and in no way anticipates the Commision's future policy in this area.

References

1. M.G. Knize, F.A. Dolbeare, K.L. Carroll, D.H. Moore and J.S. Felton, Effect of cooking time and temperature on the heterocyclic amines content of fried beef patties, *Food Chem. Toxicol.*, 1994, **32**, 595-603.

2. K. Skog, G. Steineck, K. Augustsson and M. Jägerstad, Effect of cooking temperature on the formation of heterocyclic amines in fried meat products and pan residues, *Carcinogenesis*, 1995, **16**, 861-867.
3. K. Skog, Cooking procedures and food mutagens: a literature review, *Food Chem. Toxicol.,* 1993, **31**, 655-675.
4. A. Laser Reuterswärd, K. Skog and M. Jägerstad, Mutagenicity of pan-fried bovine tissues in relation to their content of creatine, creatinine, monosaccharides and free amino acids, *Food Chem. Toxicol.*, 1987, **25**, 755-762.
5. G. A. Gross, A. Gruter and S. Heyland, Optimization of the sensitivity of high-performance liquid chromatography in the detection of heterocyclic aromatic amine mutagens, *Food Chem. Toxicol.*, 1992, **30**, 102-144.
6. A.W. Wahlefeld, G. Holz and H.U. Bergmeyer, 'Methods of Enzymatic Analysis', H. U. Bergmeyer (ed.), Verlag Chemie Weinheim & Academic Press, New York and London, 1974, Vol. 4, 1st ed., p. 1786-1790.
7. M.A.J.S. Van Boekel and P. Walstra, Use of kinetics in studying heat-induced changes in foods, Ch. 2 in IDF Monograph 'Heat-induced changes in milk', P.F. Fox (ed.), IDF (International Dairy Federation) special issue 9501, 2nd ed., Brussels, 1995, p. 22-50.
8. M.A.J.S. Van Boekel, Statistical aspects of kinetic modeling for food science problems, *J. Food Sci.*, 1996, **61**, 477-485, 489.
9. P. Arvidsson, M.A.J.S. Van Boekel, K. Skog and M. Jägerstad, Kinetics of formation of polar heterocyclic amines in a meat model system, *J. Food Sci.*, 1997, In press.
10. M. Johansson, B.F Fay, G.A Gross, K. Olsson and M. Jägerstad, Influence of amino acids on the formation of mutagenic/carcinogenic heterocyclic amines in a model system, *Carcinogenesis*, 1995, **16**, 2553-2560.
11. L.F. Bjeldanes, J.S. Felton and F.T. Hatch, Mutagens in cooked food, in 'Xenobiotics in Foods and Feeds', J.W. Finley and D.E. Schwass (eds), American Chemical Society, ACS Symposium Series, No. 234, 1983, p.149-168.
12. L.S. Jackson and W.A. Hargraves, Effects of Time and Temperature on the Formation of MeIQx and DiMeIQx in a Model System Containing Threonine, Glucose and Creatine, *J. Agric. Food Chem.*, 1995, **43**, 1678-1684.
13. B.L. Milic S.M. Djilas and J.M. Canadanovic-Brunet, Synthesis of some heterocyclic aminoimidazoazarenes, *Food Chem.*, 1993, **46**, 273-276.
14. T. Nyhammar, Studies on the Maillard reaction and its role in the formation of food mutagens, PhD Thesis, 1986, Swedish University of Agricultural Sciences, Uppsala, Sweden.
15. R.C. Jones and J.H. Weisburger, Characterization of aminoalkylimidazol-4-one mutagens from liquid-reflux models, *Mutation Res.,* 1989, **222**, 43-51.
16. M. Johansson, Influence of lipids, and pro- and antioxidants on the yield of carcinogenic heterocyclic amines in cooked foods and model systems, PhD Thesis, 1995, Lund University, Sweden.
17. K. Skog and M. Jägerstad, Effects of monosaccharides and disaccharides on the formation of food mutagens in model systems, *Mutation Res.*, 1990, **230**, 263-272.

Antioxidative Activity of Non-Enzymatically Browned Proteins by Reaction with Lipid Oxidation Products

Francisco J. Hidalgo, Manuel Alaiz and Rosario Zamora

INSTITUTO DE LA GRASA, CSIC, AVENIDA PADRE GARCÍA TEJERO, 4, 41012-SEVILLA, SPAIN

Summary

Three oxidized lipid/amino acid reaction products (OLAARP): (1-methyl-4-pentyl-1,4-dihydropyridine-3,5-dicarbaldehyde, 1-(5'-amino-1'-carboxypentyl)pyrrole, and *N*-(carbobenzyloxy)-1(3)-(1'-(formyl(methyl)-hexyl)-L-histidine dihydrate), and two browned proteins (the monomer and the dimer produced in the reaction between BSA and 4,5(*E*)-epoxy-2(*E*)-heptenal) were prepared and tested for antioxidative activity in a microsomal system in order to investigate the antioxidative function of OLAARP and non-enzymatically browned proteins in biological systems. The microsomal system consisted of freshly prepared trout muscle microsomes, which were oxidized with Cu^{2+}, Fe^{3+}/ascorbate, or Cu^{2+}/H_2O_2 at 37 °C and in the presence of the compound to be tested as antioxidant. The three OLAARP (tested at 50 µM) and the two browned proteins (tested at 40 µg/mL) efficiently protected against lipid peroxidation, assessed by the formation of thiobarbituric acid-reactive substances, and protein damage, determined by amino acid analysis. These results suggest that the formation of non-enzymatically browned proteins by reaction with lipid oxidation products may constitute an antioxidative defense mechanism, which could play a significant role *in vivo*.

Introduction

Reactive oxygen species (ROS) and free radicals are highly reactive compounds with a relatively short half-life and low specificity. They can react with most biological macromolecules, leading, in a first step, to chemical modification and loss of function, and, ultimately, to tissue damage and destruction.[1] Normally such detrimental effects are prevented by a complex and multilayered defense system in which enzymes, proteins, and low molecular mass radical scavengers interact.[2,3] Oxidative stress occurs when this antioxidant defense system is unable to cope with the amounts of oxidant present, due either to a deficiency in the defense mechanisms or an increase in the levels of oxidants. This oxidative damage is generally accepted to be involved in the etiology of a number of diseases, either as the primary cause or a secondary effect.[4]

Peroxidation of lipids is a well recognized pathway of oxidant injury, and it is always accompanied in biological systems with the formation of aldehydes and other oxidized derivatives.[5,6] These oxidized derivatives, which act as 'secondary toxic messengers', are able to interact with reactive groups in protein residues producing modified proteins, which, thereafter, are either degraded by proteolytic systems or allowed to accumulate in lipofuscin or advanced glycation endproducts.[7,8] No biological functions of these damaged proteins are known at present. However, recent studies from this laboratory have pointed out that some oxidized lipid/amino acid reaction products (OLAARP), which are present in the modified proteins,[9] exhibited antioxidative properties when tested in edible oils.[10] The objective of this study was to investigate the antioxidative function of OLAARP in biological systems and to analyze if this function is also displayed by non-enzymatically browned proteins.

Materials and Methods

Preparation of OLAARP and Non-enzymatically Browned Proteins
Three OLAARP were selected to be tested as potential oxidative stress inhibitors: 1-methyl-4-pentyl-1,4-dihydropyridine-3,5-dicarbaldehyde (DHP), 1-(5'-amino-1'-carboxypentyl)pyrrole (ACPP), and *N*-(carbobenzyloxy)-1(3)-[1'-(formylmethyl)hexyl]-L-histidine dihydrate (ZHO), which were prepared as described previously.[10-12] The three compounds were chromatographically pure, and their structures were confirmed by ^{1}H and ^{13}C nuclear magnetic resonance spectroscopy and mass spectrometry. Structures for these compounds are shown in Figure 1.

Two non-enzymatically damaged proteins, the monomer and the dimer produced in the reaction between bovine serum albumin (BSA) and 4,5(*E*)-epoxy-2(*E*)-heptenal (MBSA and DBSA, respectively), were also tested as potential oxidative stress inhibitors. MBSA and DBSA were obtained as described previously.[13] Both proteins exhibited a decrease in lysine residues of 45 and 48 % for MBSA and DBSA, respectively, and the formation of OLAARP product ε-*N*-pyrrolylnorleucine (Pnl). MBSA had 0.09 μmol Pnl/mg protein, and DBSA had 0.12 μmol Pnl/mg protein.

Preparation and Oxidation of Trout Muscle Microsomes
Muscle microsomes from freshly killed rainbow trout (*Salmo gairdnerii*) were prepared by differential centrifugation as described previously,[9] and protein concentration was adjusted to 3.0 mg/mL for experiments. Microsomes were oxidized by suspending 50 μL of the stock solution in 5 mL of Krebs-Ringer phosphate buffer and incubating at 37 °C in the presence of 5 μM Cu^{2+}, 1 mM Fe^{3+}/5 mM ascorbate, or 1 mM Cu^{2+}/10 mM H_2O_2, and the compound to be tested as antioxidant.

At different incubation times samples were tested for lipid oxidation, assessed by the formation of thiobarbituric acid-reactive substances (TBARS),[14] and protein damage, determined by amino acid analysis.[15] The incubation times were 24 h for Cu^{2+}, 3 h for Fe^{3+}/ascorbate, and 1 h for Cu^{2+}/H_2O_2, when TBARS were evaluated; and 24 h for amino acid analysis. The data given for amino acid analysis in the Results section refer to lysine, which was the amino acid mainly degraded.

For comparison purposes, a protection index (PI) was calculated according to equation [1], where *a* is the control with ROS, *b* is the control without ROS, x is the sample with antioxidant treated with ROS, and y is the untreated sample with the antioxidant added.

$$PI = 100 \times [1 - ((x - y)/(a - b))] \qquad [1]$$

Results

Antioxidative effect of OLAARP
Figure 2 shows the PIs obtained for the three OLAARP tested (DHP, ACPP, and ZHO), the amino acids used in their preparation: lysine (LYS) and *N*-(carbobenzyloxy)-L-histidine (ZH), and two common antioxidants: butylated hydroxytoluene (BHT) and α-tocopherol (TOC), when tested at 50 μM in a microsomal system. The effect of OLAARP in inhibition of TBARS production is shown in Figure 2A. The three OLAARP exhibited protection for the three systems assayed. By comparison, neither LYS nor ZH protected against TBARS production. BHT and TOC exhibited very different behavior. Thus, the

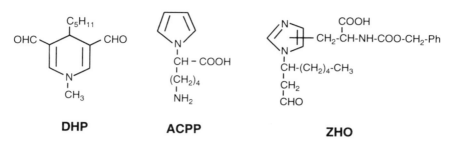

Figure 1. *OLAARP tested for antioxidative activity in this study*

protection exhibited by BHT was almost complete (PI = 97-98), and TOC did not show protection (PI = 0-7). The antioxidative activity was very similar for the three systems assayed, and the order of effectiveness observed was: BHT > ACPP ≈ ZHO ≈ DHP > ZH ≈ LYS ≈ TOC.

The PIs exhibited by the assayed compounds against protein damage, evaluated by lysine analysis, is shown in Figure 2B. The differences observed in TBARS for BHT and the OLAARP, were not observed in amino acid analysis. Although there were important differences among the different systems employed, the three OLAARP efficiently protected (PI = 40-100) the microsomal proteins against oxidative stress, and this

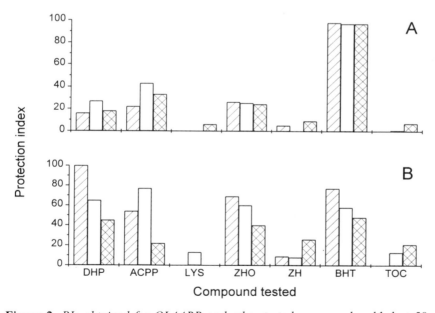

Figure 2. *PIs obtained for OLAARP and other tested compounds added at 50 µM in a microsomal system. PIs were determined for: A, lipid peroxidation, evaluated by the formation of TBARS; and B, protein damage, evaluated by lysine analysis. Three systems were used to produce ROS: Cu^{2+} (striped bars), Fe^{3+}/ascorbate (open bars), and Cu^{2+}/H_2O_2 (cross-hatched bars)*

protection was similar to that exhibited by BHT (PI = 48-77). As in the case of TBARS analysis, the protection exhibited by LYS, ZH, and TOC was lower than that exhibited by OLAARP and BHT. Therefore, the order of effectiveness obtained was: DHP ≈ ACPP ≈ BHT ≈ ZHO > ZH ≈ LYS ≈ TOC.

Antioxidative activity of non-enzymatically browned proteins
Figure 3 shows the PIs exhibited by BSA, MBSA and DBSA added at 40 µg/mL to the microsomal system. As for the OLAARP system, non-enzymatically browned proteins protected microsomes against lipid and protein damage, and this protection was higher than that exhibited by the protein used in the preparation of the damaged protein. Figure 3A shows the PIs obtained for lipid peroxidation, measured by TBARS formation (PIs were 43-61 for DBSA, 40-50 for MBSA, and 9-21 for BSA).

Analogous results were obtained for protection of protein damage when lysine was determined after acid hydrolysis. Figure 3B shows the PIs obtained for BSA, MBSA and DBSA when added at 40 µg/mL (50-83 for DBSA, 38-63 for MBSA, and 0-17 for BSA). Therefore, the order of effectiveness observer for both lipid peroxidation and protein damage was: DBSA > MBSA > BSA.

Discussion

In addition to other macromolecules, lipids are highly sensitive to ROS, producing lipid hydroperoxides in a first step, and, subsequently, secondary products. However, secondary products are not final products, and they are able to interact with reactive groups in neighboring amino acids and proteins, producing modified amino acids (OLAARP) and proteins,[16,17] which have been implicated in the etiology of many diseases. Nevertheless, the results presented in this study show that both OLAARP and OLAARP-containing proteins exhibited antioxidative activities for both lipid peroxidation and protein damage when tested in a microsomal system. Thus, the three tested OLAARP, which are

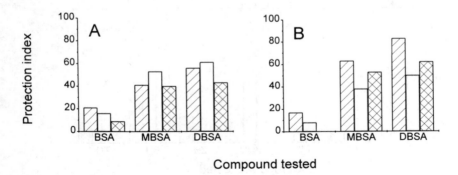

Figure 3. *PI obtained for non-enzymatically browned proteins added at 40 µg/mL to a microsomal system. PI were determined for: A, lipid peroxidation, evaluated by the formation of TBARS, and B, protein damage, evaluated by lysine analysis. Three systems were used to produce ROS: Cu^{2+} (striped bars), Fe^{3+}/ascorbate (open bars), and Cu^{2+}/H_2O_2 (cross-hatched bars)*

representative of three major groups of heterocyclic derivatives produced in oxidized lipid/amino acid reactions, exhibited antioxidative activities similar to that of BHT when protection of protein damage was considered, and lower than BHT when inhibition of lipid peroxidation was studied. In addition, the protection exhibited by OLAARP was always much higher than that exhibited by amino acids, suggesting that OLAARP formation is necessary to produce amino acids with antioxidative properties.

Similar results were obtained for the non-enzymatically browned proteins studied, which exhibited a much higher activity than the protein used in their preparation. This mechanism, by which proteins are transformed into OLAARP-containing proteins with antioxidative properties, may play a significant role in the antioxidative properties observed for many proteins. In addition, the antioxidative activity observed for the browned proteins was related to the amount of OLAARP contained in the protein. Thus DBSA, which had more Pnl than MBSA, exhibited a higher antioxidative activity than the monomeric modified protein.

These results suggest that the modification of amino acids and proteins by lipid oxidation products is not necessarily a negative consequence of oxidative stress. The antioxidative activity observed for both OLAARP and OLAARP-containing proteins seem to control, at least in a part, the oxidative stress, which is feedback inhibited, in common with many enzymatic processes. This sequence of events has been summarized in Figure 4. In addition, and according to previous studies from our laboratory,[18] all these processes takes place simultaneously. Therefore, feedback inhibition of oxidative stress should be taking place at the same time that OLAARP are being produced.

Acknowledgments
We are indebted to J. L. Navarro and M. D. García for the technical assistance. This study was supported in part by the Comisión Interministerial de Ciencia y Tecnología of Spain (Projects ALI94-0763, OLI96-2124, and ALI97-0358) and the Junta de Andalucía (Project AGR 0135).

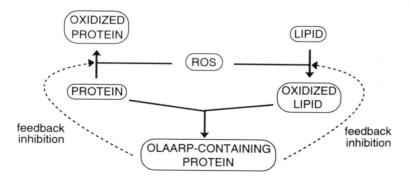

Figure 4. *Proposed mechanism by which ROS damage to lipids and proteins is feedback inhibited by OLAARP-containing proteins produced by reaction between oxidized lipids and proteins*

References

1. B. Halliwell and J. M. C. Gutteridge, 'Free Radicals in Biology and Medicine', 2nd ed., Clarendon Press, Oxford, 1989.
2. L. J. Machlin and A. Bendich, Free radical tissue damage: protective role of antioxidant nutrients, *FASEB J.*, 1984, **222**, 1-15.
3. R. J. Ellis, Chaperonins: introductory perspective, in 'The Chaperonins', R. J. Ellis (ed.), Academic Press, San Diego, 1996, pp. 1-25.
4. L. Packer, Preface, *Methods Enzymol.*, 1994, **233**, xvii-xviii.
5. H. W. Gardner, Oxygen radical chemistry of polyunsaturated fatty acids, *Free Radical Biol. Med.*, 1989, **7**, 65-86.
6. H. Esterbauer, H. Zollner and R. J. Schaur, Aldehydes formed by lipid peroxidation: mechanisms of formation, occurrence, and determination, in 'Membrane Lipid Oxidation. Vol. I', C. Vigo-Pelfrey (ed.), CRC Press, Boca Raton, 1990, pp. 239-268.
7. C. J. Dillard and A. L. Tappel, Consequences of biological lipid peroxidation, in 'Cellular Antioxidant Defense Mechanisms', C. K. Chow (ed.), CRC Press, Boca Raton, 1988, pp 103-115.
8. F. J. Hidalgo and R. Zamora, Fluorescent pyrrole products by carbonyl-amine reactions, *J. Biol. Chem.*, 1993, **268**, 16190-16197.
9. R. Zamora, J. L. Navarro and F. J. Hidalgo, Determination of lysine modification product ε-N-pyrrolylnorleucine in hydrolyzed proteins and trout muscle microsomes by micellar electrokinetic capillary chromatography, *Lipids*, 1995, **30**, 477-483.
10. M. Alaiz, R. Zamora and F. J. Hidalgo, Antioxidative activity of pyrrole, imidazole, dihydropyridine, and pyridinium salt derivatives produced in oxidized lipid/amino acid browning reactions, *J. Agric. Food Chem.*, 1996, **44**, 686-691.
11. R. Zamora and F. J. Hidalgo, Modification of lysine amino groups by the lipid peroxidation product 4,5(E)-epoxy-2(E)-heptenal, *Lipids*, 1994, **29**, 243-249.
12. M. Alaiz, R. Zamora and F. J. Hidalgo, Antioxidative activity of (E)-2-octenal/amino acid reaction products, *J. Agric. Food Chem.*, 1995, **43**, 795-800.
13. M. Alaiz, F. J. Hidalgo and R. Zamora, Antioxidative activity of nonenzymatically browned proteins produced in oxidized lipid/protein reactions, *J. Agric. Food Chem.*, 1997, **45**, 1365-1369.
14. H. Kosugi, T. Kojima and K. Kikugawa, Thiobarbituric acid-reactive substances from peroxidized lipids, *Lipids*, 1989, **24**, 873-881.
15. M. Alaiz, J. L. Navarro, J. Girón and E. Vioque, Amino acid analysis by high-performance liquid chromatography after derivatization with diethyl ethoxymethylenemalonate, *J. Chromatogr.*, 1992, **591**, 181-186.
16. R. Zamora and F. J. Hidalgo, Linoleic acid oxidation in the presence of amino compounds produces pyrroles by carbonyl amine reactions, *Biochim. Biophys. Acta*, 1995, **1258**, 319-327.
17. F. J. Hidalgo and R. Zamora, In vitro production of long chain pyrrole fatty esters from carbonyl-amine reactions, *J. Lipid Res.*, 1995, **36**, 725-735.
18. M. Alaiz, R. Zamora and F. J. Hidalgo, Contribution of the formation of oxidized lipid/amino acid reaction products to the protective role of amino acids in oils and fats, *J. Agric. Food Chem.*, 1996, **44**, 1890-1895.

Antioxidant and Prooxidant Activity of Xylose-Lysine Maillard Reaction Products

Gow-Chin Yen and Mei-Lin Liu

DEPARTMENT OF FOOD SCIENCE, NATIONAL CHUNG HSING UNIVERSITY, 250 KUOKUANG ROAD, TAICHUNG, TAIWAN, REPUBLIC OF CHINA

Summary

Maillard reaction product, prepared by heating xylose and lysine at pH 9.0 and 100°C for 3 h, was fractionated by ethyl ether and ethanol into acidic, neutral and basic low molecular weight, ethanol-soluble and ethanol-insoluble fractions. The ethanol-soluble and -insoluble fractions were the major fractions of xylose-lysine Maillard reaction product (XL MRP), contributing 79.5 and 20.1%, respectively. XL MRP inhibited linoleic acid peroxidation, initiated by the Fenton reaction, but did not inhibit liposome peroxidation catalyzed by Fe^{2+}, where it had a prooxidant action. XL MRP produced oxidative damage of deoxyribose and 2'-deoxyguanosine (2'-dG) initiated by the Fenton reaction. Ethanol-soluble and insoluble fractions also caused oxidative damage, while low molecular weight fractions had an antioxidant effect by inhibiting oxidative damage to deoxyribose initiated by the Fenton reaction. The prooxidant action of ethanol-soluble and insoluble fractions resembled unfractionated products in the 2'-dG assay. In these systems with deoxyribose, 2'-dG, linoleic acid or liposomes, the XL MRP exhibited either antioxidant or prooxidant properties, which might be due to competition between reducing power and scavenging activity against the hydroxyl radical; the low molecular weight fractions did not have prooxidant activity in these systems.

Introduction

The Maillard reaction, which is the reaction between a reducing sugar and an amino acid, is one of the most important reactions occurring during the processing, storage and cooking of foods. The antioxidant effect of Maillard reaction product (MRP) has been extensively investigated. Eichner[1] reported that the intermediate reductone compounds of MRP could break the radical chain by donation of a hydrogen atom. MRP was also observed to have metal-chelating properties and retard lipid peroxidation.[2] Hayase *et al.*[3] reported that melanoidins were powerful scavengers of reactive oxygen species. Recently, Yen and Hsieh[4] suggested that the antioxidant activity of xylose-lysine MRPs may be attributed to the combined effects of reducing power, hydrogen atom donation and scavenging of reactive oxygen species.

The prooxidant activity of some antioxidants has been increasingly studied in recent years. Plant phenolics have sometimes been found to have prooxidant properties.[5] The prooxidant activity is a result of the ability to reduce metals, (e.g., Fe^{3+}) to forms that react with O_2 or H_2O_2 to form initiators of oxidation. Aruoma *et al.*[6,7] reported that several phenolic antioxidants can accelerate oxidative damage of DNA, protein and carbohydrate *in vitro*. Therefore, it is important to assess the prooxidant activity of potential antioxidant substances in order to understand the significance and mechanism of antioxidant activity. Although the antioxidant properties of MRP are recognized, to our knowledge, the prooxidant effect of MRPs has not been examined. Thus, this work was designed to investigate the antioxidant activity and possible prooxidant action of XL MRP and its fractionation products.

Materials and Methods

Preparation of the xylose-lysine Maillard reaction product (XL MRP)
The XL MRP was prepared using reaction conditions described previously.[4] D-xylose (1 M) and L-lysine monohydrochloride (2 M) were dissolved in distilled water, adjusted to pH 9.0, and then refluxed in an oil bath at 100 °C for 3 h.

Fractionation of XL MRP
The fractionation of the XL MRP was carried out according to the method described by Kitts et al.[8] with slight modification. The reaction mixture was fractionated by subsequently ethyl ether extraction under acidic (pH 3.0), neutral (pH 7.5), and basic (pH 10.0) condition. These three fractions represented acidic, neutral and basic low molecular weight non-volatile compounds. The residual aqueous solution was re-extracted with 3 volumes of absolute ethanol. The ethanol suspension was centrifuged at 10000 rpm for 10 min to yield supernatant (ethanol-soluble) and precipitate (ethanol-insoluble) fractions. These five fractions obtained from the XL MRP were freeze-dried and the yield of each fraction was calculated.

Inhibition of linoleic acid peroxidation
XL MRP or a fraction (0.5 mL, 0.125-24 mg) was mixed with 2.5 mL 0.02 M linoleic acid emulsion. The reaction mixture was added to 0.1 mL 100 mM $FeCl_3$, 0.1 mL 100 mM ascorbate and 0.1 mL 100 mM H_2O_2 or 0.1 mL 100 mM $FeCl_2$ and 0.1 mL 100 mM H_2O_2. The final volume of reaction mixture was adjusted to 5 mL with 0.2 M phosphate buffer (pH 7.4). Reaction mixture was incubated at 37 °C for 24 h. At the end of the incubation, 1 mL 10% HCl and 1 mL 1% TBA were added and the reaction mixture heated in a water bath at 100 °C for 30 min. After cooling on ice, 5 mL chloroform was added and the reaction mixture was centrifuged at 4000 rpm for 20 min. The absorbance of the upper layer was measured at 532 nm. All data represent the average of triplicate analyses.

Inhibition of phospholipid liposome peroxidation
The ability of XL MRP and its fractions to inhibit lipid peroxidation at pH 7.4 was tested using egg phospholipid liposomes incubated with $FeCl_3$ and ascorbate as described by Aruoma et al.[7] All test data are the average of triplicate analyses. The lower the absorbance at 532 nm, the higher the antioxidant activity.

Effects of XL MRP on deoxyribose damage
To test the ability of XL MRP and its fractions to accelerate oxidative damage of deoxyribose, the Fenton reaction model containing $FeCl_3$-EDTA and H_2O_2 was used.[9] The extent of deoxyribose degradation was measured by the TBA method. The control did not caontain XL MRP or ascorbic acid.

Effect of XL MRP on oxidation of 2'-deoxyguanosine
The effect of the XL MRP and its fractions on the oxidation of 2'-deoxyguanosine (2'-dG) to 8-hydroxy-2'-deoxyguanosine (8-OH-2'-dG) were assayed, using a modification of the method of Kasai and Nishimura.[10] 2'-dG and 8-OH-2'-dG were identified by comparison of retention times with those of known standards, and quantified by peak areas from the HPLC chromatograms. The control did not contain XL MRP

or ascorbic acid. All analyses were run in triplicate.

Statistical analysis
Data were analysed using the SAS.[1] Means were compared using Duncan's multiple-range tests.

Results and Discussion

In the present study, the acidic, neutral and basic fractions in the ethyl ether extract of the XL MRP contained low molecular weight non-volatile compounds. The ethanol-soluble fraction contained Maillard reaction intermediates and low molecular weight melanoidins; and the ethanol-insoluble fraction contained high molecular weight melanoidins. The total recovery for all fractions was 90 % (w/w) of the start materials. The ethanol-soluble and ethanol-insoluble fractions were the major fractions. Two fractions contributed 79.51% and 20.08% of the total. The yields of acidic, neutral, and basic fractions recovered in the ethyl ether fractions were 0.25, 0.07, and 0.08 %, respectively.

Inhibition of linoleic acid peroxidation
The inhibition of linoleic acid peroxidation increased with increasing concentration of XL MRP whether induced by Fe^{3+}/H_2O_2/ascorbate or Fe^{2+}/H_2O_2. This finding implies that the XL MRP did not reduce Fe^{3+} to Fe^{2+} and accelerate the peroxidation of linoleic acid in the Fe^{3+}/H_2O_2/ascorbate system. The ethanol-soluble and insoluble fractions of the XL MRP revealed a similar inhibitory effect on linoleic acid peroxidation as unfractionated material. The low molecular weight fractions of XL MRP only revealed a slightly inhibitory effect on linoleic acid peroxidation induced by Fe^{3+}/H_2O_2/ascorbate but not by Fe^{2+}/H_2O_2. This weaker antioxidant effect might be due to the free-radical scavenging activity, or the weak reducing power of low molecular weight fractions, or the presence of ascorbate.

Inhibition of liposome peroxidation
The peroxidation of liposomes was increased with the increasing concentration of XL MRP in the incubation with Fe^{3+}/ascorbate. However, XL MRP showed an inhibitory effect at lower level (<0.25 mg/mL) in the absence of ascorbate, but it showed prooxidant action at levels greater than 2 mg/mL. The low molecular weight fractions of XL MRP showed no antioxidant or prooxidant action on lipsome peroxidation induced by Fe^{3+} with or without ascorbate. However, the melanoidins, ethanol-soluble and ethanol-insoluble fractions, showed similar trend as XL MRP in liposome peroxidation assays. This is not surprising in view of the fact that the ethanol-soluble and insoluble fractions are the major fractions of XL MRP.

Kanner et al.[12] discovered that the superoxide anion formed in the presence of Fe^{2+} and oxygen.. Dinis et al.[13] indicated that catalase or free radical scavengers do not inhibit lipid peroxidation induced by Fe^{2+}. Although XL MRP can scavenge the superoxide in the reaction system, it cannot inhibit the free radical chain reaction induced by Fe^{2+}.

Inhibition of ascorbate/iron-induced deoxyribose damage.
Figure 1 shows the inhibition of XL MRP on the deoxyribose damage induced by Fe^{3+}-EDTA/H_2O_2/ascorbate at low concentration (< 0.071 mg/mL). However, XL MRP

exhibited a maximum prooxidant effect at 1.14 mg/mL and then decreased with increasing concentration. In the absence of ascorbate, XL MRP showed a similar trend. Since ascorbate acts as a reducing agent to promote the production of hydroxyl radical in the reaction system, it is possible that the reducing power of XL MRP [4] could similarly promote deoxyribose damage in the Fenton reaction system. Sahu and Gay [14] reported that antioxidant and prooxidant activity of natural antioxidants resulted from competion of their ability to reducing iron ion and scavenging on hydroxyl radical in redox system.

The low molecular weight fractions of XL MRP did not promote deoxyribose damage induced by Fe^{3+}-EDTA/H_2O_2 with or without ascorbate. Moreover, the acidic low molecular fraction and neutral low molecular fraction had a protective effect on deoxyribose damage. However, the basic low molecular weight fraction only had slight inhibitory effect on Fe^{3+}-EDTA/H_2O_2 induced deoxyribose damage.

The inhibitory effect of ethanol-soluble and insoluble fractions on deoxyribose damage induced by Fe^{3+}-EDTA/H_2O_2/ascorbate increased with increasing concentration. At low concentration, there was no difference (P>0.05) between the two fractions. If the two fractions were combined in the ratio 79/20 (w/w), the inhibition of deoxyribose damage was similar to the ethanol-soluble fraction. However, in the Fe^{3+}-EDTA/H_2O_2 system, the ethanol-soluble and insoluble fractions accelerated the oxidation of deoxyribose and in a are concentration dependent manner. The prooxidant effect of the combination was similar to that of the unfractionated XL MRP.

Oxidation of 2'-deoxyguanosine
2'-dG was used to assess the potential prooxidant of XL MRP to cause DNA damage. In the Fenton reaction, 2'-dG is attacked by free radicals forming 8-OH-2'-dG. Figure 2 shows that XL MRP at low concentration (<2.86 mg/mL) exhibited a prooxidant effect on the oxidation of 2'-dG induced by Fe^{3+}-EDTA/H_2O_2/ascorbate. When the concentration was greater than 2.86 mg/mL, the reducing power of XL MRP reached a maximum and then the XL MRP appear to act as scavenger for hydroxyl radicals produced by the Fenton reaction. However, the formation of 8-OH-2'-dG from 2'-dG induced by Fe^{3+}/H_2O_2 increased with increasing concentration of XL MRP up to 11.43 mg/mL where values of 8-OH-2'-dG reached a plateau.

The formation of 8-OH-2'-dG from 2'-dG in the incubation of low molecular weight fraction with Fe^{3+}-EDTA/H_2O_2 is only 0.026 µg. Thus, it is difficult to conclude whether the low molecular weight fraction has an antioxidant or prooxidant effect. Melanoidin fractions can inhibit the oxidation of 2'-dG induced by Fe^{3+}-EDTA/H_2O_2/ascorbate and the inhibition is concentration dependent. The combination of ethanol-soluble and insoluble fractions (79/20, w/w) enhanced the formation of 8-OH-2'-dG from 2'-dG induced by Fe^{3+}-EDTA/H_2O_2. This finding is similar to unfractionated XL MRP (Figure 2). The ethanol-soluble fraction showed the strongest prooxidant effect on oxidation of 2'-dG at a concentration of 4.57 mg/mL in which the concentration of 8-OH-2'-dG was 6.3-fold higher than the control. However, this prooxidant action was decreased by increasing the concentration of ethanol-soluble fraction. The prooxidant effect of the ethanol-insoluble fraction, led to a 5.2 fold higher oxidation of 2'-dG than the control at a concentration of 1.14 mg/mL, at higher concentration a plateau formed.

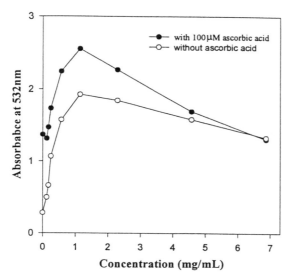

Figure 1. *Effect of xylose-lysine Maillard reaction products on Fe^{3+}-EDTA/H_2O_2-induced deoxyribose damage*

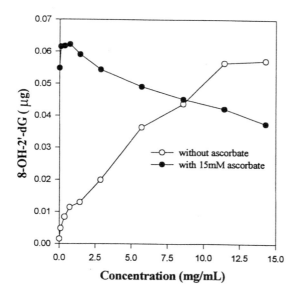

Figure 2. *Effect of xylose-lysine Maillard reaction products on oxidation of 2'-deoxyguanosine to 8-hydroxy-2'-deoxyguanosine*

Acknowledgement
This research work was supported in part by the National Science Council, Republic of China, under grant no. NSC86-2313-B005-104.

References

1. K. Eichner, Antioxidative effect of Maillard reaction intermediates, *Prog. Food Nutr. Sci.*, 1981, **5**, 441-451.
2. T. Gomyo and M. Horikoshi, On the interaction of melanoidin with metal ions, *Agric. Biol. Chem.*, 1976, **40**, 33-40.
3. F. Hayase, S. Hirashima, G. Okamoto and H. Kato, Scavenging of active oxygens by melanoidins., *Agric. Biol. Chem.* 1989, **53**, 3383-3385.
4. G.C. Yen and P.P.Hsieh, Antioxidative activity and scavenging effects on active oxygen of xylose-lysine Maillard reaction products., *J. Sci. Food Agric.*, 1995, **67**, 415-420.
5. M.J. Laguhton, B. Halliwell, P.J. Evans and J.R.S. Hoult, Antioxidant and pro-oxidant actions of the plant phenolics quercetin, gossypol and myricetin, *Biochem. Pharmacol.*, 1989, **38**, 2859-2865.
6. O.I. Aruoma, P.J. Evans, R. Aeschbach and J. Loliger, J. Antioxidant and prooxidant properties of active rosemary constituents: carnosol and carnosic acid, *Xenobiotica*, 1992, **22**, 257-268.
7. O.I. Aruoma, A. Murcia, J. Butler and B. Halliwell, Evaluation of the antioxidant and prooxidant actions of gallic acid and its derivatives, *J. Agric. Food Chem.*, 1993, **41**, 1880-1885.
8. D.D. Kitts, C.H. Wu, H.F. Stich and W.D. Powrie, Effect of glucose-lysine Maillard reaction products on bacterial and mammalian cell mutagenesis, *J. Agric. Food Chem.*, 1993, **41**, 2353-2358.
9. C. Smith, B. Halliwell and O.I. Aruoma, Protection by albumin against the pro-oxidant actions of phenolic dietary components. *Food Chem. Toxicol.* 1992, **30**, 483-489.
10. H. Kasai and S. Nishimura, Hydroxylation of deoxyguanosine at the C-8 position by ascorbic acid and other reducing agents, *Nucleic Acids Res.*, 1984, **12**, 2137-2145.
11. SAS Institute Inc., *SAS User's Guide: Statistics*, SAS Institute Inc., Cary, NC, 1985.
12. J. Kannar, S. Harel and B. Hazan, Muscle membranal lipid peroxidation by an 'iron redox cycle' system: initiation by oxy-radicals and site-specific mechanism, *J. Agric. Food Chem.*, 1986, **34**, 506-511
13. T.C.P. Dinis, V.C.M. Madeira and L.M. Almeida, Action of phenolic derivatives (acetaminophen, salicylate, and 5-aminosalicylate) as inhibitors of menbrane lipid peroxidation and as peroxyl radical scavengers, *Arch. Biochem. Biophys.*, 1994, **315**, 161-169.
14. S.C. Sahu and G.C. Gray, Interactions of flavonoids, trace metal, and oxygen: nuclear DNA damage and lipid peroxidation induced by myricetin, *Cancer Lett.*, 1993, **70**, 73-79.

Health and Disease

A Novel AGE Crosslink Exhibiting Immunological Cross-Reactivity with AGEs Formed *In Vivo*

Yousef Al-Abed and Richard Bucala

THE PICOWER INSTITUTE FOR MEDICAL RESEARCH, 350 COMMUNITY DRIVE, MANHASSET, NEW YORK 11030, USA

Summary

In a model reaction, we selected as an AGE target the dipeptide N^{α}-Z-arg-lys. The proximity of the arginine and lysine residues promotes stable intramolecular crosslink formation. Incubation of N^{α}-Z-arg-lys with 10 equivalents of glucose in 0.2 M phosphate buffer (pH 7.4) at 37 °C for five weeks produces at least 25 distinct reaction products upon fractionation of this mixture by HPLC. Each fraction was isolated, concentrated, and analysed for reactivity with a polyclonal anti-AGE antibody (RU) that has been shown previously to recognize a class of AGEs that increase as a consequence of hyperglycemia and which are inhibited from forming in human subjects by treatment with aminoguanidine. The products present within one fraction (1.5% yield) were found to block antibody binding in a dose-dependent manner. Further purification of this fraction by HPLC revealed the presence of one major immunoreactive compound (0.6% yield). Characterization of this adduct by UV, ESMS and ^{1}H-NMR spectra revealed the presence of an intramolecular arg-lys-imidazole crosslink. This crosslink is non-fluorescent and acid-labile and may represent an important class of immunoreactive AGE-crosslink that forms *in vivo*.

Introduction

The initial event in protein glycation is the reaction of a reducing sugar such as glucose with the N-terminus of a protein or the ε-amino group of a lysine to form an aldimine, or Schiff base. The Schiff's base can hydrolyse back to its reactants or undergo an Amadori rearrangement to form a more stable N^{ϵ}-(1-deoxy-1-fructosyl)lysine (Amadori product, AP). The reaction pathway leading to reactive crosslinking moieties (i.e., AGE formation) commences by further rearrangement or degradation of the AP. Possible routes leading from AP precursors to glucose-derived protein crosslinks have been suggested only by model studies examining the fate of the AP *in vitro*. One pathway proceeds by loss of the 4-hydroxyl group of the AP by dehydration, to give a 1,4-dideoxy-1-alkylamino-2,3-hexodiulose (AP-dione). An AP-dione with the structure of an amino-1,4-dideoxyosone has been isolated by trapping model APs with aminoguanidine, an inhibitor of AGE formation.[1] Subsequent elimination of the 5-hydroxy then gives a 1,4,5-trideoxy-1-alkylamino-2,3-hexulos-4-ene (AP-ene-dione), which has been isolated as a triacetyl derivative of its 1,2-enol form.[2] Both AP-diones and AP-ene-diones would be expected to be highly reactive in protein crosslinking; for instance, by serving as targets for the addition of the guanidino moiety of arginine or the ε-amino group of lysine.

Dicarbonyl-containing compounds such as methylglyoxal, glyoxal and deoxyglucosones are known to participate in condensation reactions with the side-chains of arginine and lysine. For example, the addition of methylglyoxal to the guanidino moiety of arginine leads to the formation of imidazol-4-one[3] and pyrimidinium[4] adducts. In an important study, Sell and Monnier[5] isolated from human dura collagen the fluorescent crosslink pentosidine, which is a condensation product of lysine, arginine and a reducing sugar. The mechanism of pentosidine formation remains uncertain but crosslinking requires that the lysine-bound, glucose-derived intermediate contain a dicarbonyl functional group that can react irreversibly with the guanidino group of arginine.

Several lines of evidence have established that AGEs exist in living tissue,[6] although the identity of the major AGE crosslink(s) that forms *in vivo* remains uncertain. Nevertheless, recent pharmacologically based data have affirmed the importance of the AP-dione pathway in stable crosslink formation.[7] The lack of data concerning the structure of AGEs has been attributed to the lability of AGE crosslinks to the standard hydrolysis methods employed to remove the protein backbone, and to the possible structural heterogeneity of the crosslinks themselves. Moreover, the pathologically relevant crosslinks may not themselves be fluorescent,[8] a property that has been historically associated with AGE formation and almost universally used as an indicator of the Maillard reaction *in vivo*.

Several years ago, we utilized hyperimmunization techniques directed against an AGE-crosslinked antigen to produce both polyclonal and monoclonal antibodies that recognize AGEs formed *in vivo*. These antibodies made possible the development of immunohistochemical and ELISA-based technologies that were free of the specificity and other technical problems associated with prior fluorescence-based assays, and enabled the first sensitive and quantitative assessment of advanced glycation in living systems. Remarkably, these anti-AGE antibodies were found to recognize a class of AGEs that was prevalent *in vivo* but immunochemically distinct from previously characterized structures such as 4-furanyl-2-furoyl-1H-imidazole (FFI), carboxy-methyllysine (CML), 1-alkyl-2-formyl-3,4-diglycosylpyrrole (AFGP), pentosidine and pyrraline.[9] Of importance, the AGE epitope recognized by these antibodies increased as a consequence of diabetes or protein aging in the case of several proteins such as collagen, hemoglobin, and LDL.[9-12] One polyclonal antibody species, designated 'RU', has been employed in human clinical studies in which the formation of immunoreactive AGEs was inhibited by administration of the pharmacological inhibitor, aminoguanidine;[10,12] 'RU' has also been used to provide important prognostic information in diabetic renal disease.[13]

To ascertain the epitope structure of this AGE and to obtain insight into the potential structures of important *in-vivo* crosslink(s), we used the RU anti-AGE antibody as a molecular probe to isolate novel products contained within crude AGE-containing reaction mixtures. We report herein the characterization of a non-fluorescent, imidazole-based AGE that may represent an important class of crosslink that forms *in vivo*.

MATERIALS AND METHODS

Analytical methods
NMR spectra were recorded in D_2O on a General Electric Q-300 (300 MHz) spectrometer. HPLC was conducted using a Hewlett Packard Model 1090 and Waters Instruments (Waters 626 pump and Waters 490 E multiwavelength detector). Electrospray ionization (ESI) samples were run on a Quattro triple quadrupole mass spectrometer. Loop injection sampling used an ABI model 140B syringe pump employing H_2O-CH_3CN (1:1) at a flow rate of 15 μL/ min, a Rheodyne model 7125 valve with a 10 μL loop, and a Micromass Megaflow ESI probe using nitrogen for the nebulizer/drying gas. LC/MS samples were run employing the ABI pump and Rheodyne valve with a 20 μL loop at a flow rate of 50 μL/min.

Preparation of Arg-Lys-Imidazole (ALI)
To a solution of N^α-Z-arg-lys (1g, 0.013 mmol) in 10 mL aqueous 0.2 M phosphate buffer (pH 7.4), was added D-glucose (0.13 mmol). The reaction mixture was stirred at 37 °C for five weeks. At intervals, 10 μL of the reaction mixture was analyzed by HPLC using an analytical primesphere column (5C18 MC, 5 micron, 250 x 21.2 mm, Phenomenex, Torrance, CA) and

a binary solvent gradient consisting of 0.05% trifluoroacetic acid in H_2O (solvent A), and methanol (solvent B). Solvent was delivered at a flow rate of 10 mL/min as follows. From 0-30 min: a linear gradient from A:B (95:5) to A:B (25:75); from 30-45 min: a linear gradient from A:B (25:75) to (0:100). Detection was by monitoring UV absorption at λ 254 nm. The AGE crosslink eluted as a mixture of three components at 34.0 min. Further purification of this subfraction using the same method gave the desired compound in a high purity (>95%).

AGE ELISA

HPLC fractions and purified compounds were analysed by an AGE-specific ELISA following methods described previously.[9] This ELISA employed a polyclonal anti-AGE antibody raised by hyperimmunization against a heavily AGE-crosslinked preparation of ribonuclease. Total IgG was prepared by protein-G affinity chromatography and the ribonuclease backbone specificities removed by immunoadsorption. For assaying AGE immunoreactivity, 96-well round-bottom microtitre plates (EIA/RIA plate, Costar, Cambridge, MA) were first coated with AGE-BSA (3 mg/mL, dissolved in 0.1 M sodium bicarbonate, pH 9.6) by incubation overnight at 4°C. After washing, the unbound sites were blocked with SuperBlock™ following the manufacturer's recommendations (Pierce, Rockford, IL). After washing, dilutions of test antigen together with anti-AGE IgG were added and the plates incubated at room temperature for 1 h. The plates were then washed again and incubated with a secondary antibody (alkaline phosphatase-conjugated anti-rabbit IgG) at 37°C for 1 h. The unbound antibodies were removed by extensive washing and bound antibodies were detected by incubation with *p*-nitrophenyl phosphate (pNPP) substrate for 30-60 min, and recording the optical density at 405 nm by an ELISA reader (EL309, Bio-Tek Instruments Inc., Burlington, VT). Results were expressed as B/B_0, calculated as [experimental OD - background OD (*i.e.* no antibody)] / [total OD (i.e. no competitor) - background OD].

RESULTS

We selected the dipeptide N^α-Z-arg-lys as an AGE target in our model. The close association of the arginine and lysine residues enables a proximity effect that promotes crosslink formation. Moreover, a synthetic strategy employing an arg-lys dipeptide was used successfully in the past to isolate a cyclic pentosidine in high yield.[14] We incubated N^α-Z-arg-lys (13 mmol) with glucose (130 mmol) in 0.2 M phosphate buffer (pH 7.4) for five weeks at 37°C. At least 25 distinct reaction products were identified upon fractionation of this mixture by reversed-phase HPLC. Each fraction was isolated, concentrated, and analysed for its reactivity with anti-AGE antibody by ELISA.[9] The product(s) contained within one distinct fraction, present in 1.5% yield, were found to block antibody binding in a dose-dependent manner. Further purification of this fraction by HPLC revealed the presence of one major, immunoreactive compound (0.6% yield) together with two minor ones. The UV and fluorescence spectrum of the isolated, major product was unremarkable and similar to that of the starting material. The ESMS spectrum displayed a molecular ion of m/z = 545 ($[MH]^+$), an increase of 109 daltons compared with the starting material, N^α-Z-arg-lys (MW= 436). A ^{1}H-NMR spectrum in D_2O showed, in addition to the N^α-Z-arg-lys protons, five aliphatic protons that resonate as a multiplet between 2.35-4.65 ppm (5H), and an olefinic proton that resonates at 7.3 ppm within the Z-group. Overall, these data are consistent with the structure of a cyclic, arg-lys imidazole crosslink (ALI). We propose the mechanism of formation outlined in Figure 1. The first step is dehydration of the lysine-derived AP (which we have determined to form in 18% yield under these reaction conditions) to give the obligate, AP-dione reactive intermediate. Reversible addition of the guanidino moiety to the dicarbonyl yields the 2-amino-4,5-dihydroxyimidazole adduct, which then undergoes dehydration to deliver the stable cyclic ALI. Of importance, this crosslink is non-fluorescent, acid-labile, and its formation can be inhibited by aminoguanidine.

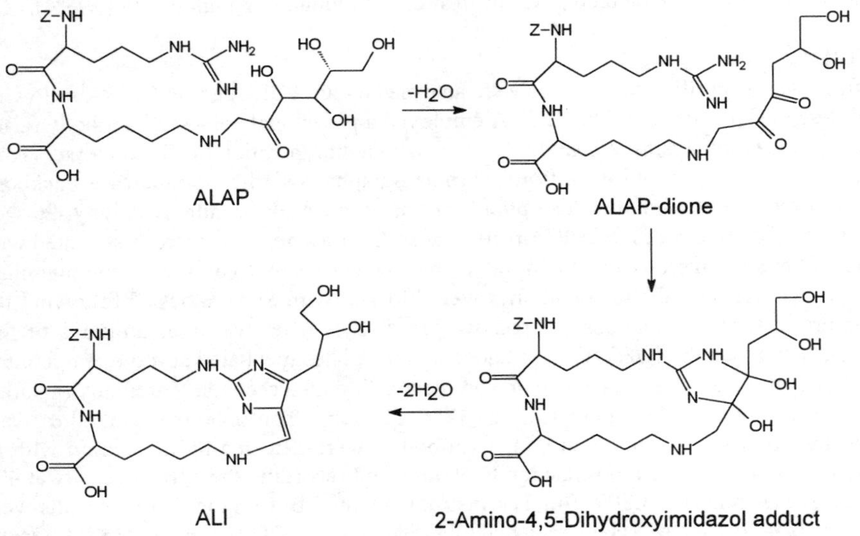

Figure 1. Proposed mechanism of formation of the cyclic arginine-lysine imidazol (ALI) crosslink from the lysine-derived Amadori product (ALAP).

Figure 2 shows the dose-dependent reactivity of synthetic ALI in the anti-AGE antibody-based ELISA. The binding curve is steep and shows 50% inhibition at 500 nmoles. For comparison purposes, we also studied the reactivity of FFI, CML, AFGP, pentosidine and cyclic pentosidine, and ligands with related epitope structure such as N^{α}-Z-arg-lys, N^{α}-Z-arg-lys-AP, an imidazolium adduct, a pyrimidinium adduct, histidine, and lys-his. With the exception of ALI, none of these compounds showed detectable cross-reactivity with the RU anti-AGE antibody shown previously to react with AGEs formed *in vivo*.

DISCUSSION

Despite the increasing body of data implicating advanced glycation in the etiology of such age- and diabetes-related conditions as atherosclerosis, renal insufficiency, and amyloid deposition, elucidation of the structure(s) of the pathologically important, AGE-crosslinks has been a challenging problem. This situation has been underscored by the recent development of high-titre specific antibodies that recognize a class of AGEs formed *in vivo* which have prognostic utility and which are inhibited from forming in patients treated with the AGE inhibitor aminoguanidine.[10,13]

In the past, investigations of AGEs that form *in vivo* have necessarily relied on chemical methods to separate the crosslinking moieties from their macromolecular backbones. These studies have led to a recognition that the major crosslinks that form *in vivo* are largely acid-labile and non-fluorescent.[5,6,8] In the present study, we have exploited the specificity of anti-AGE antibodies reactive with AGEs formed *in vivo* to identify novel crosslinking moieties contained within a synthetic mixture of AGEs. By this selection method, we found a single, immunoreactive AGE that formed in 0.6% yield in a synthetic mixture consisting of glucose and the N^α-blocked dipeptide, N^α-Z-arg-lys, as reactants. ALI was highly reactive with the anti-AGE antibody 'RU' and showed a steep binding curve at nmole amounts.

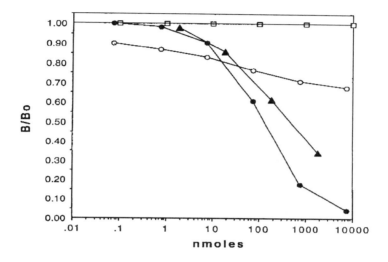

Figure 2. ELISA competition curves for the 'RU' polyclonal anti-AGE antibody "RU". Assays employed glucose-derived AGE-BSA as the adsorbed antigen. (□): FFI,[18] CML,[19] AFGP,[20] pyrraline,[5] pentosidine, cyclic pentosidine,[14] N^α-Z-arg-lys, N^α-Z-arg-lys-AP, histidine, lys-his, pyrimidinium adduct[4] and imidazolium adduct;[4,22] (○): CML-BSA; (▲): ALI; (●): AGE-BSA. nmoles: quantity of competitor added to each assay.

Among the biological consequences of AGE formation, the formation of stable crosslinks may be considered their most important pathological manifestation. We propose that ALI, an imidazole-based AGE, is a major species of the pathologically important AGE crosslinks that form *in vivo*. The mechanism of formation of ALI points to the importance of the AP-dione as a critical, reactive intermediate and further affirms prior, pharmacologically-based studies that have implicated this intermediate, as well as its dehydration product, the AP-ene-dione in irreversible, protein-protein crosslinking.[7] These intermediates have also been implicated in the formation of the cyclization product cypentodine,[15] which may display sufficient redox potential to participate in the oxidative reactions associated with phospholipid advanced glycation.[16,17]

In conclusion, the identification of the non-fluorescent crosslink ALI as an important AGE crosslink that forms *in vivo* fills an important gap in our understanding of the advanced glycation process. This finding will facilitate further clinical assessment of pathologically important AGEs and aid in the design of more specific pharmacological inhibitors of AGE formation.

REFERENCES

1. H.-J. C. Chen and A. Cerami, *J. Carbohydrate Chem.*, 1993, **12**, 731.
2. L. Estendorfer, F. Ledl and T. Severin, *Angew. Chem. Ent. Ed. Engl.*, 1990, **29**, 536.
3. T. W. C. Lo, M. E. Westwood, A. C. McLellan, T. Selwood and P. J. Thornalley, *J. Biol. Chem.*, 1994, **269**, 32299.
4. Y. Al-Abed, T. Mitsuhashi, P. Ulrich, and R. Bucala, *Bioorg. Med. Chem. Lett.*, 1996, **6**, 1577.
5. D. R. Sell and V. M. Monnier, *J. Biol. Chem.*, 1989, **264**, 21597.
6. R. Bucala and A. Cerami, *Adv. Pharm.*, 1992, **23**, 1.
7. S. Vasan, X. Zhang, X. Zhang et al., *Nature*, 1996, **382**, 275.
8. D. G. Dyer, J. A. Dunn, et al., *J. Clin. Invest.*, 1993, **91**, 2463.
9. Z. Makita, H. Vlassara, A. Cerami and R. Bucala, *J. Biol. Chem.*, 1992, **267**, 1992.
10. Z. Makita, H. Vlassara, E. Rayfield, et al., *Science*, 1992, **258**, 651.
11. B. H. R. Wolffenbuttel, D. Giordano, H. W. Founds and R. Bucala, *The Lancet*, 1996, **347**, 513.
12. R. Bucala, Z. Makita, G. Vega, et al., *Proc. Natl. Acad. Sci. U.S.A.*, 1994, **91**, 9441.
13. P. J. Beisswenger, Z. Makita, T. J. Curphey, et al., *Diabetes*, 1995, **44**, 824.
14. Y. Al-Abed, P. Ulrich, A. Kapurniotu, E. Lolis and R. Bucala, *Bioorg. Med. Chem. Lett.*, 1995, **5**, 2929.
15. X. Zhang and P. Ulrich, *Tetrahedron Lett.*, 1996, **37**, 4667.
16. R. Bucala, Z. Makita, T. Koschinsky and A. Cerami, *Proc. Natl. Acad. Sci. U. S. A.*, 1993, **90**, 6434.
17. R. Bucala, *Redox Reports*, 1996, **2**, 291.
18. S. Ponger, P. Ulrich, F. A. Bencsath, A. Cerami, *Proc. Natl. Acad. Sci. U. S. A.*, 1984, **81**, 2684.
19. M. U. Ahmed, S. R. Thorpe, and J. W. Baynes, *J. Biol. Chem.*, 1986, **261**, 4889.
20. J. G. Farmar, P. C. Ulrich and A. Cerami, *J. Org. Chem.*, 1988, **53**, 2346.
21. F. G. Njoroge, L. M. Sayre, V. Monnier, *Carbohydrate Res.*, 1987, **167**, 211.
22. E. Brinkmann, K. J. Well-Knecht, S. R. Thorpe and J. W. Baynes, *J. Chem. Soc. Perkin Trans. I*, 1995, 2817.

Modeling of Protein and Aminophospholipid Glycation Using Low Molecular Weight Analogs. A Comparative Study

O. K. Argirov, I. I. Kerina, J. I. Uzunova[1] and M. D. Argirova[1]

DEPARTMENT OF ORGANIC CHEMISTRY, UNIVERSITY OF PLOVDIV, 24 TSAR ASSEN STREET, 4000 PLOVDIV, BULGARIA; [1] DEPARTMENT OF CHEMISTRY & BIOCHEMISTRY, HIGHER MEDICAL INSTITUTE, 15A VASSIL APRILOV STREET, 4000 PLOVDIV, BULGARIA

Summary
(2-aminoethyl)phenethylphosphate and (2-aminoethyl)ethylphosphate were synthesized as models of aminophospholipids. Their glycation by glucose and fructose was compared with that of appropriate peptides (Gly-Phe, Gly-Gly, Gly-Lys) and amino compounds (2-aminoethanol, β-alanine). Formation of glycation products was studied by electron spectroscopy and reversed-phase HPLC. The results obtained confirm that amino groups in aminophospholipids are reactive enough to form glycation products under physiological conditions.

Introduction

Nonenzymatic amino-carbonyl reactions have been intensively studied over the past two decades especially in relation to their biomedical importance. It is now generally recognized that a relationship exists between glycation and some disease states like the pathology of diabetes,[1] cataract development,[2] etc.

In most studies of glycation, the amino components have been proteins, amino acids or peptides. Recently published studies,[3,4] suggest that carbohydrates can also modify aminophospholipids by reacting with their amino groups both *in vitro* and *in vivo*. Such reactions have been suggested to promote lipid oxidation as a result of glycation. As aminophospholipids are important components of cell membranes, their glycation could have a significant effect on membrane permeability, fluidity, etc.

The complex structure of natural aminophospholipids makes it difficult to study the chemical changes that accompany their glycation. In the present work, we prepared low molecular weight analogs of the natural phosphatidylethanolamines and compared their glycation with that of peptides. The compounds: (2-aminoethyl)phenethylphosphate(APP), **1**; (2-aminoethyl)ethylphosphate(AEP), **2**; Gly-Phe, **3**; Gly-Gly, **4**; 2-aminoethanol, **5**; β-alanine, **6**; and Gly-Lys, **7,** were used as model amino compounds in the present study.

The compounds **5** and **6** were included because they have structural elements in common with **1** and **2**. The peptide **7** corresponds to the N-terminus of γ-II-crystallin - one of the proteins in which glycation has been extensively studied.

Materials and Methods

Reagents
D-Glucose and D-fructose, $POCl_3$, 1,4-dioxane, triethylamine and phenethyl alcohol were purchased from Aldrich. Gly-Phe, Gly-Gly, β-alanine and 2-aminoethanol, were obtained from Sigma. 1,4-dioxane and triethylamine were dried over CaH_2 prior to use.

Analysis
HPLC separations were conducted on Eurospher C18 (100A, 5μ, 250x4mm; Knauer) column, installed on an isocratic system: pump, Well Chrom K-1000; variable wavelength monitor, Well Chrom K-2500(Knauer). The separation conditions were: eluent, 50% methanol containing 0.1% trifluoroacetic acid; detection at 254 nm; flow rate, 1 mL/min.

The spectra of the samples were measured on a spectrophotometer (Cary 1, Varian) in the region 200-500 nm in cells with an optical path length of 1 cm.

Synthesis of AEP
A solution containing 1.17 mL ethanol (20 mmol) and 2.77 mL triethylamine(20 mmol) in 10 mL dioxane was added dropwise to a stirred solution of 1.83 mL (20 mmol) $POCl_3$ in 10 mL dioxane at 4 °C. The reaction mixture was left for 4 hours at room temperature. A solution of 1.22 g 2-aminoethanol (20 mmol) and 5.54 mL triethylamine (40 mmol) in 15 mL dioxane was added and the system was left overnight at room temperature. The following day the reaction mixture was filtered under vacuum and washed with 20 mL dioxane. The filtrate obtained was concentrated *in vacuo* and 1 mL water was added to the residue. After 5 hours, the sample was dried *in vacuo* overnight. 5 mL acetone was added to the product and the crystals formed were collected by filtration. The product was recrystallized from methanol (yield, 0.55 g; mp, 226-228°C, literature:[5] 223-225 °C). The structure of the product was confirmed by infrared and ^{1}H-NMR spectral data.

Synthesis of APP
The product was obtained in an analogous way to the synthesis of AEP. Phenethyl alcohol (2.44 g, 20 mmol) was used instead of ethanol (yield, 0.81 g). The structure of the product was confirmed by ultraviolet, infrared and ^{1}H-NMR spectral data.

Model glycation experiments
Solutions containing 50 mM carbohydrate(glucose or fructose), 50 mM amino compound (**1-7**), 0.05% NaN_3, and 1 mM EDTA in 200 mM phosphate buffer (pH 6.9) were incubated for 30 days at 37°C. Periodically aliquots of 0.5 mL were taken, diluted to 4.0 mL and their UV - visible spectra were recorded. Portions (10 μl) of the reaction systems containing Gly-Phe or APP were analysed by HPLC as described above.

Results and Discussion

Synthesis of model phosphates
The model phosphates were prepared according to the following reaction scheme:

$$\text{R-OH} + \text{POCl}_3 \xrightarrow{-\text{HCl}} \text{R-O-POCl}_2 \xrightarrow{\text{H}_2\text{N-CH}_2\text{-CH}_2\text{-OH}}_{-2 \text{HCl}}$$

$$\longrightarrow \text{R-O-P}\underset{\underset{H}{N}}{\overset{O\;\;O}{\|/}} \xrightarrow{\text{H}_2\text{O}} \text{R-O}-\underset{O^-}{\overset{\overset{O}{\|}}{P}}-\text{O-CH}_2\text{-CH}_2\text{-NH}_3^+$$

$$\text{R-} = \text{CH}_3\text{-CH}_2\text{-} \, , \, \text{C}_6\text{H}_5\text{-CH}_2\text{-CH}_2\text{-}$$

The synthesis is based on an approach proposed by Arnold and Bourseaux[5]. A benzene ring was incorporated in the case of compound 1 as it is an useful spectral marker. The hydrophobic phenyl group allows for better separation by reversed-phase HPLC. In contrast to the natural aminophospholipids, AEP and APP are soluble in water, making it possible to incubate the model aminophosphates and peptides under identical conditions.

Changes in the spectra of model systems as a result of glycation
No absorbance in the region above 280 nm was seen in the spectrum of any reaction system at the beginning of reaction. The absorbance of the samples gradually increased during the course of glycation. The spectral curves obtained after 30 days of incubation are given in Figure 1.

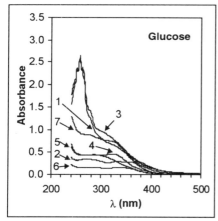
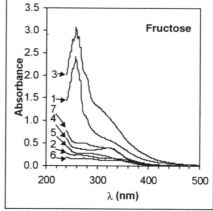

Figure 1. S*pectra of reaction mixtures obtained after 30 days of incubation. Amino compounds: 1, APP; 2, AEP; 3, Gly-Phe; 4, Gly-Gly; 5, 2-aminoethanol; 6, β-alanine; 7, Gly-Lys*

All spectral curves show a shoulder in the region 320-340 nm due to the products of glycation. If we use the absorbance in this region as a measure of the reactivity of the model compounds, we find that benzene-containing substances (APP and Gly-Phe) are among the most reactive ones toward glucose and fructose. This result is interesting because the benzene ring is not known to take part in glycation directly.

The substances AEP, β-alanine and 2-aminoethanol have similar structure close to the reacting amino group: $H_2N-CH_2-CH_2-$. The first two compounds are less reactive toward glucose and fructose, probably because of their zwitterionic structure which could diminish the reactivity of the $-NH_2$ group.

Comparison of the model aminophosphates **1** and **2** with their corresponding peptides **3** and **4** shows that both compounds form glycation products but peptides can be up to two times more reactive. The greater reactivity of peptides could be related to activation of their amino groups by the neighboring peptide carbonyl group. On the basis of this result one can expect that protein amino groups (especially N-terminal ones) will be more reactive than those of aminophospholipids. On the other hand, proteins have much higher molecular weights than aminophospholipids and therefore, the latter could be an important target for the carbohydrate molecules in living systems. In addition, the orientation of the molecules of aminophospholipids in the lipid bilayer makes their amino groups accessible for attack, while some proteins amino groups can be sterically hindered because of the tertiary structure of polypeptide chains.

Gly-Lys (**7**) reacts with fructose as quickly as Gly-Gly (**4**). In the reaction with glucose, however, it shows higher reactivity. This could be due to a cooperative interaction of the two amino groups of **7**. One can expect that the regular orientation of the aminophospholipid molecules in the lipid bilayer can favor intermolecular cooperative interactions of two amino groups, especially if the content of aminophospholipids is relatively high.

HPLC analyses of reaction mixtures containing APP (1) and Gly-Phe (3)
The chromatograms obtained after 28 days of incubation of reaction mixtures containing Gly-Phe and APP are given in Figure 2. In all cases, the main products of reaction have lower retention times than the starting amino compounds. This can be explained by the attachment of the more hydrophilic carbohydrate molecule to the amino compound. The appearance of small peaks with longer retention times than the starting amino compound can be attributed to low-molecular weight intermediates of glycation.

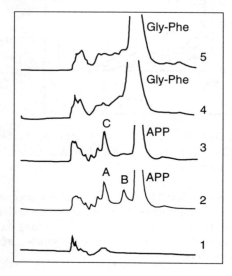

Figure 2. *HPLC chromatograms of the reaction buffer, 1 and reaction mixtures obtained after 28 days of incubation: Glc+APP, 2; Fru+APP, 3; Glc+Gly-Phe, 4; Fru+Gly-Phe, 5*

The data in Figure 2 show that complex mixtures appear as a result of peptide glycation. A number of compounds are formed with poorly resolved peaks.

The systems containing APP contain fewer products and the peaks are relatively better resolved. In the system APP+Glc there are two new main peaks marked A and B. In the system APP+Fru there is a peak marked C. The same substance might be responsible for peaks A and C. The chromatograms 2 and 3 (Figure 2) suggest it might be possible to isolate some of the aminophosphate glycation intermediates by reversed-phase chromatography.

The results obtained confirm that amino groups in aminophospholipids are reactive enough to form glycation products under physiological conditions.

Acknowledgements
This research work was supported through the European Commission program PECO, as a part of the Concerted Action Project "Role of membranes in lens ageing and cataract".

References

1. P. J. Thornalley, Advanced glycation and the development of diabetic complications. Unifying the involvement of glucose, methylglyoxal and oxidative stress, *Endocrinol. and Metab.*, 1996, **3**, 149-166.
2. E. C. Abraham, R. E. Perry, A. Abraham and M. S. Swamy, Proteins of urea-soluble high molecular weight (HMW) aggregates from diabetic cataract: Identification of in vivo glycation sites, *Exp. Eye Res.*, 1991, **52**, 107-112.
3. R. Bucala, Z. Makita, T. Hoschinsky, A. Cerami and H. Vlassara, Lipid advanced glycosylation: Pathway for lipid oxidation *in vivo*, *Proc. Natl. Acad. Sci. USA*, 1993, **90**, 6434-6438.
4. A. Ravandi, A. Kuksis, L. Marai, J.J. Myher, G. Steiner, G. Lewisa and H. Kamido, Isolation and identification of glycated aminophospholipids from red cells and plasma of diabetic blood, *FEBS Letters*, 1996, **381**, 77-81.
5. H. Arnold and F. Bourseaux, Synthese und Abbau cytostatisch wirksamer cyclischer N-Phosphamidester des Bis-(β-chloräthyl)-amins, *Angew. Chem.*, 1958, **70**, 539-544.

Protein Modifications by Glyoxal and Methylglyoxal During the Maillard Reaction of Higher Sugars

Marcus A. Glomb[1] and Ramanakoppa H. Nagaraj[2]

[1]TECHNICAL UNIVERSITY OF BERLIN, INSTITUTE OF FOOD CHEMISTRY, GUSTAV-MEYER-ALLEE 25, 13355 BERLIN, GERMANY. [2]UNIVERSITY HOSPITALS OF CLEVELAND, CENTRE OF VISION RESEARCH, DEPARTMENT OF OPHTHALMOLOGY, CLEVELAND, OHIO, USA

Summary

The modification of N^α-t-BOC-lysine, N^α-t-BOC-arginine and lens crystallins by glyoxal and methylglyoxal has been investigated under physiological conditions. In incubations involving glyoxal, a novel amide-type crosslink (GX-amide-crosslink) was identified and correlated with N^ϵ-carboxymethyllysine (CML) levels. The formation of GX-amide-crosslink should proceed via the same reaction pathways as for CML. In incubations with methylglyoxal, the formation of N^δ-(5-hydroxy-4,6-dimethylpyrimidine-2-yl)-ornithine (argpyrimidine) was established. Independent mechanistic studies point towards a synthesis via an intermediate reductone with a five-membered carbon backbone. GX-amide-crosslink and argpyrimidine were also formed from higher sugars, suggesting they may play a role in protein modification *in vivo*.

Introduction

α-Dicarbonyl compounds like glyoxal and methylglyoxal are established to be important intermediates of the Maillard reaction.[1] The main source of glyoxal formation from higher sugars should result from retro-aldol fragmentation of the glycosylamine or Schiff base. The resulting imine of glycolaldehyde is oxidized to amine-glyoxal derivatives via an intermediate pyrazine radical cation. This concept was supported by mechanistic studies on the formation of CML. Autoxidation of glucose or oxidative fragmentation of the Amadori product as a source for the formation of glyoxal was confirmed to be of minor importance.[2] As yet, there is no conclusive support for the involvement of glyoxal in protein modification *in vivo*.

Methylglyoxal has been confirmed as a product of the Maillard reaction *in vitro*. *In vivo* methylglyoxal results, in addition to protein glycation, also through various enzymatic and non-enzymatic pathways, as the enzymatic oxidation catalyzed by cytochrome P450 2E1 within the acetone metabolism. The concentration of methylglyoxal was increased in lens, kidney and blood of streptozotocin-induced diabetic rats and in blood samples of diabetic patients.[3]

Glyoxal and methylglyoxal have been shown to modify protein-bound arginine, lysine and cysteine residues. This study was undertaken to gain further insight into the resulting structures.

Materials and Methods

Sugars and amino acids were obtained from Sigma Chemical Co. Authentic standards of the GX-amide-crosslink, argpyrimidine and CML were synthesized following literature procedures.[2,4,5] High-performance liquid chromatography (HPLC) was conducted using a reversed-phase RP18 column with a methanol-water and a propanol-water gradient system.

Ion-pair reagents were heptafluorobutyric acid and sodium dodecylsulfate, respectively. For detection of argpyrimidine, the fluorescence detector was set at 320 (ex.)/380 (em.) nm. To analyse for CML and the GX-amide-crosslink the eluent was subjected to postcolumn derivatization with o-phthaldialdehyde and monitored fluorimetrically at 340 (ex.)/455 (em.) nm.[2] Incubations were carried out under sterile conditions.

Results and Discussion

Protein modification by glyoxal

The guanidino group of arginine reacts with glyoxal resulting in a single derivative (Figure 1), a dihydroxyimidazoline (1), which can be quantified after acid hydrolysis as an imidazolinone (2).[6] Glyoxal-lysine reaction products are C2-Imines (3).[2] It is speculated that these labile intermediates degrade to stable modifications such as imidazolium structure (4) and N^{ε}-carboxymethyllysine (CML).[2,7] In the case of incubations of glucose with an amine *in vitro*, the contribution of glyoxal to the formation of CML is estimated as 50%. The remaining 50% of CML is formed via oxidative cleavage of the Amadori product.[8]

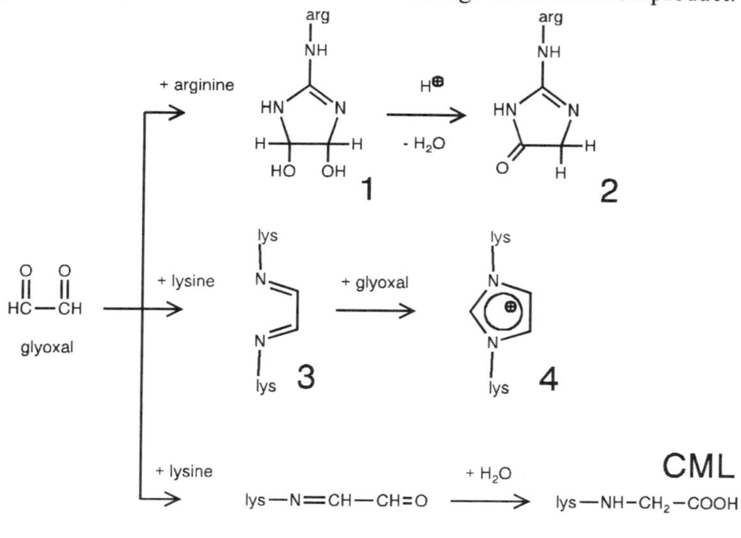

Figure 1. *Reaction of glyoxal with arginine and lysine*

An understanding of the mechanism of CML formation is important as *in vivo* CML represents the most important advanced glycation endproduct (AGE) isolated so far. Starting from the imine (5), formation of the hydrate and a subsequent hydride shift should lead to carboxymethylation (Figure 2). This Cannizzaro-type intramolecular disproportionation reaction is reminiscent of the formation of saccharinic acids under alkaline conditions. Indeed, the formation of CML is favored at high pH, while acidic conditions significantly suppress the reaction.

Importantly, CML formation starting from higher sugars is strictly dependent on oxidative conditions, as is the crosslinking of proteins during the Maillard reaction. This led to the hyphothesis that CML formation and the crosslinking phenomenon may be related. If amine structures are included in the reaction scheme (Figure 2), synthesis of CML should be accompanied by amide (6) and amidine (7) crosslink structures via the corresponding aminals. This class of molecules would fit in with the hypothesis that the main crosslink structures are non-chromophores, labile to acid and base protein hydrolysis and cannot be stabilized through sodium borohydride reduction.

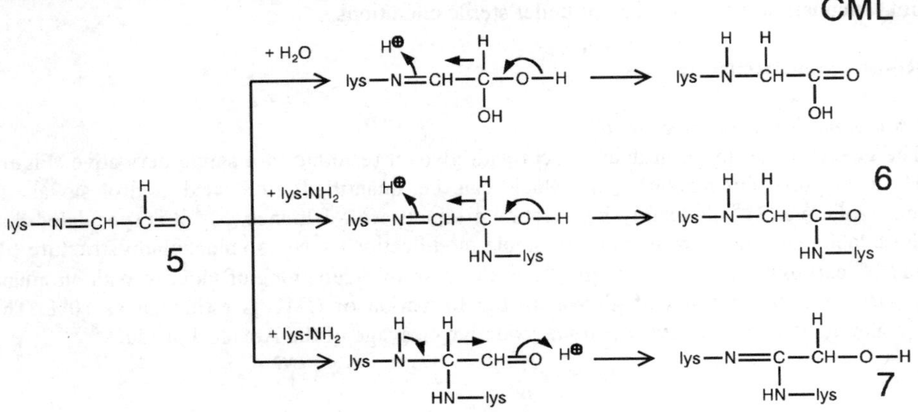

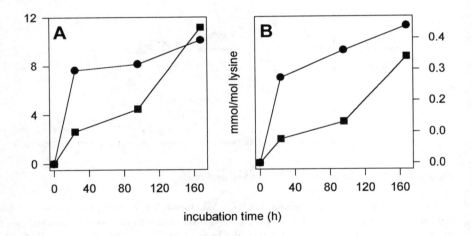

Figure 2. *Mechanism of CML, GX-amide (6) and GX-amidine (7) formation*

Figure 3. *Formation of CML (A) and the GX-amide-crosslink (B) in incubations of N^α-t-BOC-lysine (40 mM) with glyoxal (■) and glycolaldehyde (●) (both 20 mM) at 37°C in 0.2 M phosphate buffer pH 7.4. The incubation was stopped at various time points by reduction with $NaBH_4$ followed by addition of HCl.*

An authentic standard of the putative glyoxal amide crosslink (6), 2,15-diamino-7,10-diaza-8-oxo-hexadecane-1,16-dioic acid (GX-amide-crosslink), was obtained starting from

N^α-t-BOC-N^ε-CBZ-lysine in a 6-step synthesis.[4] The structure was verified by ^{1}H/^{13}C-NMR and high-resolution FAB-MS. GX-amide-crosslink could be identified and quantified in incubations of N^α-t-BOC-lysine with glyoxal, glycolaldehyde and ribose (200 mM) at 37°C and pH 7.4. The results were confirmed on two different HPLC systems using o-phthaldialdehyde postcolumn derivatization with fluorescence detection. Starting from glyoxal/glycolaldehyde the pattern of formation of GX-amide-crosslink is similar to that of CML, but at values ~20 times smaller (Figure 3). Preliminary experiments using ribose indicate a synthesis rate ~40 times less than that of CML.

Protein modification by methylglyoxal

Methylglyoxal has been shown to modify lysine residues to form carboxyethyllysine (8) and, more recently, imidazolysine (9), a lysine-lysine crosslink.[9,10] Arginine residues react to form an imidazolinone (10),[11] which has been proposed to be oxidized to a fluorescent imidazolone (11) (Figure 4).[12]

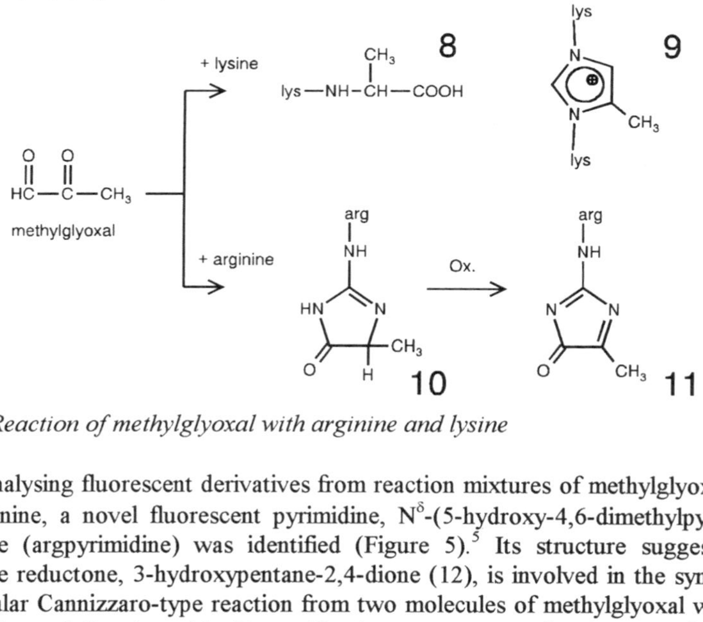

Figure 4. *Reaction of methylglyoxal with arginine and lysine*

Upon analysing fluorescent derivatives from reaction mixtures of methylglyoxal and N^α-t-BOC-arginine, a novel fluorescent pyrimidine, N^δ-(5-hydroxy-4,6-dimethylpyrimidine-2-yl)-ornithine (argpyrimidine) was identified (Figure 5).[5] Its structure suggests that an intermediate reductone, 3-hydroxypentane-2,4-dione (12), is involved in the synthesis. The intermolecular Cannizzaro-type reaction from two molecules of methylglyoxal would entail the formation of formic acid. To verify the structure, authentic argpyrimidine was synthesized from 3-O-acetyl-2,4-pentanedione (13) in high yields and compared with material obtained from incubation mixtures.

Reductones are extremely sensitive to oxidation and fragmentation reactions. Therefore, in order to support the proposed reaction mechanism, 2,4-ethoxy-3-hydroxypentane-1,4-diene (14) was synthesized as a stable precursor of reductone (12). The reductone could be freed under mild acidic conditions to yield 8 times more argpyrimidine than when the incubation was started with methylglyoxal.

Argpyrimidine was also formed in N^α-t-BOC-arginine incubations starting from glyceraldehyde and higher sugars. Yields decreased in the order glyceraldehyde, threose, ribose, ascorbic acid, with only trace amounts obtained in the case of glucose and fructose. As argpyrimidine is labile to classic protein hydrolysis, an effective enzymatic digestion was developed to probe for its formation in protein incubations (Table 1). The absorption spectra of proteins incubated with higher sugars were compatible with the spectrum of argpyrimidine. This suggests that argpyrimidine represents a major fluorescent Maillard-derived modification. The presence of aminoguanidine almost completely prevented the formation of argpyrimidine. This is not unexpected, since α-dicarbonyl compounds, like methylglyoxal, are trapped to form stable triazines.[13]

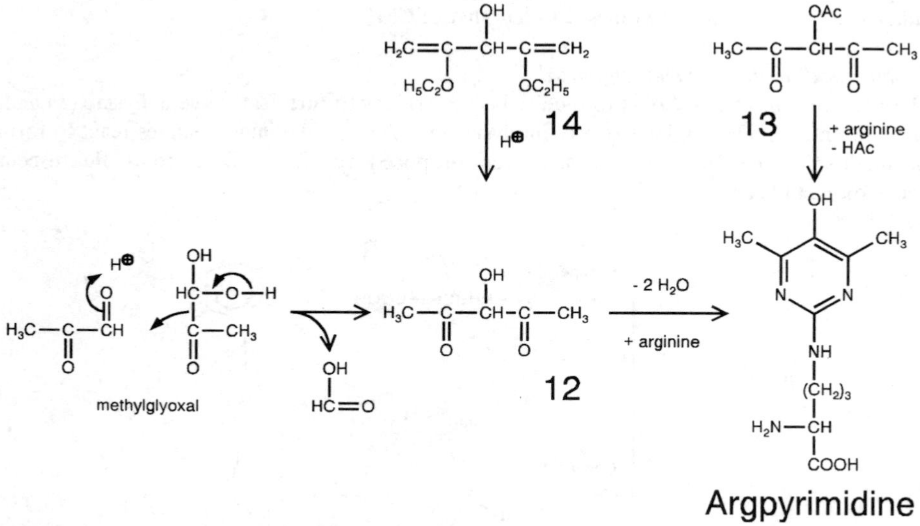

Figure 5. *Formation of argpyrimidine*

Table 1. *Formation of argpyrimidine in bovine lens water soluble crystallins (5mg/ml) incubated with 25 mM of each sugar at 37°C and pH 7.4 in phosphate buffered saline. Aliquots were withdrawn at various time points, dialysed, digested with proteolytic enzymes and analysed by HPLC.*[5]

sugar	3 days	15 days
	nmol/mg protein	
methylglyoxal	2.63	8.27
glyceraldehyde	1.53	5.94
threose	0.70	1.70
ribose	0.01	0.06

Conclusion

The reaction of N^α-t-BOC-lysine with glyoxal leads to GX-amide-crosslink formation, which is closely related to the synthesis of CML. An efficient enzymatic digestion is needed to study its significance in protein reactions and to correlate with other known crosslinks.

Argpyrimidine has been identified as a novel fluorescent protein-bound arginine modification by methylglyoxal. Its detection *in vivo* could establish a useful biomarker for protein modifications in aging and diabetes, and would confirm that enhanced intrinsic fluorescence in diabetic proteins is, in part, due to the methylglyoxal-mediated Maillard reaction.

Acknowledgements
This work was supported in part by research grants from the Deutsche Forschungsgemeinschaft (to M.A.G.), the National Eye Institute (EY 09912), the Johannsen Research Fund of Fight for Sight Research Division of Prevent Blindness America, Research to Prevent Blindness, Inc. (to R.H.N.) and the Juvenile Diabetes Foundation International.

References

1. F. Ledl and E. Schleicher, New aspects of the Maillard reaction in foods and in the human body, *Angew. Chem. Int. Ed. Engl.*, 1990, **29**, 565-594.
2. M.A. Glomb and V.M. Monnier, Mechanism of protein modification by glyoxal and glycolaldehyde, reactive intermediates of the Maillard reaction, *J. Biol. Chem.*, 1995, **270**, 10017-10026.
3. P.J. Thornalley, Pharmacology of methylglyoxal: formation of proteins and nucleic acids, and enzymatic detoxification. A role in pathogenesis and antiproliferative chemotherapy, 1996, *General Pharmacol.*, **27**, 565-573.
4. M.A. Glomb, unpublished.
5. I.N. Shipanova, M.A. Glomb and R.H. Nagaraj, Protein modification by methylglyoxal: chemical nature and synthetic mechanism of a major fluorescent adduct, *Arch. Biochem. Biophys. in press*, 1997, **344**.
6. M.A. Glomb, J. Fogarty and V.M. Monnier, unpublished.
7. K.J. Wells-Knecht, E. Brinkmann and J.W. Baynes, Characterization of an imidazolium salt formed from glyoxal and N^{α}-hippuryllysine: a model for Maillard reaction crosslinks in proteins, *J. Org. Chem.*, 1995, **60**, 6246-6247.
8. M.U. Ahmed, S.R. Thorpe and J.W. Baynes, Identification of N^{ϵ}-carboxymethyllysine as a degradation product of fructoselysine in glycated protein, *J. Biol. Chem.*, 1986, **261**, 2889-2894.
9. M.U. Ahmed, E. Brinkmann, S.R. Thorpe and J.W. Baynes, Identification of N^{ϵ}-carboxyethyllysine in tissue proteins: Evidence for formation of methylglyoxal adducts to proteins in vivo, *Diabetes (suppl.)*, 1996, **45**.
10. R.H. Nagaraj, I.N. Shipanova and F.M. Faust, Protein Cross-linking by the Maillard reaction, *J. Biol. Chem.*, 1996, **271**, 19338-19345.
11. T. Henle, A.W. Walter, R. Haessner and H. Klostermeyer, Detection and identification of a protein-bound imidazolone resulting from the reaction of arginine residues and methylglyoxal, *Z. Lebensm.-Unters. Forsch.*, 1994, **199**, 55-58.
12. T.W.C. Lo, M.E. Westwood, A.C. McLellan, T. Selwood and P.J. Thornalley, Binding and modification of proteins by methylglyoxal under physiological conditions, *J. Biol. Chem.*, 1994, **269**, 32299-32305.
13. T.W.C. Lo, T. Selwood and P.J. Thornalley, The reaction of methylglyoxal with aminoguanidine under physiological conditions and prevention of methylglyoxal binding to plasma proteins, 1994, *Biochem. Pharmacol.*, **48**, 1865-1870.

The Relative Oxidation, Glycation and Crosslinking Activity of Glucose and Ascorbic Acid in vitro

Beryl J. Ortwerth[1,2], Kwang-Won Lee[1], Amy Coots[1], and Valeri Mossine[2]

MASON EYE INSTITUTE[1] AND DEPARTMENT OF BIOCHEMISTRY[2], UNIVERSITY OF MISSOURI, COLUMBIA, MO 65212, USA

Summary

The protein crosslinking activities of glucose and ascorbic acid (ASA) were compared by the incorporation of N^α-formyl-[U-^{14}C]lysine into lens proteins. On a molar basis, ASA was 60-fold more active than glucose and 100-fold more active than fructose. ASA, but not glucose, was rapidly oxidized by the metal ions present in 100 mM phosphate buffer, but the addition of a chelating agent or even lens proteins (0.5 mg/mL) almost completely prevented this oxidation. The generation of singlet oxygen (2.0 mM) by the UVA irradiation of human lens proteins oxidized 0.5 mM ASA in 1 hr, but 4 h of irradiation had no oxidative effect on 50 µM glucose. High levels of glutathione were unable to prevent completely the oxidation of ASA by singlet oxygen. This was due to the unique products of singlet oxygen-mediated oxidation of ASA, and by the aggregate nature of the lens proteins preventing GSH access to the sites of ASA oxidation. UVA irradiation of human lens proteins caused the incorporation of [U-^{14}C]ASA into protein, even in the presence of physiological levels of glutathione. These data argue that ascorbic acid, rather than glucose, is responsible for the glycation-dependent crosslinking of lens proteins in vivo.

Introduction

Aging and cataract formation in human lens are characterized by extensive browning and protein crosslinking, which has been suggested to be the result of Maillard chemistry. Increased protein crosslinking accompanies the formation of very large protein aggregates (~10^9 Da) which can scatter light. Glucose and ascorbic acid (ASA) have both been suggested as causing this protein crosslinking,[1,2] and both are present in normal human lens at a level of 1.0 mM.[3,4] Glucose increases to only 3-7 mM in diabetes, along with increases in fructose due to the presence of the sorbitol pathway.[5] While both of these sugars are sluggish crosslinking agents, ASA crosslinking is also limited by the need for oxidation. ASA oxidation is normally prevented by the presence of 2-4 mM glutathione (GSH) in human lens, which can rapidly reduce dehydroascorbic acid back to the inactive ASA. The oxidized glutathione is in turn reduced to GSH by glutathione reductase, keeping a reducing atmosphere in the lens.

The work presented here was undertaken to evaluate the relative ability of glucose and ASA to crosslink lens proteins and to determine the relative ability of each to be oxidized by metal ions and singlet oxygen generated by UVA light. The effect of GSH on these oxidative reactions is also evaluated under conditions that mimick those in vivo.

Materials and Methods

Crosslinking activity was estimated as described previously,[6] measuring the sugar-dependent incorporation of N^α-formyl-[U-^{14}C]lysine[7] into dialyzed preparations of bovine lens proteins. The oxidation of 50 µM ASA and 50 µM [U-^{14}C]glucose in 0.1 M phosphate buffer was determined by the loss of absorbance at 265 nm and by thin layer chromatography (TLC), respectively. UVA oxidation was carried out by irradiating 2.0

mg/mL solutions of sonicated human water-insoluble proteins (WISS) with light from a 1000 W Hg/Xe lamp that had been filtered through a 5% $CuSO_4$ solution and a 338 nm cutoff filter.[8] Lamp intensities of 220 mW/cm^2 resulted in 200 J/hr of absorbed light, causing the synthesis of ~2.0 mM singlet oxygen over a 1 hr irradiation. [U-^{14}C]ASA (2.0 mCi/mMole) was synthesized by the method of Reichstein and Grusner.[9] Protein incorporation was measured in both a UVA-irradiated reaction mixture (20 mW/cm^2 for 24 hr) and a dark control consisting of 1.0 mM [U-^{14}C]ASA, 2.0 mM GSH and 2.0 mg/mL human WISS proteins in 50 mM phosphate buffer, pH 7.0 containing 50 µM diethylenetriaminepentaacetic acid (DTPA). Triplicate aliquots were taken.

Results

Protein aggregation in human lens increases with age and results in the formation of a water-insoluble (WI) protein fraction upon homogenization. Figure 1 shows the SDS-PAGE of the water-soluble (WS) and WI proteins from a 16 and 85 year human lens and a 50 yr brunescent cataract. Since >90% of the normal lens proteins are crystallins with subunit molecular weights of 20 to 30 kDa, it can be seen that the WI proteins in Fig. 1 exhibit extensive crosslinking. Assuming that such crosslinks are due to Maillard chemistry, they could arise either from glucose or ASA in the normal aged lens. These agents were tested in vitro for their crosslinking activity.

N^α-formyl-[^{14}C]lysine incorporation into lens proteins by ASA and glucose

Incubations were carried out with either 10 mM ASA or 100 mM glucose in a solution containing 5.0 mg/mL bovine lens proteins, 6.7 mM N^α-formyl-[U-^{14}C]lysine (0.3 mCi/mmol) and 100 mM phosphate buffer, pH 7.0. Aliquots were removed over 3 weeks of incubation and the incorporation determined by TCA precipitation of the proteins in filter paper discs.[6] Figure 2 shows that detectable crosslinking by 100 mM glucose was observed, which was inhibited by 2-aminoguanidine. However, it was only slightly greater that the reaction without sugar.

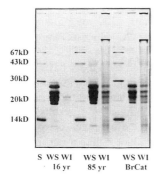

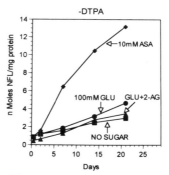

Figure 1. *SDS-PAGE separation of the water-soluble and water-insoluble proteins from human lens*

Figure 2. *Rate of incorporation of N^α-formyl-[U-^{14}C]lysine into bovine lens proteins*

Crosslinking by 10 mM ASA, on the other hand, was rapid and extensive. A comparison of the crosslinking activity of glucose, fructose and ASA are summarized in Table 1. All values are corrected for the incorporation seen in the absence of sugar. Crosslinking by all three carbohydrates was sugar-dependent and inhibited by 2-aminoguanidine (2-AG). Glucose was slightly more active than fructose, but ASA was 60-fold more active than glucose on a molar basis.

Table 1. The relative crosslinking activity of glucose, fructose and ASA[a]

Condition	100 mM Glucose	100 mM Fructose	10 mM ASA
Complete	1.7	1.0	11
Complete + 1.0 mM DTPA	1.5	0.9	11
Complete + 10 mM 2-AG	0.1	0.1	0.1

[a]Values shown are nmol N^α-formyl[U-^{14}C] lysine incorporated/mg protein/3 weeks

Glucose and ASA oxidation by metal ions and UVA irradiation of human lens proteins
Only the oxidation products of ASA are capable of glycating and crosslinking proteins.[10] When tested individually DHA, diketogulonic acid and L-threose were all active, even in the absence of oxygen.[10] Glucose readily glycates lens proteins, but a metal ion-catalyzed step is required to produce advanced glycation endproducts (AGE) which include protein crosslinks.[11] In this reaction, transition metals are thought to produce hydroxyl radicals, which fragment glucose into AGE-forming compounds, such as arabinose and glyoxal.[12] Crosslinking reactions in vitro provide the necessary transition metal ions in the 0.1M phosphate buffer used in such assays.[13]

Since both ASA and glucose require oxidation, the oxidation of each was measured in 0.1 M phosphate buffer. Figure 3 shows that 50 µM ASA was completely oxidized after 1 hr in phosphate buffer. The removal of metal ions by a prior chelex-treatment of the buffer, or the addition of DTPA prevented the oxidation over 1 h, but not over a period of days (Table 1). Significantly, the addition of only 0.5 mg/mL of lens proteins also prevented ASA oxidation, presumably by complexing metal ions. These data argue that no free Fe or Cu would exist in vivo, since the lens contains ~300 mg protein/g wet weight. Similar incubations, carried out with 50 µM [U-^{14}C] glucose, showed no oxidation products by TLC (data not shown).

Lens tissue is unique in that it is subjected to solar irradiation on a daily basis, and in that the proteins synthesized prior to birth remain present throughout life Protein-bound browning products and fluorophores accumulate with age, and are localized in the WI protein fraction. Irradiation of these proteins from aged human lens results in the formation of active oxygen species. While superoxide anion and hydrogen peroxide are generated in µM amounts, 2.0 mM singlet oxygen is produced under our in vitro irradiation conditions.[8] The ability of UVA-induced singlet oxygen to oxidize ASA is shown in Figure 4. A 1.0 mM solution of ASA was 50% oxidized over 1 hr, which is consistent with the ability of ASA to serve as a singlet oxygen scavenger. When 50 µM glucose was similarly irradiated, no alteration in the TLC profile of [^{14}C]glucose was seen after 4 hr of irradiation (Figure 5). Therefore, UVA-induced singlet oxygen in older lenses would be an oxidative threat to ASA, but not glucose.

The effect of GSH on the oxidation of ASA by singlet oxygen

Oxidative stress results in the formation of DHA, which while capable of glycation, is prevented from accumulating by chemical reduction back to ASA by GSH. Calculations of the singlet oxygen produced in the lens by 1 hr of sunlight exposure is estimated to be 30 μM based upon our in vitro experiments.[8] The ASA oxidized by this amount of singlet oxygen should be very rapidly reduced by GSH. The generation of singlet oxygen, however, largely occurs within the protein aggregates of the human lens. This location

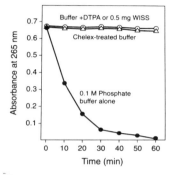

Figure 3. *Oxidation of 50 μM ASA in 0.1 M phosphate buffer ± DTPA and protein.*

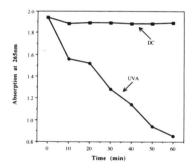

Figure 4. *Oxidation of 1.0 mM ASA by UVA-generated singlet oxygen.*

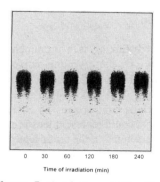

Figure 5. *Oxidation of 50 μM glucose by UVA-generated singlet oxygen.*

completely excludes GSH due its highly polar nature.[14] ASA oxidation in the presence of GSH by the UVA-irradiation of human WISS proteins and a proteinase digest of the WISS proteins is shown in Figure 6. GSH, at physiological levels (2-4 mM) allows ASA oxidation to occur, and even at levels of 10-15 mM, GSH cannot completely prevent ASA oxidation. This is partly due to the inability of GSH to penetrate to the site of oxidation, as evidenced by the increased GSH protection seen during the UV irradiation of a WISS proteolytic digest, and partly due to the ability of singlet oxygen to oxidize ASA to compounds that cannot be reduced by GSH, since 20% of the ASA oxidation could not be prevented even with the WISS digest.

UVA-dependent incorporation of ASA into protein

While UVA-generated singlet oxygen can oxidize ASA in the presence of GSH, it is possible that the oxidation products would not be glycation active. Fig. 7 shows that UVA irradiation of human WISS proteins results in the incorporation of [U-^{14}C]ASA into protein in the presence of 2.0 mM GSH. Little or no incorporation was seen with the same reaction mixture kept in the dark (DC). A similar irradiation of RNAse A as a control produced no ASA incorporation.

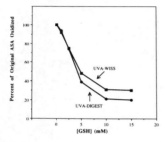
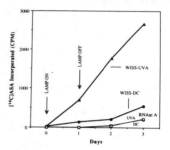

Figure 6. *The effect of added GSH on the UVA- induced oxidation of ASA.*

Figure 7. *The UVA-dependent incorporation of [U-^{14}C]ASA into lens proteins.*

Discussion

Protein crosslinking is a prominent feature of lens proteins during aging and cataract formation. Of the glycation-active compounds present in lens, glucose and ASA both caused the incorporation of N^{α}-formyl-[U-^{14}C]lysine into lens proteins. However, on a molar basis, ASA was 60-fold more effective as a crosslinking agent. These data agree with previous comparisons by SDS-PAGE.[2] The incorporation seen in the absence of added sugar could represent lens proteins preglycated in vivo, or possibly transglutaminase activity remaining in the crude lens proteins.

ASA crosslinking was assayed in air, which provided a higher level of oxygen than in the lens, and in the absence of GSH. In lens tissue this rate would be markedly decreased. Estimates of DHA levels in lens, if reliable, suggest it is present at only 5% of the levels of ASA. The presence of free Fe and especially Cu ions markedly accelerate the oxidation of ASA, but these are not likely to exist in the lens due to binding by the lens proteins. A more likely source of ASA oxidation is the generation of reactive oxygen species. The sensitizers bound to human lens proteins produce considerable singlet oxygen, mostly within the confines of the large protein aggregates present. In this milieu singlet oxygen's halflife is increased, and ASA can penetrate and be oxidized, whereas GSH cannot.[14] In addition, the enzymes that normally destroy active oxygen species are excluded from the site of generation. Consistent with this hypothesis, we were able to show a UVA-dependent incorporation (glycation) of human lens protein aggregates in the presence of 2.0 mM GSH after only 24 hours.

Evidence is increasing that the formation of AGE by glucose involves an oxidative component, likely involving the cleavage of glucose to form smaller, more reactive sugars.[12] The primary defense against oxidative stress in all tissues, however, is ASA. Oxidation of glucose by either metal ions in phosphate buffers or by reactive oxygen species could not be demonstrated with 50 µM glucose even in the absence of ASA. Any glucose oxidation in the presence of ASA would, therefore, be prohibited. The in vitro data presented here argue that the formation of ASA glycation crosslinks in lens tissue are not only possible, but may occur to the exclusion of those from glucose.

References

1. V.J. Stevens, C.A. Rouser, V.M. Monnier and A. Cerami, Diabetic cataract formation: Potential role of glycosylation of lens crystallins, *Proc. Natl. Acad. Sci. USA* 1978, **78**, 2918-2922.
2. B.J. Ortwerth, M.S. Feather and P.R. Olesen, The precipitation and crosslinking of lens proteins by ascorbic acid, *Exp. Eye Res.*, 1988, **47**, 155-168.
3. S.D. Varma, S.S. Schocket and R.D. Richards, Implications of aldose reductase in cataracts in human diabetes. *Invest. Ophthalmol. Vis. Sci.* 1979, **18**, 237-241.
4. H. Heath, The distribution and possible functions of ascorbic acid in the eye, *Exp. Eye Res.* 1962, **1**, 362-367.
5. J.A. Jedziniak, L.T. Chylack Jr., H-M. Cheng, M.K. Gillis, A.A. Kalustian and W.H. Tung, The sorbitol pathway in the human lens: aldose reductase and polyol dehydrogenase, *Invest. Ophthalmol. Vis. Sci.* 1981, **20**, 314-326.
6. M. Prabhakaram and B.J. Ortwerth, Determination of glycation crosslinking by the sugar-dependent incorporation of [^{14}C]lysine into protein, *Anal. Biochem.*, 1994, **216**, 305-312.
7. K. Hofman, E. Stutz, G. Spuhler, H. Yajima and E.T. Schwartz, Studies on polypeptides. XVI. The Preparation of N-formyl-L-lysine and its application to the synthesis of peptides. *J. Am. Chem. Soc.*, 1960, **82**, 3727-3732.
8. M. Linetsky and B.J. Ortwerth, Quantitation of the singlet oxygen produced by UVA irradiation of human lens proteins, *Photochem. Photobiol.* 1997, **65**, 522-529.
9. T. Reichstein and A. Grusner, Eine ergiebige synthese der l-Ascorbinsaure (C-vitamin), Helv. Chim. Acta 1934, 17, 311-328.
10 M. Prabhakaram and B.J. Ortwerth, The glycation-associated crosslinking of lens proteins by ascorbic acid is not mediated by oxygen free radicals, *Exp. Eye Res.*, 1991, **53**, 261-268.
11. M.X. Fu, K.J. Wells-Knecht, J.A. Blackledge, T.J. Lyons, S.R. Thorpe and J.W. Baynes Glycation, glycoxidation and crosslinking of collagen by glucose. Kinetics, mechanisms and inhibition of late stages of the Maillard reaction, *Diabetes,* 1994, **43**, 676-683.
12. K.J. Wells-Knecht, D.V. Zyzack, J.E. Litchfield. S.R. Thorpe and J.W. Baynes, Mechanism of autoxidative glycosylation: Identification of glyoxal and arabinose as intermediates in the autoxidative modification of proteins by glucose, *Biochemistry* 1995, **34**, 3702-3709.
13. G.R. Buettner, In the absence of catalytic metals ascorbate does not autoxidize at pH 7: ascorbate as a test for catalytic metals, *J. Biochem. Biophys. Meth.*, 1988, **16**, 27-40.
14 M. Linetsky, N. Ranson and B.J. Ortwerth, The scavenging of singlet oxygen within human lens protein aggregates by ascorbic acid and glutathione, in press.

Reaction Mechanisms Operating in 3-Deoxyglucosone-Protein Systems

F. Hayase, N. Nagashima, T. Koyama, S. Sagara, and Y. Takahashi

DEPARTMENT OF AGRICULTURAL CHEMISTRY, MEIJI UNIVERSITY, 1-1-1 HIGASHI-MITA, TAMA-KU, KAWASAKI, KANAGAWA 214-71, JAPAN

Summary

3-Deoxyglucosone (3DG) has been reported to be generated from Amadori compounds by non-oxidative reaction *in vitro* and *in vivo*. Several non-fluorescent imidazolone compounds were identified as arginine adducts of 3DG. Using a 3DG and N^α-benzoylarginine amide (BzArgNH$_2$) reaction system, 2-(N^α-benzoyl-N^δ-ornithylamide)-5-(2,3,4,-trihydroxybutyl)-2-imidazoline-4-one (S12), was identified as an intermediate product. A minor product formed from two molecules of 3DG and one molecule of BzArgNH$_2$ was identified as 2-(N^α-benzoyl-N^δ-ornithylamide)-5,6a-di(2,3,4,-trihydroxybutyl)-5,6-dihydroxydehydrofuro[2,3-*d*]imidazole (S11). We speculate that S11 is formed by the addition of 3DG to compound S12. S12 was also identified in a 3DG-insulin reaction system by ESI-MS. In further studies, a fluorescent product was purified from a N^α-acetyllysine-3DG reaction system and identified as 6,8-di-N^α-acetyllysyl-3,3a,8,8a-tetrahydro-3a-hydroxy-2-(1,2-dihydroxyethyl)-5-hydroxymethyl-2*H*-furo[3',2: 4,5]pyrrolo-[2,3-*c*]-pyridinium (N^α-acetyllysyl-pyrropyridine) having fluorescence at Ex$_{max}$: 376nm and Em$_{max}$: 450nm. Quantification of lysyl-pyrropyridine in glycated lysozyme revealed a significant increase in a 3DG-lysozyme system (74.3mmol/mol lysozyme) when compared with a glucose-lysozyme system (6.3 mmol/mol lysozyme) incubated at 50°C for 7 days.

Introduction

Glucose-mediated advanced glycation end products (AGEs) have been proposed to be involved the aging of tissue and diabetic complications.[1] Therefore, much attention has been focused on elucidation of chemical structures of AGEs. 3-Deoxy-D-hexos-2-ulose, also called 3-deoxyglucosone (3DG), has been reported to be generated from Amadori compounds by nonoxidative reactions *in vitro*[2] and *in vivo*.[3-6] 3DG was also shown to be formed from fructose 3-phosphate in the lens of diabetic rats,[7] and by the Maillard reaction in some foods. 3DG is known to be a highly reactive and cytotoxic compound.[8] Previous work revealed that 3DG reacted with arginine, lysine, and tryptophan residues to crosslink proteins in the solid state at 50°C and 75% relative humidity[9] and under physiological conditions at 37°C and pH 7.4.[2]

Several nonfluorescent imidazolone compounds were identified as arginine adducts of 3DG from the reaction system between 3DG and N^α-benzoylarginine amide (BzArgNH$_2$).[10,11] Major products were identified as 2-(N^α-benzoyl-N^δ-ornithylamide)-5-(2,3,4,-trihydroxybutyl)-2-imidazoline-4-one (S12) and 2-(N^α-benzoyl-N^δ-ornithylamide)-5-(2,3,4,-trihydroxybutyl)-4-imidazolone (S17) (see Figure 1). In addition, a minor product, S11, formed from one molecule of BzArgNH$_2$ and two molecules of 3DG, may be indicative of further reactions such as the formation of crosslinks.

A previous study showed that 3DG reacted with free amino groups to form a protein-bound pyrrole aldehyde, called pyrraline.[12] We also identified a pyrrolopyridinium compound, that we named pyrropyridine, as a possible fluorescent lysine-lysine cross-link generated in a butylamine-3DG or glucose reaction system.[13]

In this paper, we describe identification of S11 from 3DG-BzArgNH$_2$, S12 from 3DG-insulin, and pyrropyridine from glucose or 3DG-lysozyme reaction systems.

Materials and Methods

Lysozyme (10 mg/mL) or BzArgNH$_2$ (0.1 M) and 3DG (0.1 M for lysozyme and 0.2 M for BzArgNH$_2$) were dissolved in 0.1 M sodium phosphate buffer at pH 7.4 and incubated at 50°C. Insulin chain B oxidized (Insulin B, 5 mg/mL) was incubated with 3DG (0.1 M) in 0.2 M sodium phosphate buffer at 37°C and pH 7.6.

The reaction products from the BzArgNH$_2$-3DG system were purified by RP-HPLC with detection at 235 nm. The major fluorescent compounds formed from the reaction systems of N^α-acetyl-L-lysine (0.1 M) and 3DG (50 mM) in 0.1 M sodium phosphate buffer at 50°C and pH 7.4 were purified by RP-HPLC with fluorescence detection (Ex: 370 nm and Em: 450 nm).

Mass spectra were recorded using a JEOL SX102 mass spectrometer. The ^{13}C-NMR and ^{1}H-NMR spectra were recorded using a JEOL GSX-500 (500MHz).

Results and Discussion

3DG- N^α-benzoylarginine amide or insulin reaction systems

Similar unknown peaks were detected on an amino acid chromatogram of the acid hydrolysates of the 3DG-lysozyme as well as the glucose-lysozyme reaction systems. Unknown peaks were also detected in the acid hydrolysates of the 3DG and BzArgNH$_2$ reaction system. In this system, S17 and S12 were identified as major products in a previous paper of 3DG reaction with the guanidino group (Figure 1).[10,11] NMR data for S12 revealed that S12 was a mixture of two isomers derived from the asymmetric carbon in the imidazoline ring. Minor products (S3, S4, S5, S5', S11) were also isolated. FAB-MS data for all these minor isolated compounds revealed an [M + 1]$^+$ ion at m/z =584, indicating that these compounds are probably formed from one molecule of BzArgNH$_2$ and two molecules of 3DG by dehydration. High resolution FAB-MS data for S11 showed an [M + 1]$^+$ ion at m/z =584.2571, the molecular formula of which was estimated to be $C_{25}H_{37}N_5O_{11}$ (calcd for $C_{25}H_{38}N_5O_{11}$, 584.2568). In further work, S11 was identified as 2-(N^α-benzoyl-N^δ-ornithylamide)-5,6a-di(2,3,4,-trihydroxybutyl)-5,6-dihydroxydehydrofuro[2,3-*d*]imidazole on the basis of ^{13}C-NMR and ^{1}H-NMR spectroscopic techniques including ^{1}H-^{1}H COSY, ^{1}H-detected ^{1}H-^{13}C COSY (HMQC), and ^{1}H-detected long range ^{1}H-^{13}C COSY (HMBC).

The reaction of arginine and 3DG is extremely complex. S11 was rapidly formed after 6 h of reaction of 13.3 mM S12 incubated with 13.3 mM 3DG in a 0.1 M sodium

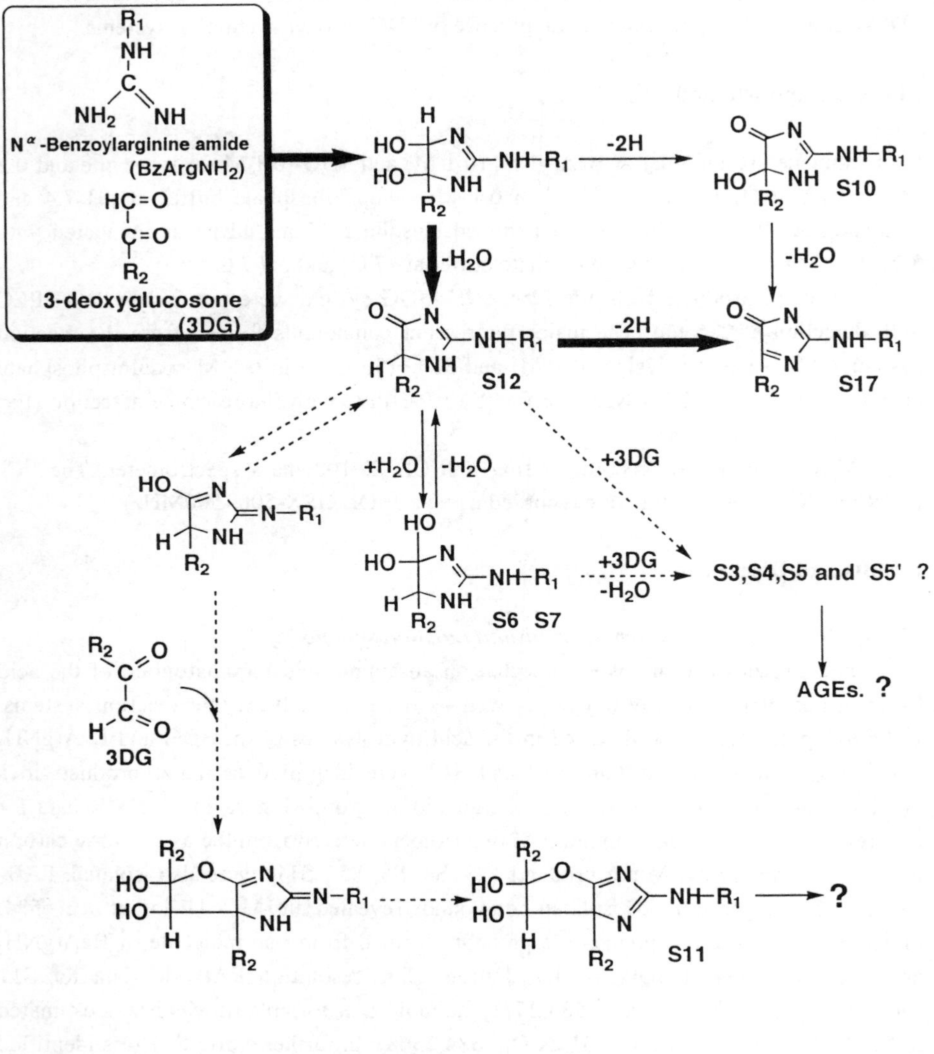

Figure 1. *Proposed scheme for the formation of compounds from 3DG and BzArgNH$_2$ reaction system (R_1=(CH$_2$)$_3$CH(NHCOC$_6$H$_5$)CONH$_2$; R_2=CH$_2$(CHOH)$_2$CH$_2$OH). Heavy arrows denote the main pathway; dotted arrow denote proposed pathway.*

phosphate buffer at 50°C and pH 7.4. Figure 1 shows the proposed formation scheme for S11 in the S12-3DG reaction system. Since S11 increased and then decreased in a BzArgNH$_2$-3DG reaction system incubated for 7.5 days, and in the S12-3DG reaction system incubated for 120 h, we propose that S11 participated in further reactions such as the formation of crosslinks. In a previous paper,[11] the formation of S17 from S12 via dehydration and subsequent oxidation was considered to be the major pathway. The formation of S17 from S10 is less important. The first step in formation of S12 is presumed to be the condensation of one molecule of 3DG with the guanidino group to give a 3,4-dihydroxy-2-imidazoline compound (Figure 1).

An arginine adduct, S12, was also identified as a major product in the 3DG-insulin B reaction system by electrospray-ionization mass spectrometry (ESI-MS). In this system, polymerization of insulin B was observed. Recently, Niwa et al.[14] have reported that amyloid tissue β_2-microglobulin is modified to contain imidazoline compounds (S12 and S17) in patients with dialysis-related amyloidosis, indicating that 3DG accumulating in uremic serum may be involved in the modification of β_2-microglobulin resulting in the formation of S12 and S17.

3DG- N^α-acetyllysine reaction system

Previous work[13] revealed that three major fluorescent compounds (FL-A, FL-B, and FL-C) were produced in a butylamine-glucose reaction system under physiological conditions at 37°C and pH 7.4. FL-A and FL-B were identified as epimers of 3,4-dihydroxy-5-(1,2,3,4-tetrahydroxybutyl)-1,7-dibutyl-1,2,3,4-terahydro-1,7-naphthyridinium, similar to compounds characterized and termed to croslines by Nakamura et al.[15] FL-C was identified as 6,8-dibutyl-3,3a,8,8a-tetrahydro-3a-hydroxy-2-(1,2-dihydroxyethyl)-5-hydroxymethyl-2H-furo[3',2: 4, 5] pyrrolo-[2,3-c]pyridinium, which we named butyl-pyropyridine. We examined the formation of the fluorescent compound, lysyl-pyropyridine (Figure 2), formed by the reaction of lysine residues with glucose. The results of HPLC analysis showed that several fluorescent compounds including FL-C1 were produced in a N^α-acetyllysine-glucose reaction system. If 3DG was added to this reaction system, FL-C1 was increased. In a N^α-acetyllysine-3DG reaction system, FL-C1 was generated as a major fluorescent compound. FL-C1 was isolated from N^α-acetyllysine-3DG reaction system and identified as N^α-acetyllysyl-pyropyridine having fluorescence at Ex$_{max}$: 376nm and Em$_{max}$: 450nm. This fluorescent compound was formed following the loss of five molecules of water from the reaction between 2 molecules of N^α-acetyllysine and 2 molecules of 3DG.

Lysyl-pyropyridine (Figure 2) was obtained by hydrolysis of N^α-acetyllysyl-pyropyridine with 6N HCl in evacuated tubes at 110°C for 22 h. The recovery of lysyl-pyropyridine was 26.7% after acid hydrolysis. However, croslines appear to have decomposed in this process.

Sell and Monnier[16] isolated a fluorescent compound (Ex$_{max}$: 335nm and Em$_{max}$: 385nm) from human extracellular matrix and identified it as pentosidine. Quantitative

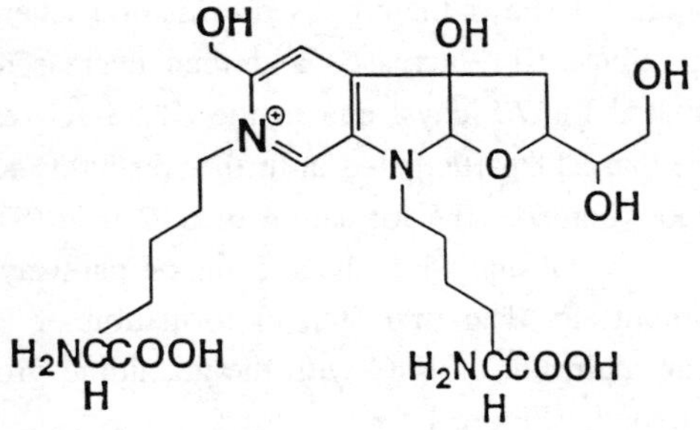

Figure 2. *Structure of lysyl-pyrropyridine*

analyses of lysyl-pyrropyridine and pentosidine were carried out by RP-HPLC. Table 1 shows amounts of lysyl-pyrropyridine and pentosidine in hydrolysates from lysozyme incubated with glucose, 3DG, or both, indicating that lysyl-pyrropyridine was the major fluorescent product in all systems. Therefore, lysyl-pyrropyridine may be generated as one of important biomarker in advanced glycated proteins *in vivo* as well as in foods.

Table 1. *Amounts (mmol/mol lysozyme) of lysyl-pyrropyridine (Lys-pyrropyridine) and pentosidine in hydrolysates from lysozyme (10mg/mL)-glucose (200mM), lysozyme-3-deoxyglucosone (50mM, 3DG), and lysozyme-glucose (200mM)-3DG (20mM) reaction systems in 0.2M phosphate buffer at pH 7.4 and 50°C.*

	Reaction system (reaction time)					
	lysozyme-glucose		lysozyme-glucose-3DG		lysozyme-3DG	
	(7 days)	(14 days)	(7 days)	(14 days)	(7 days)	(14 days)
Lys-pyrropyridine	6.3	11.2	35.2	34.6	74.3	106.5
Pentosidine	2.6	8.3	1.1	3.8	1.3	2.9

Acknowledgements

We thank Professor Vincent M. Monnier (Case Western Reserve University) for the gift of pentosidine.

References

1. V.M. Monnier, D.R. Sell, R.H. Nagaraj and S. Miyata, Mechanisms of protection against damage mediated by the Maillard reaction in aging, *Gerontology.*, 1991, **37**, 152-165.
2. D.B. Shin, F. Hayase, and H. Kato, Polymerization of proteins caused by the reaction with sugars and the formation of 3-deoxyglucosone under physiological conditions, *Agric. Biol. Chem.*, 1988, **52,** 1451-1458.
3. F. Hayase, Z.Q. Liang, Y. Suzuki, N.V. Chuyen, T. Shinoda and H. Kato, Enzymatic

metabolism of 3-deoxyglucosone, a Maillard intermediate, *Amino Acid,* 1991, **1**, 307-318.
4. K.J. Knecht, M.S. Feather and J.W. Baynes, Detection of 3-deoxyglucosone in human urine and plasma: Evidence for intermediate stages of the Maillard reaction in vivo. *Arch. Biochem. Biophys.*, 1992, **294**, 130-137.
5. T. Niwa, N. Takeda, H. Yoshizumi, A. Tatematsu, M. Ohara, S. Tomiyama, and K. Niimura, Presence of 3-deoxyglucosone, a potent protein crosslinking intermediate of Maillard reaction, in diabetic serum, *Biochem. Biophys. Res. Commun.*, 1993, **196**, 837-843.
6. H. Yamada, S. Miyata, N. Igaki, H. Yatabe, Y. Miyauchi, T. Ohara, M. Sakai, H. Shoda, M. Oimomi and M. Kasuga, Increase in 3-deoxyglucosone levels in diabetic rat plasma, *J. Biol. Chem.*, 1994, **269**, 20275-20280.
7. B.S. Szwergold, F. Kappler and T.R. Brown, Identification of fructose 3-phosphate in the lens of diabetic rats, *Science*, 1990, **247**, 451-454.
8. T. Shinoda, F. Hayase and H. Kato, Suppression of cell-cycle progression during the S phase of rat fibroblasts by 3-deoxyglucosone, a Maillard reaction intermediate, *Biosci. Biotech. Biochem.*, 1994, **58**, 2215-2219.
9. H. Kato, R.K. Cho, A. Okitani, and F. Hayase, Responsibility of 3-deoxyglucosone for the glucose-induced polymerization of proteins, *Agric. Biol. Chem.* 1987, **51**, 683-689.
10. Y. Konishi, F. Hayase and H. Kato, Novel imidazolone compound formed by the advanced Maillard reaction of 3-deoxyglucosone and arginine residues in proteins, *Biosci. Biotech. Biochem.*, 1994, **58**, 1953-1955.
11. F. Hayase, Y. Konishi and H. Kato, Identification of the modified structure of arginine residues in proteins with 3-deoxyglucosone, a Maillard reaction intermediate, *Biosci. Biotech. Biochem.*,1995, **59**, 1407-1411.
12. F. Hayase, R.H. Nagaraj, S. Miyata, F.G. Njoroge and V.M. Monnier, Aging of proteins: immunological detection of a glucose-derived pyrrole formed during Maillard reaction in vivo, *J. Biol. Chem.*, 1989, **264**, 3758-3764.
13. F. Hayase, H. Hinuma, M. Asano, H. Kato, and S. Arai, Idetification of novel fluorescent pyrrolopyridinium compound formed from Maillard reaction of 3-deoxyglucosone and butylamine, *Biosci. Biotech. Biochem.*, 1994, **58**, 1936-1937.
14. T. Niwa, T. Katsuzaki, S. Miyazaki, T. Momoi, T. Akiba, T. Miyazaki, K. Nokura, F. Hayase, N. Takemichi and Y. Takei, Amyloid β_2-microglobulin is modified with imidazolone, a novel advanced glycation end product, in dialysis-related amyloidosis, *Kidney Int.*, 1997, **51**, 187-194.
15. K. Nakamura, T Hasegawa, Y. Fukunaga, K. Ienaga, Crosslines A and B as candidates for the fluorophores in age- and diabetes-related cross-linked proteins, and their diabetes produced by Maillard reaction of α–N-acetyl-L-lysine with D-glucose, *J. Chem. Soc., Chem. Commun.* 1992, 992-994.
16. D.R. Sell, and V.M. Monnier, Structure elucidation of a senescence cross-link from human extracellular matrix, *J. Biol. Chem.*, 1989, **264**, 21597-21602.

Effects of Vitamin E on Serum Glucose Levels and Nephropathy in Streptozotocin-Diabetic Rats

M. James C. Crabbe,[1] B. Üstündag,[2] M. Cay,[3] M. Naziroglu,[3] N. Ilhan,[2] I. H. Özercan,[2] and Nihat Dilsiz[4]

[1] DIVISION OF CELL AND MOLECULAR BIOLOGY, SCHOOL OF ANIMAL AND MICROBIAL SCIENCES, UNIVERSITY OF READING, WHITEKNIGHTS, READING, RG6 6AJ, U.K.; [2] DEPARTMENT OF BIOCHEMISTRY, MEDICAL SCHOOL, FIRAT UNIVERSITY, ELAZIG, 23119, TURKEY; [3] DEPARTMENT OF PHYSIOLOGY, VETERINARY FACULTY, FIRAT UNIVERSITY, ELAZIG, 23119, TURKEY; [4] DEPARTMENT OF BIOLOGY, FACULTY OF SCIENCE, FIRAT UNIVERSITY, ELAZIG, 23119, TURKEY

Summary
We investigated the effect of vitamin E on serum glucose levels and renal changes in streptozotocin-diabetic rats. Animals were treated every other day with 50 mg/kg doses of vitamin E for a week before streptozotocin treatment. After the final streptozotocin treatment, animals were treated daily with 20 mg/kg vitamin E for three days.

The mean serum glucose levels in control diabetic rats was 387±15.35 mg/dL at day 3 and 548±47.75 mg/dL at day 15, compared with vitamin E-treated rats 326±33.76 mg/dL at day 3 and 498±13.74 mg/dL at day 15 ($p<0.005$ and $p<0.05$ respectively). There was also a significant decrease ($p<0.005$) in blood urea almost to normal levels. In histopathological studies, increased basement membrane thickness, focal glomerulosclerosis, partial tubular necrosis and tubular dilatation with 'capsular drops' in Bowman's capsules were established in control diabetic rats at day 15. Such changes were less pronounced or absent in vitamin E-treated diabetic rats.

Introduction

Hyperglycaemia is the major risk factor for the complications of diabetes mellitus;[1,2] for that reason, it is important to keep the blood glucose concentration at normal levels. In diabetes, there appears to be a close interplay between glycoxidation and oxidation in lipoproteins, with the involvement of free radicals.[3,4] Vitamin E is one of most important free-radical scavengers, or antioxidants, in the body.[5] Vitamin E acts on the unsaturated fatty acids in the phospholipid of the cell membrane to prevent or interupt oxidation reactions.[6,7] In vitamin E deficiency, cell membrane integrity is compromized due to increased lipid peroxides. Vitamin E may be an attractive candidate for supplementation because it is of low toxicity and is a readily obtained dietary supplement.
This study was designed to test the effect of vitamin E supplementation on the blood glucose level in streptozotocin-diabetic rats.

Materials and Methods

In this study, 42 female Wistar-albino rats (190-220 g body weight) were obtained from the local Animal Research Institute in Elazýõ. Rats were randomly divided into two groups, a control streptozotocin-diabetic group and a streptozotocin-diabetic group supplemented with dietary vitamin E. All rats were fed for 10 weeks on standard lab chow before the experiment. At the beginning of the experiment, all rats were weighed individually. One group was supplemented with vitamin E (100 mg/kg) three times in one week. Then both control and vitamin E-supplemented rat groups were injected with 60 mg/kg streptozotocin (Sigma Chemical Co.) to induce diabetes. The vitamin E-supplemented group was then supplemented daily with vitamin E (50 mg/kg) for three days. Rats were observed every day after beginning supplementation, and blood glucose levels measured. Rats were killed after 3 and 15 d following injection.

Blood (8-10 mL) was taken via rat tails at 3 and 15 d from lightly ether-anaesthetized and laparotomized rats; samples were centrifuged at 3,000 rpm for 15 minutes to obtain serum.

Kidneys and pancreas were obtained from rats following exsanguination to study histopathology. These organs were fixed in 37% formaldehyde for 72 h then transferred to 70% alcohol and paraffin embedded. Paraffin sections were prepared from tissues using a microtome (Reichert-Jung). Each section (4 μm) was transferred into Hematoxylin and Eosin (H&E) solution, and then examined using light microscopy.

Biochemical analyses (glucose, urea, creatinine, sodium, potassium, phosphate and calcium) were conducted on serum samples using a Technicon RA-XT autoanalyser. Results were reported as means ± SEMs, and statistical significance assessed by the t-test or Mann-Whitney U test as appropriate.

Results

The blood glucose level of the diabetic control group increased to 387 ± 15.35 mg/dL after 3 days, while it was significantly decreased to 326 ± 33.76 mg/dL in the vitamin E-supplemented group. The differences between the se two groups were significant (p<0.005). After 15 d. of supplementation, the glucose level of the control group was increased to 584 ± 47.75 mg/dL, while it was found to be 498 ± 13.7 mg/dL in the

Table 1. *The serum biochemical parameters in control diabetic and vitamin E-supplemented diabetic groups. n=10 for each column.*

	Control		Vitamin E	
	Day 3	Day 15	Day 3	Day 15
Urea (mg/dL)	63.1 ± 6.38	102.6 ± 5.05	54.4 ± 7.45*	81.4 ± 9.33***
Creatinine (mg/dL)	1.2 ± 22	2.4 ± 0.55	0.86 ± 0.24***	0.82 ± 0.12**
Sodium (meq/L)	144.8 ± 1.39	144.6 ± 1.77	143 ± 2.09*	143 ± 1.83
Potassium (meq/L)	5.6 ± 0.47	6.5 ± 0.47	4.9 ± 0.31**	5.6 ± 0.31*
Calcium (mg/dL)	9.3 ± 0.49	9.6 ± 0.62	9.8 ± 0.60	10.39 ± 0.47*
Phosphate (mg/dL)	8.3 ± 0.61	7.5 ± 0.53	7.3 ± 0.44*	7.1 ± 0.39

*, p<0.05, **, p<0.01, ***, p<0.005

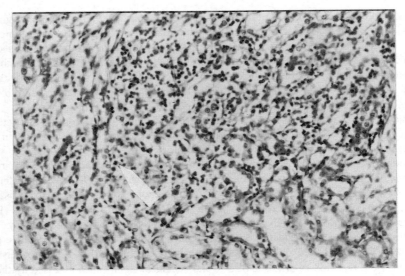

Figure 1. *Increased interstitial mononuclear cells, indicated by arrow, were observed in control diabetic rats after 15 d (H&E x 200).*

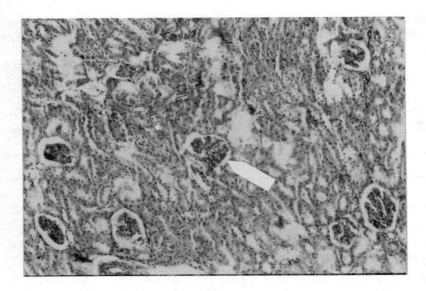

Figure 2. *Normal mesangial cells in the glomeruli of vitamin E-supplemented diabetic rats after 15 d, indicated by arrow (H&E x 200).*

vitamin E-supplemented group. This difference was also statistically significant at $p<0.05$.

The data from the biochemical analyses are shown in Table 1.

No significant differences were observed in kidney sections after three days of supplementation. After 15 d, the control group exhibited increased numbers of mesangial cells in the glomerulus, a thicker glomerular basement membrane, and some glomerulosclerosis with necrosis in the glomeruli and tubular epithelia. There was also an increased number of interstitial mononuclear cells (Figure 1). In addition, increased eosinophilic substance and 'capsular drop' lesions were found in the Bowman's capsules (data not shown). In the vitamin E-supplemented group after 15 d, there were only a few thick basement membranes, and only slight eosinophilia and infiltration (Figure 2).

Discussion

Vitamin E is an antioxidant that prevents the oxidation of cell-membrane lipids by free-radicals in diabetes.[6] Lipid peroxide levels have been found to be low in diabetic rats supplemented with high vitamin E.[8] In our study, the effect of vitamin E on the serum glucose level and related complications was determined in diabetic rats. The glucose level after 3 days supplementation was significantly lower than in the control group. A significant difference was also found after 15 days of supplementation between these two groups.

Our finding is in agreement with Mooradian *et al.*[8] Shoff *et al.*[9] found that the levels of hyperglycaemia and glycated haemoglobin were decreased in diabetic humans by administration of vitamin E. In our study, after 3 days, there were no significant effects on the kidneys of either group. However, after 15 days, various pathological changes were observed in the control group, which were less apparent in the vitamin E-supplemented diabetic group. The cellular changes noted in the control group were presumably a consequence of hyperglycaemia, as streptozotocin does not have a toxic effect on the kidney.[10-12] Vitamin E might reduce the progression of diabetic nephropathy in this rat model by influencing glycoxidation reactions and so reducing post-translational modification of proteins important in kidney function.

It is of interest that serum urea level was significantly lower in vitamin E-treated animals (Table 1). Urea is in equilibrium with cyanate, and carbamylation is a major non-enzymic post-translational modification reaction of proteins, particularly in cataract.[13] The formation of cataract is about six times more common in diabetics than in non-diabetics.[13] Carbamylation destroys the chaperone-like activity of α-B crystallin, a protein necessary for the maintenance of proteins in their correctly folded state in the lens.[14,15] The analgesic ibuprofen, a protective agent against cataract,[16] protects the chaperone-like activity of α-B crystallin against post-translational modification by carbamylation and glycoxidation, by preventing protein crosslinking.[17] It would be of interest to study the effects of both vitamin E and ibuprofen on the development of diabetic complications in model systems.

Acknowledgements
We thank the Wellcome Trust and the British Diabetic Association for support.

References

1. M.J.C. Crabbe, (ed.), Diabetic Complications, Churchill Livingstone, London, 1987.
2. M. Brownlee and A. Cerami, The Biochemistry of the Complications of Diabetes Mellitus, *Ann. Rev. Biochem.*, 1981, **50**, 385-432.
3. J.W. Baynes, The role of oxidation in the Maillard Reaction *in vivo*, In 'The Maillard Reaction', (R. Ikan, ed.), John Wiley and Sons, Chichester, UK, pp. 55-72, 1996.
4. S.P. Wolff, Free radicals and glycation theory, In 'The Maillard Reaction', (R. Ikan, ed.), John Wiley and Sons, Chichester, UK, pp. 73-88, 1996.
5. L. Packer, The role of antioxidative treatment in diabetes mellitus, *Diabetologia*, 1993, **36**, 1212-1213.
6. O. Yilmaz, S. Celik, M. Naziroglu, M. Cay and N. Dilsiz, The effects of dietary selenium and vitamin E and their combination on the fatty acids of erythrocytes, bone marrow and spleen tissue lipids of lambs, *Cell Biochem. Func.*, 1997, **15**, 1-7.
7. R.B. Brandt, G.E. Kaugars, W.T. Riley, R.A. Bei, S. Silverman, J.G. L. Lovas, B.P. Dezzutti and W. Chann, Evalution of serum and tissue levels of α-tocopherol, *Biochem. Mol. Med.* 1996, **57**, 64-66.
8. A.D. Mooradian, M. Failla, B. Hoogwert, M. Marvniuk and J.W. Rosett, Selected vitamins and minerals in diabetes, *Diabetes Care*, 1994, **45**, 464-479.
9. S.M. Shoff, J.A. Perlman, K.J. Cruickshanks, R. Klein and B.E.K. Klein, Glycosylated haemoglobin concentrations and vitamin E, vitamin C and beta-carotene intake in diabetic and non diabetic older adults, *Am. J Clin.*, 1993, **58**, 412-416.
10. G.V. Mann and L. Adams, The renal lesions associated with experimental diabetes in the rat, *Am. J Pathol.*, 1984, **50**, 565-568.
11. M.W. Steffes, D.M. Brown, J.M. Basgen and S.M. Mauer, Amelioration of mesengial volume and surface alterations following islet transplantation in diabetic rat, *Diabetes*, 1980, **29**, 509-515.
12. M.W. Steffes, R. Osterby, B. Chavers and S.M. Mauer, Mesangial Expansion as a Central Mechanism for Loss of Kidney Function in Diabetic Patients, *Diabetes*, 1989, **38**, 1077-1081.
13. J.J. Harding, Cataract, Chapman and Hall, London, 1991.
14. M.L. Plater, D. Goode and M.J.C. Crabbe, Effects of site-directed mutations on the chaperone-like activity of α-B crystallin, *J. Biol. Chem.*, 1996, **271**, 28558 - 28566.
15. J. Horwitz, Alpha crystallin behaves as a molecular chaperone, *Proc. Natl. Acad. Sci. USA*, 1992, **89**, 10449-10453.
16. K.A. Roberts and J.J. Harding, Ibuprofen, a putative anti-cataract drug, *Exp. Eye Res.*, 1990, **50**, 157-164.
17. M.L. Plater, D. Goode and M.J.C. Crabbe, Ibuprofen protects α-crystallin against post-translational modification by preventing protein cross-linking. *Ophthal. Res.*, 1997, In the press.

Diabetic Cataract : Glycation of the Molecular Chaperone α-Crystallin and its Binding to the Membrane Protein MIP26

Martinus A.M. van Boekel and Wilfried.W. de Jong

CATHOLIC UNIVERSITY OF NIJMEGEN, DEPARTMENT OF BIOCHEMISTRY, NIJMEGEN, P.O. BOX 9101, 6500 HB NIJMEGEN, THE NETHERLANDS

Summary
We investigated the influence of glycation on the chaperone-like activity of the eye lens protein α-crystallin. Formation of early glycation products *in vitro* had no consequences for the protecting properties of the protein. Late (cross-linking) glycation products had a marked negative influence on the chaperone-like activity of α-crystallin. Similar results were found for homopolymers of the α-crystallin subunits αA2 and αB2. Of all crystallins, only α-crystallin specifically binds to lens membranes. The presence of early glycation products on α-crystallins does not seem to alter their affinity for membranes *in vitro*. These processes are possibly involved in the development of diabetic cataract.

Introduction

One of the most common complications in diabetes is the early development of cataract. There are two lines of investigation that could provide an explanation for this feature. The polyol pathway is a metabolic route by which reducing sugars are converted to sugar alcohols.[1] These polyols possibly impose an osmotic stress on lens cells. This could cause membrane disruption, finally resulting in lens opacity. Cell membrane damage is a certain trademark of early cataract formation. Changes in membrane dynamics could potentially also be induced by binding of cytoplasmic proteins. α-Crystallin is able to bind specifically to lens membranes[2]; any modifications of the crystallin could, therefore, influence its association with membranes or membrane proteins.

α-Crystallin belongs to the family of small heat-shock proteins, which have the ability to protect other proteins against precipitation.[3,4] The mechanism of this so-called chaperone-like activity is still unresolved. Since lens proteins are not replaced during life, this quality of α-crystallin could play a role in the maintenance of lens transparency. This chaperone-like activity is clearly diminished in the aging lens.[5] The accumulation of post-translational modification products such as Maillard products are generally held responsible for this decay. We hypothesized that glycation of α-crystallin leads to a diminished chaperone-like activity. We therefore investigated the influence of early and late glycation products produced *in vitro* on the chaperone-like activity of α-crystallin. Glycation of α-crystallin could also be involved in the changes to membrane binding that are observed in aging and cataract.[6] We therefore investigated whether glycation of α-crystallin itself leads to changes in membrane binding.

Materials and Methods

Isolation of crystallins
We used lenses from 4-6 months old calves, from which the α- and β-crystallins were isolated by gel-permeation chromatography (Aca34, Pharmacia). The α-crystallin subunits αA1, αA2, αB1 and αB2 were separated as described before.[7] The membrane fraction, containing MIP26 (a *m*embrane *i*ntrinsic *p*rotein with a molecular weight of 26 kDa), was obtained after removal of urea-soluble material and treatment with 0.1 M NaOH.[8]

Electrophoresis
The purity of the subunit preparations was determined by flat-bed isoelectric focusing. SDS-PAGE was performed using the MiniProtean II system (Biorad). We utilized the BCA assay (Pierce) for determination of protein concentrations.

Glycation of proteins in vitro
Purified α-crystallin was incubated with 200 mM sugar (glucose-6-phosphate or ribose) in sterile 'chaperone buffer' (20 mM NaPi, 100 mM Na_2SO_4 and 1 mM EDTA, pH 6.9) for 7 to 21 days at 37°C. Levels of modification were visualized by isoelectric gel electrophoresis, scanning and data processing. For production of early glycation products, incubations were performed at up to 1 mole sugar / mole α-crystallin monomer. Prolonged incubations introduced crosslinking of subunits ('late glycation products', mainly dimers). If necessary, noncrosslinked subunits were removed by HPLC gel permeation chromatograph under denaturating conditions.

Separation of monomeric and crosslinked α-crystallin subunits
Crosslinked subunits of α-crystallin glycated *in vitro* were separated from early glycated products by gel-permeation chromatography on a Superose 6 column (Pharmacia). The elution buffer consisted of chaperone buffer with 6 M urea and 0.5% β-mercaptoethanol. Fractions containing crosslinked and monomeric subunits were collected. The proteins were reassociated from 6M urea by dialysis against chaperone buffer.

Insulin assay
Total α-crystallin, αA2-, αB2-homopolymers and glycated proteins were reassociated in chaperone buffer with 6 M urea. Samples were extensively dialyzed against chaperone buffer to remove all traces of urea followed by determination of the protein concentration. We used a spectrophotometer fitted with a thermo-controlled cell holder (Perkin-Elmer Lambda 2 UV/VIS) to maintain a constant temperature during the chaperone assay.[9] The reference cell holder was filled with chaperone buffer; the sample cell holder with 250 µg insulin and α-crystallins in a final volume of 1 mL. The samples were preincubated for 5 minutes at 37°C. We added 20 µL of a freshly prepared 1 M DTT (dithiothreitol) solution and recorded the absorption at 360 nm for 15 minutes.

Binding assay
Crystallin samples (0-200 µg) were incubated with membranes containing 25 µg of MIP26 during 4 hours at 37°C with regular homogenization. An aliquot was taken from the incubation mixture. The membranes were washed four times with chaperone buffer to remove unbound material and analyzed using SDS-PAGE. The bands were scanned (Imaging Densitometer 670, Biorad) and compared with patterns from the incubation mixture to determine the levels of bound α-crystallin (Molecular Analyst 1.2, Biorad).

Results and Discussion

Effect of glycation on the chaperone-like activity of α-crystallin
We isolated α-crystallin from the cortex of calf lenses and performed modifications *in vitro* with glucose-6-phosphate and ribose. Early glycation products were obtained by incubating until a minimal modification level of 1 mole sugar / mole α-crystallin subunit was reached, as determined on iso-electrofocussing gels. Late (crosslinked) products were obtained by removal of monomeric subunits using gel-permeation chromatography. Both fractions were reconstituted by unfolding/refolding in 6 M urea.

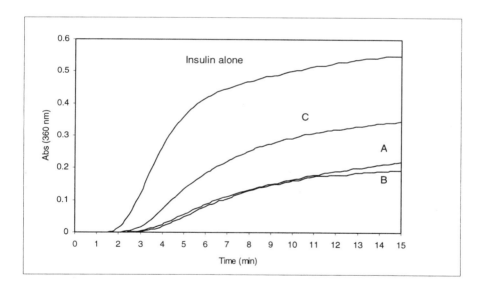

Figure 1. *Prevention of DTT-induced precipitation of insulin by α-crystallin (20 %, w/w, relative to insulin). A, control : non-modified total α-crystallin; B, early glycation products (± 1 mole ribose/ mole subunit) of α-crystallin; C, late (crosslinked) glycation products of total α-crystallin*

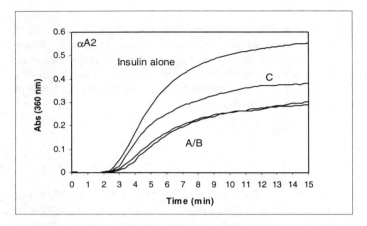

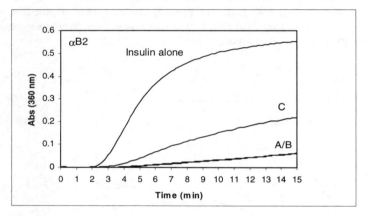

Figure 2. *Prevention of DTT-induced precipitation of insulin by αA2- or αB2-homopolymers (20 %, w/w, relative to insulin). A, controls : non-modified αA2- or αB2-complexes; B, early glycation products (± 1 mole ribose / mole subunit) of αA2- or αB2-homopolymers; C, late (crosslinked) glycation products of αA2- or αB2- complexes*

The presence of early glycation products did not lead to changes in the protective behavior of α-crystallin (Figure 1). Chaperone experiments with the insulin assay revealed a reduction in protecting ability for crosslinked α-crystallin. The chaperone assay was also performed using purified αA2- and αB2-homopolymers.The αB2-complexes were more effective in preventing insulin precipitation compared with αA2 homopolymers (Figure 2). The sugar-modified complexes revealed results similar to those found for total α-crystallin. For both homopolymers, early glycation did not lead to reduced chaperone-like activity, while crosslinking of subunits diminished their protective properties. The negligible effect of early glycation is in line with other lysine-directed modifications we performed by using cyanate or citraconic anhydride. The α-crystallin complexes thus seem to be remarkably resistant against simple modification of free amino groups.

Late glycation does have negative consequences for the functional properties of the protein. Earlier reports mentioned a reduced chaperone-like activity of α-crystallin following subunit crosslinking with dimethylsuberimidate (DMS) or dimethyl 3,3_-dithiobispropionimidate (DTBP).[10] These experiments were performed with the classical heat-denaturation assay using βlow-crystallin as a substrate. Cleavage of the DTBP-induced crosslinks resulted in a 50% recovery in chaperone-like activity underlining the restrictive role of subunit crosslinking. Our results support the view that crosslinking of crystallins, a common feature in aging and cataract, negatively affects the chaperone-like activity of α-crystallin. Since the chaperoning mechanism is unknown, one can only speculate about the nature of this effect. Crosslinking of subunits possibly limits the structural flexibility of the α-aggregates necessary for substrate binding and solubilization.

Table 1. *Binding of ribose-modified total α-crystallin to MIP26. The C-terminal domain of recombinant αB-crystallin was tested additionally*

	Effect of MIP26 Binding
Early glycation Total α–crystallin (1 mole ribose/mole Subunit)	No influence
Ribose-induced crosslinking of total α–crystallin (>90% dimerization)	No influence
C-Terminal domain α–B crystallin	No binding

α-Crystallin and MIP26

In contrast with β– and γ-crystallins, α-crystallin is able to bind specifically to lens membranes, especially to the lens membrane protein MIP26 which functions as a water-channel.[11] It is not known whether this binding reflects any metabolic or cytological function. The nature of the α-crystallin / MIP26 interaction is unclear. We tested the C-terminal domain of αB-crystallin in our binding assay but found no association with MIP26 (Table 1). The amounts of membrane-bound crystallins are higher in aging and cataractous lenses.[6] We therefore considered the possibility that glycation of α-crystallin could increase its affinity for membrane-binding. We studied the effect of glycation of calf α-crystallin *in vitro* on its binding to lens membranes. α-Crystallin was glycated with ribose up to a minimum level of 1 mole bound sugar / mole subunit and tested in a binding assay. Preliminary results with ribose revealed no difference in binding between early glycated total α-crystallin, crosslinked subunits (90% dimerization) and non-modified protein (Table 1). We are currently testing the binding of both early and late glycated α-crystallin homopolymers to the membrane fraction. Glycation of MIP26 itself, predominantly occurring at lysines 238 and 259 (Ref 12), can already have an affect on permeability as shown in reconstituted liposomes.[13] Currently, we are also developing an oocyte system which more closely mimics the lens *in vivo*. We plan to inject modified and non-modified α-crystallins and study the effect of glycation on membrane binding.

Acknowledgement
This work was supported by the *Diabetes Fonds Nederland*

References

1. J.H. Kinoshita, A thirty year journey in the polyol pathway, *Exp. Eye Res.*, 1990, **50**, 567-573.
2. W.Z. Zhang and R.C. Augusteyn, On the interaction of α-crystallin with membranes, *Curr. Eye Res.*, 1994, **13**, 225-230.
3. G.J. Caspers, J.A.M. Leunissen and W.W. de Jong, The expanding small heat-shock protein family, and structure predictions of the conserved "α-crystallin domain", *J. Mol. Evol.*,1995, **40**, 238-248.
4. J. Horwitz, α-crystallin can act as a molecular chaperone, *Proc. Natl. Acad. Sci.*, 1992, **89**, 10449-10453.
5. J. Horwitz, T. Emmons, and L. Takemoto, The ability of alpha crystallin to protect against heat- induced aggregation is age-dependent, *Curr. Eye Res.*, 1992, **11**, 817-822.
6. G. Chandrasekher and R.J. Cenedella, Protein associated with human lens "native" membrane during aging and cataract formation, *Exp. Eye Res.*, 1995, **60**, 707-717.
7. P.J.M. van den Oetelaar, P.F.H.M. van Someren, J.A. Thomson, R.J. Siezen and H.H. Hoenders, A dynamic quaternary structure of bovine alpha-crystallin as indicated from intermolecular exchange of subunits, *Biochemistry*, 1990, **29**, 3488-3493.
8. P. Russell, W.G. Robinson and J.H. Kinoshita, A new method for rapid isolation of the intrinsic membrane proteins from lens, *Exp.Eye Res.*, 1981, **32**, 511-516.
9. M.A.M. van Boekel, S.E.A. Hoogakker, J.J. Harding, and W.W. de Jong, The influence of some post-translational modifications on the chaperone-like activity of α-crystallin, *Ophthalm. Res*, 1996, **28**, 32-38.
10. K. Krishna Sharma and B.J. Ortwerth, Effect of cross-linking on the chaperone-like function of alpha crystallin, *Exp. Eye Res.*, 1995, **61**, 413-421.
11. S.M. Mulders, G.M. Preston, P.M.T. Deen, W.B. Guggino, C.H. van Os and P. Agre, Water channel properties of lens Major Intrinsic Protein, *J. Biol. Chem.*, 1995, **270**, 9010- 9016.
12. S. Swamy-Mruthhinti and K.L. Schey, Identification of in vitro glycated sites of MIP by mass spectrometry, *Invest. Ophthal. & Vis. Sci.*, 1997, **38**, S299.
13. M.S. Swamy and E.C. Abraham, Glycation of lens MIP26 affects the permeability in reconstituted liposomes, *Biochem.Biophys.Res.Comm.*, 1992, **2**, 632-638.

Aminoguanidine Prevents the Depletion of Neurons Containing Nitric Oxide Synthase in the Streptozotocin Diabetic Rat

E Roufail,[1,2] T Soulis,[3] ME Cooper[3] and SM Rees[1]

[1]DEPARTMENT OF ANATOMY AND CELL BIOLOGY, UNIVERSITY OF MELBOURNE, PARKVILLE, VICTORIA, 3052, AUSTRALIA; [2]DEPARTMENT OF OPHTHALMOLOGY, UNIVERSITY OF MELBOURNE, PARKVILLE, VICTORIA, 3052, AUSTRALIA; [3]DEPARTMENT OF MEDICINE, UNIVERSITY OF MELBOURNE, AUSTIN AND REPATRIATION MEDICAL CENTER (REPATRIATION CAMPUS) BANKSIA STREET, WEST HEIDELBERG, 3081, AUSTRALIA

Summary
Recently, we demonstrated that nitric oxide synthase immunoreactive (NOS-IR) neurons are closely associated with the retinal vasculature and have proposed that these neurons are the mechanism by which retinal blood flow and metabolism are linked. This study examined the effect of diabetes on the population of NADPH diaphorase (NADPHd)-positive (equivalent to NOS-IR) neurons in the retina.

Aminoguanidine (AG), an inhibitor of advanced glycation, has been shown to attenuate AGE accumulation as well as the progression of retinal disease in experimental diabetes. It is known that aminoguanidine also inhibits NOS *in vitro*. This study investigated the effect of diabetes and aminoguanidine (an inhibitor of advanced glycation) or N^G-monomethyl-L-arginine (L-NAME) (an inhibitor of NOS), treatment on the population of NAPDHd-positive amacrine cells in the rat retina.

Diabetic and control rats were randomized and received no treatment, aminoguanidine or L-NAME for 32 weeks. Diabetic rats had increased serum glucose levels and blood pressure, but lower body weight compared with controls. The number of NADPHd-positive neurons per retina was significantly reduced (-26%) in diabetic rats compared with control rats. Aminoguanidine treatment of diabetic rats restored the number of NADPHd-positive neurons to normal levels at 32 weeks. L-NAME treatment of diabetic animals had no effect on the number of NADPHd-positive neurons. Since the retinoprotective effect of aminoguanidine cannot be reproduced by L-NAME, it is likely to be mediated by a decrease in advanced glycation rather than by inhibition of NOS.

Introduction

Diabetic retinopathy is the most common microvascular disease affecting virtually every patient who has insulin-dependent diabetes mellitus for 15 years or longer.[1] The morphological changes in the retinal vasculature, such as the loss of pericytes and the thickening of the basal lamina have been well described.[2] The change in retinal blood flow and [I^{125}]-albumin permeation at early and late time points in the development of the disease have been documented.[3,4] The changes in vascular responses observed in experimental models and diabetic patients, including reduced endothelium-dependent relaxation and increased reactivity to vasoconstrictors such as noradrenaline, are well characterized.[5] The mechanisms for the pathogenesis of these changes have not been clearly elucidated apart from the observation that the severity of the retinopathy is associated with the degree of chronic hyperglycemia.[6] It is likely that several mechanisms are involved in the pathogenesis of diabetic vascular dysfunction and retinopathy. The observation that aminoguanidine (a potent inhibitor of inducible NO synthase (iNOS)[7] and AGE formation[8]) significantly ameliorates diabetes-induced vascular dysfunction in the retina[9]) in a similar way to N^G-monomethyl-L-arginine (L-NMMA), (a non-selective

NOS inhibitor) has indicated that there may be a relative or absolute increase in NO activity mediating the early vascular dysfunction of diabetes.[3,10]

The aim of the study was to investigate the role of AGEs and aminoguanidine treatment on the population of NOS-positive retinal neurons in long-term (32 weeks) streptozotocin (STZ)-diabetic rats. This was compared with the effect of L-NAME treatment.

Materials and Methods

Diabetes was induced in male Sprague-Dawley rats (8-9 weeks old, weighing 200-250g) by injection of streptozotocin (STZ) (Boehringer-Mannheim, Mannheim, Germany) into the tail vein after 16 hours of fasting at a dose of 55mg/kg body weight. A rat was considered to be diabetic if its morning non-fasting plasma glucose was > 15 mmol/L one day after the injection of STZ. Rats were randomized into: untreated control and diabetic groups, aminoguanidine-treated control and diabetic groups and L-NAME-treated diabetic groups at the non-pressoric dose of 5g/L in drinking water. Control treated rats were given drinking water containing 2g/L AG hydrogen carbonate and diabetic treated rats were given water containing 1g/L of AG hydrogen carbonate (Fluka Chemica,Buchs,Swizerland), a dose previously shown to be effective in preventing neurovascular dysfunction in diabetic rats.[11] Diabetic rats received 2 units of ultralente insulin (Ultratard HM, Novo Industries, Bagsvaerd, Denmark) subcutaneously every second day in order to maintain body weight and to improve survival. Non-fasting serum glucose levels were measured by the glucose oxidase technique at 32 weeks. Body weight, and blood pressure as determined by tail-cuff plethysmography, were measured every eight weeks. At the same time blood was collected for measurement of glycated hemoglobin by a high-performance liquid chromatography assay (Biorad, Richmond,CA).

Tissue preparation
Rats were killed with an overdose of sodium pentobarbital (13mg/100g body weight). The right eye from each animal was removed and the cornea perforated. The eyes were placed in 4% paraformaldehyde in 0.1M phosphate buffer (PB) for 1h at 4°C. The retina was dissected in 0.1M PB and placed in 4% paraformaldehyde for a further 2h and prepared for NADPHd histochemistry.

NADPH diaphorase staining
Dissected retinas were washed for 3x10 min in Tris Buffer (pH 7.6). They were then left overnight in 0.2% Triton X-100 in Tris buffer (pH 7.6). Stained retinal tissue was stained for NADPHd at 37°C for 2h by incubating sections in a premixed solution containing 0.25% nitroblue tetrazolium, Tris Buffer, Triton X-100 and 0.1% βNADPH. Stained retinas were mounted on slides coated with 1% gelatin solution. The retinas were compressed for 20 min and then cover slips were applied over glycerol jelly.

Analysis
Cells in all retinas were counted by eye using a projectoscope (Olympus BH-2). The number of NADPHd-positive cells per retina was assessed by counting the number of cells within or touching the upper and right borders of a grid, the area of which was equivalent to $0.45mm^2$ at 300x magnification. Counts were made at 2mm intervals by scanning systematically iacross the entire retina. From the total number of cells counted in the area of retina examined, the average density of NADPHd-positive neurons per mm^2 of retina

was determined. The total area of the retina was determined from a projected image of the whole-mount retinas using a digitizing pad interfaced to a computer. From this figure and the average density per retina, the total number of NADPHd-positive neurons per retina was determined.

All values are expressed as means ± SEM of all animals in each group. Comparison of the mean values among groups at each time point was made by ANOVA. *P-V*alues less than 0.05 were considered to be significant.

Results and Discussion

Diabetic rats had significantly elevated blood glucose and HbA_{1c} levels compared with control rats. Diabetes was also associated with reduced weight gain. Treatment of diabetic rats with either AG or L-NAME did not significantly affect body weight or glycemic control during the duration of the study. Diabetic rats had significantly elevated blood pressure throughout the study compared with controls rats. Systolic blood pressure was not affected by treatment with AG or L-NAME in the diabetic rats (Table 1).

NADPHd histochemistry

As previously reported, NADPHd histochemistry in the rat retina stains one class of amacrine cell located in the inner nuclear layer and occasionally displaced to the ganglion cell layer. These neurons are closely associated with the vasculature.[12] There was no qualitative difference in the intensity of staining in control versus diabetic animals of either the amacrine cells or the retinal vasculature. We demonstrated in untreated diabetic animals, that NOS-IR and NADPHd histochemistry were co-localized. All of the seventy cells examined all stained positively for NADPHd and NOS-IR. No NADPH diaphorase-positive neuron was observed that did not show NOS-IR or *vice versa*.

Table 1. *Glycemic control, body weight, blood pressure and NADPHd-positive cells after 32 weeks treatment of rats*

	Control	Control+AG	Diabetic	Diabetic+AG	Diabetic+ L-NAME
n	7	7	5	10	6
Body weight (g)	535±20*	523±14*	378±20	430±31	348±18
Glucose (mmol/L)#	5.5±0.2*	5.4±0.1*	26.7±1.3	26.8±1.0	25.4±0.9
HbA_{1c} (%)	†2.8±0.1*	2.9±0.1*	5.2±0.4	5.3±0.3	4.9±0.5
Blood pressure (mmHg)#	118±3*	121±2*	129±3	131±2	136±4
NADPHd cells (n)	3091±145	2595±186†	2284±148†	2838±165	2396±95†

Mean of weeks 8-32, * p<0.01 versus all diabetic groups, † p<0.01 versus control

Diabetes was associated with a reduction in the population of NADPHd-positive neurons. We have shown that there was a 26% reduction in the number of NADPHd-positive neurons. Aminoguanidine treatment resulted in a restoration of the total number of NADPHd-positive neurons to control levels after 32 weeks (Table 1). By contrast, in control rats AG treatment decreased the number of NADPHd staining cells. In diabetic rats treated with L-NAME no effect on the population of NADPHd-positive neurons was seen, the number of positive neurons remaining significantly below control levels.

This study has focused on the effects of STZ-induced diabetes on the population of neuronal NOS (nNOS) containing amacrine cells in the rat retina. We have shown that NADPHd and nNOS are 100% co-localized in the diabetic rat retina as was the case in previous studies on non-diabetic rat retina.[12,13] This is of significance since, in a number of situations, NADPHd and NOS are not co-localized and so their equivalence should be proved in all experimental situations.[12]

We have demonstrated that the reduction in NOS in retinal amacrine cells was not temporary but was consistent throughout the observation period. Furthermore, in a group of animals made euglycemic for 3 weeks after induction of diabetes there was no significant reduction in the number of nNOS-containing amacrine cells compared with their age-matched controls (data not shown). These findings support the contention that the change nNOS in diabetic rats is specific and not due to a toxic effect of streptozotocin.

There are several reports that implicate NO deficiency in the slowing of nerve conduction [11,14] and vascular relaxation[15] in the acutely streptozotocin-diabetic rat as well as other reports of NO deficiency secondary to hyperglycemia[16] Our finding of reduced numbers of NADPHd-positive neurons (nNOS equivalent) is consistent with these observations and implies a loss of nNOS activity that might contribute to retinal neurovascular dysfunction.

A number of biochemical pathways have been implicated in the diabetic alteration of NOS activity in diabetes; included among these is the process of advanced glycation. This pathway leads to irreversible crosslinking of long lived-tissue proteins.[17] This process is accentuated in diabetes and has been postulated to contribute to the development of a range of diabetic complications including nephropathy,[18] retinopathy[2], and neuropathy.[19-21] Aminoguanidine has been shown to be retinoprotective *in vivo* by retarding the long-term structural changes of retinopathy such as thickening of the capillary basal lamina and loss of retinal capillary pericytes.[2] Aminoguanidine has also been shown to protect bovine retinal pericytes from glycotoxicity *in vitro*.[22]

This study confirmed that aminoguanidine retards the progression of retinal disease as assessed by measuring the number of NADPHd-positive neurons. The lack of effect of L-NAME on the total number of NADPHd-positive neurons in the diabetic retina is consistent with the predominant action of aminoguanidine as an inhibitor of advanced glycation rather than via nitric oxide synthase inhibition *in vivo*. It is of interest that in control animals NADPHd staining was reduced by AG and it is possible that in the non diabetic context where AGE accumulation is less evident, the predominant action of AG is as an inhibitor of NOS.

The findings of this study concur with the observations of aminoguanidine treatment in preventing the long-term structural changes that occur in the diabetic retina[2] and may reflect the prevention of damage to intracellular enzyme systems by AGEs. Aminoguanidine has also been shown to prevent AGE accumulation in subendothelial collagen. It has been suggested that AGEs quench NO leading to NO deficiency.[8] Therefore, in diabetes, there would be AGE accumulation with associated NO deficiency in the retina possibly manifested by decreased NOS-containing neurons. Chronic aminoguandine treatment would presumably prevent retinal AGE accumulation and therefore reverse the depletion of NOS containing neurons in experimental diabetes. The present study has not determined if NOS depletion represents less of neurons containing this enzyme or relates to reduced expression of the enzyme in these cell. Furthermore, one cannot exclude a role for other mechanisms which were not investigated in this study such as the polyol or protein kinase C pathways.

References

1. R Klein, B.E.K. Klein, S.E. Moss, M.D. Davis and D. L. DeMets, The Wisconsin Epidemiological study of diabetic retinopathy IX. Four-year incidence and prevelance of diabetic retinopathy when age at diagnosis is less than 30 years, *Arch. Ophthalmol.*, 1989, **107**, 237-243.
2. H. P. Hammes, D. Strodter, A. Weiss, R. G. Bretzel, K. Federlin and M. Brownlee, Secondary intervention with aminoguanidine retards the progression of diabetic retinopathy in the rat model, *Diabetologia*, 1995, **38**, 656-660.
3. R. G. Tilton, K. Chang, K. S. Hasan, S. M. Smith, J. M. Tetrash, T. P. Misko, T. M. Moore, M. G. Currie, J. A. Corbett, M. L. McDaniel and J. R. Williamson, Prevention of diabetic vascular dysfunction by guanidines: Inhibition of nitric oxide synthase versus advanced glycation end-product formation, *Diabetes*, 1993, **42**, 221-232.
4. J. R. Nyengaard, K. Chang, S. Berhorst, K. M. Reiser, J. R. Williamson and R.G. Tilton, Discordant effects of guanidines on renal structure and function and on regional vascular dysfunction and collegen changes in diabetic rats, *Diabetes*, 1997, **46**, 94-106.
5. N.E. Cameron and M.A. Cotter, The relationship of vascular changes to metabolic factors in diabetes mellitus and their role in in the development of peripheral nerve complications, *Diabetes Metab. Rev.*, 1994, **10**, 189-224.
6. H. B. Chasee, W. E. Jackson, S. L. Hoops, *et al* J.Am.Med.Assoc., Glucose control under renal and retinal complications of insulin-dependant diabetes, *J. Am. Med. Assoc.*, 1989, **261**, 1155-1160.
7. T.P. Misko, W. M.Moore, T. P. Kasten, G. A. Nickols, J. A. Corbett, R. G. Tilton, M. L. McDaniel, J. R. Williamson and M. G. Currie, Selective inhibition of the inducible nitric oxide synthase by aminoguanidine, *Eur. J.Pharmacol.*, 1993, **233**, 119-125.
8. R. Bucala, K.J. Tracey and A. Cerami, Advanced glycation products quench nitric oxide and mediate defective endothelium-dependent vasodilation in experimental diabetes, *J. Clin. Invest.*, 1991, **87**, 432-438
9. K. Ido., E. Chang, W. Ostrow, C. Allison, C. Kilo and R.G. Tilton, Aminoguanidine prevents regional blood flow increases in streptozotocin-diabetic rats (Abstract), *Diabetes*, 1990, **39**, (Suppl.1) 93A.
10. J.A. Corbett, R.G. Tilton, K. Chang, K.S. Hasan, Y. Ido, J.L. Wang, M.A.Sweetland, J.R. Lancaster Jr, J.R.Williamson and M.L.McDaniel, Aminoguanidine, a novel inhibitor of nitric oxide formation, prevents diabetic vascular dysfunction, *Diabetes*, 1992, **41**, 552-556.

11. N.E. Cameron and M.A. Cotter, Rapid reversal by aminoguanidine of the neurovascular effects of diabetes in rats: modulation by nitric oxide inhibition, *Metabolism*, 1996, **45**, 1147-1152.
12. E. Roufail, M. Stringer and S. Rees, Nitric oxide synthase immunoreactivity and NADPH diaphorase staining are co-localised in neurons closely associated with the retinal vasculature in rat and human retina, *Brain Res.*, 1995, **684**, 36-46.
13. T.M.Dawson, D.S.Vredt, M. Fotuhi, P.M.Hwang, and S.H.Snyder, Nitric oxide synthase and neuronal NADPH diaphorase are identical in brain and peripheral tissue, *Proc.Natl.Acad.Sci.USA*, 1991, **88**, 7797-7801.
14. M. J. Stevens, J. Dananberg, E. L. Feldman, *et al*, The linked roles of nitric oxide, aldose reductase and (Na^+, K^+)-ATPase in the slowing of nerve conduction in the streptozotocin-diabetic rat, *J.Clin.Invest.*, 1994, **94**, 853-859.
15. V. Archibald, W.A. Cotter, A. Keegan and N.E. Cameron, Contraction and relaxation of aortas from diabetic rats: effects of chronic anti-oxidant and aminoguanidine treatments, *Naunyn-Schimiedeberg's Arch. Pharmacol.*, 1996, **353**, 584-591.
16. H. Shindo, T. P. Thomas, D. D. Larkins, A. L. Karihaloo, H. Inada, T. Onaya, M. J. Stevens and D. A. Green, Modulation of basal nitric-oxide-dependent cyclic-GMP production by ambient glucose, *Myo*-insitol, and protein kinase C in SH-SY5Y human neuroblastoma cells, *J. Clin. Invest.*, 1996, **97**, 736-745.
17. R. Bucala, A. Cerami and H. Vlassara, Advanced glycation end-products in diabetic complications: biochemical basis and prospects for therapeutic intervention, *Diabetes Rev.*, 1995, **3**, 258-268.
18. T. Soulis-Liparota, M.E. Cooper, D. Papazoglou, B. Clarke and G. Jerums, Retardation by aminoguanidine of deveplopment of albuminuria, mesangial expansion and tissue fluorescence in streptozotocin-induced diabetic rats, *Diabetes*, 191, **40**, 1328-1334.
19. N.E. Cameron, M.A. Cotter, K.C. Dines *et al.*, Effects of aminoguanidine on peripheral nerve function and polyol pathway metabolism in streptozotocin-induced diabetic rats, *Diabetalogia*, 1992, **35**, 946-950.
20. S. Yagihashi, M. Kamijo, M. Baba, N. Yagihashi and K. Nagai, Effect of aminoguanidine on functional and structural abnormalities in peripheral nerve of STZ-induced diabetic rats, *Diabet . Metab.*, **41**, 47-52.
21. J. Kihara, J. D. Schmelzer, J. F. Pudusio, G. L. Curran, K. K. Nicklander and P. A. Low, Aminoguanidine effects on blood flow, vascular permeability, electrophysiology and oxygen free radicals, *Proc. Natl. Acad. Sci. USA*, 1991, **88**, 6107-6111.
22. R. Chibber, P. A. Molinatti and J. F. K. Wong, The effect of aminoguanidine and tolrestat on glucose toxicity in bovine retinal capillary pericytes, *Diabetes*, 1994, **43**, 758-763.

Modification of Low-Density Lipoprotein by Methylglyoxal Alters its Physico-Chemical and Biological Properties

Casper G. Schalkwijk,[1] Mario A. Vermeer,[2] Nicole Verzijl,[2] Coen D.A. Stehouwer,[3] Johan te Koppele,[2] Hans M.G. Princen[2] and Victor W.M. van Hinsbergh[2]

[1]DEPARTMENT OF CLINICAL CHEMISTRY, FREE UNIVERSITY, P.O. BOX 7075, 1007 MB AMSTERDAM, THE NETHERLANDS; [3]DEPARTMENT OF INTERNAL MEDICINE AND INSTITUTE FOR CARDIOVASCULAR RESEARCH, FREE UNIVERSITY, AMSTERDAM, THE NETHERLANDS; [2]GAUBIUS LABORATORY TNO-PG, LEIDEN, THE NETHERLANDS

Summary
Nonenzymatic glycosylation of low-density lipoprotein (LDL) has been suggested to be involved in the development of atherosclerosis in patients with diabetes. Since α-dicarbonyl compounds were identified as intermediates in nonenzymatic glycosylation, we investigated the effect of the physiological α-dicarbonyl compound methylglyoxal (MG) on the physico-chemical and biological properties of LDL. MG modifies LDL in a time- and concentration-dependent manner as indicated by the production of fluorescent products and the increase in net-negative charge. Despite the production of superoxide anion radicals, LDL modification by MG is not accompanied by peroxidation of the polyunsaturated fatty acids. MG-LDL showed impaired recognition by the LDL receptor, but increased binding to scavenger receptors. However, MG-LDL did not enhance cholesteryl ester synthesis in murine macrophages.

Introduction

Modification of LDL has been suggested to be an important factor in the premature onset and rapid progression of atherosclerosis in diabetes. Several modified forms of LDL have been characterized in diabetic patients including nonenzymatic glycosylated forms of LDL (AGE-LDL).[1,2] The recognition of AGE-LDL by the LDL receptor is markedly impaired,[3] and the impaired clearance kinetics[4] may result in the higher amounts of AGE-LDL observed in diabetic patient plasma.[1,2] AGE-LDL is preferentially taken up by macrophages, leading to enhanced cholesteryl ester accumulation.[5] Although the mechanism of nonenzymatic glycosylation is not completely understood and the structures of major AGEs remain to be established, α-dicarbonyl compounds have been identified as intermediates in the nonenzymatic glycosylation and crosslinking of proteins by glucose.[6,7] Hyperglycemia is associated with the formation of the α-dicarbonyl compounds, particularly 3-deoxyglucosone[7] and MG.[8] In patients with diabetes mellitus, the concentrations of these compounds were found to be increased 2- to 6-fold.[7,9] The major source of MG is thought to be the nonenzymatic dephosphorylation of the trioses dihydroxyacetone phosphate and glyceraldehyde-3-phosphate. Formation of MG-modified proteins involves glycoxidation leading to AGE-like fluorescence characteristics.[10,11] Furthermore, MG-modified proteins have been shown to be ligands for the AGE receptor,[12] indicating that adducts in MG-modified proteins are analogous to those found in AGEs. The hydroimidazolone adduct of MG-modified arginine residues, N_δ-(5-hydro-5-methyl-4-imidazolon-2-yl)ornithine, was the receptor recognition factor for binding of proteins minimally-modified by MG to the AGE receptors.[13]

In this paper we describe modification of LDL by MG and the characterization of the chemical and biological properties of these modified LDL preparations.

Materials and Methods

The procedure for the preparation of LDL was adapted from the method described by Princen et al.[14] MG was prepared by the acid hydrolysis of 1,1-dimethoxyacetone and purified by distillation;[15] the concentration of MG was determined as described in Ref 9. LDL modified with MG was prepared by the incubation of LDL (1.0 mg/mL) with MG in PBS containing 10 µM EDTA at 37 Oc. Electrophoretic mobility of native- and modified-LDL was determined by agarose gel electrophoresis using a Beckman Paragon Lipoprotein Electrophoresis kit and barbital buffer at pH 8.6. Amino acid analysis was determined as described recently.[16] The ability of MG-LDL to generate superoxide anion radicals was determined by the reduction of cytochrome c as measured spectrophotometrically by recording the increase in absorbance at 550 nm for 30 minutes. In some experiments, superoxide dismutase was added at a concentration of 1 mg/mL to confirm the generation of the superoxide anion radical. Fatty acid composition of LDL was determined as described previously.[14] Fatty acids were quantified by peak area comparison using pentadecanoic acid (15:0) as internal standard.

LDL binding and competition experiments were conducted on human skin fibroblasts and the murine macrophage-like cell line P388D1. LDL preparations were radioiodinated with carrier-free ^{125}I-iodine. LDL (10 µg/mL) binding by cultured fibroblasts was measured after an 18-hour pre-incubation of the cells with Medium 199 containing 10% lipoprotein-deficient serum. Binding was studied at 4 °C in fresh medium. After 3 hours, the cells were washed four times with ice-cold PBS containing 0.1% human serum albumin and three times with PBS. Then the cells were solubilized in 0.5 mL of 0.3 M NaOH. Radioactivity and the amount of cellular protein in the lysate was determined. Binding was calculated by subtracting the amount of non-specific binding, as determined in the presence of a 25-fold excess of unlabeled LDL or a 25-fold excess of unlabeled MG-LDL preparations. Association of ^{125}I-Ox-LDL with P388D1 murine macrophages, cultured in 5 cm^2 dishes, was studied at 37°C at a concentration of 10 µg/mL. After 4.5 hours the medium was removed and the cells were washed four times with ice-cold PBS containing 0.1% albumin and twice with PBS. The cells were solubilized in 0.5 mL 0.3 M NaOH and the radioactivity was counted and represents association. Competitive inhibition of association was measured by the incubation of ^{125}I-Ox-LDL in the presence of a 25-fold excess of unlabeled LDL preparations.

LDL cholesterol uptake was estimated by measurement of the stimulation of ^{14}C-oleate incorporation into cholesteryl esters. Cell monolayers of P388D1 were incubated with LDL or modified LDL at a concentration of 100 µg/mL for 48 hours at 37°C in dulbecco's modified eagle medium containing 10% newborn calf serum. Then 100 µM ^{14}C-oleate complexed with 0.1% fatty acid-free bovine serum albumin was added and further incubated for another 2 hours. Cell monolayers were washed and cellular lipids were extracted according to the procedure of Bligh and Dyer. Lipids were taken to dryness and redissolved in chloroform/methanol (5:1, v/v). The lipids were separated by thin layer chromatography on silicagel 60 plates in chloroform/methanol (98:2, v/v) for 15 minutes, followed by chloroform/hexane (65:45, v/v) for 45 minutes. The ^{14}C-cholesteryl oleate was quantified on a phosphoimager with ^{14}C-cholesterol oleate as external standard.

Results and Discussion

LDL incubated for 18 hours with increasing concentrations of MG showed a dose-dependent increase in fluorescence, with an emission maximum at 420 nm at an excitation maximum of 340 nm (data not shown). The MG-induced development of fluorescence indicates interaction of MG with LDL. MG induced a dose- and time-dependent increase in net-negative charge in LDL as detected by increased mobility on agarose electrophoresis (Figure 1). MG-LDL showed major modifications of arginine residues (up to 70%) but minor modification of lysine (up to 5%). Whether MG modified both phospatidylethanoamine and the ApoB moieties of LDL, as has been shown for AGE-LDL,[2] is unknown and whether the characteristic MG-adduct-derived N^{ϵ}-[1-(1-carboxy)ethyl]-lysine and imidazolysine are formed during the reaction of LDL with MG needs to be clarified.

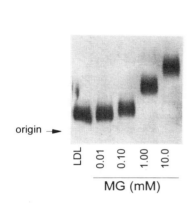

Figure 1. *Electrophoretic mobility of LDL and MG-LDL. LDL (1 mg/mL) was exposed to the indicated concentrations of MG for 24 hours at 37 °C and agarose gel electrophoresis was performed*

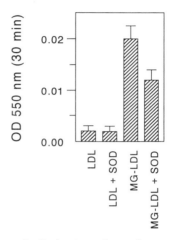

Figure 2. *Reduction of cytochrome c by MG-LDL. LDL was incubated with or without 10 mM MG for 96 hours. After dialysis the generation of superoxide production from the LDL and MG-LDL were measured with or without superoxide dismutase*

We studied whether MG-LDL is capable of generating oxidative stress. Under our *in vitro* conditions, MG-LDL generated superoxide anion radicals (Figure 2). Superoxide dismutase produced a decrease in radicals formed, indicating that superoxide was formed, although there was also significant reduction of ferricytochrome c that was not mediated by superoxide. Recently, a similar generation of free-radical species was found during the reaction of MG with amino acids.[11] Generation of superoxide radicals by MG-LDL may result in autoxidation of the LDL and peroxidation of polyunsaturated fatty acids. However, the fatty acid composition of native LDL and MG-LDL is similar, while polyunsaturated fatty acid levels were reduced in copper-induced oxidized LDL (data not shown).

Since MG was shown to modify LDL, we determined whether MG-LDL is still a ligand for the LDL receptor. In the presence of a 25-fold excess of different unlabeled MG-LDL preparations, highly methylglyoxylated-LDL failed to inhibit significantly ^{125}I-LDL binding to fibroblasts (Figure 3), demonstrating impaired binding to the LDL receptor.

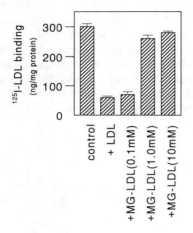 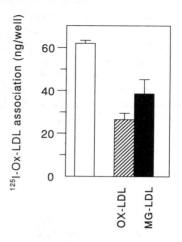

Figure 3. *Impaired binding of MG-LDL to the LDL receptor. Human fibroblasts were incubated with ^{125}I-labeled LDL (100% binding) and the competitive inhibition of binding was investigated with a 25 fold excess of unlabeled LDL and MG-LDL preparations, prepared by incubation of LDL with 0.1 mM, 1 mM, and 10 mM MG for 96 hours*

Figure 4. *Association of MG-LDL to the scavenger receptor. P3881 macrophages were incubated with ^{125}I-Ox-LDL in the absence or presence of a 25-fold excess of unlabeled Ox-LDL or MG-LDL. MG-LDL was prepared with 10 mM MG for 96 hours*

When murine macrophages, which possess scavenger receptors, were incubated with ^{125}I-LDL, unlabeled MG-LDL failed to inhibit LDL association significantly (data not shown). On the other hand, association of ^{125}I-Ox-LDL with macrophages was competitively inhibited by unlabeled MG-LDL (Figure 4), although the competition by unlabeled MG-LDL was less than the same amount of unlabeled Ox-LDL. This indicates that MG-LDL binds to scavenger receptors on macrophages. This binding to scavenger receptor(s) is in agreement with binding of other chemically modified forms of LDL with a negative charge.[5,17-19]

We studied whether association of MG-LDL with macrophages is followed by subsequent cellular esterification of cholesterol. Surprisingly, MG-LDL modified by various concentrations of MG showed significantly less cholesteryl ester synthesis compared with both LDL, Ac-LDL and Ox-LDL, and even less than control LDL in P388D1 murine macrophages (Figure 5). It has been proposed that binding of LDL to macrophages is not only dependent on the charge of LDL but also on the conformation of the LDL particle and, therefore, it may be that MG induces conformational changes with consequences for uptake and cholesterol ester synthesis by macrophages.

In conclusion, MG modifies LDL *in vitro*, with the formation of a fluorescent LDL particle with a net negative charge which generates superoxide anions and, thus, oxidative stress. The uptake of MG-LDL by the LDL receptor is impaired and therefore MG-LDL may become prone to further modification and, therfore, may contribute to the hyperglycemia-induced accelerated development of atherosclerosis in patients with diabetes. However, MG-

LDL does not lead to cholesterol-ester accumulation in murine macrophages. Further studies will be necessary to determine whether MG-LDL and its derivatives exist *in vivo* and whether they contribute to the development of atherosclerosis via activation of macrophages and endothelial cells, similar to those found for AGE-adducts.[20,21]

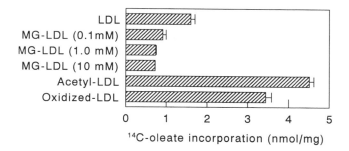

Figure 5. *Effect of methylglyoxal-LDL on cholesteryl ester synthesis in P338D1 murine macrophages. The ^{14}C-oleate incorporation into cholesteryl esters in P338D1 murine macrophages in the absence of LDL was 0.40 nmol/mg protein. The MG-LDL preparations were prepared by incubation with 0.1 mM, 1.0 mM and 10 mM MG for 96 hours*

Acknowledgments
We wish to thank W. van Duyvenvoorde for assistance with the fatty acid analysis and Dr H. ten Brink for the synthesis of MG. This project is financial supported by the Praeventiefonds (grant 28-2622-2), the Netherlands.

References

1. E. Schleicher, T. Deufel and O.H. Wieland, Non-enzymatic glycosylation of human serum lipoproteins, *FEBS Lett.*, 1981, **129**, 1-4.
2. R. Bucala, Z. Makita, T. Koschinsky, A. Cerami and H. Vlassara, Lipid advanced glycosylation: pathway for lipid oxidation in vivo, *Proc. Natl. Acad. Sci. USA*, 1993, **90**, 6434-6438.
3. J.L. Witzum, E.M. Mahoney, M.J. Branks, M. Fisher, R. Elam and D. Steinberg, Nonenzymatic glucosylation of low-density lipoprotein alters its biologic activity *Diabetes*, 1982, **31**, 283-291.
4. R. Bucala, Z. Makita, G. Vega, S. Grundy, T. Koschinsky, A. Cerami and H. Vlassara, Modification of low density lipoprotein by advanced glycation end products contributes to the dyslipidemia of diabetes and renal insufficiency, *Proc. Natl. Acad. Sci. USA*, 1994, **91**, 9441-9445.
5. M.F. Lopes-Virella, R.L. Klein, T.J. Lyons, H.C. Stevenson and J.L. Witzum, Glycosylation of low-density lipoprotein enhances cholesteryl ester synthesis in human monocyte-derived macrophages, *Diabetes*, 1988, **37**, 550-557.
6. M.C. Wells-Knecht, D.V. Zyzak, J.E. Lichtfield, S.R. Thorpe and J.W. Baynes, Mechanism of autoxidative glycosylation: identification of glyoxal and arabinose as intermediates in the autoxidative modification of proteins by glucose, *Biochemistry*, 1995, **34**, 3702-3709.
7. H.I., Yamada, S. Miyata, N. Igaki, H. Yatabe, Y. Miyauchi, T. Ohara, M. Sakai, H.

Shoda, M. Oimomi and Kasuga, Increase in 3-deoxyglucosone levels in diabetic rat plasma, *J. Biol. Chem.* 1994, **269**, 20275-20280.
8. P.J. Thornalley, Monosaccaride autoxidation in health and disease, *Environ. Health Perspect.* 1995, **64**, 297-307.
9. A.C. McLellan, S.A. Phillips and P.J. Thornalley, The assay of methylglyoxal in biological systems by derivatization with 1,2-diamino-4,5-dimethoxybenzene, *Anal. Biochem.*, 1992, **206**, 17-23.
10. T.W.C. Lo, M.E.Westwood, A.C. McLellan, T. Selwood and P.J.Thornalley, Binding and modification of proteins by methylglyoxal under physiological conditions, *J. Biol. Chem.*, 1994, **269**, 32299-32305.
11. H-S. Yim, S-O. Kang, Y-C. Hah, P.B.Chock and M.B Yim, Free radicals generated during the glycation reaction of amino acids by methylglyoxal, *J. Biol. Chem.*, 1995, **270**, 28228-28233.
12. M.E. Westwood, A.C. McLellan P.J. and Thornalley, Receptor-mediated endocytic uptake of methylglyoxal-modified serum albumin, *J. Biol. Chem.*, 1994, **269**, 32293-32298.
13. M.E. Westwood, O.K. Argirov, E.A. Abordo and P.J. and Thornalley, Methylglyoxal-modified arginine residues- a signal for receptor-mediated endocytosis and degradation of proteins by monocytic THP-1 cells. *Biochem. Biophys. Acta* 1997, **1356**, 84-94.
14. H.M.G. Princen, W. Van Duyvenvoorde, R. Buytenhek, A.Van der Laarse, G. Van Poppel, J.A. Gevers Leuven and V.W.M.Van Hinsbergh, Supplementation with low doses of vitamin E protects low density lipoprotein from lipid peroxidation, *Arterioscler. Thromb. Vasc. Biol.*, 1995, **15**, 325-333.
15. A.C. McLellan and P.J. Thornalley, Synthesis and chromatography of 1,2-diamino-4,5-dimethoxybenzene, 6,7-dimethoxy-2-methylquinoxaline and 6,7-dimethoxy-2,3-dimethylquinoxaline for use in a liquid chromatographic fluorimetric assay of methylglyoxal, *Anal. Chim. Acta.*, 1992, **263**, 137-142.
16. R.A. Bank, E.J. Jansen, B. Beekman and J.M. te Koppele, Amino acid analysis by reverse-phase high-performance liquid chrmomatography: improved derivatization and detection conditions with 9-fluorenylmethyl chloroformate, *Anal. Biochem.*, 1996, **240**, 167-176.
17. I.A. Sobenin, V.V. Tertov, and A.N. Orekhov, Atherogenic modified LDL in diabetes, *Diabetes*, 1996, **45**, S35-S39.
18. M.S. Brown, S.K. Basu, J.R. Falck, Y.K. Ho and J.L. Goldstein, The scavenger cell pathway for lipoprotein degradation: specificity of the binding site that mediates uptake of negatively-charged LDL by macrophages, *J. Supramol. Struct.*, 1980, **13**, 67-81.
19. G.Camejo, A. Lopez, F. Lopez and J. Quinones, Interaction of low density lipoproteins with areterial proteoglycans: the role of charge and sialic acid content *Atherosclerosis* 1988, **55**, 93-105.
20. C. Esposito, H. Gerlach, J. Brett, D. Stern and H. Vlassara, Endothelial receptor-mediated binding of glucose-modified albumin is associated with increased monolayer permeability and modulation of cell surface coagulant properties, *J. Exp. Med.*, 1989, **170**, 1386-1407.
21. A.M. Schmidt, O. Hori, J.X. Chen, J.F. Li, J. Crandall, J. Zhang, R. Cao, S.D. S.D. Yan, J. Brett and D. Stern, Advanced glycation endproducts interacting with their endothelial receptor induce expression of VCAM-1 in cultured human endothelial cells and in mice, J. Clin. Invest. 1995, **96**, 1395-1403.

Production and Metabolism of 3-Deoxyglucosone in Humans

Sundeep Lal, Benjamin S. Szwergold, Michael Walker, William Randall, Francis Kappler, Paul J. Beisswenger[1] and Truman Brown

DEPT. OF NMR, FOX CHASE CANCER CENTER, 7701 BURHOLME AVE., PHILADELPHIA, PA 19111, USA; [1] DEPT. OF ENDOCRINOLOGY, DARTMOUTH-HITCHCOCK MEDICAL CENTER, LEBANON, NH 03756, USA

Summary

3-deoxyglucosone (3DG), a reactive dicarbonyl sugar involved in nonenzymatic glycation of proteins is found in elevated concentrations in diabetic patients and has been implicated in the pathogenesis of diabetic complications. In this study, we provide ex-vivo and in-vivo evidence for production of 3DG via the fructose-3-phosphokinase pathway. Evidence of further metabolism of this carbohydrate by oxidative and reductive pathways to 2-keto-3-deoxygluconate (DGA) and 3-deoxyfructose (3DF), respectively, is presented. Levels of these three sugars, 3DG, DGA and 3DF, in blood components and fluids from normoglycemic and diabetic humans are summarized. A comparison of the relative levels of 3DG and 3DF in urine from normoglycemics and diabetics suggests a possible impairment of the detoxification of 3DG to 3DF in diabetics.

Introduction

Over the past decade, significant experimental support has been gathered in support of the hypothesis that the Maillard reaction plays a role in the pathogenesis of diabetic complications[1]. In diabetics, these reactions lead to increased modification of proteins and elevated levels of advanced glycation endproducts (AGEs)[2] which affect protein structure and function. Previous studies have shown the appearance of AGEs to be coincident with the earliest signs of complications[2] or predictors of their future development.[3] It is generally recognized that the key intermediates in such reactions and in the production of AGEs are the reactive dicarbonyls 3-deoxyglucosone (3DG) and methyl glyoxal.[1,4] In order to characterize the production and metabolism of 3DG and determine its contribution to the pathogenesis of diabetic complications, we have studied the production of this compound as a decomposition product of fructose-3-phosphate (F3P)[5] in human erythrocytes. Its further metabolism to its reductive and oxidative detoxification products, 3-deoxyfructose (3DF)[6] and 2-keto-3-deoxygluconate (DGA)[7] was also investigated (Figure 1). These ex-vivo results were extended to man by measurement of 3DG, 3DF, DGA and F3P, in plasma, erythrocytes and urine, in diabetic and normoglycemic subjects.

It has been suggested that the ability to detoxify 3DG may be an important variable in the development of diabetic complications.[1] To investigate this possibility, we measured concentrations of 3DG and 3DF in urine and plasma from three groups of volunteers, normoglycemics, type II diabetics and type I diabetics. Results from this study demonstrate that, while both 3DG and 3DF are elevated in diabetics relative to normal subjects, the relationship between 3DG and 3DF from the three groups of volunteers differ dramatically suggesting a possible impairment of detoxification of 3DG to 3DF. Our studies support a new conceptual framework for the role of the 3-phosphokinase in the production of 3DG and the metabolic ramifications of its activity.

Material and Methods

Materials
[U-^{13}C] standards of 3DG, DGA and 3DF were synthesized as described previously.[8, 9, 10] 2,3-Diaminonaphthalene and phenylenediamine were obtained from Sigma (St. Louis) and Aldrich (Milwaukee), respectively. Tri-Sil reagent (TMS) was purchased from Pierce Co. (Rockford, Il.).

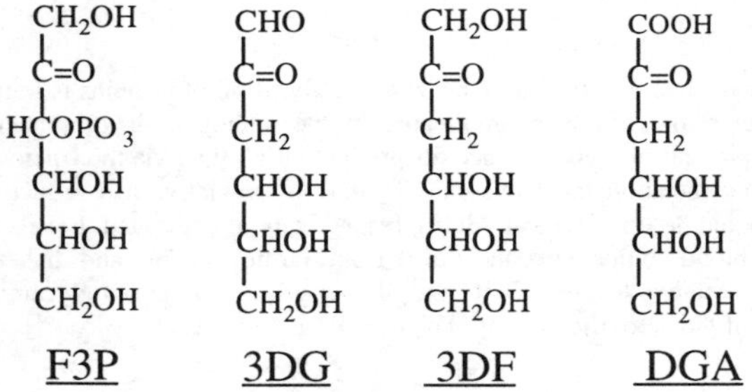

$$\begin{array}{cccc}
CH_2OH & CHO & CH_2OH & COOH \\
| & | & | & | \\
C=O & C=O & C=O & C=O \\
| & | & | & | \\
HCOPO_3 & CH_2 & CH_2 & CH_2 \\
| & | & | & | \\
CHOH & CHOH & CHOH & CHOH \\
| & | & | & | \\
CHOH & CHOH & CHOH & CHOH \\
| & | & | & | \\
CH_2OH & CH_2OH & CH_2OH & CH_2OH \\
F3P & 3DG & 3DF & DGA
\end{array}$$

Figure 1. *Chemical structures of fructose-3-phosphate (F3P), and it's metabolic products 3-deoxyglucosone (3DG), 3-deoxyfructose (3DF), 2-keto-3-deoxygluconate (DGA)*

Collection of blood, urine and description of the patient population
Approximately 8-10 mL of second voided urine and blood samples were collected after an overnight fast from normoglycemic volunteers and patients with type I and type II diabetes. These populations are described in Ref. 11. Blood was collected in heparinized tubes and was separated into plasma and erythrocytes prior to storage at -70 °C.

Incubations of erythrocytes
Erythrocytes from normoglycemic volunteers were incubated in a basal medium buffered with bicarbonate and HEPES between 2-5 % hematocrit. The media contained 5 mM glucose and were supplemented with 30 mM fructose and 2 % antibiotics (penicillin-streptomycin-neomycin). This mixture was incubated under a 5 % CO_2 atmosphere. Periodic aliquots were drawn and red cells were washed and stored at -40 °C until processed. The corresponding media samples were also stored for further analysis.

Processing of plasma and urine for quantification of 3-deoxyglucosone
Plasma samples were processed as described previously.[12] To a 0.35 to 0.5 mL aliquot of a urine sample, 20 mL of a 10 µM 3-[U-^{13}C]DG internal standard was added. This mixture was passed through 3 mL of cation-exchange resin, AG 50W-X4, 200-400 mesh hydrogen form (Bio-Rad, CA) to remove fructoselysine which would interfere with the assay. The column was washed with 3 mL of de-ionized water. The wash was collected and was mixed with 1 mL of 1 mM 2,3-diamino-naphthalene solution containing 100 mM HEPES and 5 mM phenol red (for pH adjustment). The pH of this mixture was adjusted to 7.0 and 3DG converted to silyl ethers as described previously.[12]

GC-MS analyses
The derivatized sample was analysed using a HP 5890 GC equipped with a HP 5971 mass-selective detector and HP 7673 automatic sampler using a method described previously.[12]

Quantification of 3-deoxyfructose
Samples were processed by passing a 0.3 mL aliquot of the test sample through an ion-exchange column containing 0.15 mL of AG 1-X8 and 0.15 mL of AG 50W-X8 resins. The columns were then washed twice with 0.3 mL deionized water, aspirated to remove free liquid and filtered through a 0.45 mm Millipore filter.

50 µL injections of the treated samples were analyzed using a Dionex DX 500 chromatography system. A carbopac PA1 anion-exchange column was employed with an eluant consisting of 16% sodium hydroxide (200 mM) and 84% deionized water. 3DF was detected electrochemically using a pulsed amperometric detector. Standard 3DF solutions spanning the anticipated 3DF concentrations were run both before and after each unknown sample.

Quantification of 2-keto-3-deoxygluconate
A previously described synthetic procedure[9] for synthesis of quinoxalinol derivatives of DGA was adapted for its assay by GC-MS. A 50 µL aliquot of a perchloric acid extract of erythrocytes was incubated overnight with 250 µL of a solution of phenylenediamine (6 mg/mL in 1 M acetic acid). The DGA-phenylenediamine adduct was dried under reduced pressure in a centrifugal evaporator. The dried sample was converted to its trimethyl silyl ethers by addition of 250 mL of TMS. These silyl derivatives were then analysed by GC-MS using the following temperature program: hold for 2 min at 150°C, ramp to 320°C at 10 °C/min., hold for 5 min. at 320 °C. The oven was set at 150°C, injector at 300°C and the transfer line at 290 °C. The mass spectral data acquisition was performed using selected ion monitoring at m/z of 333, 358 and 448 for the sample DGA and the corresponding ions of m/z at 337, 363, and 448 for the 3-[U-13C]DGA internal standard. Quantification of erythrocytic DGA was done using m/z fragment at 333 and the corresponding 3-[U-13C]DGA fragment at 337 as this was the most sensitive and selective mass fragment

Measurement of urine creatinine
Urine creatinine concentrations were determined by the end-point colorimetric method (Sigma Diagnostic kit 555-A) modified for use with a plate reader.

Results & Discussion

Erythrocyte incubations
Previous studies in this laboratory have demonstrated a rapid production of F3P in erythrocytes incubated with 30 mM fructose over a duration of a few hours.[13] Incubation of human erythrocytes with 30 mM fructose over a similar duration resulted in substantial intracellular production of 3DG (Figure 2). To confirm that this 3DG was directly produced from fructose, an identical incubation with [U-13C]-fructose was conducted which yielded 3-[U-13C]DG. Control incubations with glucose did not show such an elevation of 3DG. Concomitant to the increase in 3DG, its reductive and oxidative detoxification products, 3DF and DGA, respectively, also increased over the course of the experiment (Figure 2).

Correlation of erythrocytic F3P with 3DG
Figure 3 shows the correlation between plasma levels of free 3DG with intracellular F3P measured in type I diabetic volunteers (R=0.58; $p<10^{-4}$). These data support the thesis that, in man, F3P is a systemic source of plasma 3DG. Since 3DG is cleared from plasma at the

glomerular filtration rate (GFR), the product of [3DG] and GFR (mL/min) represents the total amount of 3DG cleared from plasma per minute. Therefore, with the knowledge of [3DG]$_{plasma}$, GFR, [F3P]$_{RBCs}$, and assuming that the average volume of the erythrocyte compartment in man is 2000 mL, the *in-vivo* lifetime of F3P (τ) in man was estimated to be 38 hours using the following expression:

[F3P]RBCs / τ x 2000 mL = [3DG]plasma x GFR

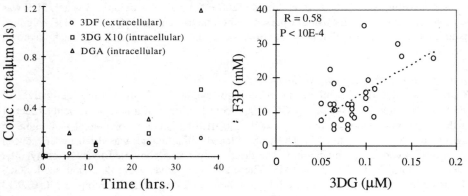

Figure 2. *Production of 3DG, 3DF and DGA in erythrocyte incubation with fructose (5% hematocrit in basal medium)*

Figure 3. *Correlation between plasma levels of 3DG and RBC levels of F3P in type I diabetics*

Summary of levels of 3DG and 3DF in plasma and urine
Table I summarizes the concentrations of 3DG & 3DF in the plasmas of type I and type II diabetics. DGA was measured only in erythrocytes from type I diabetics and normoglycemics. While DGA is also detectable in plasma and urine, an assay for its measurements in these fluids is still being refined. The mean plasma concentration of free 3DG in individuals with type I and type II diabetes was elevated compared with normoglycemics. There were significant differences in the levels of these metabolites in

Table 1. *Summary of levels of 3DG and 3DF in plasma and urine*

Measurement	Normal Subjects	Diabetics Type II	Diabetics Type I
3DG, Plasma (nM)	48 ± 9 (n=30)	79 ± 24 (n=67)	78 ± 24 (n=47)
3DF, Plasma (µM)	1.08 ± 0.25 (n=30)	1.5 ± 0.5 (n=64)	2.1 ± 1.4 (n=41)
DGA, RBCs (µM)	11.5 ± 3.2 (n=11)	NA	31.4 ± 14 (n=17)
3DG, Urine (µmol/gm Creat)	2.7 ± 0.9 (n=19)	5.76 ± 4.3 (n=66)	4.6 ± 3.6 (n=28)
3DF, Urine (µmol/gm Creat)	92 ± 15 (n=20)	166 ± 122 (n=66)	125 ± 71 (n=28)

Means ± Standard deviations; NA, Not Analyzed

normoglycemics compared with type II and type I diabetics, using a two-tailed t-test (p< 10^{-14} and p<10^{-11} for 3DG and p<10^{-7} and p<10^{-5} for 3DF, respectively). The difference between the level of 3DG in type I and type II patients was not significant, although the difference in the 3DF level was significant at p < 0.01.

The urine concentrations of 3DG and 3DF were normalized to the amount of creatinine (Table 1). The differences between normoglycemics and type II and type I diabetics, using a two-tailed t-test, were also significant (p<0.003 and p<0.008 for 3DG and p<0.002 and p<0.015 for 3DF). These data further demonstrate that while the plasma levels of 3DG and 3DF in normoglycemic individuals are tightly regulated, in diabetics there is a considerable variation of the levels of these metabolites among individual patients.

Detoxification of 3DG to 3DF in normoglycemic individuals and diabetics

When plasma and urine 3DG levels were correlated with corresponding 3DF levels in the three groups, a dramatic difference in the slope of these correlations was observed when diabetics were compared to normoglycemic group (Figure 4). These results are particularly striking in the case of urine where there was a significant decline in the slope of this correlation in the order: normoglycemic individuals > type II > type I diabetics. By comparison, in the plasma, the slope of the 3DG to 3DF correlation increased from normoglycemic individuals and type IIs to type I diabetics. Taking into account the clearance of 3DG and 3DF into urine and the ratios of 3DG/3DF in urine and plasma, ~ 35 % of plasma 3DG is unaccounted for in urine. This 3DG, is either detoxified in the kidney, is

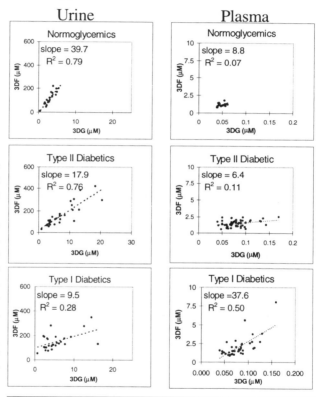

Figure 4. *Relationship between plasma and urine levels of 3DG and 3DF in three populations: normoglycemics, type II and type I diabetics*

recirculated back to the plasma or possibly interacts with the proteins on the lumen surface as it passes through the kidney. From our data it appears that this process appears to be more active in type I diabetics than in type II diabetics or normoglycemic individuals.

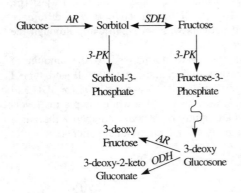

Figure 5. *Schematic of the expanded aldose reductase pathway*

Conclusions

Our findings provide evidence for a mechanism where by fructose produced by the aldose reductase pathway can be converted to a potent dicarbonyl, 3DG, with F3P as a relatively short lived intermediate. This is summarized in Figure 5, which integrates the aldose-reducates pathway with 3-phosphokinase and the 3DG detoxification reactions. Our data suggests that increased production of AGEs in diabetes maybe a result of two mutually reinforcing processes: increased production of 3-deoxyglucosone, and impairment of its detoxification.

References

1. M. Brownlee, Glycation and diabetic complications, *Diabetes*, 1994, **43**, 836-841.
2. Z. Makita, S. Radoff, E.J. Rayfield, Z. Yang, E. Skolnik, V. Delaney, E.A. Friedman and A. Cerami, Advanced glycosylation end products in patients with diabetic nephropathy, *New. Engl. J. Med.*, 1991, **325**, 836-842.
3. P. J. Beisswenger., Z. Makita, T. J. Curphey, L. L. Moore, S. Jean, T. Brinck-Johnsen, R. Bucala and H. Vlassara, Formation of immunochemical advanced glycosylation end products precedes and correlates with early manifestations of renal and retinal disease in diabetes, *Diabetes* 1995, **44**, 824-829.
4. P. J. Thornalley, M. L. Westwood and A.C. McLellan, Formation of methylglyoxal-modified proteins in vitro and in vivo and their involvement in AGE-related processes, *Contributions to Nephrology*, 1995, **112**, 24-31.
5. S. Lal, B. S. Szwergold, A.H. Taylor, W. C. Randall, F. Kappler, K. Wells-Knecht, J. W. Baynes and T. R. Brown, Metabolism of fructose-3-phosphate in the diabetic rat lens. *Arch. Biochem. Biophys.*, 1995, **318**, 191-199.
6. K.J. Knecht, M.S. Feather and J.W. Baynes, Detection of 3-deoxyfructose and 3-deoxyglucosone in human urine and plasma. *Arch. Biochem. Biophys.* 1992, **294**, 130-137.
7. E. Fujii, H. Iwase, I. Ishii-Karakasa, Y. Yajima and K. Hotta, Quantitation of the glycation intermediate 3-deoxyglucosone by oxidation with rabbit liver oxoaldehyde dehydrogenase to 2-keto-3-deoxygluconic acid followed by high performance liquid chormatography, *J. Chromatogr.*, 1994, **660**, 265-270.
8. M.A. Madson and M.S. Feather, An improved preparation of 3-deoxy D-erythro-hexose-2-ulose via the bis(benzoyl hydrozone) and some related constitutional studies. *Carbohydr. Res.*, 1981, **94**, 183-191.
9. M.A. Madson, M.S. Feather, The oxidation of 3-deoxy-D-erythro-hexos-2-ulose ('3-deoxyglucosone') to 3-deoxy-D-erythro-2-hexulosonic acid ('2-keto-3-deoxy-D-gluconate') by D-glucose oxidase, *Carbohydr. Res.*, 1983, **115**, 288-291.

10. W. L. Dills, 3-Deoxy-D-erythro-hexulose: a convenient synthesis and its interaction with the enzymes of fructose metabolism, *Carbohydr. Res.* 1990, **208**, 276-279.
11. P. J. Beisswenger, S. Howell, R. Stevens, A. Seigal, S. Lal, W.R. Randall, B.S. Szwergold, F. Kappler and T.R. Brown, The role of 3-deoxyglucosone and the activity of its degradative pathways in the etiology of diabetic microvascular disease. *Proceedings of the Sixth Maillard Symposium*, 1997.
12. S. Lal, F. Kappler, M. Walker, T. J. Orchard, P. J. Beisswenger, B. S. Szwergold and T. R. Brown, Quantitation of 3-Deoxyglucosone Levels in Human Plasma, *Arch. Biochem. Biophys.* 1997, **342**, 254-260.
13. A. Petersen, F. Kappler, B.S. Szwergold and T.R. Brown, Fructose metabolism in the human erythrocyte phosphorylation to fructose 3-phosphate, *Biochem. J.*, 1992, **284**, 363-366.

The Role of 3-Deoxyglucosone and the Activity of its Degradative Pathways in the Etiology of Diabetic Microvascular Disease

Paul J. Beisswenger[1], Scott Howell[1], Rosalind Stevens[1], Alan Siegel[1], Sundeep Lal[2], William Randall[2], Benjamin S. Szwergold[2], Francis Kappler[2], and Truman Brown[2]

[1]DARTMOUTH MEDICAL SCHOOL AND DARTMOUTH-HITCHCOCK MEDICAL CENTER, LEBANON, NH 03756, USA
[2]DEPT. OF NMR, FOX CHASE CANCER CENTER, PHILADELPHIA, PA 19111, USA

Summary

To address the role of 3-deoxyglucosone (3DG) and the activity of its major degradative pathways in diabetic nephropathy and retinopathy we have measured 3DG and its degradation products [3-deoxyfructose (3DF) and 3-deoxy-2-keto-gluconic acid (DGA)] in plasma and RBCs. In addition to these compounds, we have quantified HbA_{1c}, renal dysfunction [glomerular filtration rate (GFR) and urinary albumin excretion(UAE)] and diabetic retinal sequelae over 2-3 years in 25 IDDM subjects with minimal complications during the earliest stages of diabetic microangiopathy. They were also measured in 58 subjects with NIDDM and 30 non-diabetic subjects.

Plasma 3DG, 3DF and erythrocyte DGA were significantly elevated in diabetic subjects relative to controls. We also found a highly significant association between plasma 3DG and HbA_{1c} concentrations, indicating that glycemic control is an important determinant of 3DG levels. An increased flux of 3DG to DGA and 3DF was associated with greater degrees of hyperglycemia, since both degradation products correlated with HbA_{1c} values.

Increasing rates of GFR also correlated with plasma 3DG levels and 3DG levels increased with increasing urinary albumin excretion (UAE). By contrast, there was an inverse correlation between plasma 3DF and UAE. Subjects showing progression of retinopathy also showed reduced plasma levels of the 3DG product 3DF relative to nonprogressors. As retinal perfusion decreased, we also observed increased 3DG levels.

Introduction

Nonenzymatic glycation is considered to be a major factor in the etiology of diabetic complications. The highly reactive α dicarbonyl 3-deoxyglucosone(3DG) has been postulated to play an important role in glycation and in diabetic complications due to its intrinsic toxicity and its role as a precursor for AGEs.

3DG has been shown to be produced from the Amadori product, fructoselysine (FL)[1], from autoxidation of glucose[2,3] or from fructose-3-phosphate (F3P)[4] which is produced from fructose by a specific 3-phosphokinase. Because of its toxic potential, 3DG is converted to innocuous products by several mechanisms. Both oxidative and reductive pathways exist for the breakdown of 3DG.[5,6] The reductive pathway results in the production of 3-deoxyfructose (3DF) while a less well characterized oxidative pathway leads to the production of 3-deoxy-2-keto-gluconic acid (3-DGA).

3DG levels have been shown to be elevated in diabetes and are further increased by end-stage renal disease.[7,8] Direct quantification of 3DG and its metabolites has not yet been performed in diabetic populations that have been carefully evaluated for early diabetic complications. To evaluate the role of 3DG and its breakdown pathways in the development of diabetic complications, we have developed new assays or modified existing assays to quantify these products.

Materials and Methods

Subjects
Twenty five IDDM subjects were studied for 2 consecutive years. To study the role of 3-deoxyglucosone and its detoxification pathways during early vascular dysfunction and to avoid the effect of decreased renal function on clearance from plasma, we have studied subjects with normal renal function. Subjects were chosen with IDDM duration of ≤ 15 years and age ≤ 40 years. At entry, they did not have significant microalbuminuria(<40 mg/24h) and had minimal or no background retinopathy. Renal function was quantified by determining glomerular filtration rate (by Tc^{99}DTPA clearance)[9] and 24h albumin excretion by RIA. Retinal status was determined on 25 IDDM subjects by 7 field-retinal photographs before and after fluorescein infusion and by vitreous fluorophotometry. Seven field stereo retinal photographs with and without fluorescein infusion were read blind each year by two ophthalmologists. Retinal microaneurysms (MA) were quantified using a grid according to a standard protocol[10] with progression defined as an increase of ≥ 5 MA/year. Preretinal fluorescein concentration was measured by vitreous fluorophotometry [VF], and the results were expressed as the Penetration Ratio (PR), to correct for plasma fluorescein concentration.[11] Glycemic control was monitored by determination of HbA_{1c} at four-month intervals throughout the study.

We also studied 58 subjects with NIDDM. Inclusion criteria for study subjects were: Type II diabetes, age 20-70, no renal impairment (serum creat.<1.6 mg/dl), no significant liver disease, and stable cardiac status. Each subject underwent a complete clinical and laboratory evaluations on the same day. This included determination of the duration of diabetes, the level of retinopathy by dilated ophthalmoscopy and the degree of atherosclerosis as graded based by standard clinical criteria. Blood sampling was performed in the fasting state for determination of glucose, urea nitrogen, creatinine, electrolytes, lipids, and HbA_{1c}. A 24-hour urine collection was performed over the preceding day for determination of UAE and creatinine clearance. 30 non-diabetic controls were also studied.

Sampling of blood and urine
Blood samples were collected in the mid-morning on a yearly basis for each IDDM subject. The NIDDM and control subjects were sampled once following an overnight fast. Plasma and red blood cells were separated and immediately frozen at -80° C for storage until analysis. The metabolites measured were glucose, fructose, 3DG, 3DF, and DGA. DGA was measured in erythrocytes only. All plasma samples were deproteinized either by ultrafiltration or perchloric acid (PCA) extraction. RBCs also were extracted with PCA. Appropriate internal standards for quantification of metabolites were added prior to any sample processing.

Analytical
Determination of 3DG, 3DF, glucose and fructose in plasma: 3DG was assayed by a modification[12] of an assay described previously.[8] Concentrations of 3DF, glucose and fructose in samples was determined through HPLC analysis of ultrafiltered, deionized samples.[12]
Determination of DGA in erythrocytes: A previously described synthetic procedure for making phenylene diamine derivatives of DGA[13] was adapted for assay by GC-MS.[12]
Determination of Hemoglobin A_{1c}: HbA_{1c} was determined by the Diamat HPLC method (normal range, 4.8-6.1 %).

Results

Characteristics of study populations

The mean age of the IDDM subjects was 30±6 years and the duration of diabetes was 6.6±5.1 years. Mean GFR was 132.9±22.5 mL/min, mean UAE was 6.5±5.3 mg/g creatinine/24 h, and mean HbA_{1c} was 8.2±1.7%. The NIDDM study population consisted of 33 men and 25 women and the mean age and known duration of diabetes were 62.5±7.5 and 13.6±8.3 years, respectively. Mean GFR was 126.6±36.5 mL/min by creatinine clearance; UAE for the 50 normoalbuminuric subjects was 10±7.2 ml/min; and HbA_{1c} was 8.2±1.2%. The mean age of the control group was 41±14 years (20-76) and their mean HbA_{1c} was 5.2±0.5% (range, 3.9-6.2).

3DG, 3DF and DGA in diabetics and controls

There was a significant elevation of 3DG in the plasma of diabetic subjects when compared with normal controls (p<0.0001 for IDDM and NIDDM). We also observed a significant elevation of 3DF (P<0.0001 for IDDM and NIDDM) and DGA (P<0.0001 for IDDM) in diabetics relative to controls (Table 1).

Table 1. Levels of 3DG and its Detoxification Products in Diabetes and Controls

Blood	Controls	IDDM	NIDDM
3DG (nM) Plasma	48±9	76±24*	79±22*
3DF (µM) Plasma	1.1±0.25	2.0±1.1*	1.5±0.48*
DGA (µM) RBC's	11.5 ± 3.2	31.4 ±14*	ND

* Significantly different from controls, P<0.0001

Relationship of 3DG to its detoxification products 3DF and DGA

Plasma levels of 3DG and its major reductive product, 3DF, were highly correlated in both IDDM and NIDDM subjects. (IDDM: R=0.60, p<0.0001; NIDDM: R=0.34, P=0.006).

A significant relationship was also found between 3DG and its major oxidative product, DGA (R=0.69, p=0.005).

Glycemic control and plasma 3DG, 3DF and DGA

A highly significant relationship was found between plasma 3DG and the Hemoglobin A_{1c} value for IDDM (R=0.52, p=0.0004) (Figure 1) and NIDDM subjects (R=0.46, P=0.0001)(not shown). We also observed highly significant relationships between HbA_{1c} and the major 3DG detoxification products, 3DF (NIDDM) (R=0.50, P<0.0001) and DGA (R=0.82, P<0.0001), although we found a weaker relationship between 3DF and HbA_{1c} for subjects with IDDM (R=0.28, P=0.07). These relationships persisted when adjusted for age and duration of diabetes by multiple regression analysis.

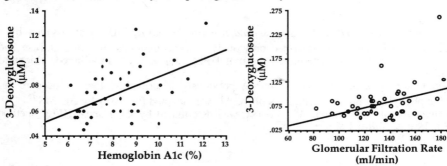

Figure 1. 3-Deoxyglucosone and Glycemic Control.

Figure 2. 3-Deoxyglucosone and Glomerular Filtration Rate.

3DG levels and glomerular filtration rate

As shown in Figure 2, a significant positive correlation was found between plasma 3DG levels and GFR in the IDDM population (R=0.40, p=0.009). We did not observe a significant relationship between 3DG and GFR (by creatinine clearance) in the NIDDM population. The significance of this relationship persisted when adjusted for age, duration of diabetes and glycemic control by multiple regression analysis.

Urinary albumin excretion, 3DG, and 3DG metabolites

For the normoalbuminuric NIDDM subjects (N=50) (Figure 3) we found a significant positive relationship between 3DG and urinary albumin excretion (R=0.34, P=0.02). When we investigated the relationship between increasing levels of urinary albumin excretion and the plasma levels of 3DF in our IDDM population (Figure 4), we found a weak inverse correlation (R=0.3, p=0.06). All of the NIDDM and IDDM subjects had urinary albumin levels within the normal or minimally elevated range (<40 mg/24 h).

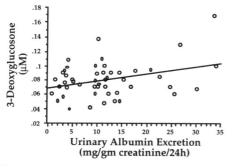
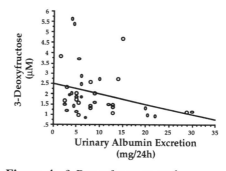

Figure 3. *Urinary Albumin Excretion and 3-Deoxyglucosone levels.*

Figure 4. *3-Deoxyfructose and Urinary Albumin Excretion.*

Progression of diabetic retinopathy and 3DG Detoxification products

Of 25 IDDM subjects studied, 7 showed progression of retinopathy at year two (Progressors). We found that progressors had significantly lower plasma levels of 3DF than nonprogressors (NP) (1.44±0.59 vs 2.45±1.61µM; P=0.05).

3DG levels and retinal perfusion

We also observed a significant inverse correlation between plasma 3DG and post-injection preretinal fluorescein concentrations as determined by vitreous fluorophotometry (R=0.46, p=0.005) (Figure 5).

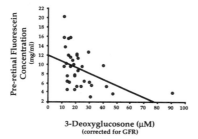

Figure 5. *3-Deoxyglucosone levels and Vitreous Fluorescein Content.*

Discussion

Utilizing populations with IDDM, NIDDM and non-diabetic controls, we have studied the levels of 3DG as well as its precursor and detoxification products in plasma. Our findings confirm those of Knecht and Yamada [7,8] and show that diabetes is associated with significant elevation in plasma levels of 3DG and its major detoxification products. Since the populations studied were documented to have normal renal function, the levels of these compounds in plasma and RBCs should accurately reflect their production and detoxification rates.

Our understanding of 3DG production has recently become more clear. Production of 3DG has generally been thought to occur via glucose, either by its autoxidation or by decomposition of fructoselysine. [2,3,14] As discussed by Lal et al ,[12] another pathway for production has been investigated where production of 3DG occurs via decomposition of F3P. [15,16] On the basis of this work, the phosphokinase pathway is likely to be the dominant source of 3DG in vivo.

Studies to date have shown a high degree of individual variation in rates of dicarbonyl detoxification among diabetic subjects. This was initially suggested by the high coefficient of variation of ± 96% for tissue levels of the major detoxification product 3DF, particularly in subjects with IDDM, indicating a differential capacity of individuals to produce and/or detoxify 3DG. Studies of the activity of pathways for degrading 3DG have shown highly predictable production of 3DF from 3DG in non-diabetic subjects. Our studies also show significant associations between plasma levels of 3DG and its two major detoxification products, 3DF and DGA, in both IDDM and NIDDM subjects. Thus it might appear that dicarbonyls are detoxified in a rapid and predictable manner as they are formed. Studies by Lal et al, [12] comparing urinary levels of 3DG and 3DF in diabetic populations have shown a different picture. In these studies variable 3DG detoxification is apparent in diabetic subjects. This variable and overall reduced rate of detoxification of 3DG appears to be characteristic of NIDDM and IDDM and could play a role in enhancing dicarbonyl accumulation and subsequent toxicity in some individuals. Such reduced detoxification could be a function of elevated glucose levels, altered redox status and the genetic expression of enzymes that degrade 3DG and methylglyoxal.

Our data suggest a strong association between tissue levels of 3DG and diabetic nephropathy. For instance, studies during the early stages of diabetic renal dysfunction showed that glomerular hyperfiltration, an early manifestation of diabetic nephropathy, correlates with plasma 3DG levels and the relationship remains when adjusted for age, duration of diabetes and glycemic control. We also observed a positive relationship between urinary albumin excretion, another early marker of diabetic glomerulopathy, and plasma levels of 3DG in subjects with NIDDM. These data suggest that elevation of 3DG levels may be related to early renal dysfunction, and may play a causal role in the development of diabetic nephropathy.

Our data also suggest that reduced detoxification rates of 3DG may play a role in the pathogenesis of diabetic nephropathy. This is supported by our observation that subjects with early diabetic nephropathy demonstrated an inverse correlation between plasma levels of 3DF and urinary albumin excretion, suggesting that reduced 3DG detoxification may play a direct role in early glomerular damage.

We have observed that progression of early diabetic retinopathy, as measured by increasing microaneurysm counts, is associated with significantly lower levels of the 3DG product 3DF relative to nonprogressors. This suggests that worsening of diabetic retinopathy is associated with decreased conversion of 3DG to its major degradation product. In these studies we also observed significant inverse correlations between preretinal fluorescein levels, as measured by vitreous fluorophotometry, and levels of 3DG. Although the significance of this finding is not clear , it could suggest that 3DG plays a role in progressive retinal capillary loss in diabetes if reduced dye concentrations reflect retinal capillary density. To investigate this possibility, sensitive methods to measure retinal perfusion will have to be carried out in our study population.

Our preliminary data and the data of Lal *et al* demonstrate a pattern that suggests a role for 3DG and reduced activity of its detoxification pathways in the pathogenesis of diabetic microangiopathy. Additional studies are clearly needed to assess the role of 3DG and its degradative pathways in the development of diabetic complications, but these results appear sufficiently provocative and encouraging to justify such studies.

References

1. D. Edelstein and M. Brownlee, Mechanistic studies of advanced glycosylation end product inhibition by aminoguanidine, *Diabetes* , 1992, **41**, 26-29.
2. S. P. Wolff and R. T. Dean, Glucose autoxidation and protein modification, *Biochem. J* , 1987, **245**, 243-250.
3. D. Zyzak, J. Richardson, S. Thorpe and J. Baynes, Formation of reactive intermediates from Amadori compounds under physiological conditions, *Archi. Biochem. Biophy.* , 1995, **316**, 547-554.
4. B. Szwergold, P. Beisswenger, J. Cavender, T. Brinck-Johnsen, A. Siegel, T. Brown, F. Kappler and S. Lal, Assessment of potential new erythrocyte markers of diabetic nephropathy, *Diabetes (Supplement 1)* , 1995, **44**, 222A.
5. E. Fujii, H. Iwase, I. Ishii-Karakasa, Y. Yajima and K. Hotta, The prescence of 2-keto-3-deoxygluconic acid and oxoaldehyde dehydrogenase activity in human erythrocytes, *Biochem. biophys. res. commmun.*, 1995, **210**, 852-857.
6. K. Wells-Knecht, T. Lyons, D. McCance, S. Thorpe, M. Feather and J. Baynes, 3-Deoxyfructose concentrations are increased in human plasma and urine in diabetes, *Diabetes* , 1994, **43**, 1152-1156.
7. K. Knecht, M. Feather and J. Baynes, Detection of 3-deoxyfructose and 3-deoxyglucosone in human urine and plasma: evidence for intermediate stages of the Maillard reaction in vivo, *Arch. Biochem. Biophys* , 1992, **294**, 130-137.
8. H. Yamada, S. Miyata, N. Igaki, H. Yatabe and *et al*, Increase in 3-deoxyglucosone levels in diabetic rat plasma, *J. Biol. Chem.* , 1994, **269**, 20275-20280.
9. J. J. Goates, K. A. Morton, W. W. Whooten, H. E. Greenberg and *et al* , Comparison of methods for calculating glomerular filtration rate:technetium-99m-DTPA scintigraphic analysis, protein-free and whole-plasma clearance of technetium-99m-DTPA, and iodine-125-Iothalamate clearance, *J. Nucl. Med.* , 1990, **31**, 424-429.
10. B. E. Klein, M. D. Davis, P. Segal and *et al*, Diabetic retinopathy: assessment of severity and progression, *Ophthalmology* , 1984, **91**, 10-17.
11. H. Lund-Andersen, B. Krogsaa, M. La Cour and *et al*, Quantitative vitreous fluorophotometry applying a mathematical model of the eye, *Invest. Ophthalmol. Vis. Sci.*, 1985, **26**, 698-710.
12. S. Lal, B. Szwergold, M. Walker, W. Randall, F. Kappler, P. Beisswenger and T. Brown, Production and metabolism of 3-deoxyglucosone in humans, *Proceedings of the 6th International Symposium on the Maillard Reaction, London England* , 1997, .
13. M. Madson and M. S. Feather, The oxidation of 3-deoxy-d-erythro-hexo-2-ulose to 3-deoxy-d-erythro-2-hexulosonic acid by glucose oxidase, *Carbohydrate Res.* , 1983, **115**, 288-291.
14. S. P. Wolff, Z. A. Basca and J. V. Hunt, Autoxidative glycosylation: Free radicals and glycation theory, *The Maillard Reaction in Aging,Diabetes, and Nutrition*, 1989, Editors, Alan R. Liss,New York, 259-275.
15. B. S. Szwergold, F. Kappler and T. R. Brown, Identification of fructose 3-phosphate in the lens of diabetic rats, *Science* , 1990, **247**, 451-454.
16. A. Petersen, B. S. Szwergold, F. Kappler, M. Weingarten and T. R. Brown, Identification of sorbitol 3-phosphate and fructose 3-phosphate in normal and diabetic human erythrocytes, *J. Biol. Chem.* , 1990, **265**, 17424-17427.

Role of Maillard Reaction in Cataractogenesis and Altered α-Crystallin Chaperone Function in Diabetes

E.C. Abraham, S. Swamy-Mruthinti, M. Cherian and H. Zhao

DEPARTMENT OF BIOCHEMISTRY AND MOLECULAR BIOLOGY, MEDICAL COLLEGE OF GEORGIA, AUGUSTA, GA 30912-2100, USA

Summary
Effect of aminoguanidine (AG), an inhibitor of advanced glycation, on the development of cataract was studied in streptozotocin-diabetic rats. Advanced glycation end products (AGEs) were estimated immunochemically and lens opacity by Scheimpflug densitometric analysis. AG treatment resulted in a significant decrease in AGE levels and a concomitant delay in cataract development, indicating that AGEs play an important role in the development of cataract in this animal model. The function of α-crystallin as a molecular chaperone, *i.e.* its ability to protect other lenticular proteins from denaturation and aggregation, was significantly compromised in the diabetic rat lenses. Glycation studies *in vitro* showed that AG alone did not improve the hyperglycemia-induced loss in the chaperone function. However, in the presence of catalase and conditions that prevent metal-catalyzed oxidation, chaperone function was recovered. This suggests that reactive oxygen species like hydroxy radicals, generated by glycoxidation or metal-catalyzed glucose autoxidation, are involved in the loss of α-crystallin chaperone activity.

Introduction

Earlier studies have suggested that Maillard reaction products or advanced glycation end products (AGEs) are contributory factors in the development of diabetic cataracts.[1,2] Prior studies from our laboratory have shown a strong correlation between lens opacity (quantified by Scheimpflug densitometry) and lens crystallin and membrane intrinsic protein (MIP) glycation.[3] It is quite possible that lens protein crosslinking, mediated by advanced glycation, can influence aggregation and insolubilization of lens cytosolic proteins. However, direct evidence to support the view that AGEs play a significant role in the development of diabetic cataracts is lacking. Our approach was to study the effect of a known inhibitor of AGE formation, namely aminoguanidine (AG), on cataractogenesis and AGE formation. We have reported recently that AG delays cataract development in moderately diabetic rats but not in severely diabetic rats.[4]

Another factor that can influence aggregation of proteins in the diabetic lens is protein modification of α-crystallin by glycation. α-Crystallin is known to prevent denaturation and aggregation of other proteins in the lens by its chaperone like activity[5,6] and posttranslational modifications of α-crystallin could affect its function as a molecular chaperone.[6-8] We studied the effect of glycation *in vivo* (in diabetic rats) as well as glycation *in vitro* of α-crystallin on its chaperone function.

Materials and Methods

Streptozotocin-diabetic rats
One-month-old male Sprague-Dawley rats were made diabetic by a single *i.v.* injection of streptozotocin (STZ) (45 mg kg^{-1} body weight). One half of the diabetic rats received daily *i.p.* injections of aminoguanidine-HCl (25 mg kg^{-1} body weight) starting from the day of STZ injection. Age-matched normal animals served as controls. Progression of cataracts was

evaluated by Scheimpflug slit-imaging and densitometric analysis in each animal every 30 days. On the 90th day, all the animals were killed to obtain lenses and blood.

Immunochemical measurement of AGEs
For preparation of the anti-AGE antibody, AGE-proteins were generated by incubating bovine serum allumin (BSA) with 1M glucose at $37^{0}C$ for 90 days in dark, and the resultant BSA-AGE was used to immunize rabbits, as described by Horiuchi et al.[9] AGEs were estimated by enzyme-linked immunosorbent assay (ELISA) using the above polyclonal anti-AGE antibody. Lens water-soluble and water-insoluble (urea-soluble) protein fractions were prepared as described before.[1] For estimation of AGEs formed *in vivo* in rat lenses and during glycation of bovine α-crystallin *in vitro* noncompetitive ELISA was used. Each well was coated with 0.1 mL (1 µg protein) of sample, blocked with 0.25 mL of 0.5% gelatin and incubated with 0.1 mL of 1: 1000 diluted antiserum. The well was washed and reacted with 0.1 mL of second antibody (monoclonal anti-rabbit immunoglobins-horse raddish peroxidase conjugate) followed by reaction with 0.2 mL of tetramethyl benzidine (TMB) substrate system (Sigma). The absorbance at 650 nm was read on a THERMO max microplate reader and the level of AGEs were expressed as A_{650} units.

Purification of α-crystallin from rat and bovine lenses
To show whether elevated concentrations of glucose, as in diabetes, affect the function of α-crystallin as a molecular chaperone, α-crystallin was purified from streptozotocin-diabetic as well as from normal rat lenses (not treated with AG) by preparative gel permeation chromatography (Sephacryl S-300-HR gel). For *in vitro* glycation studies, α-crystallin was purified from calf lenses.

In vitro glycation of α-crystallin
Glycation of calf α-crystallin *in vitro* was done under the following conditions: α-Crystallin (5 mg/mL) was incubated with 25 mM erythrose for 24 hours at $37^{0}C$, alone or in combination with 100 units of catalase (scavenger of hydroxy radicals), 50 mM AG (inhibitor of AGE formation), and 25 mM diethylenetriaminepentaacetic acid (DPTA) (inhibitor of metal catalyzed oxidation). α-Crystallin was also treated with 50 mM AG alone. α-Crystallin was examined by SDS-PAGE after the above treatments.

Molecular chaperone assay
Heat-induced denaturation and aggregation of calf $β_L$-crystallin, the target protein, was done in the presence or absence of α-crystallin as described earlier.[7,8] α-Crystallin from control rats, diabetic rats, or calf α-crystallin glycated *in vitro* were used for these studies. Forty micrograms of α-crystallin were mixed with 400 µg of $β_L$-crystallin in a cuvette and made up to a final volume of 1 mL with 50 mM phosphate buffer, pH 7.0. The cuvette was heated at 55°C and scanned for turbidity over a time period of 50 min (in some cases heated to 58°C and scanned for 30 min) at 360 nm in a Shimadzu UV/160 spectrophotometer equipped with a temperature regulated cell holder.

Results and Discussion

Scheimpflug photography and subsequent densitometric scans of Scheimpflug images were used to quantify the progression of overall lens opacification. Area of the total lens including epithelium, cortex, and nucleus was used, and opacity expressed in arbitrary units. An earlier study from our laboratory has shown that lens opacity progressed in a biphasic manner in diabetic rats, an initial slow increase for 60 days followed by a steep increase for the next 30 days.[3] Since the level of opacity attained by 90 days was close to maximum (mature cataract) all the studies reported here were done on the 90th day, post-STZ injection, and the results are summarized in Table I. In the diabetic rats, lens opacity increased 20-fold compared with age-matched control rats. Aminoguanidine treatment did result in a significant reduction (34%; $p<0.001$) in opacity. However, it should be pointed out that when a similar study was done in severely diabetic rats (mean plasma glucose 896 ± 74 mg dL-1), AG had no effect on cataractogenesis,[4] and it was concluded that the beneficial effect of AG seems to be overwhelmed by excessive accumulation of AGEs.

As expected, control rats maintained a stable level of plasma glucose levels (89.8 ± 4.5 mg dL^{-1}) whereas a three-fold increase was seen in diabetic rats (Table 1). AG treatment did not alter the plasma glucose levels. AGEs in the water-soluble and urea-soluble fractions were determined immunochemically by ELISA. In the diabetic rats, the level of AGEs in the water-soluble fraction was increased by 1.5-fold and in the urea-soluble fraction by 3.2-fold; AGEs in the urea-soluble fraction were two-fold higher than in the water-soluble fraction. However, we have reported earlier a ten-fold increase in AGE related fluorescence in the urea- soluble fraction of diabetic rats.[4] This suggests that all the fluorescent products (detected at emission.440 nm/excitation 370 nm) were not AGEs or our AGE antibody is specific for only a group of AGE products. Treatment with AG, a known inhibitor of advanced glycation, showed ~10% and ~20% decreases in the levels of AGEs in water-soluble and urea-soluble fractions, respectively. Interestingly, our recent study reveals that carboxymethyllysine (CML) is a major AGE in diabetic rat lenses and AG does inhibit the formation of CML (Matsumoto, Ikeda, Horiuchi, Zhao, and Abraham, unpublished).

The present study confirms an earlier report of Kumari et al.[10] who showed a delay in the development of cataracts in alloxan-diabetic rats with mean glucose levels of 350 mg dL^{-1}. However, the mechanism seems to be inhibition of the formation of AGEs rather than

Table 1. *Effect of Aminoguanidine (AG) treatment for 90 days on plasma glucose, AGE, and lens opacity*

Animal Group	Lens opacity (arbitrary units)	Plasma glucose (mg dL^{-1})	AGE-ELISA (A_{650} units)	
			Lens water-Soluble fraction	Lens urea-soluble fraction
Control rats (n=6)	0.54±0.02	89.8±4.5	0.255±0.004	0.241±0.002
Diabetic rats (n=6)	9.54±0.20	263.8±56.5	0.383±0.01	0.774±0.007
AG treated Diabetic rats (n=6)	6.32±0.81	269.8±74.1	0.344±0.01	0.630±0.006

aldose reductase inhibition as suggested by Kumari et al.[10] Lens opacification is a complex phenomenon and AGEs seem to play a major role in the lens opacification in this animal model. Since AG treatment only delayed the development of mature cataracts, it can be concluded that other factors are also important. However, it should be pointed out that with the dosage of AG used in this study, only about 20% inhibition of AGEs was accomplished; a higher level of inhibition might lead to a further delay of cataractogenesis. Unlike the animal model, human cataractogenesis is relatively slow and cataracts tend to be more brunescent and highly cross-linked, presumably due to accumulation of AGEs. Thus, inhibition of AGE formation by AG should be more relevant in human cataractogenesis.

Table 2. *Influence of in vivo glycation or diabetes on α-crystallin function as a molecular chaperone*

Sample	Chaperone function* (A_{360} units)
β_L-crystallin	0.91
β_L-crystallin + α-crystallin from 90-day control rats	0.216
β_L-crystallin + α-crystallin from 90-day diabetic rats	0.516

*Average of duplicate determinations

Lens protein aggregation and insolubilization are known to increase during aging and diabetes[1,11] which may be, at least in part, due to the α-crystallin loosing its chaperone-like function, (i.e. its ability to protect other lenticular proteins from denaturation and aggregation). We have shown earlier that oxidation and glycation of α-crystallin can lead to significant loss in chaperone function.[7] Since diabetic lenses are known to undergo more oxidation and glycation than age-matched normal lenses, it is possible that in diabetic lenses the chaperone function of α-crystallin may be compromised. To investigate this, we have isolated α-crystallin from the water-soluble fraction of streptozotocin-diabetic and control rats, and the chaperone-like function was assessed by the β_L-crystallin thermal denaturation assay; the results are summarized in Table 2. β_L-crystallin heated alone underwent complete thermal denaturation, whereas in the presence of α-crystallin from 90-day control rats, β-crystallin aggregation was reduced substantially. However, when α-crystallin from 90-day diabetic rats was used, only ~50% protection of β_L-crystallin aggregation was observed. The levels of lens protein glycation (Amadori products) were 5.4% and 15.4% in 90-day control and 90-day diabetic rats, respectively. As discussed earlier, levels of AGEs were also significantly elevated in the diabetic rats (Table 1). Thus, glycation appeared to be the cause of altered chaperone function. However, the following glycation studies led to a somewhat different conclusion.

Glycation of bovine α-crystallin *in vitro* was achieved using erythrose as the glycating sugar. An earlier study showed that erythrose could glycate and undergo advanced glycation relatively fast.[14] α-Crystallin lost about 54% of its chaperone activity after incubation with erythrose (Table 3). When the incubation with erythrose was done in the presence of catalase (scavenger of hydroxy radicals) or DPTA+N_2 (prevents metal catalyzed oxidation) the chaperone activity was almost fully recovered. Neither of these two treatments had any effect on AGE production, which was increased substantially by erythrose. As judged by ELISA (Table 3) and SDS-PAGE (data not given), AG showed almost normalization of AGE levels, but without any improvement of chaperone activity. Surprisingly, when α-crystallin was treated with AG alone, ~ 35% loss of chaperone activity was seen. Based on these observations, it is not possible to conclude whether AGEs formed on α-crystallin are the cause for the decline in chaperone function. However, the oxidation by the hydroxy radicals, generated during glycation, seems to be the main cause for the altered chaperone function. There are, at least, two well-recognized mechanisms that may explain the generation of reactive oxygen species like hydroxy radicals: metal catalyzed glucose autoxidation[12] and glycoxidation.[13]

Table 3. *Effect of glycation in vitro on α-crystallin chaperone function*

Treatment of α-crystallin	Chaperone function* (A_{360} units)	AGE-ELISA* (A_{650} units)
Untreated Control	0.313±0.02	0.250±0.03
Erythrose	0.482±0.06	1.030±0.03
Erythrose + AG	0.555±0.03	0.292±0.02
AG	0.422±0.02	-
Erythrose + catalase	0.280±0.03	0.977±0.13
Erythrose + DPTA+N_2	0.270±0.09	0.710±0.08

*Mean of 3 determinations

We have shown that inhibition of AGE formation by AG delays the development of mature cataracts in diabetic rats. AG, on the other hand, did not improve α-crystallin function. It is not known whether AG has any effect other than inhibiting the formation of AGEs. Antioxidants were also shown to have a delaying effect on sugar cataract formation,[15] which strongly suggests an important role for reactive oxygen species in the development of diabetic cataracts.

Acknowledgements
This study was supported NIH Grants EY07394 and EY11352 (E.C.A.) and EY10219 (S.S.M.).

References

1. R.E. Perry, M.S. Swamy and E.C. Abraham, Progressive changes in lens crystallin glycation and high-molecular-weight aggregation leading to cataract development in streptozotocin-diabetic rats, *Exp. Eye Res.*, 1987, **44**, 269-282.
2. M.S. Swamy, A. Abraham and E.C. Abraham, Glycation of human lens proteins:

Preferential glycation of αA subunits, *Exp. Eye Res.*, 1992, **54**, 337-345.
3. S. Swamy-Mruthinti, K. Green and E.C. Abraham, Scheimpflug densitometric analysis of cataracts in diabetic rats: Correlation with glycation, *Ophthalmic Res.*, 1996, **28**, 230-236.
4. S. Swamy-Mruthinti, K. Green and E.C. Abraham, Inhibition of cataracts in moderately diabetic rats by aminoguanidine, *Exp. Eye Res.*, 1996, **62**, 505-510.
5. J. Horwitz, α-Crystallin can function as a molecular chaperone, *Proc. Natl. Acad. Sci. U.S.A.*, 1992, **89**, 10449-10453.
6. J. Horwitz, T. Emmons and L. Takemoto, The ability by lens α-crystallin to protect against heat-induced aggregation is age dependent, *Current Eye Res.*, 1992, **8**, 817-822.
7. M. Cherian and E.C. Abraham, Decreased molecular chaperone property of α-crystallins due to posttranslational modifications, *Biochem, Biophys. Res. Commun.*, 1995, **208**, 675-679.
8. M. Cherian and E.C. Abraham, Diabetes affects α-crystallin chaperone function, *Biochem. Biophy. Res. Commun.*, 1995, **212**, 184-189.
9. S. Horiuchi, N. Araki and Y. Morino, Immunochemical approach to characterize glycation end products of the Maillard reaction, *J. Biol. Chem.*, 1991, **266**, 7329-7332.
10. K. Kumari, S. Umar, V. Bansal and M.K. Sahib, Inhibition of diabetes associated complications by nucleophilic compounds, *Diabetes*, 1991, **40**, 1079-1084.
11. M.S. Swamy and E.C. Abraham, Lens protein composition, glycation and high molecular weight aggregation in aging rats, *Invest. Ophthalmol. Vis. Sci.*, 1987, **28**, 1693-1701.
12. S.P. Wolff and R.T. Dean, Glucose antoxidation and protein modification, *Biochem. J.*, 1987, **245**, 243-250.
13. M. Fu, K.J. Wells-Knecht, J.A. Blackledge, T.A. Lyons, S.R. Thorpe and J.W. Baynes, Glycation, glycoxidation, and cross-linking of collagen by glucose, *Diabetes*, 1994, **43**, 676-683.
14. M.S. Swamy, C. Tsai, A. Abraham and E.C. Abraham, Glycation mediated lens crystallin cross-linking by various sugars and sugar phosphates, *Exp. Eye Res.*, 1993, **56**, 177-185.
15. S.K. Srivastava and N. Ansari, Prevention of sugar induced cataractogenesis in rats by BHT, *Diabetes*, 1993, **37**, 1505-1508.

Immunochemical Approaches to AGE-Structures —Characterization of Anti-AGE Antibodies

Kazuyoshi Ikeda, Ryoji Nagai, Tamami Sakamoto, Takayuki Higashi, Yoshiteru Jinnouchi, Hiroyuki Sano, Kenshi Matsumoto, Masaki Yoshida, Tomohiro Araki,[1] Shoichi Ueda and Seikoh Horiuchi

DEPARTMENT OF BIOCHEMISTRY AND UROLOGY, KUMAMOTO UNIVERSITY SCHOOL OF MEDICINE, KUMAMOTO, JAPAN; [1] FACULTY OF AGRICULTURE, KYUSHU TOKAI UNIVERSITY, KUMAMOTO, JAPAN

Summary
Recent immunological approaches have greatly helped broaden our understanding of the biomedical significance of AGEs (advanced glycation end-products) in aging and age-enhanced disease processes. We previously prepared a monoclonal anti-AGE antibody (6D12) that recognized a common AGE-structure(s) as a major immunochemical epitope. Subsequently, N^ε-(carboxymethyl)lysine (CML), one of the glycoxidation products of AGEs, was demonstrated to be a major immunological epitope among AGEs, and 6D12 turned out to recognize CML as an epitope. In the present study, 13 different polyclonal anti-AGE antibodies were characterized in order to obtain the other epitope structure(s), other than CML (non-CML). We used CML-bovine serum albumin as an authentic CML-protein and AGE-lysozyme as an authentic non-CML-protein. The results indicated that these antibodies were classified into 3 groups (Group I, II & III). Group I was specific for CML, but both Group II and Group III were unreactive to CML. Group II, but not Group III, recognized AGE-lysozyme, suggesting Group II and III were specific for non-CML but different epitopes. The epitope of Group II was formed much earlier than that of Group III during incubation of BSA with glucose *in vitro*. Furthermore, we made two monoclonal anti-AGE antibodies (M-1 and M-2) whose epitope structures appeared to be identical or closely similar to Group III and Group II, respectively. These results indicate that AGE-proteins express two major non-CML epitopes in addition to CML.

Introduction

Immunological approaches[1] have been used to examine the presence of AGEs in several human tissues[2-5] and to determine their major structure(s). Using AGE-bovine serum albumin (BSA) as an antigen, we previously prepared a mouse monoclonal anti-AGE antibody (6D12) and a rabbit polyclonal anti-AGE antibody.[1] Characterization of these antibodies led us to an interesting observation: both antibodies reacted with AGE-samples obtained from proteins, peptides, lysine derivatives and monoaminocarboxylic acids, suggesting the presence of a common AGE-structure(s) in these AGE-preparations. Determination of this common AGE-structure, i.e. the epitope(s) of anti-AGE antibodies, was a major focus of research until Reddy *et al.*[6] showed that N^ε-(carboxymethyl)lysine (CML), one of the glycoxidation AGE-products,[7,8] served as a major immunological epitope among AGE-structures. Our subsequent study identified the epitope of 6D12 as CML, whose carbonyl group played an important role in recognition by the antibody.[9] Furthermore, separation of a CML-specific antibody population from our polyclonal anti-AGE antibody suggested the presence of a structure(s) other than CML (non-CML) as a major immunological epitope.[9] To further test this notion, the present study was undertaken to characterize 13 polyclonal anti-AGE antibodies prepared by different laboratories. The results provided evidence for the presence of at least two major non-CML epitopes in addition to CML in AGE-proteins.

Materials and Methods

Preparation of AGE-proteins used as immunogens
AGE-BSA (Batch E) was prepared by incubation for 12 weeks with glucose[1] and used as immunogen for preparing antibodies PA-1, PA-2 and PA-5. Similar AGE-BSA

Table 1. List of anti-AGE antibodies examined in the present study

No.	Immunogen	Species	monoclonal/polyclonal
S-1	AGE-BSA	mouse	monoclonal
S-2	CML-BSA	rabbit	polyclonal
S-3	AGE-BSA	rabbit	polyclonal
S-4	AGE-BSA	rabbit	polyclonal
PA-1	AGE-BSA	rabbit	polyclonal
PA-2	AGE-BSA	rabbit	polyclonal
PA-3	AGE-BSA	rabbit	polyclonal
PA-4	AGE-BSA	rabbit	polyclonal
PA-5	AGE-BSA	mouse	polyclonal
PA-6	AGE-KLH	guinea pig	polyclonal
PA-7	AGE-KLH	guinea pig	polyclonal
PA-8	AGE-KLH	guinea pig	polyclonal
PA-9	AGE-KLH	rabbit	polyclonal
PA-10	AGE-KLH	mouse	polyclonal
PA-11	AGE-KLH	rabbit	polyclonal
PA-12	AGE-collagen	rabbit	polyclonal
PA-13	AGE-RNase	rabbit	polyclonal
M-1	AGE-RNase	mouse	monoclonal
M-2	AGE-RNase	mouse	monoclonal

preparations obtained by incubation for 9 and 16 weeks were used for PA-3 and PA-4, respectively. AGE-keyhole limpet hemocyanin (KLH) was prepared as described previously and used to obtain PA-6 to PA-11 (Ref. 10). AGE-collagen was prepared by incubating collagen with glucose in the same manner as AGE-BSA and used for raising PA-12 in a rabbit. AGE-ribonuclease (RNase), prepared as described by Makita et al.,[11] was used to obtain PA-13. AGE-RNase was prepared by incubating 50 mg/mL RNase with 1 M glucose in 0.2 M sodium phosphate buffer (pH 7.5) for 12 weeks at 37°C, followed by a further 12 weeks incubation without glucose. This preparation was used to make two monoclonal antibodies in mice (M-1 and M-2). AGE-lysozyme was prepared as described previously.[12] After being glycated in 5% glucose solution, lysozyme was lyophilized and stored at 75% humidity for 10 days at 50°C.

Preparation of anti-AGE antibodies
As listed in Table 1, four standard antibodies (S-1, S-2, S-3 & S-4) were used. S-1 was 6D12,[1] specific for the CML-protein adduct.[9] S-2 was an affinity-purified antibody monospecific for CML. S-3 was separated from our polyclonal anti-AGE antibody (PA-2) by a CML-BSA affinity column as the adsorbed fraction, and thus monospecific for CML, whereas S-4 was separated as the non-adsorbed fraction, thus specific for non-CML structure(s).[9] PA-3 was a gift from Dr Noriyuki Sakata. PA-6 and PA-7 were from Dr Hidetaka Nakayama. To make antibodies against non-CML epitopes, two monoclonal anti-AGE antibodies (M-1 and M-2) were prepared as follows. BALB/c mice were immunized with AGE-RNase and their spleen cells were fused with P3U1 cells. The hybridomas were screened and two cell lines, M-1 and M-2, positive to AGE-BSA (Batch E) but negative to both of CML-BSA and RNase were selected through successive subcloning.

Enzyme-linked immunosorbent assay
Assays were performed at room temperature for both competitive ELISA and noncompetitive ELISA.[1,9] In competitive ELISA, results were expressed as the ratio B/B0 in which B represents the amount of binding of the antibody in the presence of a competitor and B0 represents that in the absence of it.

Time-course study of formation of AGE-epitopes
Each tube containing 50 mg/mL of BSA in 1.0 mL of 0.5 M sodium phosphate buffer (pH 7.4) was incubated at 37°C with 2.0 M glucose for indicated intervals up to 62 days, dialyzed against PBS, and subjected to immunoreaction with antibodies tested.

Analytical procedures
Hippuryl-CML was prepared by incubating 40 mg/mL of hippuryllysine with 0.13 M glyoxylic acid in the presence of 0.65 M $NaCNBH_3$ in 0.5 mL of 0.1 M sodium carbonate buffer (pH 10.0) overnight at room temperature.[6,9] CML contents of modified proteins were determined by amino acid analyses after hydrolysis of each protein in 6 N HCl for 24 h at 110°C (Ref. 9).

Results

Immunological and analytical characterization of AGE-antigens
As shown in Figure 1, all these standard antibodies (S-1 to S-4) commonly recognized AGE-BSA. Their reactions to AGE-BSA were completely inhibited by CML-BSA in S-1, S-2 and S-3 but not in S-4, indicating the immunospecificity of S-1, S-2 and S-3 for CML and S-4 for structure other than CML (non-CML). CML contents of AGE-BSA and CML-BSA measured by amino acid analysis were 7.69 and 9.25 mol/mol of BSA, respectively. None of these anti-CML antibodies (S-1, S-2 or S-3) reacted with AGE-lysozyme. The CML content of AGE-lysozyme was less than a detectable level (<0.01 mol/ mol of protein) by amino acid analysis. AGE-HSA prepared in the same way as AGE-lysozyme neither contained CML nor reacted with these anti-CML antibodies (data not shown). Therefore, these immunological and analytical data suggest that CML is unlikely to be generated by the method of Cho et al.[12] Thus, AGE-lysozyme was used as an AGE-protein lacking CML (non-CML) in the following experiments.

Immunoreactivity of polyclonal anti-AGE antibodies to CML and non-CML
Figure 2 shows the immunoreactivity of 13 different polyclonal anti-AGE antibodies towards CML-BSA and AGE-lysozyme. No antibodies reacted with native proteins, such as BSA or lysozyme without glucose-modification (data not shown). The immunoreactivity of these antibodies to AGE-BSA was inhibited completely by CML-BSA in the case of PA-1, PA-6, PA-8, PA-10, PA-11 and PA-12, but partially in PA-2, PA-5, PA-7, PA-9 and PA-13. A similar immunoreactivity was observed with hippuryl-CML, a low molecular weight CML-peptide (data not shown). In a sharp contrast, CML-BSA did not react with PA-3 and PA-4. The immunoreactivity to AGE-lysozyme differed from one antibody to another. The immunoreactivity of these antibodies to AGE-BSA was inhibited by AGE-lysozyme completely in PA-3 and PA-4, but partially in PA-9 and PA-13. The other 9 antibodies did not recognize AGE-lysozyme at all. These results appeared to suggest a high diversity of immunoreactivity of these anti-AGE antibodies; these antibodies could be classified into three groups according to their reactivity to CML-BSA and AGE-lysozyme. Group I, including PA-1, PA-6, PA-8, PA-10, PA-11 and PA-12, are monospecific for CML-BSA. Group II, including PA-3 and PA-4, does not react with CML-BSA, but shows a positive reaction to AGE-lysozyme, suggesting they are specific for non-CML. Group III, which is similar to Group II, is specific for non-CML, but unreactive to AGE-lysozyme whose typical example is S-4 (Figure 1). These results not only confirm the notion that CML is one of the major immunological epitopes on AGE-proteins, but also suggest the presence of non-CML epitope in AGE-proteins.

Health and Disease 313

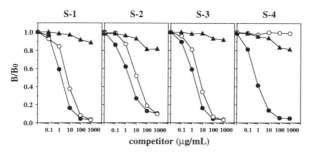

Figure 1. Immunoreactivities of standard anti-AGE antibodies against AGE-BSA (●), CML-BSA (○) and AGE-lysozyme (▲) examined by competitive ELISA

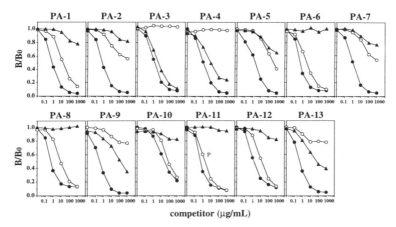

Figure 2. Immunoreactivities of 13 polyclonal anti-AGE antibodies against AGE-BSA (●), CML-BSA (○) and AGE-lysozyme (▲) examined by competitive ELISA

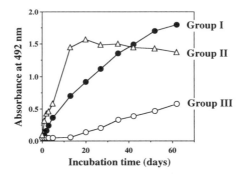

Figure 3. Time-course of CML and non-CML epitope formation in the Maillard reaction in vitro

Time-course study of formation of the AGE-epitopes
To evaluate the formation of CML and non-CML *in vitro* during the Maillard reaction, the time-course of the immunoreactivities were examined by the noncompetitive ELISA using anti-AGE antibodies of Groups I-III (Figure 3). The immunoreactivity of the glucose-modified BSA with Group I antibody became detectable on day 1 and increased slowly but progressively until day 62, exhibiting a saturating tendency. In contrast to Group I, the immunoreactivity to one of Group III antibodies (S-4) was not detectable before day 5, but became detectable on day 13 and increased proportionately with time. Immunoreactivity of the glucose-modified BSA with Group II became detectable on day 1 and rapidly increased to the plateau-level until day 13. These results showed that epitopes of Group I and Group II were formed relatively earlier than that of Group III during the Maillard reaction *in vitro*.

Cloning of two monoclonal anti-AGE antibodies specific for non-CML epitopes
To obtain a much clearer picture of epitopes of non-CML structure, we attempted to prepare monoclonal antibodies specific for non-CML structure(s). AGE-RNase was used to immunize mice and two different clones (M-1 and M-2) were obtained, which were specific for non-CML epitopes. M-1 did not react with CML-BSA or AGE-lysozyme, whereas M-2 did not react with CML-BSA but did react with AGE-lysozyme. The time-course study of their epitopes during the Maillard reaction revealed that the M-2 reactivity was similar to that of Group II, whereas the M-1 reactivity was similar to Group III (data not shown), suggesting that M-1 and M-2 could be classified into Group III and Group II, respectively.

Discussion

According to the immunoreactivity towards CML-BSA and AGE-lysozyme, polyclonal anti-AGE antibodies could be categorized into 3 groups. Group I was specific for CML, Group II was specific for non-CML and recognized AGE-lysozyme. Group III was also specific for non-CML, but did not recognize AGE-lysozyme. AGE-lysozyme used in the present study was prepared under unique conditions.[12] After being glycated in 5% glucose solution, lysozyme was lyophilized and stored at 75% humidity for 10 days, resulting in browning, polymerization and fluorophore formation. In spite of these physicochemical features of AGE-proteins, no CML was generated in the AGE-lysozyme preparation. Since CML formation is catalysed by transition metal ions and inhibited under antioxidative conditions,[7,8,13] it is likely that metal-catalysed oxidation might be minimized in this method of preparation. Thus, AGE-structures formed on AGE-lysozyme might be considered as oxidation-independent.

Immunochemical kinetic studies showed that the epitope of Group II is formed as early as CML (early non-CML) and that of Group III appears much later (late non-CML) (see Figure 3). We successfully prepared two monoclonal antibodies specific for non-CML epitopes (M-1 and M-2) and demonstrated that M-1 is specific for the late non-CML epitope and M-2 is specific for the early non-CML epitope. Therefore, the present study made it clear that at least three different epitopes are commonly expressed in AGE-proteins, CML, early non-CML and late non-CML.

In conclusion, it is suggested that the early non-CML epitope recognized by Group II and M-2 is oxidation-independent, whereas the late non-CML epitope recognized by Group III and M-1 is oxidation dependent. Chemical structures of these non-CML epitopes seem to differ from Amadori products or known AGE-structures such as pyrraline and pentosidine because neither of them was recognized by PA-2 (Ref.1) and pyrraline and pentosidine conjugated with BSA were not recognized by all antibodies tested in the present study (data not shown). Further studies are needed to elucidate chemical structures of these epitopes.

Acknowledgment

We are grateful to Dr Noriyuki Sakata (Department of Pathology, School of Medicine, Fukuoka University, Japan) for providing a polyclonal anti-AGE antibody (PA-3) and to Dr Hidetaka Nakayama (Manda Hospital, Sapporo, Japan) for PA-6 and PA-7.

References

1. S. Horiuchi, N. Araki and Y. Morino, Immunochemical approach to characterize advanced glycation end products of the Maillard reaction, *J. Biol. Chem.*, 1991, **266**, 7329-7332.
2. N. Araki, N. Ueno, B. Chakrabarti, Y. Morino and S. Horiuchi, Immunochemical evidence for the presence of advanced glycation end products in human lens protein and its positive correlation with aging, *J. Biol. Chem.*, 1992, **267**, 10211-10214.
3. S. Kume, M. Takeya, T. Mori, N. Araki, H. Suzuki, S. Horiuchi, T. Kodama, Y. Miyauchi and K. Takahashi, Immunohistochemical and ultrastructural detection of advanced glycation end products in atherosclerotic lesions of human aorta with a novel specific monoclonal antibody, *Am. J. Pathol.*, 1995, **147**, 654-667.
4. S. Horiuchi, Advanced glycation end products (AGE)-modified proteins and their potential relevance to atherosclerosis, *Trends Cardiovasc. Med.*, 1996, **6**, 163-168.
5. K. Mizutari, T. Ono, K. Ikeda, K. Kayashima and S. Horiuchi, Photo-enhanced modification of human skin elastin in actinic elastosis by N^ε-(carboxymethyl)lysine, one of the glycoxidation products of the Maillard reaction, *J. Invest. Dermatol.*, 1997, **108**, 797-802.
6. S. Reddy, J. Bichler, K.J. Wells-Knecht, S.R. Thorpe, J.W. Baynes, N^ε-(carboxymethyl)lysine is a dominant advanced glycation end product (AGE) antigen in tissue proteins, *Biochemistry*, 1995, **34**, 10872-10878.
7. J.W. Baynes, Role of oxidative stress in development of complications in diabetes, *Diabetes*, 1991, **40**, 405-412.
8. R. Nagai, K. Ikeda, T. Higashi, H. Sano, Y. Jinnouchi, T. Araki and S. Horiuchi, Hydroxyl radical mediates N^ε-(carboxymethyl)lysine formation from Amadori product, *Biochem. Biophys. Res. Commun.*, 1997, **234**, 167-172.
9. K. Ikeda, T. Higashi, H. Sano, Y. Jinnouchi, M. Yoshida, T. Araki, S. Ueda and S. Horiuchi, N^ε-(carboxymethyl)lysine protein adduct is a major immunochemical epitope in proteins modified with advanced glycation end products of the Maillard reaction, *Biochemistry*, 1996, **35**, 8075-8083.
10. H. Nakayama, T. Mitsuhashi, S. Kuwajima, S. Aoki, Y. Kuroda, T. Itoh and S. Nakagawa, Immunochemical detection of advanced glycation end products in lens crystallins from streptozocin-induced diabetic rat, *Diabetes*, 1993, **42**, 345-350.
11. Z. Makita, H. Vlassara, A. Cerami and R. Bucala, Immunochemical detection of advanced glycosylation end products in vivo, *J. Biol. Chem.*, 1992, **267**, 5133-5138.
12. R.K. Cho, A. Okitani and H. Kato, Chemical properties and polymerization ability of the lysozyme monomer isolated after storage with glucose, *Agric. Biol. Chem.*, 1984, **48**, 3081-3089.
13. M.-X. Fu, K.J. Wells-Knecht, J.A. Blackledge, T.J. Lyons, S.R. Thorpe and J.W. Baynes, Glycation, glycoxidation, and cross-linking of collagen by glucose, *Diabetes*, 1994, **43**, 676-683.

Immunolocalization of the Glycoxidation Product N^ε-(carboxymethyl) Lysine in Normal and Inflamed Human Intestinal Tissues

Erwin D. Schleicher, Klaus E. Gempel,[1] Eva Wagner[1] and Andreas G. Nerlich[2]

DEPARTMENT FOR INTERNAL MEDICINE, DIVISION OF ENDOCRINOLOGY, METABOLISM AND PATHOBIOCHEMISTRY, EBERHARD-KARLS-UNIVERSITÄT TÜBINGEN, OTFRIED-MÜLLER-STR. 10, D-72076 TÜBINGEN, GERMANY; [1]INSTITUTE FOR DIABETES RESEARCH, KÖLNER PLATZ 1, D-80804 MÜNCHEN, GERMANY; [2]DEPARTMENT OF PATHOLOGY; LUDWIG-MAXIMILIANS-UNIVERSITÄT MÜNCHEN, THALKIRCHNERSTR. 36, D-80337 MÜNCHEN, GERMANY

Summary

It has been suggested that N^ε-(carboxymethyl)lysine (CML), the major product of oxidative degradation of glycated proteins represents an integrative marker of oxidative stress in aging, arteriosclerosis and diabetes. Our previous immunohistochemical analysis using a CML-specific antiserum provided evidence for an age- and diabetes-dependent increase in CML-accumulation in the extracellular matrix. Intracellular accumulation of CML was found in macrophages and foam cells of arteriosclerotic plaques. In this study we immunohistochemically investigated the occurrence of CML in normal intestinal tissues. Furthermore, we analysed the immunolocalization of CML in chronically inflamed intestinal tissues (inflammation associated with M. Crohn or colitis ulcerosa) as a "model system" for a chronic inflammatory process. In these lesions we found widespread, intense, positive CML-staining in the macrophages. However, since in the normal intestinal wall we observed also positively labelled macrophages, it appears that exogenous CML uptake and phagocytosis in macrophages, e.g. from diet, may contribute to this staining. Furthermore, the formation of CML was studied *in vitro* to elucidate the origin of the unexpectedly high extent of intracellular CML. We found that CML was more rapidly formed from arachidonic acid than from glycated HSA. In summary, our observations provide strong evidence that CML-modified proteins may be formed by both, glycoxidation and lipid peroxidation, i.e., in situations of enhanced oxidative stress and that the CML-content of intestinal cells may result from phagocytosed AGE-modified proteins of endo-or exogenous origin.

Introduction

With advancing age, insoluble collagen undergoes marked physicochemical changes including decreased solubility, elasticity, protease sensitivity, and increased thermal stability.[1,2] These alterations are accelerated in diabetic patients in whom collagen behaves as though prematurely aged. The changes in physical properties are paralleled by biochemical alterations. These include increases in advanced glycosylation end-products (AGE) particularly pentosidine, pyralline, carboxymethyllysine and other less well-defined products of the Maillard reaction.[2] The identification of specific Maillard products has allowed us to investigate the different routes, i.e. oxidative and non-oxidative, leading to the post-ribosomal modifications occuring in long-lived proteins in the human body. Recently the major product resulting from oxidative cleavage of glycated proteins has been characterized. Glycated proteins are oxidatively cleaved between carbon 2 and 3 resulting in the formation of N^ε-(carboxymethyl)lysine (CML) in a yield of approximately 40% and erythronic acid.[3,4] The formation of CML modifications in proteins is irreversible. Therefore, it has been suggested that CML may be an integrative biomarker for the accumulated oxidative stress the respective tissue had been exposed to. However, recent *in-vitro* studies revealed that CML may also be formed by oxidative cleavage of polyunsaturated fatty acids.[5] Since the molecular structure of CML has been unequivocally characterized by chemical methods we produced an antiserum to this well-defined AGE-product recognizing CML-modified proteins in tissues.[6] We found an age-dependent increase in CML content in extracellular dermal tissue which was further increased in diabetic patients confirming earlier biochemical determinations.[7] In addition, we found an age-dependent increase in CML-staining of normal arterial walls. Arteriosclerotic lesions were intensely stained both in the extracellular matrix and intracellularly, particularly in macrophages. These data indicated that macrophages may take up CML-modified proteins.

It was our aim to study the occurrence of CML in intestinal cells and to investigate the origin of intracellular CML staining. In particular, we wanted to study if normal intestinal macrophages are also labelled to find out, if possibly exogenous AGE-products are taken up by the body. Furthermore, we analysed the effect of inflammation on cellular CML content to estimate the contribution of the glycation reaction in comparison with lipid peroxidation reaction to the formation of CML in human tissues. We also analysed the *in vitro* kinetic of CML formation from preglycated HSA and arachidonic acid in order to compare the rate of CML formation from glycoxidation and lipid peroxidation, respectively.

Methods

Generation and characterization of CML antibodies

The generation and characterization of CML-antibodies has been described recently.[6] Briefly, rabbits were immunized with CML-modified keyhole limpet hemocyanin and the antiserum was evaluated using CML-modified bovine serum albumin. CML-antibodies were obtained by immunoadsorption.[6] Specificity of the antiserum/antibody was assessed by competition with structurally related compounds.[6]

Formation of CML from glycated proteins and from polyunsaturated fatty acids or glyoxal in vitro

Glycation of human serum albumin (HSA) and incubation in air and oxygen-free atmosphere was performed as previously described.[6] To evaluate the contribution of polyunsaturated fatty acids we incubated HSA (20 mg/mL) with 0.4 M arachidonic acid at 37°C in the presence or absence of 20 µg/mL vitamin E. Similarly, HSA (5 mg/mL) was incubated with 4 M glyoxal. Samples from the incubation mixtures were taken at the time points indicated and immediately dialysed against phosphate-buffered saline. The formation of CML-HSA was determined by an enzyme-immuno-assay using the CML-antibody as previously described.[6] Briefly, 5 µg of the CML-modified protein were coated to microtiter plates and incubated overnight at 4°C. After several washings the microtiter plates were blocked and 0.1 mL CML-antiserum (1:1,000) was added and incubated for 2 h at room temperature. Then 0.1 mL of goat anti-rabbit IgG conjugated with peroxidase (1:8,000) was added to each well and incubated for 2 h. After eight washings 0.1 mL of phenylenediamine substrat solution (1.68 mg/mL) was added and the plate was incubated for 30 min in the dark. The absorbance was measured at 450 nm by a microplate reader and the data are expressed as mE (450nm).

Selection of tissue samples

Formalin-fixed and paraffin-embedded tissues obtained at autopsy or from biopsy material were used for immunohistochemical analysis. We analysed samples obtained from skin, lung, heart, kidney, bone and intestine from apparently healthy individuals at different ages. According tissues were analysed from diabetic patients and from patients with inflammatory intestinal diseases.

Immunohistochemistry
The immunohistochemical localization of CML-modified proteins was performed using the avidin-biotin complexed peroxidase staining method (ABC method; Vector Laboratories, Inc., Burlingame, USA) for paraffin-embedded tissue as established.[6] 3 µm tissue sections were deparaffinized and incubated with 0.1% pronase in phosphate-buffered saline at 37°C for 15 min and for further 15 min at room temperature. After washing and blocking steps, slides were incubated with CML-antibody (1:100) at 20°C for 1h. Development of the sections was performed as described previously.[6] The specificity of the immune reaction was evaluated by competition with CML-modified keyhole limpet hemocyanine (5 µg/mL).

Results

In-vitro formation of CML
Incubation of HSA with arachidonic acid in the presence of air yielded rapid formation of CML modification within 3 days with only little further increase after 2 to 4 weeks (Table 1A). Addition of vitamin E (20 µg/mL) had only a minor effect on CML formation. Similarly incubation with glyoxal caused rapid formation of CML (Table 1B). Furthermore, our studies revealed linear formation of CML modification with time from glycated proteins under normal atmosphere (oxygen) while no CML was formed under unaerobic (argon) conditions (Table 1C) The data indicate a slower CML formation from glycated proteins when compared to lipid peroxidation.[5,6]

Table 1. *In-vitro formation of CML-HSA by coincubation of HSA with arachidonic acid or glyoxal*

A	condition	0	3	7	14	28	days
	arachidonic acid (ara)	38	408	623	807	515	mE (450 nm)
	ara + vitamin E	50	230	592	554	500	mE (450 nm)

B	condition	0	3	6	13	-	days
	glyoxal	43	582	843	1121	-	mE (450 nm)

C	condition	0	6	10	20	23	days
	gHSA + O$_2$	41	91	113	179	238	mE (450 nm)
	gHSA + Ar	42	40	43	39	41	mE (450 nm)

HSA was incubated with 0.4 M arachidonic acid without or with vitamin E. (A). Similarly HSA was incubated with glyoxal (B). HSA was preglycated (gHSA) and the CML formation was followed under aerobic (O_2) and anaerobic (Ar) conditions (C). At the time points indicated samples were taken and the CML content of HSA was analyzed by enzyme immuno assay as described in the method section.

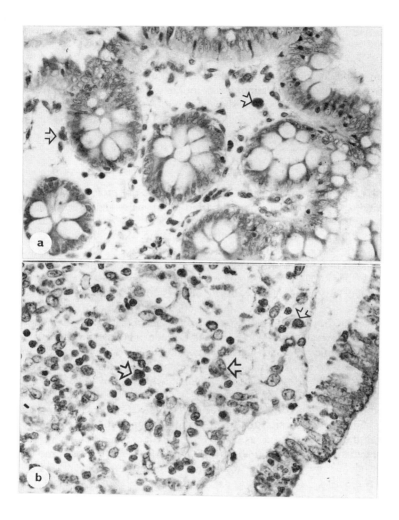

Figure 1. *Localization of CML-modified proteins in human intestinal mucosa;*
(a) In the mucosa of a normal individual (38 years) few scattered macrophages show cytoplasmatic staining (arrows). Note the slight staining of the brush border of the enterocytes;
(b) In the chronic colitis (33 year old individual, chronic non-active ulcerative colitis) the number of CML-labelled macrophages is significantly increased (arrows). The apical enterocytes are also more intensely stained

Localization of CML in normal and inflamed intestinal tissue

The immunostaining for CML in intestinal tissue showed in the normal mucosa a slight superficial staining of the epithelial cytoplasm and scattered positively labelled interstitial cells, particularly macrophages (Figure 1a). In intestinal mucosa with chronic inflammation (chronic active and non-active ulcerative colitis and M.Crohn) the amount of positively labelled cells was significantly increased. An enhanced CML staining was seen intracellularly in apical enterocytes, but also in interstitial cells. In particular, macrophages, which were markedly increased in number, showed a significant cytoplasmic staining, so that the overall macrophage staining was strongly enhanced (Figure 1b). In addition, plasma cells, but also granulocytes, showed a positive reaction. In our restricted series, however, no significant correlation could be observed between the CML-staining intensity and the activity of the inflammatory reaction.

Discussion

Our *in-vitro* data confirmed that CML modification in proteins may originate from glycoxidation of glycated proteins or from lipid peroxidation in the presence of polyunsaturated fatty acids as previously described.[5] Of note, CML formation from incubations with arachidonic acid was much more rapid when compared to formation from glycated HSA; however only in the first few days. Presence of glyoxal, an intermediate of lipid peroxidation of arachidonic acid, caused even faster formation of CML-HSA: These data indicate that CML modification of proteins may also occur via oxidation of polyunsaturated fatty acids. This pathway appears likely to occur in inflammatory and phagocytic processes where an activation of both polyunsaturated fatty acids and oxygen radicals is present. Since we found also positively stained macrophages in normal e.g. non-inflammatory intestinal sections, we cannot exclude that intestinal uptake of Maillard products with the diet and subsequent phagocytosis by macrophages contributes to this staining. This assumption is corroborated by the observation that normal enterocytes show a very faint superficial cytoplasmic CML-staining, which seems to be enhanced in chronic intestinal inflammation.

No reports on immunohistochemical detection of AGE-products in intestine appeared up to now, although the localization of AGE-products in arteriosclerotic plaques has been studied.[8,9] In these studies extra- as well as intracellular staining of atherosclerotic lesions has been found similarly to our findings with the CML-specific antiserum.[6] The similarity may be explained by the recent result that AGE-antisera recognize CML as major epitope.[10] The intracellular location of CML in macrophages indicates that these cells may specifically recognize and take up CML-modified proteins. However, extensive studies by our laboratory using CML-modified LDL indicated that CML modification is not recognized by macrophages and that the CML-modified LDL species are not recognized by the LDL-specific receptor on fibroblasts.[11] Noteworthy, we also found increased intra- and extracellular CML staining in intervertebral discs with early "degenerative" lesions suggesting that CML modification occurs during repair processes.[12]

In sum, our data indicate an age-dependent appearance of CML modification in specific extracellular matrix compounds and an ubiquitous staining of macrophages in lesioned tissues. While CML in the matrix is probably formed from glycated precursors, intracellular CML content may originate from lipid peroxidation.

References

1. E. Schleicher and A. Nerlich, The role of hyperglycemia in the development of diabetic complications, *Horm. Metabol. Res.*, 1996, **28**, 367-378.
2. R. Bucala and A. Cerami, Advanced glycosylation: chemistry, biology and implications for diabetes in aging, *Adv. Pharmacol.*, 1992, **23**, 1-34.
3. S. Reddy, J. Bichler, K.J. Wells-Knecht, S.R. Thorpe and J.W. Baynes, N^ε-(carboxymethyl)lysine is a dominant advanced glycation end product (AGE) in tissue proteins, *Biochemistry*, 1995, **34**, 10872-10878.
4. M.U. Ahmed, S.R. Thorpe and J.W. Baynes, Identification of N^ε-(carboxymethyl)lysine as a degradation product of fructose lysine in glycated protein, *J. Biol. Chem.*, 1986, **261**, 4889-4894.
5. M.X. Fu, J.R. Requena, A.J. Jenkins, T.J. Science, J.W. Baynes and S.R. Thorpe, The advanced glycation endproduct, N^ε-(carboxymethyl)lysine, is a product of both lipid peroxidation and glycoxidation reactions, *J. Biol. Chem.*, 1996, **271**, 9982-9986.
6. E.D. Schleicher, E. Wagner and A.G. Nerlich, Increased accumulation of the glycoxidation product N^ε-(carboxymethyl)lysine in human tissues in diabetes and aging, *J. Clin. Invest.*, 1997, **99**, 457-468.
7. D.G. Dyer, J.A. Dunn, S.R. Thorpe, K.E. Baillie, T.J. Lyons, D.R. McCance and J.W. Baynes, Accumulation of Maillard reaction products in skin collagen in diabetes and aging, *J. Clin. Invest*, 1993, **91**, 2463-2469.
8. Y. Nakamura, Y. Horii, T. Nishino, H. Shiiki, Y. Sakaguchi, T. Kagoshima, K. Dohi, Z. Makita, H. Vlassara and R. Bucala, Immunohistochemical localization of advanced glycosylation end products in coronary atheroma and cardiac tissue in diabetes mellitus, *Am. J. Pathol.*, 1993, **143**, 1649-1656.
9. S. Kume, M. Takeya, T. Mori, N. Araki, H. Suzuki, S. Horiuchi, T. Kodama, Y. Miyauchi and K. Takahashi, Immunohistochemical and ultrastructural detection of advanced glycation end products in atherosclerotic lesions of human aorta with a novel specific monoclonal antibody, *Am. J. Pathol.*, 1995, **147**, 654-667.
10. K. Ikeda, T. Higashi, H. Sano, Y. Jinnouchi, M. Yoshida, T. Araki, S. Ueda and S. Horiuchi, N^ε-(carboxymethyl)lysine protein adduct is a major immunological epitope in proteins modified with advanced glycation end products of the Maillard reaction, *Biochemistry*, 1996, **35**, 8075-8083.
11. K. E. Gempel, K. D. Gerbitz, B. Olgemöller and E. D. Schleicher, In-Vitro Carboxymethylation of low density lipoprotein alters its metabolism via the high affinity receptor, *Horm. Metabol. Res.*, 1993, **25**, 250-252.
12. A. Nerlich, E. D. Schleicher and N. Boos, Immunohistochemical markers for age related changes of human lumbar intervertebral discs, *Spine* (in press)

Pharmacological Reversal Of AGE-Related Protein Crosslinking with Agents that Cleave α-Dicarbonyls

Peter Ulrich

CERAMI CONSULTING CORPORATION, 765 OLD SAW MILL RIVER ROAD, TARRYTOWN, NY 10591, USA

Summary
Substituted thiazolium salts, designed to cleave α-dicarbonyl-containing precursors of AGE-mediated protein crosslinking, have been found to be able to cleave a substantial proportion of the AGE crosslinks themselves as formed *in vitro* and *in vivo*. This finding has important implications as to the predominant mechanisms of protein crosslinking by AGEs (advanced glycation end-products) *in vivo*, and shows the potential for pharmacological reversal of AGE-mediated diabetic complications.

Introduction

What if an important circulating energy source could chemically damage proteins in the body? What if such damage even included the inappropriate crosslinking of proteins and other biomolecules to each other, leading to deleterious tissue changes? What if the body had to devote resources to remove this damage throughout the lifespan of the organism? It might seem paradoxical that evolution would have produced such a scenario.

And yet, at least two examples of this situation occur. Damage by active oxygen species is the best known of such processes. Appreciation has been slower outside the Maillard and diabetes research communities of the importance *in vivo* of another process, the non-enzymatic reaction of glucose with protein amino groups, or advanced glycation end-product (AGE) formation. Recognition of the significance of AGE formation *in vivo* has not been helped by the fact that the dominant mechanisms by which AGEs produce protein crosslinking and other deleterious changes under physiological conditions are still poorly understood.

Agents that cleave α-dicarbonyls can break AGE protein-protein crosslinks
In collaboration with other scientists we recently designed a new class of inhibitors of AGE formation[1] based on the assumption that α-dicarbonyl species like methylglyoxal, 3-deoxyglucosone, and 1,4-dideoxy-1-lysino-fructosone are important mediators of protein crosslinking. The new inhibitors are thiazolium salts such as *N*-phenacylthiazolium bromide (PTB), which are designed to cleave chemically α-dicarbonyl species between the two carbonyl groups. These compounds were found to be excellent inhibitors of AGE formation. More significantly, they were found to cleave a major portion of protein-protein AGE crosslinks that had already formed under physiological conditions, *in vitro* and *in vivo*.[1] The well-known and highly effective inhibitor of AGE formation, aminoguanidine,[2] can neither cleave α-dicarbonyls nor break such pre-existing AGE crosslinks. These findings indicate the possibility that a major class of heretofore unrecognized AGE crosslinking structures contain an α-dicarbonyl unit integral to the covalent linkage. I will discuss here the current status and future implications of these findings.

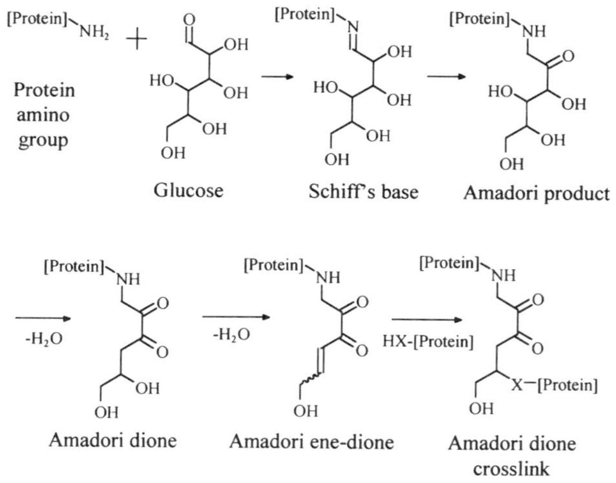

Figure 1 *Hypothetical formation of an α-dicarbonyl containing crosslink via the Amadori product*

Amadori product dehydration as a source of α-dicarbonyl AGE crosslinks
It has already been noted by others that currently known AGE crosslinking structures cannot explain the bulk of AGE crosslinking observed under physiological conditions.[3] Cleavage of AGE crosslinks by specific α-dicarbonyl-cleaving compounds led us to consider the possibility that a major proportion of physiological AGE crosslinks may be due to a simple but little studied potential route which could produce crosslinks containing α-dicarbonyl linkages. This route involves dehydration of the Amadori product (Figure 1). Loss of the hydroxyl at carbon 4 via β-elimination gives a 1,4-dideoxy-1-alkylamino-2,3-hexodiulose,[4] which we call the Amadori dione. A second β-elimination of water from carbon 5 gives an ene-dione which we call the Amadori ene-dione; this has been isolated as a triacetylated enol derivative.[5] Conjugate addition of a nucleophilic protein residue to the carbon-carbon double bond of the Amadori ene-dione would produce a protein-protein crosslink containing an α-diketone, which we call the Amadori dione crosslink. The process would actually be more complex than shown in Figure 1, because many of these intermediates probably exist preferentially in a 2,6-pyranose form.

Mechanism of α-dicarbonyl cleavage by N-*phenacylthiazolium bromide (PTB)*
Simple N-substituted thiazolium salts can cleave α-dicarbonyls at high pH.[6] The mechanism is similar to enzymatic cleavages using thiamine as a cofactor. In order to allow the cleavage to proceed at physiological pH without an enzyme, we designed PTB to have a second nucleophilic carbon center at the α position of the side chain. This allows nucleophilic additions by PTB to both carbonyls of the α-dicarbonyl compound, forming a cyclic adduct in which both of the former ketone carbons have assumed the tetrahedral sp^3 geometry necessary for carbon-carbon bond cleavage to occur. In Figure 2 this process is depicted for the hypothetical Amadori dione crosslink; for simplicity, this figure arbitrarily shows only one of the two possible orientations of PTB addition. After cleavage of the carbonyl-carbonyl bond, one of the crosslink fragments would detach via hydrolytic

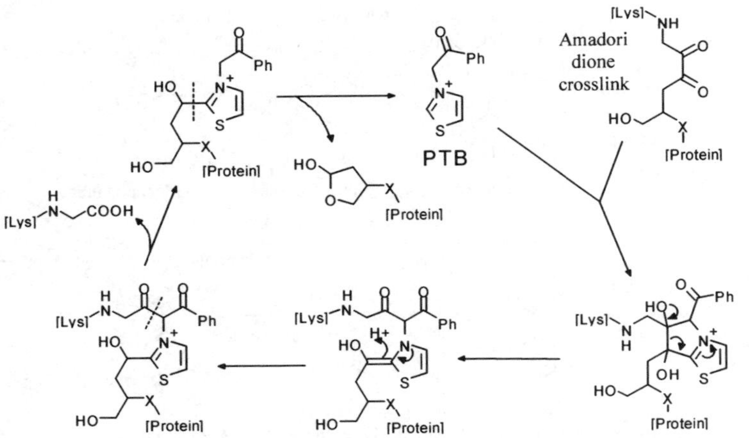

Figure 2 *Proposed mechanism of AGE crosslink cleavage by PTB in a potentially catalytic cycle*

deacylation to give a carboxylic acid, effecting cleavage of the crosslink. Ideally, the other crosslink fragment would also detach, regenerating the thiazolium cleavage agent to allow a catalytic cycle to occur, but the efficiency of this second step at physiological pH is not yet established. Future drug design efforts based on PTB will include increasing the rate of this second step as an important criterion.

Effects of PTB on AGE protein crosslinks in vitro and in vivo
PTB rapidly cleaves a model diketone, 1-phenyl-1,2-propanedione, with release of benzoic acid. Of more importance are its effects in model systems of AGE crosslinking, and on AGE crosslinks *in vivo*.[1]

AGE-BSA was allowed to crosslink to collagen which had been pre-coated onto microtiter wells. Unreacted AGE-BSA was washed away. PTB was found to release the BSA that had been crosslinked to the collagen in a dose- and time-dependent manner. Aminoguanidine is not able to release crosslinked BSA in this assay.

On digestion with cyanogen bromide (CNBr), normal rat tail tendon collagen yields fragments small enough to enter an electrophoresis gel. By comparison, tail collagen from diabetic rats releases almost no such fragments on CNBr treatment, because of extensive additional crosslinking due to AGE formation. This failure to release fragments is almost completely reversed on pre-incubation of the isolated collagen with PTB prior to digestion with CNBr. This normalization of fragment release is not seen on pre-treatment with aminoguanidine.

In diabetic rats IgG becomes tightly bound to red cell membrane proteins, presumably through covalent AGE crosslinking reactions. However, release of such bound IgG, reaching a level of 60-70% cleavage, occurred in diabetic rats that received PTB at 10 mg/kg/day for up to 4 weeks. Treatment of diabetic rat red cells *in vitro* with PTB causes IgG binding to revert to control levels, an effect not seen with aminoguanidine.

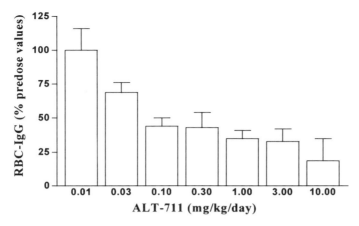

Figure 3 *RBC-IgG levels after 8 days oral dosing: EC50 = 0.06 mg/kg/day*

ALT-711, a derivative of PTB with improved properties
It was found that PTB, like many thiazolium salts including thiamine[7], undergoes hydrolytic thiazolium ring opening at physiological pH. Screening of a number of thiazolium compounds revealed a methyl-substituted derivative of PTB, designated ALT-711, which has much improved hydrolytic stability. ALT-711 has proven to have high activity in cleaving AGE crosslinks in a number of assays in vitro and in vivo. For example, in the assay described above for cleavage of AGE crosslinking of IgG to red blood cells, ALT-711 was found to have an ED50 of 0.06 mg/kg (60 µg/kg) per day after 8 days of dosing (Figure 3).[8] ALT-711 is currently in pre-clinical development.

Future directions in thiazolium-mediated reversal of AGE crosslinking.
Much work has been done in establishing a causative association between the formation of covalent AGE protein-protein crosslinks and many of the pathological sequelae observed in diabetes.[9,10] A pharmacological method of reversing this process, rather than just slowing or preventing it as with aminoguanidine, would have important therapeutic potential for treating AGE-related diabetic complications, and even for ameliorating AGE-related changes in aging.

If this potential is to be realized, it is important to increase our understanding of the details of AGE crosslink cleavage by PTB and related agents. Confirmation of our proposed mechanism in Figure 2 would require a reliable source of Amadori ene-dione, which is unstable and difficult to purify. A novel synthetic precursor, a 4,5-di-*O*-tosylate of an Amadori product, was recently developed for this purpose.[11] Availability of this chemical tool for studying the Amadori dione crosslink pathway will allow more detailed investigation as to the validity of this pathway as an explanation of the AGE crosslink breaking activity of thiazolium agents in vivo.

It is our hope that the development currently underway of new analogues of PTB with improved chemical and pharmacological properties, such as ALT-711, will lead to a clinically useful AGE-crosslink cleavage agent, which would represent a major advance in the battle against diabetic complications.

References

1. S. Vasan, X. Zhang, X.N. Zhang, A. Kapurniotu, J. Bernhagen, S. Teichberg, J. Basgen, D. Wagle, D. Shih, I. Terlecky, R. Bucala, A. Cerami, J. Egan and P. Ulrich, An agent cleaving glucose-derived protein crosslinks in vitro and in vivo, Nature, 1996, **382**, 275-278.
2. M. Brownlee, H. Vlassara, A. Kooney, P. Ulrich and A. Cerami, Aminoguanidine prevents diabetes-induced arterial wall protein cross-linking, Science, 1986, **232**, 1629-1632.
3. D.G. Dyer, J.A. Blackledge, B.M. Katz, C.J. Hull, H.D. Adkisson, S.R. Thorpe, T.J. Lyons and J.W. Baynes, The Maillard reaction in vivo, Zeitschr. Ernahrungswiss., 1991, **30**, 29-45.
4. B. Huber and F. Ledl, Formation of 1-amino-1,4-dideoxy-2,3-hexodiuloses and 2-aminoacetylfurans in the Maillard reaction, Carbohyd. Res., 1990, **204**, 215-220.
5. S. Estendorfer, F. Ledl and T. Severin, Formation of an aminoreductone from glucose, Angew. Chem. Int. Ed. Engl., 1990, **29**, 536-537.
6. A.I. Vovk, I.V. Murav'eva and A.A. Yasnikov, Mechanism of the cleavage of acetylbenzoyl in methanol catalyzed by thiazolium salts, Ukr. Khim. Zh., Russ. Ed., 1985, **51**, 521-525.
7. J.A. Zoltewicz and G. Uray, Thiamin - A critical evaluation of recent chemistry of the pyrimidine ring, Bioorganic Chemistry, 1994, **22**, 1-28.
8. S. Vasan, J. Egan, personal communication.
9. S. Horiuchi, Advanced glycation end products (AGE)-modified proteins and their potential relevance to atherosclerosis. Trends Cardiov. Med., 1996, **6**, 163-168.
10. W. Palinski, T. Koschinsky, S.W. Butler, E. Miller, H. Vlassara, A. Cerami and J.L. Witztum, Immunological evidence for the presence of advanced glycosylation end products in atherosclerotic lesions of euglycemic rabbits, Arteriosc. Thromb. Vasc. Biol., 1995, **15**, 571-582.
11. X.N. Zhang and P. Ulrich, Directed approaches to reactive Maillard intermediates: Formation of a novel 3-alkylamino-2-hydroxy-4-hydroxymethyl-2-cyclopenten-1-one ("cypentodine"), Tetrahedron Lett., 1996, **37**, 4667-4670.

Role of Protein-Bound Carbonyl Groups in the Formation of Advanced Glycation Endproducts

Jason Liggins, Nicola Rodda, Jim Iley * and Anna Furth

THE OPEN UNIVERSITY, OXFORD RESEARCH UNIT, BOARS HILL, OXFORD OX1 5HR, UK. * THE OPEN UNIVERSITY, DEPARTMENT OF CHEMISTRY, MILTON KEYNES, MK7 6AA, UK

Summary
Several mechanisms have been postulated for the formation of advanced glycation endproducts (AGEs) from glycated proteins; they all feature protein-bound carbonyl intermediates. Using 2,4-dinitrophenylhydrazine (DNPH), we have detected these intermediates on bovine serum albumin (BSA), lysozyme and β-lactoglobulin after glycation *in vitro* by glucose or fructose. Carbonyls were formed in parallel with AGE-fluorophores, via oxidative Maillard reactions. Surprisingly, neither Amadori nor Heyns products contributed to the DNPH reaction. Fluorophore and carbonyl yields were much enhanced in lipid-associated proteins, but both groups could also be detected in lipid-free proteins. When pre-glycated proteins were incubated in the absence of free sugar, carbonyl groups were rapidly lost in a first-order reaction, while fluorescence continued to develop beyond the 21 days of incubation. Some carbonyl groups appeared resistant to inhibition by 25 mM aminoguanidine (AG), although 4 mM AG could fully inhibit reactions leading to AGE-fluorescence. It is suggested that carbonyls acting as fluorophore precursors react readily with AG, while others are resistant because they are involved in ring closure to give hemiketals. Factors influencing the relative rates of acyclisation and hydrazone formation are discussed.

Introduction

Aminoguanidine (AG) inhibits the formation of AGEs *in vitro*, and is on clinical trial for its effectiveness against diabetic complications *in vivo*. Yet its exact molecular targets are still unclear. As a hydrazine it would be expected to block AGE formation by reacting with carbonyl intermediates, and indeed the expected triazine products have been isolated[1] from the reaction of AG with deoxyglucosone *in vitro*. However, other protein-bound carbonyl intermediates have long been postulated in the proposed routes to AGEs such as carboxymethyllysine,[2] pentosidine[3] and pyrraline-derived crosslinks.[4] Carbinolamines are among protein-bound carbonyl intermediates in the Namiki route to AGEs.[5]

We investigated the role in AGE formation of these other potential targets for AG. (The route to AGEs may not always involve simple glycation. The Schiff base may be oxidized directly, bypassing the Amadori product (AP), or protein may be attacked by sugar oxidation products such as glyoxal. We are interested in the role of protein-bound carbonyls in any of these routes, which are all referred to as Maillard reactions.)

We have used 2,4-dinitrophenylhydrazine (DNPH) to demonstrate the formation of protein-bound carbonyls following glycation by glucose or fructose. We attach particular emphasis to fructose, a potent glycating agent whose serum levels can rise to 0.5 mM in health and 2 mM in cirrhosis, causing fructation of human haemoglobin and lens protein *in vivo* (reviewed in Ref. 6).

Materials and Methods

Glycation in vitro
Proteins (BSA at 40 mg/mL; lysozyme and β-lactoglobulin at 10 mg/mL) were incubated in glucose or fructose at 37° and pH 7.4, in 0.1 M Na phosphate buffer with 3 mM Na azide. (Delipidated BSA and other proteins came from Sigma.)

Re-incubation of pre-glycated BSA
BSA was pre-glycated to different extents by incubating for different periods in 0.05 M fructose at 37°. After exhaustive dialysis at 4° over 4 days against phosphate buffer (to remove free sugar) the protein was immediately re-incubated at 37° in buffer alone.

Periodate assay for AP and other periodate-positive material
This was our standard microassay.[7]

Assay of AGE-fluorophores in glycation solutions
Wavelengths optimum for pentosidine were used: ex/em = 325/375 nm.

DNPH assay for protein-bound carbonyl groups
The method of Levine et al.[8] was adapted (see Ref. 9). Briefly, DNPH-hydrazones were formed by reaction of protein with DNPH in 2 M HCl. After precipitation in TCA, the protein was extensively washed in ethanol/ethyl acetate to remove free DNPH, then re-dissolved in 6 M guanidine-HCl. Carbonyl concentration was calculated from the net absorbance at 379 nm using a molar absorption coefficient of 22000 M^{-1} cm^{-1} (see Ref. 8).

Results and Discussion

Reaction with DNPH is a standard way of detecting carbonyl groups, and can be used to follow changes on protein oxidation.[8] However the carbonyls detected under our conditions are clearly the result of sugar-induced reactions, since there is little DNPH-reactivity from proteins incubated in buffer alone.

Carbonyl and fluorescent groups were produced on glycated BSA, from both glucose and fructose (Figure 1). Yields were considerably higher with fructose than glucose, and were greatly enhanced by the presence of lipid. Glycation of delipidated BSA gave comparatively few fluorescent or carbonyl groups even with fructose, and Table 1 suggests that carbonyl groups are more readily formed on lipid-associated proteins. The metal chelator DETAPAC was a potent inhibitor, reducing yields of both carbonyl and fluorescent groups by 80%. Hence both groups appear to result from metal-catalysed glycoxidation reactions, enhanced by lipid and catalysed here by the transition metal in phosphate buffer.

Table 1. *Effect of lipid on output of protein-bound carbonyl and fluorescent AGEs (7 days glycation in 0.1 M fructose)*

Protein	mol carbonyl/mol protein	fluorescence/mol protein
native BSA	0.35	366
delipidated BSA	0.06	15
β-lactoglobulin	0.31	310
lysozyme	0.06	35

When BSA samples pre-fructated to different extents were re-incubated in sugar-free buffer, there was a steady increase in fluorescence and a sharp drop in carbonyl content (Figure 2). (This accentuated changes already begun at 4° during the dialysis.) The carbonyl groups were lost (transformed to other products) in a first-order reaction that appeared to be independent of initial carbonyl concentration. This is consistent with their role as transient intermediates. Whatever the extent of previous fructation, all samples had a half-life of around 4 days under our conditions. Of the many AGEs into which carbonyls may be converted, we followed the rise of fluorophores only. These constitute a very minor fraction of AGEs,[24] and arise several days before crosslinks.[10,11] Some AGE-fluorophores may be derived directly from the carbonyl intermediates as suggested for pentosidine,[3] but since fluorophore build-up continues long after carbonyl decay is complete, there must be further 'slow' reactions on the fluorophore pathway.

How far can these DNPH-reactive carbonyl compounds be identified with known Maillard intermediates? Small dicarbonyl compounds like 3-deoxyglucosone and glyoxal will be removed by precipitation and washing of the protein DNPH-hydrazone. Acyclic forms of AP and Heyns product might be expected to react with DNPH, but they appeared to contribute little to the reaction here. Firstly, the build-up of AP was unaffected by lipid (Figure 3), with native and delipidated BSA giving similar yields of periodate positive material (PPM). This is in sharp contrast to the effect of lipid on protein-bound carbonyls (Table 1). Secondly, AP yields were enhanced from 60-600% by adding 1 mM DETAPAC, whereas carbonyl content as measured by DNPH was reduced by 80%.[9]

The periodate assay for glucated protein is generally assumed to measure AP, and gives quantitative data comparable to the thiobarbituric acid and other assays.[12] Therefore our work strongly suggests that the DNPH assay was not detecting AP. It also suggests that cyclic and acyclic forms of AP may not be in rapid equilibrium, an apparent anomaly supported by the work of Fischer and Winterhalter;[13] they found little reaction between glycated haemoglobin and phenylhydrazine, and concluded that the proportion of acyclic AP is very low. In contrast, DNPH readily forms hydrazones with the AP of glyceraldehyde and haemoglobin, where the sugar backbone is too short to cyclize.[14]

It is difficult to see whether the acyclic Heyns product reacts with DNPH, since it cannot be assayed independently. (The periodate assay measures only HCHO from *cis* diol or aminol groups, neither of which appear in the major Heyns product.[12]) Lipid significantly enhanced the yield of PPM in fructated BSA, but not in glucated BSA (Figure 3), suggesting that different factors affect the formation of PPM from the two sugars. Hence the small amount of PPM formed by fructation is not the fructose equivalent of AP, but more likely some post-Heyns Maillard product with a periodate-reactive grouping.

Table 1 suggests that most of the carbonyls reacting with DNPH derive from lipid. Although the figures in Table 1 must be treated as approximate, it is clear from comparing native and delipidated BSA, that lipid can greatly enhance the yield of carbonyl groups. Similarly, the lipid-free lysozyme gives fewer carbonyls than β-lactoglobulin, another small protein with at least one mole of lipid per mole. Antibody-reactive AGEs are found on both lipid and apoprotein fractions of lipoprotein glycated *in vivo*.[15] Our lipid carbonyls may arise partly from glycation of phospholipids with primary amine groups, and partly from lipid peroxidation. This last can be initiated by AGEs, and would augment lipid carbonyls when the resulting hydroperoxides fragment to aldehydes.[15,16] Hence our lipid-enhanced

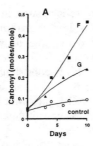 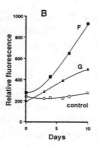

Figure 1. *Formation of protein-bound carbonyl groups (A) and fluorophores (B) during glycation of BSA by 0.05 M fructose (F), 0.2 M glucose (G) or buffer alone (controls)*

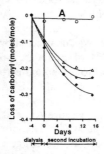 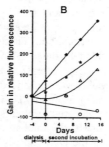

Figure 2. *Loss of carbonyl groups (A) and gain of fluorophores (B) during re-incubation of pre-glycated BSA in the absence of free sugar. Preglycation was at 37° for 7 days (triangles), 14 days (stars) and 21 days (solid diamonds) - see Methods. Unbound sugar was removed by dialysis at 4°. Samples were immediately re-incubated at 37° in buffer alone, starting at "0 days". In controls (open circles) fructose was omitted throughout*

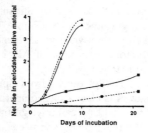 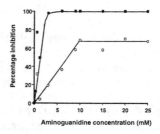

Figure 3. *Yield of periodate-positive material on glycation of native BSA (full lines) and delipidated BSA (dotted lines) in 0.2 M glucose (triangles), 0.05 M fructose (squares)*

Figure 4. *Effect of aminoguanidine concentration on output of AGE fluorophores (squares) and protein-bound carbonyls (open circles), during glycation of BSA in 0.1 M glucose for 8 days*

carbonyls appear to result directly or indirectly from Maillard reactions as defined earlier. No protein gave significant readings in the absence of sugar.

So what is the source of DNPH-reactive carbonyls in lipid-free proteins? Lysozyme (Table 1) and delipidated BSA (Table 1 and Ref. 17) contribute both protein-bound carbonyls and fluorophores, although yields are considerably lower than for lipid-associated proteins. Having discounted Amadori (and Heyns) products as contributors to the DNPH reaction, we are left with some unidentified Maillard intermediate. Protein-bound sugar derivatives, both mono- and di-carbonyl, feature in most of the reaction schemes proposed for AGE formation.[2-5] It is one or more of these carbonyls that we appear to be detecting. There may also be some amino acid sidechain carbonyls,[17,24] resulting from AGE-initiated oxidation.

Carbonyls susceptible to DNPH would be expected to react with aminoguanidine, since this is another hydrazine. However AG was never fully effective against protein-bound carbonyls (Figure 4), and its inhibitory effect tailed off rapidly above 10 mM AG, leaving a significant percentage of carbonyls still apparently unblocked. In contrast, AGE-fluorescence was fully inhibited in only 4 mM AG. Fructose gave similar results.[9] Requena et al.[18] found AP carbonyls resistant to AG concentrations that fully inhibited fluorescence, while Fujii et al.[19] found the release of 3-deoxyglucosone from glucated BSA to be highly sensitive to AG concentrations below 8 mM; thereafter the release tailed off very slowly, reaching zero only in 50 mM AG.

All this suggests there may be two classes of protein-bound carbonyl groups, some less reactive towards hydrazines because they are involved in forming cyclic hemiacetals. It is noteworthy that glucose itself reacts very slowly with AG[20] and this would explain how the hydrazine is effective in a large molar excess of glucose. Aminoguanidine may also be a rather weak hydrazine. If acyclization is slower than hydrazone formation, then acyclization becomes the rate-determining step; Harding[21] makes a similar point. Kaanane and Labuza[22] have pointed out in the context of glycation, that acyclisation rates of sugars and their derivatives would be expected to vary with pH and ionic composition. Phosphate anions have been shown to catalyse acyclization of the small model AP, deoxyfructosyl hippuryllysine.[23] So microenvironment could also be critical for acyclization, causing some carbonyls to be less reactive than others. It would clearly be useful to have more information on what influences hydrazine reactivity and acyclization rates in sugars and their Maillard derivatives.

References

1. J. Hirsch, V.V. Mossine and M.S. Feather, Detection of some dicarbonyl intermediates arising from degradation of Amadori compounds, *Carbohydr. Res.*, 1995, **273**, 171-177.
2. T. Sakurai and S. Tsuchiya, Superoxide production from nonenzymatically glycated protein, *FEBS Lett.*, 1988, **236**, 406-410.
3. S.K. Grandee and V.M. Monnier, Mechanism of formation of the Maillard protein crosslink pentosidine, *J. Biol. Chem.*, 1991, **266**, 11649-11653.
4. F. Hayase, R.H. Nagaraj, S. Miyata, F.G. Njoroge and V.M Monnier, Aging of proteins: Immunological detection of a glucose-derived pyrrole formed during Maillard reaction in vivo, *J. Biol. Chem.*, 1989, **263**, 3758-3764.

5. M.A. Glomb and V.M. Monnier, Mechanism of protein modification by glyoxal and glycolaldehyde, intermediates of the Maillard reaction, *J. Biol. Chem.*, 1995, **270**, 10017-10026.
6. A.J.Furth, Glycated proteins in diabetes, *Brit. J. Biomed. Sci.*, 1997, in press.
7. N. Ahmed and A.J. Furth, A microassay for protein glycation based on the periodate assay, *Anal. Biochem.*, 1991, **192**, 109-111.
8. R.L. Levine *et al.*, Determination of carbonyl content in oxidatively modified proteins, *Meth. Enzymol.*, 1990, **186**, 464-478.
9. J. Liggins and A.J.Furth, Role of protein-bound carbonyl groups in formation of advanced glycation endproducts, *Biochim. Biophys. Acta*, 1997, in press.
10. J.D. McPherson, B.H. Shilton and D.J. Walton, Role of fructose in glycation and crosslinking of proteins, *J. Biol. Chem.*, 1988, **27**, 1901-1907.
11. J. Liggins and A.J. Furth, Mixed dimers formed by crosslinking of native and glycated proteins in the absence of free sugar, *Biochem. Biophys. Res. Commun.*, 1996, **219**, 186-190.
12. N. Ahmed and A.J. Furth, Failure of common glycation assays to detect glycation by fructose, *Clin. Chem.*, 1992, **38**, 1-3.
13. R.W. Fischer and K.H. Winterhalter, The carbohydrate moiety in hemoglobin A_{1c} is present in the ring form, *FEBS Lett.*, 1981, **135**, 145-147.
14. A.S. Acharya and J.M. Manning, Amadori rearrangement of glyceraldehyde-haemoglobin Schiff base adducts, *J. Biol. Chem.*, 1980, **255**, 7218-7224.
15. R. Bucala, Z. Makita, T. Koschinsky, A. Cerami and H. Vlassara, Lipid advanced glycosylation: pathway for lipid oxidation *in vivo*, *Proc. Natl. Acad. Sci.*, 1993, **90**, 6434-6438.
16. J.W. Baynes, Role of oxidative stress in development of complications in diabetes, *Diabetes*, 1991, **40**, 405-412.
17. Y. Takagi, A. Kashiwaga, Y. Tanaka, T. Asahina, R. Kikkawa and Y. Shigeta, Significance of fructose-induced protein oxidation and formation of advanced glycation endproduct, *J. Diab. Comp.*, 1995, **9**, 87-91.
18. J.R. Requena, P. Vidal and J. Cabezas-Cerrato, Aminoguanidine inhibits protein browning without extensive Amadori blocking, *Diab. Res. Clin. Pract.*, 1993, **19**, 23-30.
19. E. Fujii, I. Hitoo, I. Ishii-Karakasa, Y. Yajima and K. Hotta, Quantitation of the glycation intermediate 3-deoxyglucosone by oxidation with rabbit liver oxoaldehyde dehydrogenase, *J. Chromatog. B.*, 1994, **660**, 265-270.
20. M.S. Feather, in 'Thermally Generated Flavours', *American Chemical Society Symposium Series*, 1994, **543**, pp. 127-141.
21. J. Harding, 'Cataract: biochemistry, epidemiology and pharmacology', Chapman and Hall, 1991, p. 243.
22. A. Kaanane and T.P. Labuza, in 'The Maillard Reaction in Aging, Diabetes and Nutrition', J.W. Baynes and V.M. Monnier (eds), Alan R. Liss, New York, 1989, pp. 301-327.
23. P.R. Smith and P.J. Thornalley, Influence of pH and phosphate ions on kinetics of enolisation and degradation of fructosamines, *Biochem. Int.*, 1992, **28**, 429-439.
24. E.R. Stadtman, Protein oxidation and aging, *Science*, 1992, **257**, 1220-1224.

Role of Glucose Degradation Products in the Generation of Characteristic AGE Fluorescence in Peritoneal Dialysis Fluid?

Anne Dawnay, Anders P. Wieslander[1], and David J. Millar

THE RENAL RESEARCH LABORATORY, DEPARTMENT OF CLINICAL BIOCHEMISTRY, ST BARTHOLOMEW'S HOSPITAL, LONDON EC1A 7BE, UK; [1]BIOLOGICAL AND MEDICAL RESEARCH, GAMBRO GROUP, PO BOX 10101, S-220 10 LUND, SWEDEN

Summary
Recent evidence suggests that components of peritoneal dialysis fluid other than glucose contribute substantially to the generation of characteristic protein-linked advanced glycation endproduct fluorescence. These compounds may be glucose degradation products generated during the heat-sterilization of dialysis fluid. We therefore compared protein glycation and the generation of protein-linked fluorescence in peritoneal dialysis fluids *in vitro* using albumin as a model protein. Compared with a filter-sterilized control, the use of heat-sterilized fluid caused a rapid increase in fluorescence which was most marked in the first five days of incubation. This effect of heat-sterilization could be attenuated by separating the glucose from other components of the fluid during the sterilization process. The mode of sterilization had no effect on protein glycation rate. We suggest that increased formation of glucose degradation products such as methylglyoxal and 3-deoxyglucosone during conventional heat-sterilization are most likely to be responsible for the accelerated formation of fluorescent advanced glycation endproducts which are likely to be detrimental to the long-term functional integrity of the peritoneal membrane.

Introduction

Peritoneal dialysis (PD) fluids contain high concentrations of glucose (75-214 mM), which acts as an osmotic agent to remove fluid from patients with end-stage renal failure. Patients treated chronically by PD display histological changes in the peritoneal membrane resembling those found in basement membranes in patients with diabetes. The most notable include thickening and duplication of the basal lamina of the peritoneal capillaries and mesothelium which worsen with duration of treatment.[1,2] Advanced glycation endproducts (AGEs) have been detected immunohistochemically in the mesothelium, connective tissue and vascular wall of the peritoneal membrane of patients on PD[3] and in the effluent PD fluid from patients.[4] The accumulation of AGEs in the peritoneal membrane is likely to be detrimental to its functional integrity in the longer term contributing to treatment failure in a significant number of patients.[5,6]

Although glucose has been assumed to be the cause of peritoneal AGE formation, studies *in vitro* on effluent PD fluids from patients have shown accelerated generation of characteristic protein-bound AGE fluorescence compared with control solutions containing the same concentrations of glucose.[7] The promotion of AGE formation was greatest in unused PD fluid and diminished with increasing dwell time in the patient suggesting that the causative agents were either being absorbed into the patient or were being bound locally in the peritoneal membrane. It was speculated that reactive dicarbonyl glucose degradation products, such as 3-deoxyglucosone and methylglyoxal, which are known intermediates in the Maillard reaction and generate fluorescent AGEs, were responsible for the accelerated AGE formation over and above that attributable to glucose.[6,7] These degradation products are present in glucose-containing parenteral fluids as a consequence of the heat-sterilization process.[8,9] The aim of the present study was to determine the effect of heat-sterilization of PD fluid containing glucose on AGE formation in an *in vitro* system using human serum albumin as a model protein.

Materials and Methods

Preparation of materials for experiment 1
PD fluid was prepared in our laboratory containing 75 mM glucose, 132 mM sodium, 1.75 mM calcium, 0.75 mM magnesium, 102 mM chloride and 35 mM lactate, pH 5.5. The solution was filter-sterilized (0.2μm filter) and a portion heat-sterilized at 121.5°C for 35 min to reproduce typical manufacturing conditions. Protein glycation and AGE formation in the filter-sterilized and heat-sterilized solutions were evaluated as described below.

Preparation of materials for experiment 2
The PD fluids contained 1.5%(83 mM) glucose and were donated by Gambro Lundia AB. PD-Bio was heat-sterilized in a two-compartment bag comprising a small compartment containing highly concentrated glucose (50%, pH 3.2) and a separate larger compartment containing lactate and electrolytes (pH 6.7) as described elsewhere.[8] The separation of glucose at a lower pH from the other components during the heat-sterilization process substantially reduces the generation of some glucose degradation products.[8] After sterilization and just before use, a frangible pin was broken and the solutions in the two compartments mixed. Protein glycation and AGE formation in PD-Bio and in a conventionally heat-sterilized (single-compartment bag) PD fluid of the same composition[8] were evaluated as described below.

Methods
Samples of the PD fluids described above were adjusted to pH 7.4 by the addition of sodium phosphate buffer to a final concentration of 50 mM. Sterile filtered (0.2μm filter) aliquots containing sodium azide (1 g/L) and human serum albumin (1 g/L) were incubated for 0, 1, 2, 5, 10, 20 and 30 days at 37°C in the dark and then frozen at -70°C until analysis. Albumin was separated from other components of the incubation mixture by ultrafiltration (MW cut-off of 30,000). Albumin glycation was assessed by boronate affinity chromatography[10] as described previously.[11] Albumin concentration was measured by radioimmunoassay.[12] AGE formation was detected as albumin-linked fluorescence (Ex, 350nm; Em, 430nm) as described previously[7] except that, following ultrafiltration, retentates were diluted to a volume of 5 mL with the equilibration buffer for the affinity columns. Fluorescence intensity was expressed as arbitrary units/g/L albumin. Data were analysed statistically using the Wilcoxon matched-pairs signed ranks test.

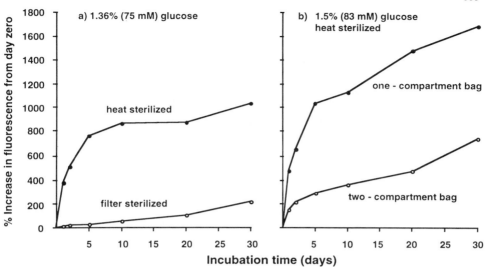

Figure 1. *The formation of characteristic AGE-fluorescence following incubation of albumin in (a) heat- or filter-sterilized PD fluid containing 75 mM glucose; and (b) PD fluid containing 83 mM glucose which was heat-sterilized either in a conventional one-compartment bag or in a novel two-compartment bag (PD-Bio) in which glucose was kept separate from other components of the solution*

Results

Experiment 1
The generation of characteristic AGE-fluorescence was significantly greater ($p < 0.05$) in the case of albumin that had been incubated in heat-sterilized PD fluid than in the filter-sterilized control solution of identical glucose concentration (Figure 1a). Even after 30 days, AGE formation in the filter-sterilized fluid was less than that found after one day of incubation in the heat-sterilized solution. After 5 days of incubation, the rate of increase in AGE formation in the heat-sterilized fluid declined and resembled that seen in the filter-sterilized control. On average, 75% of the fluorescence generated in the presence of the heat-sterilized fluid arose as a consequence of the sterilization process. The glycation of albumin increased progressively with time up to 92% after 30 days and was identical in the two fluids (data not shown).

Experiment 2
There was a rapid increase in AGE formation when albumin was incubated in the PD fluid that had been heat-sterilized in a conventional one-compartment bag; this was significantly attenuated ($p < 0.05$) in the PD fluid heat-sterilized in the novel two-compartment bag (PD-Bio) (Figure 1b). The five-fold increase in fluorescence after one day of incubation of albumin in the one-compartment-bag PD fluid was not achieved in the two-compartment-bag solution until 20 days of incubation. The difference in the rate of AGE formation between the two fluids was most pronounced in the first five days of incubation, after which rates were comparable. The glycation of albumin increased progressively with time and was identical in the two fluids (data not shown).

Discussion

The significantly greater and more rapid generation of AGEs observed when albumin was incubated in conventional heat-sterilized PD fluid compared with a filter-sterilized control could not be attributed to either glucose or to glycated albumin since levels of both were identical in the two fluids. It is likely that the generation of glucose degradation products as a consequence of the heat-sterilization process and which are intermediates in the formation of AGEs accounted for the difference in both the speed and magnitude of AGE formation. This conclusion is further supported by the lower rate of generation of AGEs in PD fluid heat-sterilized in a two-compartment rather than a one-compartment bag.

Methylglyoxal and 3-deoxyglucosone are the glucose degradation products generated during the heat-sterilization of PD fluid[8,9] which are most likely to give rise to accelerated fluorescent AGE formation since they are intermediates in the Maillard reaction and are more reactive than glucose.[13,14] In support of this, methylglyoxal concentrations were substantially reduced (from 23 to <2.8 µM) in PD fluid heat-sterilized in a two-compartment rather than one-compartment bag[8] and we found a significant reduction in both the speed and magnitude of fluorescent AGE formation. 3-Deoxyglucosone concentrations in the two-compartment bag have not been reported, but would be predicted to be lower since the greater acidity of the glucose solution (pH 3.2 vs pH 5.4 in the one-compartment bag)[8] should favour formation of hydroxymethylfurfural and diminish 3-deoxyglucosone concentrations.[9] Certainly, the concentration of hydroxymethylfurfural was increased in the two-compartment bag (8.3 µM vs 2.2 µM in the one compartment bag).[8]

The concentration of methylglyoxal in heat-sterilized PD fluid is some 100-fold higher than that found in the circulation of healthy individuals and ten-fold higher than in patients with diabetes.[15] The development of diabetic microvascular complications is related to methylglyoxal concentrations,[16] and albumin minimally modified with 20 µM methylglyoxal has altered biological activity (e.g., the stimulation of pro-inflammatory cytokine release from human monocytes).[17] These data suggest that the concentration of methylglyoxal in PD fluid is sufficient to be of pathological relevance.

Although the concentration of 3-deoxyglucosone in PD fluid is not known, it is likely to be present in similar micromolar quantities to other glucose degradation products. Serum 3-deoxyglucosone levels average 1.7 µM in healthy people, rising to 4.3 µM in diabetics, 9 µM in non-diabetic patients with chronic renal failure and 12 µM in diabetic patients with chronic renal failure.[13] There is increasing evidence that such concentrations may contribute to the development of dialysis-related amyloidosis in patients with renal failure[13] and to imidazolone formation in diabetic kidneys and atherosclerotic aortas.[18] Similar concentrations in PD fluid would therefore be expected to be detrimental to peritoneal function.

Patients on PD are exposed to between 3,000 and 4,000 litres of PD fluid per year based on an average of four exchanges of fluid, each of two to three litre volume, per day. The contribution to AGE formation of reactive dicarbonyl glucose degradation products, such as methylglyoxal and 3-deoxyglucosone, as opposed to glucose is likely to be even greater than suggested by our *in vitro* experiments in which albumin was incubated in the same fluid for up to 30 days. The rapid rate of AGE formation, which declined after five days of incubation, presumably due to consumption and/or instability of the degradation products, would be expected to be maintained *in vivo* due to frequent replenishment with fresh PD fluid.

The accumulation of AGEs in the peritoneal membrane, which is constantly exposed to PD fluid, may contribute to an increased susceptibility to infection[19] and increased solute permeability leading to a loss of ultrafiltration,[7] the two most important complications limiting continuation of PD. The relative importance of glucose and its degradation products and their precise role in the development of these complications requires further study.

References

1. N. Di Paolo and G. Sacchi, Peritoneal vascular changes in continuous ambulatory dialysis (CAPD): an *in vivo* model for the study of diabetic microangiopathy, *Perit. Dial. Int.*, 1989, **9**, 41-45.
2. J.W. Dobbie, Morphology of the peritoneum in CAPD, *Blood Purif.*, 1989, **7**, 74-85.
3. K. Yamada, Y. Miyahara, K. Hamaguchi, *et al.*, Immunohistochemical study of human advanced glycosylation end-products (AGE) in chronic renal failure, *Clin. Nephrol.*, 1994, **42**, 354-361.
4. M.A. Friedlander, Y.C. Wu, A. Elgawish and V.M. Monnier, Early and advanced glycosylation end products. Kinetics of formation and clearance in peritoneal dialysis, *J. Clin. Invest.*, 1996, **97**, 728-735.
5. A. Dawnay, Advanced glycation end products in peritoneal dialysis, *Perit. Dial. Int.*, 1996, **16(suppl.1)**, S50-S53.
6. A. Dawnay and D.J. Millar, Glycation and advanced glycation end-product formation with icodextrin and glucose, *Perit. Dial. Int.*, 1997, **17**, 52-58.
7. E.J. Lamb, W.R.Cattell and A.B.St.J. Dawnay, In vitro formation of advanced glycation end products in peritoneal dialysis fluid, *Kidney Int.*, 1995, **47**, 1768-1774.
8. A.P. Wieslander, R. Deppisch, E. Svensson, G. Forsback, R. Speidel and B. Rippe, *In vitro* biocompatibility of a heat-sterilised, low toxic, and less acidic fluid for peritoneal dialysis, *Perit. Dial. Int.*, 1995, **15**, 158-164.
9. R.J. Ulbricht, S.J. Northup and J.A. Thomas, A review of 5-hydroxymethyl-furfural (HMF) in parenteral solutions, *Fundam. Appl. Toxicol.* 1984, **4**, 843-853.
10. A.C. Silver, E. Lamb, W.R. Cattell and A.B.St.J. Dawnay, Investigation and validation of the affinity chromatography method for measuring glycated albumin in serum and urine, *Clin. Chim. Acta*, 1991, **202**, 11-22.
11. E. Lamb, W.R. Cattell and A. Dawnay, Glycated albumin in serum and dialysate of patients on continuous ambulatory peritoneal dialysis (CAPD), *Clin. Sci.* 1993, **84**, 619-626.
12. A. Silver, A. Dawnay, J. Landon and W.R. Cattell, Immunoassays for low concentrations of albumin in urine, *Clin. Chem.*, 1986, **32**, 1303-1306.
13. T. Niwa, T. Katsuzaki, T. Momoi, *et al.*, Modification of β2m with advanced glycation end products as observed in dialysis-related amyloidosis by 3-DG accumulating in uraemic serum, *Kidney Int.*, 1996, **49**, 861-867.
14. P.J. Thornalley, Pharmacology of methylglyoxal: formation, modification of proteins and nucleic acids, and enzymatic detoxification - a role in pathogenesis and

anti-proliferative chemotherapy, *Gen. Pharmacol.*, 1996, **27,** 565-573.
15. A.C. McLellan, P.J. Thornalley, J. Benn and P.H. Sonksen, The glyoxalase system in clinical diabetes mellitus and correlation with diabetic complications, *Clin. Sci.,* 1994, **87,** 21-29.
16. P.J. Thornalley, A.C. McLellan, T.W.C. Lo, J. Benn and P.H. Sonksen, Negative association of red blood cell reduced glutathione concentration with diabetic complications, *Clin. Sci.,* 1996, **91,** 575-582.
17. E.A. Abordo and P.J. Thornalley, Pro-inflammatory cytokine synthesis by human monocytes induced by proteins minimally modified by methylglyoxal, *in these proceedings*.
18. T. Niwa, T. Katsuzaki, S. Miyazaki, *et al.*, Immunohistochemical detection of imidazolone, a novel advanced glycation end product, in kidneys and aortas of diabetic patients, *J. Clin. Invest.,* 1997, **99,** 1272-1280.
19. Y.M. Li, A.X. Tan and H. Vlassara, Antibacterial activity of lysozyme and lactoferrin is inhibited by binding of advanced glycation-modified proteins to a conserved motif, *Nature Med.,* 1995, **1,** 1057-1061.

Increased Formation of Pentosidine and N^ε–(Carboxymethyl)lysine in End Stage Renal Disease: Role of Dialysis Clearance

Miriam A. Friedlander, Yu Ching Wu, Christopher P. Randle, Gavin P. Baumgardner, Peter B. DeOreo, and Vincent M. Monnier[1]

DEPARTMENT OF MEDICINE, AND [1]INSTITUTE OF PATHOLOGY, CASE-WESTERN RESERVE UNIVERSITY, CLEVELAND, OHIO 44106 USA

Summary

Advanced glycation end-products (AGEs) accumulate to extremely high levels on tissue and plasma proteins of uremic patients. In order to define differences between uremia and health, we modeled the rates of formation of Amadori product (as furosine) and the chemically defined AGEs pentosidine and N^ε–(carboxymethyl)lysine(CML) in vitro. Both pentosidine and CML formed at an accelerated rate in serum from uremic patients, an effect slowed most effectively by chelating agents. Clinical factors which affected circulating levels of protein-bound CML and pentosidine included dialysis clearance and dialyzer membrane pore size. These data point to a dynamic equilibrium between dialyzable low molecular weight promotors of AGE formation (or circulating fragments of AGEs) and protein-bound AGEs. We hypothesize that an important promotor of AGE formation in uremia may consist of a low molecular weight protein-metal complex.

Introduction

Advanced glycation end-products (AGEs) accumulate to extremely high levels on tissue and plasma proteins of patients with end-stage renal disease (ESRD).[1, 2] As determined by ELISA techniques, the majority of circulating AGEs in these patients are low molecular weight AGE peptides. In contrast to AGE peptides, 98% of plasma pentosidine in patients with ESRD is bound to high molecular weight proteins, with less than 2% circulating in low molecular weight or free form.[3, 4] Levels of AGE peptides return to normal after a successful kidney transplantation.[5] However the pentosidine content of plasma proteins does not completely normalize after transplant,[6] and tissue levels of pentosidine remain elevated or decrease only slightly.[7] These apparently disparate observations suggest that a dynamic equilibrium may exist between AGEs that form on proteins too large for renal or dialysis clearance, and dialyzable promotors of AGE formation or circulating fragments of AGEs.

The first goal of this study was to model the rate of formation in vitro of Amadori product (as furosine) and the chemically defined AGEs pentosidine and N^ε–(carboxymethyl)lysine(CML) in order to define differences between dialysis modalities. Various inhibitors of AGE formation were also tested. The second goal was to describe clinical factors that modify the accumulation of pentosidine and CML in patients with ESRD treated by both peritoneal dialysis (PD) and hemodialysis (HD).

Materials and Methods

In-vitro incubations

Serum was collected from stable patients treated by HD (n=5) or PD (n=5) and from healthy control individuals (n=5) and pooled. A pool of spent peritoneal dialysate was created from the first morning drain of 20 uninfected patients treated by PD. The glucose level was determined in each pooled serum and in the pooled spent dialysate. Glucose was then added to a final concentration of 1M. The protein level of the dialysate pool was also measured, and normal human serum albumin (HSA) added to bring the protein content to 70 mg/mL. Serum or pooled spent peritoneal dialysate was incubated at 37°C for up to 28 days. At each time point, an aliquot of serum or dialysate was processed for the measurement of Amadori product or AGEs as described below.

To determine the mechanism of accelerated formation of AGEs in uremic serum and pooled spent peritoneal dialysate, the incubation experiments were repeated with the addition of chelating agents (DTPA (1 mM) and phytic acid (1 mM)), aminoguanidine (5 mM), catalase (500 units added daily) or heat inactivation.

Because serum from patients with ESRD contains high levels of free pentosidine as well as protein-bound AGE, experiments were conducted to ascertain if free pentosidine forms a non-covalent linkage to plasma proteins. HSA (50 mg/ml) was incubated with 1000 nM free pentosidine for 24 hr. Protein was separated (not precipitated) from solution using a 10,000 MW cut off filter (Microcon 10) prior to processing for HPLC analysis.

Patients
Patients treated by HD were recruited at the outpatient Center for Dialysis Care, Cleveland, Ohio USA: 116 patients with ESRD due to diabetes mellitus and 95 with ESRD from other causes, treated by either F8 'high efficiency' or F80 'high-flux', polysulfone dialyzers, (Fresenius). Patients treated by PD (n=30) were recruited from the PD unit at University Hospitals of Cleveland (UHC). Patients with diabetes and diabetic nephropathy were recruited from the renal clinic at UHC (n=24). Healthy control subjects were recruited in response to a posting in work areas at UHC (n=36). All study subjects gave written consent. Serum was obtained from patients before the HD treatment or PD exchange.

Assays
Serum and dialysate glucose were measured by the clinical laboratory at UHC using an autoanalyzer method. Serum and dialysate total protein was assayed in a microassay modification of the Bradford method using Coomassie brilliant blue.[3,8] Amadori product or AGEs were measured by HPLC as previously described: furosine, a stable hydrolysis product of the Amadori compound fructose-lysine, according to the method of Wu;[8] pentosidine, a lysine-arginine cross-link formed during glucose condensation and oxidation (glycoxidation) of Amadori product according to a modification of the method of Odetti;[3,9] and CML, formed from both lipid peroxidation and glycoxidation reactions according to the method of Glomb.[10,11]

Calculations and Statistical Methods
Measures of dialysis adequacy were calculated by standard formulae for both patients on HD[12] and patients on PD.[13] Between-group comparisons of parametric data were performed by 2-way ANOVA. Multiple regression analysis and Pearson's pairwise comparisons were used to determine relationships between AGEs and other parameters. Statistical significance was deemed to be attained for $p<.05$.

Results

Pentosidine formed at an accelerated rate in serum from patients on HD and PD, compared to healthy subjects (Figure 1B). Spent dialysate from patients on peritoneal dialysis (known to contain low molecular weight AGE fragments) promoted the formation of CML, but not pentosidine, on serum proteins (Figure 1C).

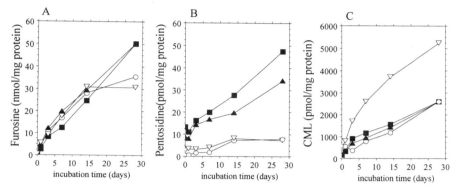

Figure 1. *The pattern of increase in the content of furosine (A), pentosidine (B) and CML (C) is shown after incubation of serum proteins or spent pooled peritoneal dialysate(with normal serum albumin added) in 1M glucose, -○- healthy, -□-HD, -▲- PD, -▽- dialysate.*

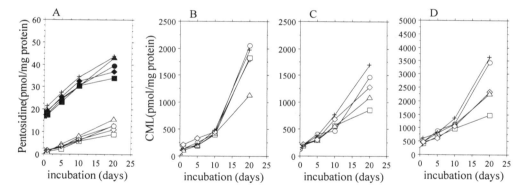

Figure 2. *Effect of inhibitors on the formation of pentosidine in HD (filled symbols) and healthy (open symbols) serum(A). Effect of inhibitors on CML formation in healthy serum (B), HD serum (C) and pooled spent peritoneal dialysate (D).). -○-control, -□-chelators, -△- aminoguanidine, -◇- catalase, -+- heat inactivated.*

Chelators were the most effective suppressors of the formation of pentosidine in serum from a patient on HD and, to a lesser extent, in serum from healthy controls (Figure 2). In both uremic serum and pooled spent dialysate, the most dramatic suppression is caused by chelators, with a lesser effect by aminoguanidine and catalase. In contrast, in healthy control serum, CML formation is suppressed by aminoguanidine.

Incubation of normal HSA with 1000 nM free pentosidine did not result in an increase in the pentosidine content of the >10,000 MW fraction or a decrease in the concentration of pentosidine in the low MW fraction over a time course of 30 min, 2, 6 and 24 h. (Data not shown).

Circulating pentosidine and CML levels were highest in serum from patients dialyzed against an HD membrane with a smaller pore size (F8) compared with the large pore size (F80). The presence or absence of diabetes mellitus had no significant effect on levels of these AGEs in any dialysis treatment group (p=0.97, pentosidine, p=0.69 CML). Comparison groups are healthy subjects and diabetics with normal kidney function.

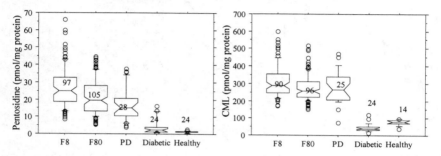

Figure 3. *Whisker plots demonstrate the 10th, 25th, 50th, 75th, and 90th % range of the data. Outliers = open circles. Pentosidine: F8 vs F80, p<0.0001; F8 vs PD, p<0.0001; F80 vs PD, p<0.05; all dialysis vs healthy or diabetic with normal renal function, p<0.0001; diabetic vs healthy, p=0.37. CML: F8 vs F80, p<0.003; F8 vs PD, p=0.09; F80 vs PD, p=0.80; all dialysis vs healthy or diabetic, p<0.0001 diabetic vs healthy, p=0.24. Significance by ANOVA.(The number of observations is indicated for each group.)*

There was no effect of race, sex, or age on serum AGE levels as judged by multiple regression analysis. Pairwise Pearson's correlation showed that CML and pentosidine were highly correlated (R=.535, p<.0001) and that duration of ESRD had significant positive correlation with both pentosidine (R=.45, p<.0001) and CML (R=.19, p<.005). No relationship between commonly used markers of dialysis adequacy (Kt/V or PCR) and AGE levels was found in the patients treated by HD. However in the patients treated by PD, the greater the dialysis clearance (as Kt/V, weekly creatinine clearance (Figure 4), or dialysate volume) the lower the serum level of AGEs.

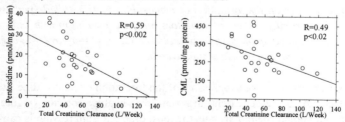

Figure 4. *Correlations between a marker of PD clearance (Total creatinine clearance) and pentosidine/CML in the subgroup of patients treated by PD.*

Discussion

The formation of Amadori product (as furosine) occurs at equivalent rates on serum proteins from healthy subjects and those with uremia. In contrast, uremic serum demonstrates a dramatic increase in autoxidation resulting in the formation of pentosidine and CML. Thus, uremic serum may contain promotors of glycoxidation reactions, or decreased quantities of inhibitor substances. Spent peritoneal dialysate contains many of the 'uremic toxins' cleared from the body, including AGE fragments.[3] The lack of acceleration of pentosidine formation by spent peritoneal dialysate suggests no additional Amadori precursors in dialysate compared with those that form in 1M glucose. Moreover, there is no evidence that AGE fragments can form non-covalent bonds with proteins in the in-vitro incubations of free pentosidine with HSA or in spent dialysate.

However the spent dialysate did contain an apparent promotor of CML formation. The nature of this promotor activity is more fully defined in the experiments involving inhibitors. Aminoguanidine slowed formation of both pentosidine and CML in healthy control serum, implicating glyoxal as an intermediate.[10] In uremic serum and pooled spent peritoneal dialysate, chelating agents were most effective. Because catalase was also effective, these results suggest that CML formation in normal serum stems from lipid peroxidation,[14] whereas sugar autoxidation, enhanced by a chelatable promotor, appears to play an important role in uremic fluids (serum and spent dialysate).

To our knowledge, these data are the first to demonstrate an effect of HD membrane pore size or dialysis clearance on circulating levels of protein-bound pentosidine and CML. Although increased circulating levels of free pentosidine[3, 4] and AGE fragments[1, 15] have been demonstrated in uremic patients, more than 90% of circulating AGEs are bound to high molecular weight proteins including albumin.[3, 4] Nonetheless, these fragments may have pathophysiologic importance. Low molecular weight AGE-peptides have been shown to 'catalyze' the formation of high molecular weight AGE on calf collagen in-vitro.[1] Therefore subtle differences in dialysis clearance of low molecular weight AGE fragments might result in differences in levels of protein-bound AGEs in the patients with ESRD.

By inference, the chelatable promotor of auto-oxidation is likely to be present in the dialyzable low molecular weight fractions of serum. In health, most reactive transition metals exist tightly bound to carrier protein. However alterations induced by the uremic environment (such as low pH) might result in dissociation of such a metal from binding proteins[16]. Alternatively, metal ions might complex with other proteins or low molecular weight fragments of AGEs. In summary, the nature of the dialyzable promotor of AGE formation present in uremia is not yet defined. In patients with ESRD, a dynamic equilibrium exists between clearance of a low molecular weight fraction by dialysis and formation of AGEs on high molecular weight proteins. These data suggest that the promotor of AGE formation in uremia may consist of a low molecular weight protein-metal complex.

Acknowledgments: Supported in part by a grant from the National Institutes of Diabetes and Digestive and Kidney Diseases DK-45619. The authors thank Ms. Kelly Weigel, RN, BSN, Nephrology Study Coordinator and the nurses at the Center for Dialysis Care, Cleveland, and the PD unit at UHC.

References

1. Z. Makita, R. Bucala, E.J. Rayfield, E.A. Friedman, A.M. Kaufman, S.M. Korbet, R.H Barth, J.A. Winston, H. Fuh, K.R. Manogue, A. Cerami and H. Vlassara, Reactive glycosylation endproducts in diabetic uremia and treatment of renal failure, *Lancet*, 1994, **343**, 1519-22
2. M. Friedlander, Y. Wu, J. Schulak, V. Monnier and D. Hricik, Influence of dialysis modality on plasma and tissue concentrations of pentosidine in patients with end-stage renal disease *Am. J. Kidney Dis.*, 1995, **25**, 445-451
3. M. Friedlander, Y. Wu, A. Elgawish and V. Monnier, Early and Advanced Glycosylation End Products: Kinetics of formation and clearance in peritoneal dialysis, *J. Clin. Invest.*, 1996 **97**, 728-735
4. T. Miyata, Y. Ueda, T. Shinzato, Y. Iida, S. Tanaka, K. Kurakawa, C. van Ypersele de Strihou and K. Maeda, Accumulation of albumin-linked and free-form pentosidine in the circulation of uremic patients with end stage renal failure: renal implications in the pathophysiology of pentosidine, *J. Am. Soc. Nephrol.*, 1996, **7**, 1198-1206
5. Z. Makita, S. Radoff, E.J. Rayfield, Z. Yang, E. Skolnik, V. Delaney, E.A. Friedman, A Cerami and H. Vlassara, Advanced glycosylation end products in patients with diabetic nephropathy, *New Engl. J. Med.*, 1991, **325**, 836-842
6. D. Hricik, J. Schulak, D. Sell, J. Fogarty and V. Monnier, Effects of kidney or kidneypancreas transplantation on plasma pentosidine, *Kidney Int.*, 1993, **43**, 398-403
7. D. Hricik, Y. Wu, J. Schulak and M. Friedlander, Disparate changes in plasma and tissue pentosidine levels after kidney and kidney-pancreas transplantation, *Clin. Transplant.*, 1996 **10**, 568-573
8. Y. Wu, V. Monnier and M. Friedlander, Reliable determination of furosine in human serum and dialysate proteins by high-performance liquid chromatography, *J Chromatog B*, 1995 **667**, 328-332
9. P. Odetti, J. Fogarty, D.R. Sell and V.M. Monnier, Chromatographic quantitation o plasma and erythrocyte pentosidine in diabetic and uremic subjects, *Diabetes*, 1992, **41**, 153-9
10. M. Glomb and V. Monnier, Mechanism of protein modification by glyoxal and glycoaldehyde, reactive intermediates of the Maillard reaction, *J. Biol. Chem.*, 1995, **270**, 10017-10026
11. A.H. El Gawish, M. Glomb, M. Friedlander and V.M. Monnier, Involvement of hydroger peroxide in collagen crosslinking by high glucose in vitro and in vivo, *J. Biol. Chem.*, 1996, **271** 12964-12971
12. F. Gotch, Kinetic Modeling in Hemodialysis, in 'Clinical Dialysis', A. Nissenson, R. Fine and D. Gentile, (ed). Appleton and Lange: Norwalk. 1990, pp. 118-146
13. P. Blake, K. Somboslos, G. Abraham, J. Weissgarte, R. Pemberton, G. Chu and D. Oreopoulos, Lack of correlation between urea kinetic indices and clinical outcomes in CAPD patients, *Kidney Int.*, 1991, **39**, 700-706
14. M. Fu, J. Requena, A. Jenkins, T. Lyons, J. Baynes and S. Thorpe, The advanced glycation end product, N^{ε}-(carboxymethyl)lysine, is a product of both lipid peroxidation and glycoxidation reactions., *J. Biol. Chem.*, 1996, **271**, 9982-9986
15. S. Korbet, Z. Makita, C. Firanek and H. Vlassara, Advanced glycosylation end products in continuous ambulatory peritoneal dialysis patients, *Am. J. Kidney Dis.*, 1993, **22**, 588-591
16. M. Cooper, B. Buddington, N. Miller and A. Alfrey, Urinary iron speciation in nephrotic syndrome, *Am. J. Kidney Dis.*, 1995, **25**, 314-319

Reducing Sugars Induce Apoptosis in Pancreatic β-Cells by Provoking Oxidative Stress via a Glycation Reaction

H. Kaneto, J. Fujii, T. Myint, N. Miyazawa, K. N. Islam, Y. Kawasaki, K. Suzuki and N. Taniguchi

DEPARTMENT OF BIOCHEMISTRY, OSAKA UNIVERSITY MEDICAL SCHOOL, 2-2 YAMADAOKA, SUITA, OSAKA 565

Summary

Reducing sugars brought about apoptosis in isolated rat pancreatic islet cells as well as in a pancreatic β-cell-derived cell line, HIT. This apoptosis was characterized biochemically by internucleosomal DNA cleavage and morphologically by nuclear shrinkage, chromatic condensation, and apoptotic body formation. N-acetyl-L-cysteine and aminoguanidine inhibited the apoptosis. Proteins in β-cells were actually glycated by the binding with an antibody that can specifically recognize the protein glycated by fructose. The FACS analysis using dichlorofluorescin diacetate showed that reducing sugars increased intracellular peroxide levels preceding the induction of apoptosis. Levels of carbonyl and malondialdehyde were also increased. These results suggest that reducing sugars trigger oxidative modification and apoptosis in pancreatic β-cells by provoking oxidative stress mainly via a glycation reaction, which may explain the deterioration of β-cells under diabetic conditions.

Introduction

Reactive oxygen species (ROS) have been implicated in a wide range of biological functions, but they can be both essential to cellular homeostasis and highly toxic.[1] Several conditions are known to disturb the balance between the production of ROS and cellular defense, resulting in cellular destruction and dysfunction. An imbalance between pro- and antioxidant factors plays an important role in many disease processes including diabetes mellitus. Reducing sugars are known to produce ROS mainly through the glycation reaction.[2] D-Ribose and 2-deoxy-D-ribose, which rank at the top of reducing capacity among reducing sugars, can induce cell death in mononuclear cells.[3] Under diabetic conditions, glucose is converted to fructose through the polyol pathway. Fructose has a stronger reducing capacity than glucose and the glycation reaction is easily induced by fructose. Thus, fructose is believed to play an important role in diabetes-induced deterioration of various organ systems. Incubation of β-cells with a large amount of reducing sugars results in injury of β-cells[4] resulting in inhibition of insulin secretion, but the mechanism by which the injury of β-cells occurs is not known. The aim of the present study was to explore DNA damage, morphological change, and oxidative modification after exposure to reducing sugars and investigate the mechanism by which β-cells are damaged.

Materials and Methods

Cell culture
HIT cells (hamster pancreatic β-cell-derived cell line) were grown in RPMI 1640 medium supplemented with 10% (v/v) fetal calf serum, 100 units/mL penicillin and 0.1 mg/mL streptomycin sulfate in a humidified atmosphere of 5% CO_2.

Determination of internucleosomal DNA cleavage
After incubation with the reagents described Internucleosomal DNA cleavage was detected as we reported recently.[5]

Fluorescence microscopy
Morphological change was examined by fluorescence microscopy. After incubation with the reagents, fluorescence microscopy was carried out as described.[4]

Immunological detection of proteins glycated by fructose
We examined whether cytosolic proteins were actually glycated in cells after incubation with fructose, by using an antibody specific for protein glycated by fructose, as shown by Western blot analysis. The antibody was established (Miyazawa, N., Kawasaki, Y., Fujii, J., Theingi, M., Hoshi, A., Hamaoka, R., Matsumoto, A., Uozumi, N., Teshima, T. and Taniguchi, N. submitted for publication), following a procedure similar to that used when we purified a hexitol lysine antibody that recognized proteins glycated by glucose, but not by fructose or ribose.[6]

Results and Discussion

Internucleosomal DNA cleavage and morphological change in β-cells induced by reducing sugars
After incubation of HIT cells with 100 mM fructose or 25 mM ribose for 3 days, cleavage of DNA into nucleosomal fragments was observed, suggesting apoptosis (Figure 1A). To investigate the mechanism by which DNA is damaged in β-cell nuclei, several reducing sugars were employed. DNA cleavage was not detected after 3 days incubation with 100 mM glucose; after incubation with sorbitol, a non-reducing sugar, it was not observed at all. These results suggested that the DNA cleavage was dependent on the reducing capacity of the sugars. Nuclear fragmentation, a characteristic change in apoptosis, was observed after 3 days of treatment with 100 mM fructose or 25 mM ribose. The percentage of cells undergoing apoptosis was measured after 1-7 days of treatment with 100 mM fructose or 25 mM ribose (Figure 1B). Treatment with fructose or ribose resulted in lysis of some cells. Thus, the cell number was counted after 1-7 days of treatment with 100 mM fructose or 25 mM ribose (Figure 1C).

Immunological determination of glycated proteins in β-cells treated with fructose

After incubation with 100 mM fructose for more than 2 days, fructated proteins in HIT cells were detected by ELISA with the anti-fructated lysine antibody (Figure 2A). The extent was increased after 3 days of incubation. Fructated proteins were detected after incubation with 25 mM fructose, but the extent was more remarkable after incubation with 50 mM or 100 mM fructose. These changes were suppressed by the addition of 1 mM aminoguanidine, an inhibitor of glycation (Figure 2B).

Increase of intracellular peroxide levels produced by reducing sugars

We assessed intracellular peroxide level by flow cytometric analyses using a peroxide sensitive fluorescence probe 2',7'-dichlorofluorescin diacetate (H2DCF-DA). The production of intracellular peroxide in HIT cells was observed after 3 days of incubation with 100 mM fructose or 25 mM ribose. Thus, intracellular peroxide may be a factor in the DNA cleavage induced by fructose or ribose (data not shown).

Increase of carbonyl and malondialdehyde levels induced by reducing sugars

After 3 days of incubation with 100 mM fructose, carbonyl formation was increased. The increase of carbonyl formation was suppressed by the addition of 1 mM aminoguanidine or 10 mM N-acetyl-L-cysteine (NAC). Furthermore, to confirm that β-cells were modified by oxidative stress after treatment with reducing sugars, we measured malondialdehyde levels. Malondialdehyde is a major aldehyde product of lipid peroxidation and is used as an index of peroxidation. After 3 days of incubation with 100 mM fructose, malondialdehyde level was increased. The increase in malondialdehyde levels was suppressed by the addition of 1 mM aminoguanidine or 10 mM NAC. These results indicate that reducing sugars induce oxidative modification by provoking oxidative stress via the glycation reaction.

Involvement of oxidative stress in apoptosis induced by reducing sugars

In order to elucidate the mechanism by which reducing sugars induce apoptosis, experiments were performed with several possible inhibitors. It is possible that fructose or ribose may interfere with the redox status of the cell. Because the increase of oxidative stress, including the production of hydrogen peroxide was implicated in the internucleosomal DNA cleavage by reducing sugars, we have tested whether the induction of apoptosis could be suppressed by NAC which can raise intracellular GSH levels and, thereby, protect cells from the effects of ROS. Addition of 10 mM NAC inhibited internucleosomal DNA cleavage induced by fructose or ribose. These results suggest that oxidative stress is involved in the DNA cleavage. In order to confirm the participation of the glycation reaction in the DNA cleavage, aminoguanidine, an inhibitor of the glycation reaction, was employed. Addition of 1 mM aminoguanidine to the incubation medium suppressed internucleosomal DNA cleavage. All

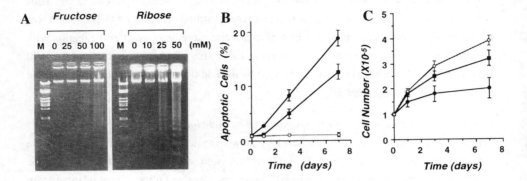

Figure 1. *Internucleosomal DNA cleavage and morphological change of β-cells treated with reducing sugars*
(A) After incubation of HIT cells with several concentrations of fructose (0-100 mM) or ribose (0 - 50 mM) for 3 days, DNA from the cells was extracted and subjected to electrophoresis on a 1.5% agarose gel. Lane M, molecular weight marker. (B) After incubation of HIT cells with 100 mM fructose or 25 mM ribose for 1-7 days, the cells were stained with 4',6-diamidino-2-phenylindole hydrochloride and the percentage of cells undergoing apoptosis was measured. Open circle, control; closed square, 100 mM fructose; closed circle, 25 mM ribose. (C) After incubation of HIT cells with 100 mM fructose or 25 mM ribose for 1-7 days, the number of cells per 6-well plate was counted. Open circle, control; closed square, 100 mM fructose; closed circle, 25 mM ribose.

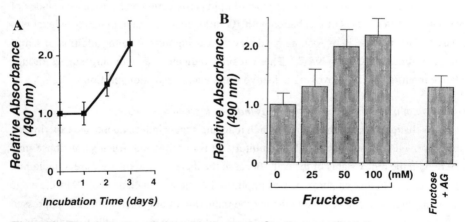

Figure 2. *Determination of glycated proteins in β-cells treated with fructose*
(A) After incubation of HIT cells with 100 mM fructose for 1-3 days, glycated proteins were detected by ELISA with the antibody specific for the glycated proteins by fructose. (B) After incubation with several concentrations of fructose (0 - 100 mM) or 100 mM fructose plus 1 mM aminoguanidine (AG) for 3 days, glycated proteins were detected by ELISA with the antibody.

of these results were consistent with the idea that oxidative stress following the glycation reaction with reducing sugars was responsible for internucleosomal DNA cleavage (data not shown). In diabetes the production of several reducing sugars is increased through glycolysis or polyol pathway and oxidative stress is provoked. We have reported that DNA cleavage as well as protein fragmentation is induced by ROS produced through glycation reaction.[7,8] An involvement of oxidative stress in sugar-induced apoptosis is supported by the inhibitory effect of NAC.

Internucleosomal DNA cleavage triggered by reducing sugars in isolated rat pancreatic islets
The pancreatic β-cell-derived cell line, HIT, is known to possess some β-cell functions such as insulin secretion, but may differ from normal β-cells in other respects. So we performed the experiments with freshly isolated rat pancreatic islets in order to validate the significance of sugar-induced apoptosis in deterioration of β-cells. Islets were isolated from Sprague-Dawley rats by collagenase digestion and Ficoll gradient. After incubation of isolated pancreatic islets with 100 mM fructose or 25 mM ribose for 3 days, cleavage of DNA into nucleosomal fragments was observed, confirming observation on the HIT cell line. Furthermore, addition of 10 mM NAC to the incubation medium suppressed the DNA cleavage (data not shown).

Morphological evidence for apoptosis by electron microscopy
We examined morphological changes by electron microscopy (data not shown). Examination by scanning electron microscopy showed untreated HIT cells were covered with microvilli and ruffles. After 3 days of treatment with 100 mM fructose or 25 mM ribose, numbers of microvilli and ruffles were reduced. Instead, extensive surface blebbings and granular protrusions were often observed. Spherical bodies with smooth surface were sometimes found to be attached to the cells. These structures are thought to correspond to apoptotic bodies. Such morphological changes were suppressed by the addition of 10 mM NAC. Furthermore, after incubation with fructose or ribose, some vacuoles, lots of small vesicles, and nuclear fragmentation with condensed nuclear chromatin were observed by transmission electron microscopy, although these morphological changes were not observed in untreated cells. These morphological changes were also suppressed by the addition of 10 mM NAC. These results suggest that oxidative stress as well as the DNA damage is involved in the morphological changes.

References

1. B. Halliwell, J. M. C. Gutteridge and C. E. Cross, Free radicals, antioxidants, and human disease: where are we now? *J. Lab. Clin. Med.*, 1992, **119**, 598-620.

2. T. Sakurai and S. Tsuchiya, Superoxide production from nonenzymatically glycated protein, *FEBS Lett..*, 1988, **236**, 406-410.
3. D. Barbieri, E. Grassilli, D. Monti, S. Salvioli, M. G. Franceschini, A. Franchini, E.Bellesia, P. Salomoni, P. Negro, M. Capri, L. Troiano, A. Cossarizza and C. Francechi, D-ribose and deoxy-D-ribose induce apoptosis in human quiescent peripheral blood mononuclear cells, *Biochem. Biophys. Res. Commun.*, 1994, **201**, 1109-1116.
4. H. Kaneto, J. Fujii, T. Myint, K. N. Islam, N. Miyazawa, K. Suzuki, Y. Kawasaki, M. Nakamura, H. Tatsumi, Y. Yamasaki and N. Taniguchi, Reducing sugars trigger oxidative modification and apoptosis in pancreatic β-cells by provoking oxidative stress through the glycation reaction, *Biochem. J.*, 1996, **320**, 855-863.
5. H. Kaneto, J. Fujii, H. K. Seo, K. Suzuki, T. Matsuoka, M. Nakamura, H. Tatsumi, Y. Yamasaki, T. Kamada and N. Taniguchi, Apoptotic cell death triggered by nitric oxide in pancreatic β-cells, *Diabetes*, 1995, **44**, 733-738.
6. T. Myint, S. Hoshi, T. Ookawara, N. Miyazawa, K. Suzuki and N. Taniguchi, Immunological detection of glycated proteins in normal and streptozotocin-induced diabetic rats using anti hexitol-lysine IgG, *Biochem. Biophys. Acta*, 1995, **1272**, 73-79.
7. T. Ookawara, N. Kawamura, Y. Kitagawa and N. Taniguchi, Site-specific and random fragmentation of Cu,Zn-superoxide dismutase by glycation reaction. Implication of reactive oxygen species, *J. Biol. Chem.*, 1992, **267**, 18505-18510.
8. H. Kaneto, J. Fujii, K. Suzuki, H. Kasai, R. Kawamori, T. Kamada and N. Taniguchi, DNA cleavage induced by glycation of Cu,Zn-superoxide dismutase, *Biochem. J.*, 1994, **304**, 219-225.

Low Density Lipoprotein Carboxymethylated *in vitro* does Not Accelerate Cholesterylester Synthesis in Mouse Peritoneal Macrophages

Tamiko Sakurai,[1] Yorihiro Yamamoto,[2] Makiko Shimoyama[3] and Minoru Nakano[3]

[1] TOKYO UNIVERSITY OF PHARMACY AND LIFE SCIENCE, SCHOOL OF PHARMACY, HACHIOJI, TOKYO 192-03, JAPAN;   [2] RESEARCH CENTER FOR ADVANCED SCIENCE AND TECHNOLOGY, UNIVERSITY OF TOKYO 153, JAPAN;   [3] DEPARTMENT OF PHOTON AND FREE RADICAL RESEARCH, IMMUNORESEARCH LABORATORIES, TAKASAKI 370, JAPAN

Summary
One of the main endproducts of advanced glycation through metal-catalysed oxidation is carboxymethylated lysine residues in a protein. Carboxymethylation of the low density lipoprotein (LDL) *in vitro* resulted in an increased negative charge on the apoprotein B-100 comparable to acetylation and succinylation of lysine residues. However, carboxymethylated LDL did not induce an accumulation of lipids in mouse peritoneal macrophages assesed by oil red O stain, even when 60 % of lysine residues were modified, which is completely different from the effect of acetylated and succinylated LDLs. An experimental system for incorporation of ^{14}C-oleate into macrophages confirmed that carboxymethylated LDL did not accelerate cholesterylester synthesis, suggesting that the polyanionic character resulting from carboxymethylation is not recognized by the macrophage scavenger receptor. Oxidation levels of modified LDLs assessed by cholesterylester hydroperoxide/cholesterylester were in the order: carboxymethylated LDL< control LDL< acetylated LDL< succinylated LDL<< oxidized LDL.

Introduction

Glycated proteins produced nonenzymatically by reactions of proteins and reducing sugars have been considered as a model of protein aging, as advanced reactions form comlex fluorescent materials termed advanced glycation endoproducts (AGEs) during long incubation periods under physiological conditions. Indeed, AGEs have been detected immunohistochemically in human tissues or plasma by using anti-AGE-antibodies. Recently, attention has been focused on carboxymethylated lysine residues among known structures of AGEs, because a dominant epitope of anti-AGE-antibodies is carboxymethylated lysine.[1]   In our earlier studies,[2] glycated human serum albumin treated with ferric ion showed a shift of isoelectric point towards acidic; i.e., an increase in net negative charge, with an increase in carboxymethylated lysine residues. It is believed that oxidized LDL and acetylated LDL are recognized by the macrophage scavenger receptor because of the increase in their negative charges. The present work was undertaken to compare the carboxymethylated LDL with the aforementiond atherogenic modified LDLs.

Materials and Methods

LDL and its modifications
A 10 mL sample of human blood was collected into a tube containing disodium EDTA(10 mg). The LDL fraction obtained by sequential ultracentrifugation, was dialysed against 150 mM phosphate buffer   (pH 7.4) containing 1 mM EDTA at   4° C for 20 h. This was referred to as control LDL (c-LDL) and used for preparation of modified LDL. *Carboxymethylation:*   LDL (2.4 mg /mL, 1.75 mM lysine equivalent) was reacted with glyoxylic acid (GOX) and then NaCNBH$_3$ (GOX: NaCNBH$_3$=1:4 mol/mol) in 1 mL of 150

mM PBS (pH 7.8) and 4° C for 20 h. The concentrations of GOX used were 0.88, 1.75, 5.25, and 8.75 mM (GOX/lysine residue of apoprotein B=0.5, 1, 3, 5, respectively). These LDLs were referred to as CM-LDL and lysine residues as CM-lysine. As a control, LDL was reacted with 35 mM $NaCNBH_3$ but without GOX. *Acetylation:* One volume of LDL (2.4 mg/mL) was mixed with one volume of saturated sodium acetate (pH 8.0) under ice-cold conditions, and then treated with acetic anhydride (equal to the weight of apoprotein) for 1 h. *Succinylation:* One mL of LDL (2.4 mg/mL) was adjusted to pH 8.0 by adding a few drops of 5 M NaOH. Solid succinic anhydride (3 mg, ten times lysine equivalent) was added slowly in small portions at 5°C for 1 h. *Oxidized-LDL:* LDL (2.4 mg/mL) was treated with 10 μM $CuSO_4$ at 37°C for 5 h. Modified LDL after dialysis was applied to agarose gel electrophoresis (Ciba Corning: Universal Gel) and stained by Fat red 7 B.

Assay of CM-lysine content by amino acid analysis
Modified LDLs were dialyzed against distilled water and extracted with cold acetone n-butanol (1:1) to remove lipids. Apoprotein (1 mg) was hydrolyzed in 6 N HCl *in vacuo* at 110° C for 72 h. Synthetic CM lysine[2] was used as an external standard. An Hitachi L-8500 amino acid analyzer was used to determine CM-lysine content.

Uptake of modified LDLs by mouse peritoneal macrophages
Female BALB/C mice (24-30) were used. After injection of 8 mL of tissue culture medium (RPMI-1640, Sigma) in the abdominal cavity, the peritoneal cells were harvested using standard methods. The cells were then added to culture plates (Cell wells 25820, Corning), at 2×10^6 cells/plate and incubated at 37° C in 5% of CO_2 in air for 2 h. Macrophages adherent to the plates were used for experiments. After exposure of macrophages to LDL and incubation at 37° C for 5 h, lipids in the cells were stained by oil red-O. The incorporation of ^{14}C-oleate (25.9 Gbq/mmol, Daiichi Kagaku Ltd, Japan) in macrophages was measured using the method of Gianturco et al..[3] Modified LDL preparations together with ^{14}C-oleate was added to the macrophages. After incubation at 37°C for 24 h, the lipids in the macrophages were extracted with 2mL of hexane:i-propanol (3:2). The extracts were applied to a TLC plate (Kiesekgel 60, Merck) and exposed to iodine gas. The cholesterylester section was cut out and analyzed using a liquid scintillation counter. Each sample was assayed using 4 wells.

Oxidation levels of modified LDLs
Phosphatidylcholine hydroperoxide (PC-OOH) and cholesterylester hydroperoxide (CE-OOH), free cholesterol, cholesterylesters(CE: arachidonate, Ch20:4; linoleate, Ch18:2; oleate, Ch18:1) were assayed by HPLC equipped with UV and chemiluminescence detectors as described previously.[4]

Measurement of conformational features
Circular dichroism spectra of modified LDLs were measured. Adsorption properties of modified LDLs were assayed on an immunoaffinity gel bound anti-apoprotein B-100 monoclonal antibody which recognizes the amino acid residues 2291-2318 of apoprotein B 100.

Results and Discussion

Modified LDLs and cholesterylester synthesis
Carboxymethylation was carried out by the addition of GOX and reduction by $NaCNBH_3$. Table 1 shows the levels in CM-lysine of modified LDLs at various concentrations of GOX. At a concentration of GOX 3 times that of lysine residues, 60 % of lysine residues of apoprotein were modified. Figure 1a shows the mobility of modified LDLs on agarose gel. At high levels of CM-lysine residues, the migration of modified LDLs to the anode became faster, indicating an increase in net negative charge on the LDL particle. LDL modified to 60 % of its lysine residues was more negatively charged than Cu-oxidized LDL. Uptake of the modified LDLs by macrophages as assayed by oil red O stain,

Table 1. Carboxymethyl lysine contents in CM-LDLs

Sample		CM-lysine (mole/apo B*)	% modified
LDL:GOX**	1 : 0.5	69	19
	1 : 1	168	47
	1 : 3	214	60
	1 : 5	208	59
c-LDL		present	

* mole of apoprotein B
**glyoxylic acid/Lys in apo B

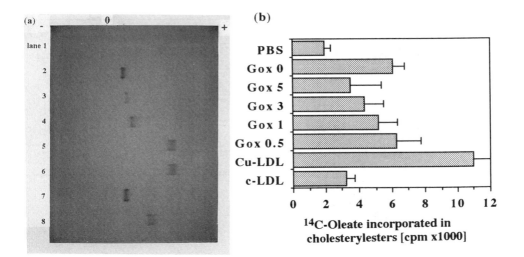

Figure 1. *Agarose gel electrophoresis (a) and effects of modified LDLs on cholesterylester synthesis of macrophages (b); lanes 1 and 7, c-LDL; lane 2, GOX 0.5; lane 3, GOX 1; lane 4, GOX 3; lane 5, GOX 5; lane 6, GOX 0; lane 8, Cu-LDL. PBS means phosphate buffer saline without LDL. Bars represent means ± SD.*

however, was not observed in CM-LDL but only in Cu- oxidized LDL (data not shown). Figure 1b shows the effects of modified LDLs on cholesterylester synthesis in mouse macrophages. This experimental system was to study accumulation of lipoprotein-derived cholesterylester by macrophages, established by Goldstein et al..[5] They showed that mouse macrophages take up only a small amount of normal human LDL, whereas the cells take up large amounts of negatively charged LDL. The cholesterylesters of LDL particles taken up by macrophages are hydrolysed by lysozomal enzymes and the resultant free cholesterol is then re-esterified. In the present system, exogenous ^{14}C-oleate was used to esterify the free cholesterol to cholesteryl ^{14}C-oleate in the cytoplasm where it accumulates as lipid droplets. In Figure 1b, Cu-induced oxidized LDL accelerated cholesterylester synthesis 4-fold compared with c-LDL. In contrast, highly negatively charged CM-LDL did not accelerate the incorporation of ^{14}C-oleate into cholesteryl ^{14}C-oleate compared with LDL treated with NaCNBH$_3$ only. A modest increase in LDL treated with NaCNBH$_3$ only may be due to the effect of reductant on apoprotein B 100.

Figures 2a and 2b show the results of other chemical modifications of LDL: acetyl-LDL, succinyl-LDL and glycated LDL treated with ferric ion. Acetyl-LDL and succinyl-LDL showed faster migration towards the anode than CM-LDL. Glycated LDL treated with ferric ion also migrated faster than c-LDL. As expected, acetyl-LDL and succinyl-LDL accelerated the cholesterylester synthesis 8-fold compared with c-LDL. Glycated LDL treated with ferric ion showed a modest increase in ^{14}C-oleate incorporation.

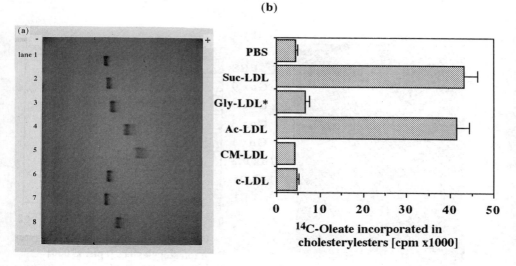

Figure 2. *Agarose gel electrophoresis (a) and effects of modified LDLs on cholesterylester synthesis of macrophages (b); lane 1, acetyl LDL; lanes 2 and 7, c-LDL; lane 3, glycated LDL; lane 4, glycated LDL treated with ferric ion; lanes 5 and 6, succinyl LDL; lane 8, CM-LDL. * glycated LDL treated with ferric ion (200 mM glucose, 3 days, and then incubated with 60 µM ferric ion for 20 h). Bars represent means ± SD*

Table 2. Contents* of lipids and lipid peroxides of modified LDLs**

sample LDL:GOX	PC-OOH (mM)	CE-OOH (mM)	FC (mM)	Ch20:4 (mM)	Ch18:2 (mM)	Ch18:1 (mM)	CE (mM)	CE/FC	TC (mM)	CE-OOH/CE ($\times 10^3$)
1:0.5	0.184	0.240	0.73	0.24	1.31	0.52	2.07	2.86	2.80	0.12
1:1	0.187	0.124	0.73	0.25	1.35	0.53	2.13	2.91	2.86	0.06
1:3	0.111	0.095	0.72	0.25	1.36	0.54	2.15	2.98	2.87	0.04
1:5	0.167	0.175	0.74	0.24	1.27	0.52	2.02	2.74	2.75	0.09
1:0***	0.096	0.030	0.72	0.25	1.35	0.54	2.14	2.98	2.86	0.01
c-LDL	1.749	2.028	0.79	0.27	1.46	0.55	2.29	2.90	3.07	0.89
ox-LDL	76.301	164.522	0.73	0.01	0.28	0.36	0.65	0.88	1.38	253.97
Ac-LDL	23.819	38.019	0.65	0.15	1.08	0.46	1.69	2.58	2.34	22.54
Suc-LDL	86.611	106.312	0.89	0.14	1.23	0.54	1.91	2.15	2.79	55.81

*mean values of two assays
** Concentration of LDLs was assayed by Lowry Method and was adjusted to 1 mg/mL.
*** LDL reacted with $NaCNBH_3$ without GOX.

Oxidation of modified LDLs
It is widely believed that oxidation of LDL produces lipid hydroperoxides and their degradation products react with lysine residues of the apoprotein rendering the LDL particle more negatively charged; the increased anionic nature of the protein is recognized by the macrophage scavenger receptors. Therefore, the extent of oxidation of modified LDLs was assayed and the results are shown in Table 2. Oxidation levels of modified LDLs when assessed by CE-OOH/CE were in the order: CM-LDL< c-LDL < acetyl-LDL < succinyl-LDL << Cu-oxidized LDL. The levels of PC-OOH and CE-OOH/CE in any CM-LDL preparation were lower than c-LDL, due to the reduction of hydroperoxides to alcohol by $NaCNBH_3$. An increase in negative charge on CM-LDLs is apparently attributable only to CM-lysine residues. On the other hand, acetylation and succinylation processes provoked lipid peroxidation. These data suggest that macrophages do not always recognize polyanionic character, but oxidative modifications are always recognized.

Differences between CM-LDL and acetyl-LDL or succinyl-LDL
If the macrophages recognize an increased anionic character of acetyl-LDL or succinyl-LDL, what is the difference between CM-LDLs and acetyl-LDL or succinyl-LDL? It is possible that micro-environmental conformation may be altered by acetylation and succinylation. Both residues, acetyl lysine(Lys-NH-CO-CH_3) and succinyl lysine (Lys-NH-CO-CH_2-CH_2-COOH) on the LDL particle result in the loss of positive charge on the amino groups; the microenvironment of the former becomes no charge and the latter negatively charged. In contrast CM-lysine residues (Lys-NH-CH_2-COOH) do not result in the loss of positive charge on the amino group. Circular dichroism spectra of acetyl-LDL and Cu-oxidized LDL were changed at the wavelength region 210-230 nm. When the affinity of modified LDL preparations on an immunoaffinity gel bound anti-apoprotein B-100 monoclonal antibody was assayed, succinyl-LDL and acetyl-LDL was decreased (data not shown), suggesting conformational changes.

Whilst submitting this paper, we read a related study entitled 'In-Vitro Carboxymethylation of Low Density Lipoprotein Alters its Metabolism Via the High-Affinity Receptor' by Gempel et al.[6] They have shown the same results as our data concerning CM-LDL on cholesterylester synthesis of mouse macrophages.

References

1 S.Reddy, J.Bichler, K.J.Wells-knecht, S.R.Thorpe and J.W.Baynes, N^ϵ-Carboxymethyl)lysine is a Dominant Advanced Glycation End Product (AGE) Antigen in Tissue Proteins, *Biochemistry*, 1995, **34**, 10872-10878.
2. M. Takanashi, T. Sakurai and S.Tsuchiya, Identification of the Carboxymethyllysine Residue in the Advanced Stage of Glycated Human Serum Albumin, *Chem. Pharm. Bull.*, 1992, **40**, 705-708.
3. S.H. Gianturco, W.A. Bradley, A.M. Jr. Gotto, J.P. Morisett and D.L. Peavy, Hypertriglycemic Very Low Density Lipoproteins Induce Triglyceride Synthesis and Accumulation in Mouce Peritoneal Macrophages, *J. Clin. Invest.*, 1982, **70**, 168-178.
4. Y. Yamamoto and E. Niki, Presence of Cholesterylester Hydroperoxide in Human Blood Plasma, *Biochem. Biophys. Res. Commun.*, 1989, **165**, 988-993.
5. J.L.Goldstein, Y.K. Basu and M.S.Brown, Binding Site on Macrophages That Mediates Uptake and Degradation of Acetylated Low Density Lipoprotein, Producing Massive Cholesterol Deposition, *Proc. Natl. Acad. Sci.*, U.S.A., 1979, **76**, 333-337.
6. K.E.Gempel, K.D.Gerbitz, B.Olgemoller and E.D.Schleicher, In-Vitro Carboxymethylation of Low Density Lipoprotein Alters its Metabolism Via the High-Affinity Receptor, *Horm. metab. Res.*,1993, **25**, 250-252.

Pro-Inflammatory Cytokine Synthesis by Human Monocytes Induced by Proteins Minimally-Modified by Methylglyoxal

Evelyn A. Abordo and Paul J. Thornalley

DEPARTMENT OF BIOLOGICAL AND CHEMICAL SCIENCES, UNIVERSITY OF ESSEX, WIVENHOE PARK, COLCHESTER, ESSEX CO4 3SQ, UK

Summary
Methylglyoxal is a reactive α-oxoaldehyde, physiological metabolite and potent glycating agent. It reacts with lysine and arginine residues irreversibly to form advanced glycation endproducts (AGE): N_ε-(1-carboxyethyl)lysine, 4-methylimidazolium crosslinks, hydroimidazolone, N_δ-(5-hydro-5-methyl-4-imidazolon-2-yl)ornithine and imidazolone N_δ-(5-methyl-4-imidazolon-2-yl)ornithine residues. The induction of pro-inflammatory cytokines by AGE-modified proteins has been suggested as a mechanism by which AGE accumulation *in vivo* may contribute to the development of diabetic complications and other disease mechanisms. Proteins typically have only minimal modification by AGE (1-3 AGE per protein molecule) in vivo, even in pathological states. We therefore studied the induction of cytokines in human monocytes *in vitro* by minimally advanced glycated proteins. Human serum albumin minimally modified by methylglyoxal stimulated the synthesis and secretion of interleukin-1β, tumour necrosis factor-α and macrophage-colony stimulating factor. Human serum albumin minimally- or highly-modified by glucose-derived AGE gave a much weaker response under the same conditions. This suggests that the modification of proteins by methylglyoxal produced a potent pharmacophore for activation of the cytokine response. Cytokine induction *in vivo* provides a mechanism by which the accumulation of AGE, particularly methylglyoxal-modified proteins, may contribute to the development of diabetic complications, chronic renal insufficiency, Alzheimer's disease and aging.

Introduction

Non-enzymatic modification of proteins by glucose to form end-stage adducts, advanced glycation endproducts (AGE), *in vivo* has been linked to the development of chronic clinical complications associated with diabetes mellitus - retinopathy, neuropathy and nephropathy, macrovascular disease, Alzheimer's disease and dialysis-related amyloidosis.[1] Glucose-derived AGE formation was originally proposed as 'a specific signal for recognition and degradation of senescent macromolecules'.[2] Glucose-derived AGE compounds are: N_ε-carboxymethyllysine, N_ε-lactatolysine, pentosidine and pyrraline. Proteins highly modified by glucose-derived AGE underwent receptor-mediated endocytosis by monocytes, macrophages and endothelial cells *in vitro*, stimulating monocyte chemotaxis and transendothelial movement, proliferation of macrophages, and chemotaxis and angiogenesis of human endothelial cells.[1]

The hypothesis that glucose-derived AGE contribute to disease mechanisms via receptor-mediated binding of AGE modified-proteins has been exemplified by studies *in vitro* using proteins with a much higher extent of modification than found *in vivo*. The extent of protein modification by AGE has become of increasing interest since receptor-mediated responses induced by glucose-derived AGE-modified proteins were limited to proteins with high modification. Only proteins modified by incubation with $\geq$0.5 M glucose for 2 months were bound and internalized by cells.[3] Human serum albumin (HSA) highly modified by glucose-derived AGE used in such studies had 37 glucose-derived AGE per molecule.[4] These proteins have a high content of N_ε-carboxymethyllysine, modified arginine residues and a high net negative charge - a characteristic recognition factor of scavenger receptors.[5,6] Other AGE-binding receptors have been characterized: for example, Receptor for Advanced Glycosylation Endproduct (RAGE),[7] oligosaccharyltransferase-48 complex (OST-48) and a

80 kDa protein kinase C substrate (80K-H).[8] RAGE and OST-48 were found in monocytes, endothelial cells, mesangial cells and astrocytes.[7,8]

The extent of modification of proteins by glucose-derived AGE *in vivo* is typically 1-3 AGE residues per molecule.[9] In healthy human subjects, increases in molecular mass of HSA corresponded to less than one glucose moiety and, even in pathological conditions, rarely exceeded 1-3 glucose moieties. Proteins with this minimal extent of glucose-derived AGE modification had very low affinity for AGE receptors. The dichotomy between the requirement of high modification of glucose-derived AGE-proteins for functional activity and the presence of only proteins minimally modified by AGE *in vivo* is an important disparity in the hypothesis which implicates AGE in disease mechanisms. Receptor binding of proteins and functional responses induced by proteins minimally-modified by AGE are likely to be of greater physiological importance than the binding of proteins highly modified by AGE. Moreover, minimally modified proteins are likely to be involved in the early stages of pathological mechanisms. An amended form of this hypothesis is now emerging, initiated by observations that the physiological α-oxoaldehyde methylglyoxal is an important physiological precursor of AGE,[10] and proteins minimally-modified by methylglyoxal had high affinity for cell surface receptors on human monocytic cells.[11]

Methylglyoxal is formed by the fragmentation of triosephosphates, by the cytochrome P450 2E1-catalyzed oxidation of acetone from ketone bodies, and the catabolism of threonine.[12] Methylglyoxal is a potent glycating agent.

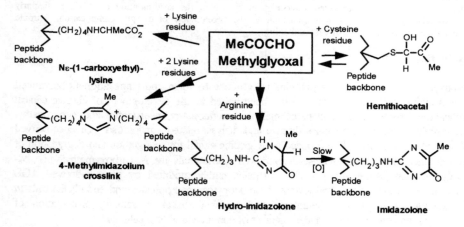

Figure 1. *Reaction of proteins with methylglyoxal*

It binds reversibly cysteine residues to form hemithioacetal adducts but undergoes irreversible reactions with lysine and arginine residues forming stable adducts.[13] Modification of lysine residues forms N_ε-(1-carboxyethyl)lysine and 4-methylimidazolium crosslinks, and modification of arginine residues forms hydroimidazolone and fluorescent oxidized imidazolone.[1] Our recent studies indicated that the hydroimidazolone derivative of methylglyoxal-modified proteins was the recognition factor for binding to AGE receptors of human monocytes. HSA minimally modified with methylglyoxal (MG_{min}-HSA) was bound by cell surface receptors of human monocytic THP-1 cells *in vitro at* 4°C: the binding constant K_D value was 377 ± 35 nM and there were $5.9 \pm 0.2 \times 10^5$ receptors per cell.[11]

The binding of proteins highly modified by glucose-derived AGE to cell surface receptors on monocytes and macrophages induced the synthesis and secretion of interleukin-1β (IL-1β),[14] tumour necrosis factor-α (TNF-α)[14] and macrophage colony-stimulating factor (M-CSF)[15] and was implicated in tissue remodelling and vascular dysfunction associated with microvascular complications of diabetes mellitus.[14,15] To test if proteins modified by glucose-derived AGE and methylglyoxal to physiological extents were competent inducers of cytokines, we measured the secretion of IL-1β, TNF-α and M-CSF proteins by human monocytes stimulated with HSA minimally modified by methylglyoxal (MG_{min}-HSA) and by glucose-derived AGE (AGE_{min}-HSA) - 2-3 AGE moieties per molecule. HSA highly modified by glucose-derived AGE (AGE-HSA) and unmodified HSA were used as controls.

Materials and Methods

Materials
RPMI 1640 medium was purchased from Gibco BRL (Paisley, Scotland). Polymyxin B, HSA, Harris haematoxylin stain and bacterial lipopolysaccharide (LPS) were purchased from Sigma Chemical Company Ltd (Poole, Dorset, UK). Quantikine™ ELISA kits for the assay of IL-1β, TNF-α and M-CSF were purchased from Research and Development Systems (Abingdon, Oxon, UK). A QC-100 limulus endotoxin assay kit was purchased from Biowhittaker (Wokingham, UK).

Methods
Peripheral mononuclear cells were isolated from heparinized human blood (50 ml) by density gradient centrifugation.[15] Mononuclear cells were washed and resuspended in RPMI 1640 with 5 μg/ml of polymyxin B at a density of 2×10^6 cells/mL (IL-1β and M-CSF experiments) or 1×10^6 cells/mL (TNF-α experiments). Staining with Wright's stain indicated the presence of $12.5 \pm 0.2\%$ monocytes, and with trypan blue indicated a cell viability of >99%. Cell suspension (1 mL) was added to each well of a 24-well plate (Costar, High Wycombe, UK) and incubated for 60 min. at 37°C when the monocytes had adhered to the well surface. Non-adherent cells were removed by washing with RPMI 1640 with 5 μg/mL of polymyxin B, and the cells were finally immersed in RPMI 1640 with 5 μg/mL of polymyxin B with 5% autologous plasma and HSA derivative. The monocytes were incubated with and without 20 μM HSA derivatives for 24 h (IL-1β and M-CSF experiments) or 2 h (TNF-α experiments) at 37°C. The concentrations of IL-1β, TNF-α and M-CSF in the extracellular medium were measured by ELISA. The extracellular medium was also assayed for L-lactic dehydrogenase activity as an indicator of cell viability. MG_{min}-HSA, AGE_{min}-HSA and AGE-HSA were prepared and characterized as described.[15] Mean extents of AGE modification per protein molecule were: MG_{min}-HSA, 2.4; AGE_{min}-HSA, 1.9; and AGE-HSA, 36.7. Endotoxin content of HSA derivatives was determined by chromogenic limulus assay (QC-1000 kit) by the manufacturer's instructions.

Results

When human monocytes were incubated for 2 h with AGE_{min}-HSA, the concentration of TNF-α in the extracellular medium was increased 77%, relative to control incubations with HSA. The extracellular concentration of TNF-α increased more, 2-fold, in the presence of AGE-HAS but potent secretion of TNF-α was induced in the presence of MG_{min}-HSA where the extracellular concentration of TNF-α was increased approximately 19-fold (Figure 2a). The extracellular concentration of TNF-α with 20 μM MG_{min}-HSA was 0.60 ± 0.06 ng/mL (n = 3) which is sufficient to induce characteristic functional responses: expression of vascular adhesion molecules, increased monocyte adhesion to glomerular mesangial cells and vascular smooth-muscle cells; enhanced superoxide formation, degranulation and chemotactic response of neutrophils; and increased expression of heparin-binding epidermal growth factor and monocyte chemoattractant protein-1 in vascular endothelial cells.

The secretion of IL-1β protein was also increased markedly by MG_{min}-HSA. Incubation of human monocytes with AGE_{min}-HSA for 24 h induced a 61% increase in the extracellular concentration of IL-1β; with AGE-HSA there was a 3-fold increase but with MG_{min}-HSA the extracellular concentration of IL-1β was increased 18-fold (Figure 2b). With 20 μM MG_{min}-HSA, the extracellular concentration of IL-1β was 0.96 ± 0.01 ng/mL (n = 3) which is sufficient to induce functional responses: expression of vascular adhesion molecules, production of cytokines/growth factors, chemokines, nitric oxide, oxygen free radicals and neutral proteinases in mesangial cells, vascular endothelial cells and smooth muscle cells.

When human monocytes were incubated for 24 h with AGE_{min}-HSA, the concentration of M-CSF in the extracellular medium was 0.13 ± 0.02 ng/mL (n = 3). The secretion of M-CSF from monocytes further increased in the presence of AGE-HSA, indicating that the secretion of M-CSF from human monocytes was dependent on the extent of modification of HSA by glucose-derived AGE. When human monocytes were incubated for 24 h with MG_{min}-HSA, however, the secretion of M-CSF into the extracellular medium was markedly higher than that induced by AGE_{min}-HSA and by AGE-HSA: the concentration of M-CSF in the extracellular medium was 1.22 ± 0.01 ng/mL (n = 3). At this concentration, M-CSF induced characteristic functional responses: growth and differentiation of bone marrow progenitor cells to mononuclear phagocytes, the expression of the scavenger receptor, synergism with TNF-α to stimulate macrophage proliferation, and synergism with interferon-γ to induce increased expression of IL-1β and TNFα, and expression of vascular cell adhesion molecule-1 and intercellular adhesion molecule-1.[15] Increased expression of M-CSF may contribute to the development of diabetic complications - particularly glomerulosclerosis associated with nephropathy, proliferative vitreoretinopathy and atherosclerosis associated with macrovascular disease.

Although cytokine induction was studied in the presence of polymyxin B to suppress effects of contaminating endotoxin, the concentration of LPS in HSA derivatives was determined. The endotoxin contamination of HSA, AGE_{min}-HSA, AGE-HSA and MG_{min}-HSA was determined by a chromogenic limulus assay and values were (ng LPS/mg protein): 0.38 ± 0.09, 2.22 ± 0.12, 2.09 ± 0.06 and 2.54 ± 0.09 (n = 3), respectively. Stimulation of human monocytes with a similar amount of LPS as added with 20 μM MG_{min}-HSA (3.35 ng) produced TNF-α, IL-1β and M-CSF secretion (ng/mL): 0.05 ± 0.01, 0.04 ± 0.01 and 0.010 ± 0.003 (n = 3). The stimulation of TNF-α concentration in incubations of monocytes with HSA derivatives could therefore not be attributed to endotoxin contamination of the HSA derivatives.

Modification of proteins by methylglyoxal forms a particularly potent pharmacophore for the secretion of pro-inflammatory cytokines by human monocytes. This may underlie the link of methylglyoxal metabolism with the development of diabetic complications[16] where cytokines may contribute to microvascular dysfunction. Moreover, proteins modified by methylglyoxal (and other α-oxoaldehydes) may be involved in other disease mechanisms where glucose-derived AGE have been implicated - macrovascular disease, Alzheimer's disease, chronic renal insufficiency, and the production of pro-inflammatory cytokines by monocytes and other cells (microglia, for example) may be important mediators of these processes.

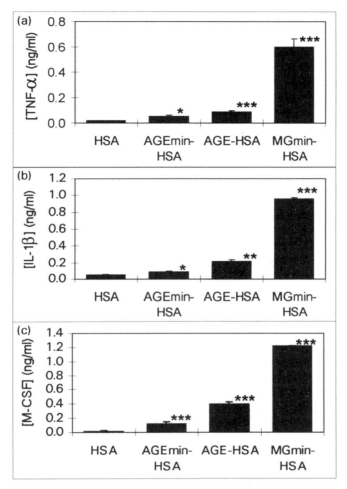

Figure 2. *Effect of HSA derivatives modified by AGE and methylglyoxal on the secretion of (a) TNF-α, (b) IL-β and (c) by human monocytes. Data are means ±S.D. (n = 3). Significantly different from control: *, P<0.05, ** P<0.01, *** P<0.001 (t-test).*

References

1. P.J. Thornalley, Advanced glycation and the development of diabetic complications. Unifying the involvement of glucose, methylglyoxal and oxidative stress, *Endocrinol. & Metabolism*, 1996, **3**, 149-166.
2. H. Vlassara, M. Brownlee and A. Cerami, High-affinity-receptor-mediated uptake and degradation of glucose-modified proteins: A potential mechanism for the removal of senescent macromolecules, *Proc. Natl. Acad. Sci. USA*, 1985, **82**, 5588-5592.
3. S.M. Shaw and M.J.C. Crabbe, Non-specific binding of advanced glycosylation endproducts to macrophages outweighs specific receptor-mediated interactions,

Biochem. J., 1994, **304,** 121-129.
4. M.E. Westwood and P.J. Thornalley, Molecular characteristics of methylglyoxal-modified bovine and human serum albumins. Comparison with glucose-derived advanced glycation endproducts-modified serum albumins, *J. Prot. Chem.*, 1995, **14,** 359-372.
5. T. Shinoda, F. Hayase, N. Van Chuyen and H. Kato, Uptake of proteins modified with 3-deoxyglucosone, a Maillard reaction intermediate, by the type I macrophage scavenger receptor, *Biosci. Biotech. Biochem.*, 1993, **57,** 1826-1831.
6. N. Araki, T. Higashi, T. Mori, R. Shibayama, Y. Kawabe, T. Kodama, K. Takahashi, M. Shichiri and S. Horiuchi, Macrophage scavenger receptor mediates the endocytic uptake and degradation of advanced glycation end-products of the Maillard reaction, *Eur. J. Biochem.*, 1995, **230,** 408-415.
7. A.M. Schmidt, M. Vianna, M. Gerlach, J. Brett, J. Ryan, J. Kao, C. Esposito, H. Hegarty, W. Hurley, M. Clauss, F. Wang, Y.E. Pan, T.E. Tsang and D. Stern, Isolation and characterization of two binding proteins for advanced glycation endproducts from bovine lung which are present on the endothelial cell surface, *J. Biol. Chem.*, 1992, **267,** 14987-14997.
8. Y.M. Li, T. Mitsuhashi, D. Wojciechowicz, N. Shimizu, J. Li, A. Stitt, C. He, D. Bannerjee and H. Vlassara, Molecular identity and cellular distribution of advanced glycation endproduct receptors: relationship of p60 to OST-48 and p90 to 80K-H membrane proteins, *Proc. Natl. Acad. Sci. USA*, 1996, **93,** 11047-11052.
9. A. Lapolla, D. Fedele, R. Seraglia, S. Catinella, L. Baldo, R. Aronica and P. Traldi, A new effective method for the evaluation of glycated intact plasma proteins in diabetic subjects, *Diabetologia*, 1995, **38,** 1076-1081.
10. M. Shinohara, I. Giardino and M. Brownlee, Overexpression of glyoxalase I inhibits intracellular advanced glycation endproduct (AGE) formation, *Diabetes*, **45,** 1996, 126A(Abstract)
11. M.E. Westwood, O.K. Argirov, E.A. Abordo and P.J. Thornalley, Methylglyoxal-modified arginine residues - a signal for receptor-mediated endocytosis and degradation of proteins by monocytic THP-1 cells, *Biochim. Biophys. Acta.*, 1997, **1356,** 84-94.
12. P.J. Thornalley, The glyoxalase system in health and disease, *Molecular Aspects of Medicine*, 1993, **14,** 287-371.
13. T.W.C. Lo, M.E. Westwood, A.C. McLellan, T. Selwood and P.J. Thornalley, Binding and modification of proteins by methylglyoxal under physiological conditions. A kinetic and mechanistic study with N_α-acetylarginine, N_α-acetylcysteine, N_α-acetyl-lysine, and bovine serum albumin., *J. Biol. Chem.*, 1994, **269,** 32299-32305.
14. H. Vlassara, M. Brownlee, K.R. Manogue, C.A. Dinarello and A. Pas, Cachectin/TNF and IL-1 induced by glucose-modified proteins: Role in normal tissue remodelling, *Science*, 1988, **240,** 1546-1548.
15. E.A. Abordo, M.E. Westwood and P.J. Thornalley, Synthesis and secretion of macrophage colony stimulating factor by mature human monocytes and human monocytic THP-1 cells induced by human serum albumin derivatives modified with methylglyoxal and glucose-derived advanced glycation endproducts, *Immunol. Lett.*, 1996, **53,** 7-13.
16. P.J. Thornalley, A.C. McLellan, T.W.C. Lo, J. Benn and P. Sonksen, Negative association of red blood cell reduced glutathione with diabetic complications, *Clin. Sci.*, 1996, **91,** 575-582.

Detection of AGE-Lipids *In Vivo*: Glycation and Carboxymethylation of Aminophospholipids in Red-Cell Membranes

J.R. Requena, M.U. Ahmed, S. Reddy, C. W. Fountain, T.P. Degenhardt, A.J. Jenkins,[1] B. Smyth,[1] T.J. Lyons[1] and S.R. Thorpe

DEPARTMENT OF CHEMISTRY AND BIOCHEMISTRY, UNIVERSITY OF SOUTH CAROLINA, COLUMBIA SC 29208, USA AND [1]DIVISION OF ENDOCRINOLOGY, MEDICAL UNIVERSITY OF SOUTH CAROLINA, CHARLESTON SC 29425, USA

Summary
The study of the Maillard reaction in vivo has emphasized reactions between glucose and amino groups of proteins. The present report extends this work to reactions of aminophospholipids with glucose, and with products of glucose and polyunsaturated fatty acid autoxidation. Glycation of aminophospholipids under oxidative conditions in vitro, resulted in the formation of both carboxymethylethanolamine and carboxymethylserine. These compounds have now been quantified in the lipid fraction of red-cell ghost membranes. Glycated and carboxymethylated lipids represent another class of compounds produced during the Maillard reaction in vivo, and should be useful for assessing the impact of aminophospholipid glycoxidation and/ or lipoxidation in disease states.

Introduction

The study of the Maillard reaction *in vivo* has focused primarily on reactions between sugars and protein. A number of early and advanced glycation end-products (AGEs) have been quantified in human tissues, and their concentrations correlated with aging and disease, including pathology in diabetes, renal failure and Alzheimer's disease.[1] Pentosidine and N^ε-(carboxymethyl)lysine (CML) are two chemically characterized products formed *in vitro* by reaction of proteins and glucose (and other sugars) under oxidative conditions. Both are increased in tissues during normal aging and their age-adjusted levels in collagen correlate positively with the severity of diabetic complications.[1] CML can be produced by the oxidative cleavage of the Amadori compound, fructose-lysine, *in vitro*. CML can also be generated *in vitro* following autoxidation of glucose, yielding glyoxal, which in turn can react directly with lysine residues to form CML. Recent findings indicate that there is a third route to formation of CML via autoxidation of polyunsaturated fatty acids (PUFAs), independent of carbohydrate.[2] Overall, these results demonstrate a complex interplay between oxidation of sugars and lipids during the Maillard reaction, including the formation of common dicarbonyl intermediates.

From a theoretical point of view, it is likely that the amino group of aminophospholipids will also react with glucose and initiate many of the same reactions that occur with proteins. However, scant data are available on glycation and glycoxidation of lipids. Using a chemical assay, Pamplona *et al.*[3] described the formation of early glycation products on amino-phospholipids in livers of control rats and their elevation in livers from diabetic rats. Ravandi *et al.*[4] also showed the presence of glycated phosphatidylethanolamine (PE) in human red blood cell membranes, but the sugar appeared to be present largely in Schiff's base form. Using antibodies prepared against protein modified by glucose under oxidative conditions, Bucala et al. [5] described the immunologic detection of AGE-lipids during reaction of glucose with PE but not phosphatidylcholine, and developed an ELISA technique to quantify them. These authors found that much more AGE immunoreactivity developed in the lipid fraction of LDL, compared with the apoprotein fraction, when LDL was incubated with glucose; elevated levels of circulating AGE-lipids were also reported in diabetic patients vs healthy individuals.[6]

The studies described below were initiated to develop sensitive and specific gas chromatography-mass spectrometry (GC-MS) assays for glycated and carboxymethylated (CM) phospholipids. These assays will be useful for quantifying the extent of glycation of aminophospholipids and the usefulness of CM-lipids as markers of oxidative stress *in vivo*, and their possible contribution to pathology in aging and vascular disease. Scheme 1 shows various routes to the formation of carboxymethyl-ethanolamine (CME) and -serine, (CMS) analogous to those known for CML.

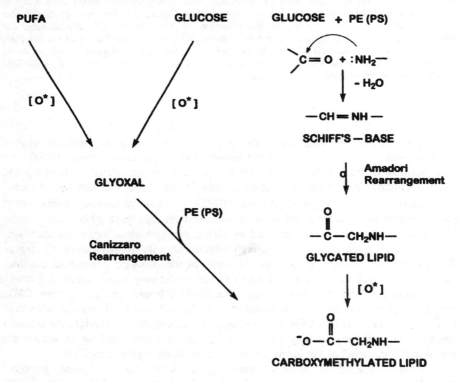

Figure 1. *Scheme for the formation of CM-aminophospholipids. Direct reaction of glucose with an aminophospholipid to form a Schiff's-base, followed by subsequent Amadori rearrangement would yield a glycated lipid; oxidative cleavage of the Amadori adduct would then yield the CM-lipid. Alternatively, autoxidation of either glucose or of PUFA in lipoproteins can result in the formation of glyoxal, which can react with lipid amino groups via a Canizzaro reaction, to yield the CM-lipid. PE = phosphatidylethanolamine; PS = phosphatidylserine; [O*] = reactive oxygen species*

Materials and Methods

Glucitolethanolamine (GE) and aminophospholipids were obtained from Sigma (St Louis, MO). CME and d_4-CME (CMS and $^{13}C_3$-CMS) were synthesized by reaction of ethanolamine or d_4-ethanolamine (serine or $^{13}C_3$-serine) and glyoxylic acid in the presence of $NaBH_3CN$. Phospholipids were incubated at 37°C for two weeks in the presence or absence of 500 mM glucose dissolved in 0.2 M phosphate buffer, pH 7.4, under aerobic or anti-oxidative conditions (1 mM DTPA, nitrogen atmosphere). Lipids were extracted with chloroform: methanol (2:1), reduced with 100 mM $NaBH_4$ in methanol for 4 h at 25°C then hydrolyzed for 3 h at 110°C in 3N HCl. Samples were converted to their acetylmethylester (AME)[7] or trifluoroacetylmethylester (TFAME)[8] derivatives for analysis by GC-MS. Red blood cell ghosts were prepared by the method of Dodge et al.[9]; the ghost membranes were dried and lipids were extracted and treated as above.

Results and Discussion

Incubation of dioleoyl-PE (DO-PE), or dioleoyl-PS (DO-PS) with glucose under aerobic conditions resulted in the formation of several new products detected by GC-MS. In the case of DO-PE, two products, CME and GE, were both identified by comparison with elution times and fragmentation patterns of authentic standards. Surprisingly, despite using several experimental protocols to exclude oxygen, including argon or nitrogen atmospheres and metal ion chelators, CME was still detectable in antioxidative incubations of DO-PE and glucose, but at only 10-15% the levels under oxidative conditions. CME was not detected in the starting DO-PE, or in DO-PE incubated under oxidative conditions in the absence of glucose. CMS was formed on DO-PS in the presence of glucose under oxidative conditions only. Finally, incubation of linoleoyl-palmitoyl-PE under aerobic conditions, but in the absence of glucose, also resulted in the formation of CME, indicating that products of autoxidation of PUFA can produce CME on phospholipid. CME was not detected in the PUFA-containing phospholipid incubated under antioxidative conditions. The yields of CME in phospholipids treated under glycoxidative (glucose, air) or lipoxidative (PUFA, air) conditions were comparable, ranging from 9-18 mmoles CME/ mol ethanaolamine.

The chromatograms shown in Figure 2 are typical selected-ion traces for red cell membrane lipids isolated from healthy individuals. The respective concentrations of CME and CMS were 0.128 ± 0.08 mmol CME/ mol ethanolamine and 0.044 ± 0.007 mmol CMS/ mol serine (means $\pm$ SD, n = 4), which is consistent with the approximate 3:1 ratio of ethanolamine:serine in the red-cell membrane. In these same samples the amount of GE was 21.9 ± 11.8 mmol GE/ mol ethanolamine. In a separate study,[10] we have also measured red-cell membrane protein modification by carboxymethylation and glycation. CML and fructose-lysine represented $\sim$ 0.2 and 2 mmol/mol lysine, respectively. Compared with the results for modified lipids in the present study, these data suggest that there is a similar extent of carboxymethylation of the membrane lipid and protein amino groups. At the same time, there appears to be somewhat more glycation of the aminophospholipids than of the protein. This conclusion is tentative because the methodology used to measure lipid glycation, unlike that for lysine glycation, does not distinguish between glucose bound as a Schiff's base and that bound as an Amadori adduct. Experiments are in progress to clarify this issue.

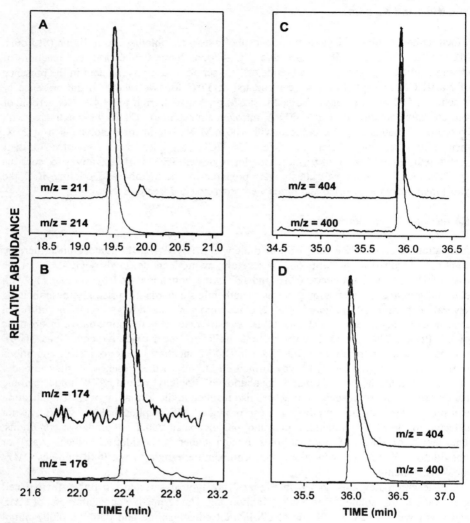

Figure 2. *Selected-ion monitoring traces of modified aminolipids in red-cell membrane lipid extract. Chromatograms are for lipids extracted from ghosts isolated from the blood of a single individual. (A) traces m/z = 211 and 214, correspond to TFAME derivative of CME and d_4-CME internal standard, respectively; (B) traces m/z = 174 and 176, correspond to AME derivative of CMS and $^{13}C_3$-CMS internal standard, respectively. (Because of interference, the AME derivative was not suitable for quantification of CME; thus, in order to measure both CME and CMS two different derivatization procedures were needed.) (C) and (D) traces show m/z = 400 and 404 ions for GE in red cell lipids (C) and authentic GE (D). Elution times and similarity in ion-ratios for red-cell extracts and authentic GE were used to confirm the presence of GE; quantification was by comparison with a standard curve constructed from the ratio of authentic GE and $^{13}C_3$-serine*

These data provide the first evidence for carboxymethylation of aminophospholipids *in vivo*. We know little at this point about the origin of CME *in vivo*. Both carbohydrate- and lipid-derived dicarbonyl intermediates may be involved. It is tempting to speculate that PUFAs, relatively prone to oxidation, and physically close to the amino groups of ethanolamine and serine may be the primary sources of the lipid modification. It also seems likely that a substantial fraction of the previously described AGE-lipids[5,6] may, in fact, be CME and/ or CMS. In this respect, CME was only somewhat less effective than CML in a competitive ELISA assay,[10] using an anti-AGE protein antibody to detect AGE-rabbit serum albumin; we have previously shown that this anti-AGE antibody recognizes CML as a primary epitope. Furthermore, reports of immunoreactive AGEs in frozen sections of athero sclerotic plaque from normoglycemic individuals may reflect reaction of both CM-protein and lipid moieties.[11] Finally, the low levels of red-cell CM-lipids suggest that they are biomarkers, rather than important effectors of pathogenicity. At the same time, measurement of CME and CMS should prove useful to assess the impact of aminophospholipid glycoxidation and/or lipoxidation in disease states.

Abbreviations
AGE, advanced glycation end-product; AME, acetylmethylester; CM-, carboxymethylated; CME, carboxymethylethanolamine; CML, N^ε-(carboxymethyl)lysine, CMS, carboxymethylserine; DO-PE, dioleoyl-PE; DO-PS dioleoyl-PS; GC-MS, gas chromatography-mass spectrometry; GE, glucitolethanolamine; PE, phosphatidylethanolamine; PS, phosphatidylserine; PUFAs, polyunsaturated fatty acids; TFAME, trifluoroacetylmethylester

Acknowledgements
This work was supported by USPHS grants AG11472 (SRT) and 5R29 EY10697 (TJL). CWF is supported by a research grant from the South Carolina Affiliate of the American Heart Association and TJL by an American Diabetes Association research grant. JRR is a recipient of a Juvenile Diabetes Foundation postdoctoral fellowship; AJJ is a recipient of a Lions Sight First Training Grant and Juvenile Diabetes Foundation research grant 195052.

References

1. S.R.Thorpe and J.W.Baynes, Role of the Maillard reaction in diabetes mellitus and diseases of aging, *Drugs & Aging*, 1996, **9**, 69-77.
2. M.X.Fu, J.R.Requena, A.J.Jenkins, T.J.Lyons, J.W.Baynes and S.R.Thorpe, The advanced glycation end-product, N^ε(carboxymethyl)lysine, is a product of both lipid peroxidation and glycoxidation reactions, *J. Biol. Chem.*, 1996, **271**, 9982-9986.
3. R.Pamplona, M.J.Bellmunt, M. Portero, D.Riba and J.Prat, Chromatographic evidence for Amadori product formation in rat liver aminophospholipids, *Life Sci.*, 1995, **57**, 873-879.
4. A. Ravandi, A.Kuksis, L.Marai, J.J.Myher, G.Steiner, G. Lewis and H. Kamido, Isolation and identification of glycated aminophospholipids from red cells and plasma of diabetic blood, *FEBS Lett.,*1996, **381**, 77-81.
5. R.Bucala, Z.Makita, T.Koschinsky, A.Cerami and H.Vlassara, Lipid advanced glycosylation: pathway for lipid oxidation *in vivo*, *Proc. Natl. Acad. Sci., USA,* 1993, **90**, 6434-6438.

6. R. Bucala, Z. Makita, G.Vega, S.Grundy, T.Koschinsky, A.Cerami and H.Vlassara, Modification of low density lipoprotein by advanced glycation end products contributes to the dyslipidemia of diabetes and renal insufficiency, *Proc. Natl. Acad Sci. USA*, 1994, **91**, 9441-9445.
7. D.R.Knapp, 'Handbook of Analytical Derivatization Reactions', John Wiley & Sons, New York, 1979, p. 254.
8. K.J.Knecht, J.A.Dunn, K.F.McFarland, D.R.McCance, T.J.Lyons, S.R.Thorpe and J.W.Baynes, Effect of diabetes and aging on carboxymethyllysine levels in human urine, *Diabetes*, 1991, **40**, 190-196.
9. J.T.Dodge, C.Mitchell and D.J.Hanahan, The preparation and chemical characterization of hemoglobin-free ghosts of human erythrocytes, *Arch. Biochem. Biophys.*,1963, **100**,119-130.
10. J.R.Requena, M.U.Ahmed, C.W.Fountain, T.P.Degenhardt, S.R.Reddy, C. Perez, T.J.Lyons, A.J.Jenkins, J.W.Baynes and S.R. Thorpe, Carboxymethylethanolamine: a biomarker of phospholipid modification during the Maillard reaction *in vivo*, *J. Biol. Chem.*, 1997, **272**, 17473-17479.
11. S.Kume, M.Takeya, T.Mori, N.Araki, H.Suzuki, S.Horiuchi, T.Kodama, M.Miyauchi and K. Takahashi, Immunohistochemical and ultrastructural detection of advanced glycation end products in atherosclerotic lesions of human aorta with a novel specific monoclonal antibody, *Am. J. Pathol.*, 1995, **147**, 654-667.

Cellular Receptors for N^ε-Fructosyllysine: Potential Role in the Development of Diabetic Microangiopathy

S. Krantz, Rowena E. Brandt and Ramona Salazar

INSTITUTE OF BIOCHEMISTRY, ERNST MORITZ ARNDT UNIVERSITY,
D-17487 GREIFSWALD, GERMANY

Summary
Evidence has been obtained that cellular receptors for the Amadori product fructosyllysine are important for the development of diabetic vascular complications. Binding studies with glycated albumin and glycated LDL and competition experiments have demonstrated that the binding proteins (100 and 200 kDa) recognize fructosyllysine as binding epitope rather than an Amadori-modified amino acid sequence or AGE. The 100 kDa receptor protein is nucleolin-like. Ligand binding to monocytes induced a release of the cytokines IL-1ß and TNF-α. Receptor expression varies from individual to individual. But it is neither age nor sex dependent. It is associated with some indices of diabetic microangiopathy, such as capillary basement membrane thickening.

Introduction

Increased glycation, resulting in Amadori-modified proteins and accumulation of advanced glycation end products (AGE), is a major factor of glucose toxicity and has been linked to the development of diabetic vascular complications. Over the past few years much attention has been focused on the role of AGE in the pathogenesis of diabetic sequelae and the possible importance of Amadori adducts has been ignored. However, evidence has accumulated showing that Amadori-modified proteins affect functions of endothelial and mesangial cells. Recent work has established that these cells express receptors that recognize an Amadori-modified sequence in glycated albumin. Ligand-receptor interactions induced activation as well as suppression of transcription of collagen and other matrix protein genes.[1,2]

We report that glycated proteins can be bound to fructosyllysine-specific binding sites on different cell types, such as monocytes, macrophages, fibroblasts and endothelial cells. The main investigations have been performed with macrophages, monocytes and the monocyte-like cell line U937.

Binding Studies with ^{125}I-Labelled Glycated Albumin

Macrophages, monocytes, U937 and MonoMac 6 cells bound glycated albumin (1 mol fructosyllysine/mol protein, purified by boronate affinity chromatography) in a dose-dependent and saturable manner. The number of binding sites amounted to 10,000 to 20,000 per cell with an affinity constant of 10^7 M^{-1}. Glycated albumin was proved to be free of AGE using binding studies with a bovine aortic endothelial cell line, which only expressed AGE-receptors and did not bind in vitro short-term glycated albumin.

Specific binding of glycated albumin was inhibited by glycated IgG, glycated LDL, glycated polylysine and fructosyllysine, to a lower extent by fructosyl-ß-alanine, but not by hexitollysine, native albumin or IgG, mannose-neoglycoprotein albumin, maleylated albumin and chondroitinsulfates. Amadori-modified albumin did not compete with binding of AGE-albumin.

Two main binding proteins of 100-110 and 190-200 kDa have been identified by ligand blotting and ligand receptor cross-linking in U937 and MonoMac 6 cell membrane preparations with ^{125}I-labelled glycated albumin as a ligand.[3,4]

The 100 kDa Fructosyllysine-Specific Binding Protein is Nucleolin-Like

The 100 kDa fructosyllysine-specific binding protein was purified from U937 cell membranes after solubilization with octylglucoside by use of DEAE-ion exchange chromatography, fructosyllysine-agarose affinity chromatography and SDS-polyacrylamide gel electrophoresis. Amino acid sequence analysis by Edman degradation revealed a single N-terminal sequence which was identical with the positions 2 to 7 of nucleolin. The N-terminal residue was blocked. Two tryptic peptides with 14 and 15 amino acids were identical with the positions 349 to 362 and 610 to 624 of the nucleolin amino acid sequence (see Table 1).[5]

Ligand blotting experiments with nuclear extracts from U937 and RINm5F (an insulinoma cell line) cells showed no binding of glycated albumin. This indicated that nucleolin which is a major soluble nuclear protein has amino acid sequences in common with the 100 kDa fructosyllysine-specific membrane protein, but is not completely identical to it.

RT-PCR (40 cycles) of cDNA from U937 cells using primers deduced from amino acid sequences in common with nucleolin and the 100 kDa fructosyllysine-specific receptor protein failed to produce amplificates different from nucleolin cDNA. One explanation might be that the mRNA of the nucleolin-like receptor protein is of low abundance.

Table 1. *Partial amino acid sequences of the 100 kDa fructosyllysine-specific receptor protein and nucleolin*

	Binding protein	Nucleolin[*]
N-terminal end	-VKLAKA	MVKLAKA
Peptide 1	FGYVDFESAEDLEK	FGYVDFESAEDLEK
		(positions 349 to 362)
Peptide 2	KGFGFVDFNSEEDAK	KGFGFVDFNSEEDAK
		(positions 610 to 624)

[*], Ref. 6.

Binding Studies with ^{125}I-Labelled Glycated LDL

Binding of glycated LDL (8 to 9 mol fructosyllysine/mol apo B) to macrophages and U937 cells involved fructosyllysine-specific sites as well as LDL and scavenger receptors and nonspecific binding. Fructosyllysine, glycated albumin, native LDL and maleylated albumin competed for binding of ^{125}I-labelled glycated LDL to macrophages. Scatchard analysis of binding data from studies with U937 cells, which lack scavenger receptors,[7] showed a linear plot, from which a K_a of 2.6 x 10^7 M^{-1} could be calculated. Bound LDL was internalized and degraded, as could also be demonstrated with glycated albumin.

On U937 cells only the 200 kDa fructosyllysine-specific binding protein and the 160 kDa LDL receptor were involved in binding glycated LDL as evidenced by ligand blotting.

Receptor proteins with the ability to bind glycated albumin were defined on endothelial and mesangial cells by Cohen and co-workers.[1,2] Glycated LDL did not compete with binding of glycated albumin in contrast to our studies, where also fructosyllysine was competitive. Based on these results, it seems unlikely that the binding sites on endothelial and mesangial cells are only fructosyllysine-specific, but recognize an Amadori-modified amino acid sequence in glycated albumin.[1,2]

Cytokine Production by MonoMac 6 Cells

MonoMac 6 monocyte-like cells were most sensitive in response towards glycated albumin which was as efficient a stimulant of IL-1 and TNF-α secretion as AGE-albumin (Table 2). The pathways by which ligand binding leads to cytokine release are still unknown[4].

Table 2. *IL-1β and TNF-α secretion of MonoMac 6 cells exposed to glycated albumins*

Ligand	IL-1β (pg/mL)	TNF-α (pg/mL)
LPS (30 ng/mL)	30.9±0.9	236.0±21.0
AGE-albumin (0.25 mg/mL)	14.2±2.0 p< 0.01	161.7±18.6
Native albumin (0.25 mg/mL)	9.7±1.9	162.6±19.6
Native albumin (0.5 mg/mL)	11.6±3.5	178.3±22.2
Glycated albumin (0.25 mg/mL)	19.7±7.0 p< 0.025	203.4± 4.0 p< 0.001
Glycated albumin (0.5 mg/mL)	60.8±1.5 p< 0.001	233.9± 1.4 p< 0.001
No ligands	10.3±1.2	158.2±15.6

Data are means ± S.D. from four experiments; p was estimated using a t-test (compared with secretion in the absence of ligands). LPS: bacterial lipopolysaccharide.

Different Individual Expression of Fructosyllysine-Specific Receptors

Binding of glycated albumin and LDL via fructosyllysine-specific sites by rat macrophages could only be demonstrated in 40% of the 335 animals investigated. A differing individual expression of these receptors has also been found on monocytes of 90 insulin-dependent diabetic patients and 101 healthy control subjects. The degree of receptor expression was neither age- nor sex-dependent. However, in the diabetic group it correlated significantly with the severity and age of onset of diabetic microangiopathy. A positive, statistically significant correlation between receptor expression, determined as %

receptor-positive monocytes, and an empirically defined microangiopathy score from microvascular complications could be found (r = 0.305; p = 0.0035; χ^2 = 17.1; p = 0.002).

Table 3. *Multivariate stepwise discriminant analysis. Forward selection summary*

Step	Variable	Number in	Partial R^2	F (statistic)	Probability > F
1	HbA$_1$	1	0.1125	5.512	0.0056
2	Duration of diabetes	2	0.0709	3.284	0.0422
3	% Positive monocytes	3	0.0604	2.732	0.0708
4	Age at onset of diabetes	4	0.0180	0.799	0.4530

The altitude of F-values determines the priority of variables.

Multivariate analysis resulted in forward selection of HbA$_1$, duration of diabetes and fructosyllysine receptor expression as the main quantitative variables defining the distribution among the score classes (low, medium, severe complications)(Table 3).

To interpret the results of the human study, spontaneous diabetic and non-diabetic BB/OK rats were used to estimate tissue content of glucose-modified proteins and capillary basement membrane thickness in relation to receptor expression on macrophages. In non-diabetic and diabetic rats no correlation was found between receptor expression and tissue content (i.e. aorta, nerve) of fructosyllysine and fluorescent AGE. However, animals that expressed the fructosyllysine receptor showed a greater increase in muscle capillary basement membrane thickness (Table 4).[8]

Table 4. *Changes in basement membrane thickness in BB/OK rats*

	Basement membrane thickness (nm)	
	Control rats	Diabetic rats
Receptor negative	70.30±30.83	98.64± 27.87[a]
Receptor positive	78.30±28.84[b]	161.42± 54.42[c]

Results are means±S.D. for 6 rats ; [a] p < 0.001 *vs* receptor negative control animals, [b] p < 0.01 *vs* receptor negative diabetic animals, [c] p < 0.001 receptor negative diabetic animals and receptor positive controls.

Conclusions

Evidence has accumulated that glycation of body proteins results in tissue damage. Interactions with specific receptors, such as for AGE (AGE-R$_{1-3}$, RAGE),[9] an Amadori-modified sequence in glycated albumin^{1,2} and for fructosyllysine affect numerous cell functions. Studies with Amadori-modified proteins focus attention back to the role that these proteins may play in the development of late diabetic complications.

The fructosyllysine-specific binding proteins are different from receptors specific for AGE and an Amadori-modified sequence in glycated albumin. This raises the possibility that different rates of development of complications in diabetic patients with similar degrees of hyperglycaemia may be a result of genetic variability in the expression of fructosyllysine-specific sites. There are indications that this receptor expression is positively associated with indices of diabetic complications, such as microangiopathy and capillary basement membrane thickening.

Acknowledgement
These studies were supported by the Deutsche Forschungsgemeinschaft and the Bundesministerium für Bildung, Wissenschaft, Forschung und Technologie, Bonn, Germany. We are greatful to Dr. R. Jack for critically reading the manuscript and making valuable comments.

References

1. M.P. Cohen, E. Hud, V.-Y. Wu and F.N. Ziyadeh, Glycated albumin modified by Amadori adducts modulates aortic endothelial cell biology, *Mol. Cell. Biochem.,* 1995, **143**, 73-79.
2. M.P. Cohen and F.N. Ziyadeh, Role of Amadori-modified nonenzymatically glycated serum proteins in the pathogenesis of diabetic nephropathy, *J. Am. Soc. Nephrol.*, 1996, **7**, 183-190.
3. S. Krantz, R. Brandt and B.Gromoll, Binding sites for short-term glycated albumin on peritoneal cells of the rat, *Biochim. Biophys. Acta,* 1993, **1177**, 15-24.
4. R. Salazar, R. Brandt and S. Krantz, Expression of fructosyllysine receptors on human monocytes and monocyte-like cell lines, *Biochim. Biophys. Acta,* 1995, **1266**, 57-63.
5. S. Krantz, R. Salazar, R. Brandt, J. Kellermann and F. Lottspeich, Purification and partial amino acid sequencing of a fructosyllysine-specific binding protein from cell membranes of the monocyte-like cell line U937, *Biochim. Biophys. Acta,* 1995, **1266**, 109-112.
6. M. Srivastava, P.J. Fleming, H.B. Pollard and A.L. Burns, Cloning and sequencing of the human nucleolin cDNA, *FEBS Lett.*, 1989, **250**, 99-105.
7. D.P. Via, H.A. Dresel and A.M. Gotto, Isolation and assay of the Ac-LDL receptor, *Methods Enzymol.*, 1986, **109**, 216-226
8. R. Brandt, C. Landmesser, L. Vogt, B. Hehmke, R. Hanschke, J. Kasbohm, K. Hartmann, B. Jäger, S. Krantz and D. Michaelis, Differential expression of fructosyllysine-specific receptors on monocytes and macrophages and possible pathophysiological significance, *Diabetologia,* 1996, **39**, 1140-1147.
9. H. Vlassara, Pathogenesis of diabetic nephropathy. Advanced glycation and new therapy, *Med. Klinik,* 1997, **92** suppl. I, 29-34.

The Receptor for Advanced Glycation End Products Mediates the Chemotaxis of Rabbit Smooth Muscle Cells

Takayuki Higashi,[1,2] Hiroyuki Sano,[1] Kenshi Matsumoto,[1] Tetsuto Kanzaki,[3] Nobuhiro Morisaki,[3] Heikki Rauvala,[4] Motoaki Shichiri[2] and Seikoh Horiuchi[1]

[1]DEPARTMENTS OF BIOCHEMISTRY AND [2]METABOLIC MEDICINE, KUMAMOTO UNIVERSITY SCHOOL OF MEDICINE, HONJO, 2-2-1, KUMAMOTO 860, JAPAN; [3]DEPARTMENT OF INTERNAL MEDICINE, CHIBA UNIVERSITY SCHOOL OF MEDICINE, INOHANA, 1-8-1, CHIBA, JAPAN; [4]INSTITUTE OF BIOTECHNOLOGY, UNIVERSITY OF HELSINKI, FINLAND

Summary

We recently demonstrated immunologically the intracellular accumulation of advanced glycation end products (AGEs) in foam cells derived from smooth muscle cells (SMCs) in advanced atherosclerotic lesions. To understand the mechanism of AGE-accumulation in these foam cells, the interaction of AGE-proteins with rabbit cultured arterial SMCs was studied in the present study. In experiments at 4°C, ^{125}I-AGE-bovine serum albumin (AGE-BSA) showed dose-dependent saturable binding to SMCs with an apparent dissociation constant (Kd) of 4.0 µg/mL. In experiments at 37°C, AGE-BSA underwent receptor-mediated endocytosis and subsequent lysosomal degradation. The endocytic uptake of ^{125}I-AGE-BSA was effectively inhibited by unlabeled AGE-proteins, but not by acetylated low density lipoprotein (LDL) and oxidized LDL, well-known ligands for the macrophage scavenger receptor (MSR). Moreover, the binding of ^{125}I-AGE-BSA to SMCs was affected neither by amphoterin, a ligand for one type of the AGE receptor named RAGE, nor by 2-(2-Furoyl)-4(5)-(2-furanyl)-1H-imidazole-hexanoic acid-BSA (FFI-BSA), a ligand for the other AGE receptors called p60 and p90, indicating that the endocytic uptake of AGE-proteins by SMCs is mediated by an AGE receptor distinct either from MSR, RAGE, p60 or p90. To examine the functional role of this AGE receptor, the effects of AGE-BSA on the migration of SMCs were tested. Incubation with 1-50 µg/mL of AGE-BSA resulted in significant dose-dependent cell migration. The AGE-BSA-induced SMCs migration was chemotactic in nature, and was significantly inhibited (~ 80%) by an antibody against transforming growth factor-β (TGF-β), and the amount of TGF-β secreted into the culture medium from SMCs by AGE-BSA was 7-fold higher than that of control, indicating that TGF-β is involved in the AGE-induced SMCs chemotaxis.

Introduction

Data have accumulated recently that support the notion that protein modifications by advanced glycation end products (AGEs) play a causative role in aging processes, and disease processes such as diabetic complications and atherosclerosis.[1] This role of AGE has mainly been attributed to the AGE-binding proteins or AGE receptors, by which cellular binding of AGE-proteins is known to elicit several responses such as cytokine induction, macrophage growth, the induction of fibronectin and type IV collagen synthesis from mesangial cells via Platelet-derived growth factor (PDGF) synthesis, and enhancement of angiogenesis in endothelial cells.[2] Three different AGE receptors have been reported: RAGE, Galectin-3 and MSR. Recent immunohistochemical studies using anti-AGE antibodies demonstrated AGE-modification of extracellular matrix proteins in human atherosclerotic lesions. More interestingly, AGE-accumulation is much more prominent in foam cells derived from monocytes/macrophages in early atherosclerotic lesions (diffuse intimal thickening and fatty streaks) as well as those derived from smooth muscle cells (SMCs) in advanced atherosclerotic lesions (atherosclerotic plaques).[3] These results suggest the interesting possibility that this intracellular AGE-accumulation is due largely to active endocytic uptake of extracellular AGE-modified proteins by

AGE receptor(s).[1] Related to this point, Brett et al.[4] showed that the anti-RAGE antibody reacted with SMCs from bovine aorta. As a logical extension of these lines of experimental evidence, it seems reasonable to postulate that the AGE receptor, if it exists in SMCs, may be involved in atherosclerotic processes such as cell migration and proliferation of SMCs. To evaluate this possibility, we examined whether SMCs from rabbit arterial medial layers expressed the functional AGE receptor.[5]

Materials and methods

Preparation of ligands
AGE-BSA and AGE-human hemoglobin were prepared as described previously.[6] FFI was synthesized and covalently coupled to BSA (FFI-BSA); the amount of FFI incorporated was 7.6 mol/mol BSA. Recombinant amphoterin was prepared by the baculovirus expression method as described previously.[7]

Cell culture and preparation of rabbit SMCs
Primary cultures of SMCs were established from the medial layer of the aorta of Japanese white rabbits by the explant method described previously.[8] Cells were subcultured in Dulbecco's modified Eagle's medium (DMEM) supplemented with 50 µg/mL gentamicin and 10% fetal calf serum (FCS) and the cells at the 3rd to 6th passages were used for the experiments.

Cellular assays
For cellular experiments, 3×10^4 cells were seeded in each well of a 24-well culture plate (15.5 mm diameter, Corning) in DMEM containing 10% FCS (medium A), and cultured for 2 days to subconfluence. The uptake study was carried out for various times at 37°C as described previously.[6] For the binding study, cells were incubated for 2 h at 4°C

Cell migration assay
This assay was performed by a slight modification of the method as described previously.[9] The migratory activity of SMCs was determined in a Transwell cell-culture chamber (Costar, MA).

TGF-β bioassay
To prepare the conditioned medium, SMCs were incubated with AGE-BSA (10 or 100 µg/mL) or BSA (10 or 100 µg/mL) for 18 h at 37°C. TGF-β production was determined in the medium by bioassay as described previously.[10]

DNA synthesis
This assay is the same as that described previously.[11]

Results

The AGE receptor of rabbit SMCs mediates endocytic uptake of AGE-BSA
The cellular binding of ^{125}I-AGE-BSA to SMCs increased in a dose-dependent manner, which was inhibited more than 90% by excess unlabeled AGE-BSA. The specific binding, obtained by subtracting the non-specific binding from the total binding, gave a saturation pattern. The Scatchard analysis of this specific binding disclosed a binding site with an apparent Kd of 4.0 µg/mL and a

maximal surface binding of 2.2 μg/mg cell protein, indicating the presence of a high affinity binding site for AGE-BSA on SMCs.

The events that occurred after AGE-BSA bound to these cells were then studied. During incubation at 37°C, the amount of cell-associated ^{125}I-AGE-BSA increased with time and was effectively reduced by unlabeled AGE-BSA. The specific cellular association of AGE-BSA reached a plateau at 1.25 μg/mg cell protein after 24 h of incubation. In parallel with this increase in cell-association, TCA-soluble radioactivity derived from ^{125}I-AGE-BSA was detected in the medium, and also increased with time, indicating that cell-associated ^{125}I-AGE-BSA underwent intracellular degradation. The degradation of ^{125}I-AGE-BSA was completely inhibited by excess unlabeled AGE-BSA. The specific degradation increased with time and reached 0.27 μg/mg cell protein after 24-h incubation.

The AGE receptor of rabbit SMCs differs in ligand specificity from known AGE receptors
To characterize the AGE-receptor on rabbit SMCs, several ligands were tested for their effect on the endocytic uptake of AGE-BSA. The cell-association of ^{125}I-AGE-BSA with, and its degradation by these cells was effectively inhibited by AGE-BSA and AGE-human hemoglobin, whereas unmodified proteins had no effect, suggesting that the receptor recognizes some common AGE structure(s), as reported previously using rat liver sinusoidal cells and peritoneal macrophages.[12] Furthermore, cellular association and degradation were effectively inhibited by polyanions such as dextran sulfate and polyinosinic acid, but not by polycytidylic acid, indicating that ligand binding to the AGE receptor of rabbit SMCs is sensitive to polyanions, as has been shown with other cells.[6]

We recently demonstrated that the MSR is responsible for the endocytic uptake of AGE-proteins.[6] Available data have suggested that the MSR is not expressed on medial SMCs.[13] However, there are some reports that showed low level expression of MSR on medial SMCs transformed with SV 40.[14] Therefore, we determined the contribution of MSR to the endocytic uptake of AGE-proteins by SMCs. Acetyl-LDL and oxidized LDL, well-known ligands for MSR, had no effect on the endocytic uptake and degradation of ^{125}I-AGE-BSA by SMCs, whereas unlabeled AGE-BSA had a marked and dose-dependent inhibitory effect on both processes. On the other hand, we did not observe significant endocytic degradation of ^{125}I-acetyl-LDL by these SMCs. These data indicate that, in a sharp contrast to macrophages, binding sites of rabbit SMCs for AGE-proteins are different from those for acetyl-LDL and oxidized LDL, and the endocytic uptake of AGE-proteins by these SMCs might be mediated mainly by a receptor other than MSR. Previous studies have demonstrated the expression of an AGE receptor called RAGE in SMCs.[4] So we examined whether RAGE was involved in the recognition of AGE-BSA by these SMCs. By taking advantage of the recent report that amphoterin, a 30-kDa protein expressed in the developing rat central nerve system, is recognized by RAGE as an effective ligand,[15] we examined the effect of amphoterin on the binding of AGE-BSA to these SMCs. Although the binding of ^{125}I-AGE-BSA to these cells was effectively inhibited by AGE-BSA, amphoterin and acetyl-LDL had no effect on it. Furthermore we examined the effect of FFI-BSA, a ligand for other AGE receptors(p60

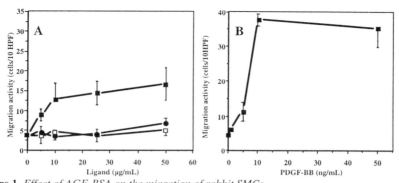

Figure 1. *Effect of AGE-BSA on the migration of rabbit SMCs*
A; The cell suspension (3 x 10^5 cells/mL, 100 mL) was placed in the upper compartment of the Transwell cell-culture chamber, and 0.6 mL of PBS containing various concentrations of AGE-BSA (■), BSA (□) or acetyl-LDL (●) was placed in the lower chamber. The chamber was incubated at 37•C in 5% CO_2 for 14 h. The number of cells per 400 x high-power field (HPF) that had migrated to the lower surface of the filters was counted microscopically. Migration activity is expressed as the mean number of migrated cells seen in ten HPFs. B; PDGF-BB at 2.5-50 ng/mL was used as a positive control. Data represent the mean of three separate experiments. Error bars represent S.D., * $p < 0.05$ comparing AGE-BSA with BSA and acetyl LDL

and p90).[16] However, it had no effect on the binding of ^{125}I-AGE-BSA to these SMCs. These findings indicate that the AGE receptor of rabbit SMCs possesses ligand specificity different from that of MSR, RAGE, p60 or p90, suggesting that it may belong to a unique class of AGE receptor family.

Induction of chemotactic migration of rabbit SMCs by AGE-BSA

To determine a functional aspect of the AGE receptor of SMCs, we examined the migratoryactivity of AGE-proteins for SMCs. AGE-BSA (1 to 50 µg/mL) stimulated a much stronger cell migration activity of SMCs than BSA or acetyl-LDL (Figure 1A). The cell migration activity produced by AGE-BSA was 3 to 4-fold higher than that of the control, whereas 2.5 - 50 ng/mL of PDGF-BB, a well-known chemotactic factor for SMCs, could elicit the cell migration 8 to 9-fold higher than control (Figure 1B). Checkerboard analysis was performed to determine if the migration activity of AGE-BSA was chemotactic or chemokinetic or both. There was no significant AGE-BSA-induced SMC migration unless AGE-BSA was placed in the lower chamber. Thus, it is likely that SMC migration induced by AGE-BSA is mainly due to chemotaxis, rather than to chemokinesis. To determine a factor(s) responsible for the AGE-BSA-induced SMC chemotaxis, we studied the effect of TGF-β, a chemotactic factor for SMCs, on the migration of rabbit SMCs. Incubation with 25 pg/mL of TGF-β significantly induced SMC migration from 8.2 to 17.5 cells/10 HPF. This TGF-β-induced SMC migration was neutralized to 11.1 cells/10 HPF (~ 80%) by an anti-TGF-β antibody. Under identical conditions, without exogenous TGF-β, 50 µg/mL of AGE-BSA was able to induce the SMC migration from 8.2 to 16.5 cells/10 HPF and this AGE-BSA-induced migration was suppressed to 9.5 cells/10 HPF (~ 80%) by the anti-TGF-β antibody. These results suggest that TGF-β is involved in the SMC migration induced by AGE-BSA.

To further elucidate the involvement of TGF-β in AGE-BSA-induced SMC chemotaxis, the TGF-β concentration in the culture medium after incubation with AGE-BSA was determined by TGF-β bioassay. Incubation of SMCs with AGE-BSA resulted in a dose-dependent secretion of total TGF-β (both the latent and active forms) in the culture medium, with a maximal level of 700 pg/mL. In contrast, the corresponding TGF-β level in the medium from cells incubated with control BSA was negligible (<100 pg/mL). It is evident from these results that AGE-BSA is able to induce the secretion of TGF-β from rabbit SMCs, suggesting a possible participation of TGF-β in the AGE-induced SMC chemotaxis.

AGE-BSA does not stimulate the growth of rabbit SMCs

Finally we determined the mitogenic activity of AGE-BSA for SMCs. Exogenous PDGF-BB (10 ng/mL) stimulated significant DNA synthesis for these SMCs, whereas AGE-BSA had no effect up to 100 μg/mL. The parallel experiments by cell-counting assay showed the same result. Thus, it is likely that AGE-proteins stimulates the chemotaxis of these rabbit SMCs, but not their growth.

Conclusions

The present study has clearly shown that cultured rabbit SMCs express the novel AGE receptor that mediates the endocytic uptake of AGE-proteins. Functionally, this receptor may mediate the SMC chemotaxis induced by AGE-proteins. Taken together, these *in vitro* data implicate that extracellular accumulation of AGE in atherosclerotic lesions of arterial walls is involved in SMC migration from the medial layer into the intimal layer and subsequent SMC conversion to foam cells.

References

1. S. Horiuchi, AGE-modified proteins and their potential relevance to atherosclerosis, T*rends in Cardiovascular Medicine,* 1996, **6**, 163-168.
2. H. Vlassara, R. Bucala,and L Striker, Pathogenic effects of advanced glycosylation: biochemical, biologic, and clinical implications for diabetes and aging, *Lab. Invest.,* 1994, **70** 138-151.
3. Kume, M. Takeya, T. Mori, N. Araki, H. Suzuki, S. Horiuchi, T. Kodama, Miyauchi, and K. Takahashi, Immunohistochemical and ultrastructural detection of advanced glycation end products in atherosclerotic lesions of human aorta using a novel specific monoclonal antibody, *Am. J. Pathol.,* 1995, **147**, 654-667.
4. Brett, A.M. Schmidt, S.D. Yan, Y.S. Zou, E. Weidman, D. Pinsky, R. Nowygrod, M. Neeper, C. Przysiecki, A. Shaw, A. Migheli, and D. Stern, Survey of the distribution of a newly characterized receptor for advanced glycation end products in tissues, *Am. J. Pathol.,* 1993, **143**, 1699-1712.
5. Higashi, H. Sano, T. Saishoji, K. Ikeda, Y. Jinnouchi, T. Kanzaki, N. Morisaki, H. Rauvala, M. Shichiri and S. Horiuchi, The receptor for advanced glycation end products mediates the chemotaxis of rabbit smooth muscle cells, *Diabetes,* 1997, **46,** 463-472.

6. Araki, T. Higashi, T. Mori, R. Shibayama, Y. Kawabe, T. Kodama, K. Takahashi, M. Shichiri and S. Horiuchi, Macrophage scavenger receptor mediates the endocytic uptake and degradation of advanced glycation end-products of the Maillard reaction, *Eur. J. Biochem.*, 1995, **230**, 408-415.
7. Parkkinen, E. Raulo, J. Merenmies, R. Nolo, E.O. Kajander, M. Baumann and H. Rauvala, Amphoterin, the 30-kDa protein in a family of HMG1-type polypeptides, *Biol. Chem.*, 1993, **268**, 19726-19738.
8. Fischer-Dzoga, R.M. Jones, D. Vesselinovitch and R.W. Wissler, Ultrastructural and immunohistochemical studies of primary cultures of aortic medial cells, *Exp. Mol. Pathol.*, 1976, **18**, 162-176.
9. Koyama, T. Koshikawa, N. Morisaki, Y. Saito and S. Yoshida, Bifunctional effects of transforming growth factor-β on migration of cultured rat smooth muscle cells, *Biochem. Biophys. Res. Commun.*, 1990, **169**, 725-729.
10. T. Ikeda, M.N. Lioubin and H. Marquardt, Human transforming growth factor type β2: production by a prostatic adenocarcinoma cell line, purification, and initial characterization, *Biochemistry*, 1987, **26**, 2406-2410.
11. Morisaki, M. Kawano, N. Koyama, T. Koshikawa, K. Umemiya, Y. Saito and S. Yoshida, Effect of transforming growth factor-β1 on growth of aortic smooth muscle cells, *Atherosclerosis*, 1991, **88**, 227-234.
12. Takata, S. Horiuchi, N. Araki, M. Shiga, M. Saitoh and Y. Morino, Endocytic uptake of nonenzymatically glycosylated proteins is mediated by a scavenger receptor for aldehyde-modified proteins, *J. Biol. Chem.*, 1988, **263**, 14819-14825.
13. N. Morisaki, K. Yokote, K. Takahashi, M. Otabe, Y. Saito, S. Yoshida and S. Ueda, Role of phospholipase A_2 in expression of the scavenger pathway in cultured aortic smooth muscle cells stimulated with phorbol 12-myristate 13-acetate, *Biochem. J.* 1994, **303**, 247-253.
14. P.E. Bickel and M.W. Freeman, Rabbit aortic smooth muscle cells express inducible macrophage scavenger receptor messenger RNA that is absent from endothelial cells, *J. Clin. Invest.*, 1992, **90**, 1450-1457.
15. O. Hori, J. Brett, T. Slattery, R. Cao, J. Zhang, J.X. Chen, M. Nagashima, E.R. Lundh, S. Vijay, D. Nitecki, J. Morser, D. Stern and A.M. Schmidt, The receptor for advanced glycation end products (RAGE) is a cellular binding site for amphoterin, *J. Biol. Chem.*, 1995, **270**, 25752-25761.
16. H. Vlassara, Y.M. Li, F. Imani, D. Wojciechowicz, Z. Yang, F.T. Liu, A. Cerami, Identification of Galectin-3 as a high-affinity binding protein for advanced glycation end products (AGE): a new member of the AGE-receptor complex, *Molecular Med.* 1995, **1**, 634-646.

Macrophage Scavenger Receptor Mediates The Endocytic Uptake of Advanced Glycation End Products (AGEs)

Seikoh Horiuchi,[1] Takayuki Higashi,[1,2] Hiroyuki Sano,[1] Kenshi Matsumaoto,[1] Ryoji Nagai,[1] Hiroshi Suzuki,[3,4] Tatsuhiko Kodama,[4] Motoaki Shichiri[2]

[1]DEPARTMENTS OF BIOCHEMISTRY AND [2]METABOLIC MEDICINE, KUMAMOTO UNIVERSITY SCHOOL OF MEDICINE, HONJO 2-2-1, KUMAMOTO 865, JAPAN; [3]CHUGAI PHARMACEUTICAL CO. LTD., 1-135 KOMAKADO, SHIZUOKA 412, JAPAN; [4]DEPARTMENT OF MOLECULAR BIOLOGY AND MEDICINE, RESEARCH CENTER FOR ADVANCED SCIENCE AND TECHNOLOGY, THE UNIVERSITY OF TOKYO, 4-6-1 KOMABA, MEGURO, TOKYO 153, JAPAN

Summary

Cellular interactions of AGEs are mediated by AGE receptors. The AGE receptors so far reported are RAGE, galectin-3 and MSR (macrophage scavenger receptor). Macrophages or macrophage-derived cells are known to show the highest endocytic activity for AGE-proteins. Our recent study using CHO (Chinese Hamster Ovary) cells overexpressing MSR clearly showed that the endocytic uptake of AGE-proteins by macrophages is mediated by MSR. To strengthen this contention, the present study was undertaken to examine the interaction of AGE-proteins with peritoneal macrophages from MSR-deficient mice (MSR (-/-)). In experiments at 37°C, thioglycolate-induced peritoneal macrophages from MSR (-/-) showed a marked decrease (more than 80%) in the endocytic degradation capacity for ^{125}I-acetylated low-density lipoprotein (acetyl-LDL). Under parallel conditions, the degradation activity of ^{125}I-AGE-bovine serum albumin (BSA) by these MSR-deficient macrophages was less than 20%. The remaining endocytic capacity of ^{125}I-AGE-BSA by these MSR-deficient macrophages was not inhibited by acetyl-LDL, but was inhibited significantly by AGE-BSA, AGE-hemoglobin or polyanions such as dextran sulfate and polyinosinic acid. These results indicate that ~80 % of the endocytic uptake of AGE-proteins by macrophages is mediated by MSR, while the remaining part is mediated by other AGE receptors.

Introduction

Recent demonstration, using anti-AGE antibodies, of AGEs in several human tissues suggests a functional link of AGE-modification to aging processes[1] and age-enhanced disease processes such as diabetic complications,[2] atherosclerosis,[3,4] hemodialysis-related amyloidosis,[5,6] Alzheimer's disease[7] and photo-enhanced skin lesions of actinic elastosis.[8] One functional consequence of AGEs is explained by the presence of AGE receptors, by which cellular binding of AGE-proteins elicits several responses. These include the induction of cytokines such as TNF and IL-1, cell growth in macrophages, the induction of fibronectin and type IV collagen synthesis from mesangial cells via PDGF synthesis, and enhancement of angiogenesis in endothelial cells.[9] Two different AGE receptors have been characterized. One is from rat liver with Mr=60 kDa and Mr=90 kDa (called p60 and p90); galectin-3 was recently identified as a component of p90[10]. The second receptor has been isolated from bovine lung with Mr=35 kDa (called RAGE[11]). In addition to these receptors, we have been interested in the macrophage scavenger receptor (MSR) because the MSR is known to mediate endocytic uptake of chemically modified low-density lipoprotein (LDL) such as acetylated LDL (acetyl-LDL) and oxidized LDL. In 1995, we succeeded in producing over-expression of MSR in Chinese hamster ovary (CHO) cells (we named them MSR-CHO cells). Our previous studies using MSR-CHO cells showed that endocytic uptake of ^{125}I-AGE-bovine serum albumin (BSA) by these transfected cells was significantly enhanced, in parallel with that of ^{125}I-acetyl-LDL, compared with wild-type CHO cells[12]. Furthermore, the endocytic uptake of ^{125}I-acetyl-LDL by these cells was competitively inhibited by AGE-BSA and the endocytic uptake of ^{125}I-AGE-BSA was significantly inhibited by acetyl-LDL. These results strongly suggest that endocytic uptake of AGE-proteins by macrophages is mediated mainly by the MSR. The present study was undertaken to examine the interaction of AGE-proteins with peritoneal macrophages obtained from MSR-deficient mice.

Materials and Methods

Chemicals
BSA and fetal calf serum (FCS) were purchased from Sigma. Na^{125}I was purchased from Amersham. Tissue culture medium was from Gibco Laboratories. Other chemicals were of the best grade available from commercial sources.

Ligand preparation
LDL (d=1.019-1.063 g/mL) was isolated by sequential ultracentrifugation of fresh plasma from normolipidemic subjects after overnight fasting, and dialyzed against 0.15 M NaCl and 1 mM EDTA (pH 7.4).[13] Acetyl-LDL was prepared by chemical modification of LDL with acetic anhydride as described previously.[13] Oxidized LDL was prepared by incubation of LDL with 5 µM CuSO$_4$ for 20 h at 37°C as described previously.[13] AGE-BSA and AGE-human hemoglobin were prepared as described previously.[14] Briefly, 2.0 g of BSA or 0.5 g of human hemoglobin was dissolved in 10 mL of 0.5 M sodium phosphate buffer (pH 7.4) with 3.0 g of D-glucose. Each sample was sterilized by ultrafiltration, incubated at 37°C for 40 weeks, and dialysed against 20 mM sodium phosphate buffer (pH 7.4). The resulting AGE-proteins showed typical absorption and fluorescent spectra patterns that were indistinguishable from those reported previously,[15] as well as a significant immunoreactivity to the anti-AGE antibody.[16]

Cellular assays
MSR-knockout mice (heterozygous MSR (-/+) and homozygous (MSR (-/-)) were prepared as described previously.[17] Peritoneal macrophages were collected from the mice 3 days after intraperitoneal injection of thioglycolate. Peritoneal macrophages were also collected from wild-type littermates (MSR (+/+)) and used as a control. All assays were performed at 37°C in a humidified atmosphere of 5% CO$_2$ in air. For uptake studies, 5 x 10^5 cells were seeded in each well of a 12-well culture plate (22.1 mm diameter, Corning) in 1.0 mL of DMEM containing 10% FCS, and cultured for 3 h. Cells were washed once with 1.0 mL of PBS and the medium was replaced with 1.0 mL of pre-warmed DMEM containing 3% BSA and then used for the following experiments. The cells in each well were incubated for 18 h with 5 µg/mL of ^{125}I-AGE-BSA, ^{125}I-acetyl-LDL or ^{125}I-ox-LDL with or without an excess of unlabeled ligand. The cells were centrifuged and 0.75 mL of the culture medium was taken from each well and mixed with 0.3 mL of 40% trichloroacetic acid (TCA) in a vortex mixer. To this solution was added 0.2 mL of 0.7 M AgNO$_3$, followed by

centrifugation. The resulting supernatant (0.5 mL) was used to determine TCA-soluble radioactivity, which was taken as an index of cellular degradation. The remaining cells were washed 3 times with PBS (pH 7.4) containing 1% BSA, and another 3 times with PBS. The cells were lysed at 37°C for 30 min with 1.0 mL of 0.1 N NaOH and the cell-associated radioactivity and cellular proteins were determined by BCA protein assay reagent.

Results and Discussion

We compared the endocytic capacity of acetyl-LDL and oxidized LDL by macrophages derived from MSR (-/-), MSR (-/+) and MSR (+/+) mice. The specific cell-association of ^{125}I-acetyl-LDL with, and its specific degradation by MSR (-/-) macrophages was less than 10% of the values for MSR (+/+) macrophages. The endocytic capacity of MSR (-/+) macrophages was reduced to about a half. The specific cell-association of ^{125}I-oxidized LDL with, and its specific degradation by MSR (-/-) macrophages was reduced to 20% and to a negligible level, respectively. The endocytic degradation of ^{125}I-AGE-BSA by macrophages from MSR (-/-) mice was determined under identical conditions. As shown in Figure 1, the specific degradation of ^{125}I-AGE-BSA by MSR (+/+) macrophages exhibited a dose-dependent saturation pattern. Parallel experiments in MSR (-/-) macrophages showed reduced endocytic degradation of AGE-BSA (less than 20% of the value for MSR (+/+) macrophages), indicating that ~80% of the endocytic uptake of AGE-BSA by macrophages is mediated by the MSR and the remaining portion is mediated by other pathway(s).

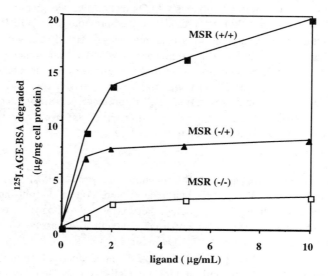

Figure 1. Endocytic degradation of ^{125}I-AGE-BSA by peritoneal macrophages from MSR (-/-), MSR (-/+) and MSR (+/+) mice. Peritoneal macrophages were collected from these mice 3 days after intraperitoneal injection of thioglycolate. Cells (5 x 10^5) were incubated for 18 h with selected concentrations of ^{125}I-AGE-BSA, as shown, with or without excess unlabeled AGE-BSA. TCA-soluble radioactivity in the medium was

measured as described in 'Materials and Methods'. The level of specific degradation was obtained by subtracting the nonspecific degradation from the total degradation.

To further characterize the recognition site for AGE-BSA on MSR (-/-) macrophages, competitive experiments were performed (Figure 2). Degradation of ^{125}I-AGE-BSA was inhibited by more than 80% using both unlabeled AGE-BSA and AGE-hemoglobin in MSR (-/-) macrophages, suggesting that the receptor recognizes some common AGE-structure(s), as reported previously for rat liver sinusoidal cells and peritoneal macrophages.[14] The effect of unlabeled dextran sulfate (DS) on this system was also significant but polycytidylic acid had no effect. As expected from the above data, the effect of unlabeled acetyl-LDL was prominent in MSR (+/+) macrophages, whereas it was not prominent in MSR (-/-) macrophages. These results strongly suggest that AGE-BSA is actively taken up by MSR (-/-) macrophages through the non-MSR pathway, which has the polyanion sensitivity.

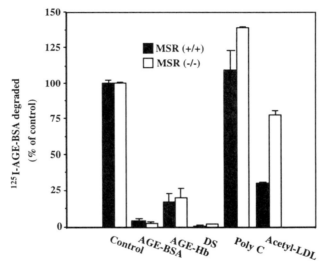

Figure 2. *Effects of several ligands on the degradation of ^{125}I-AGE-BSA by macrophages from MSR (-/-) mice and from MSR (+/+) mice. Peritoneal macrophages were collected from these mice 3 days after intraperitoneal injection of thioglycolate and incubated for 18 h with 5 µg/mL of ^{125}I-AGE-BSA with or without a 50-fold excess of the unlabeled ligands to be tested (AGE-BSA, AGE-hemoglobin, dextran sulfate, polycytidylic acid (Poly C) and acetyl-LDL)*

Conclusions

Two conclusions may be drawn from the present study. The first is that the MSR plays a major role in endocytic uptake of AGE-proteins by macrophages or macrophage-derived cells (~80% of their total endocytic uptake). The second is that the remaining portion of endocytic uptake is mediated by an AGE receptor other than MSR (non-MSR receptor). It is unknown whether this non-MSR receptor is identical to known AGE receptors such as RAGE or Galectin-3 or whether an unknown receptor(s) is involved. Experiments of endocytic uptake of AGE-proteins by MSR (-/-) macrophages show that the non-MSR receptor is able to recognize the common AGE-structure(s) as a ligand and is sensitive t polyanions, being inhibited by dextran sulfate not by polycytidylic acid.

Shaw and Crabbe[18] emphasized the non-specific binding of AGE to macrophages Their data were obtained from binding experiments with cell suspension systems, wherea our results were obtained with cells that adhered tightly to the culture plate at 37°C. In fac a large proportion of ^{125}I-AGE-BSA bound to these cells in suspension corresponded t non-specific binding from our experience. However, the cellular association an degradation of ^{125}I-AGE-BSA in adherent cell culture system showed a specific associatio and degradation with a typical dose dependent saturation curve. Therefore the marke difference of the present results from those reported by Shaw and Crabbe might lie in th experimental systems.

References

1. Araki, N. Ueno, B. Chakrabarti, Y. Morino and S. Horiuchi, Immunochemical evidence for the presence of advanced glycation end products in human lens proteins and its positive correlation with aging, *J. Biol. Chem.*, 1992, **267**, 10211-10214.
2. Makino, K. Shikata, K. Hironaka, M. Kushiro, Y. Yamasaki., H. Sugimoto, Z. Ota, N. Araki and S. Horiuchi, Ultrastructure of nonenzymatically glycated mesangial matrix in diabetic nephropathy, *Kidney Internat.*, 1995, **48**, 517-526.
3. Kume, M. Takeya, T. Mori, N. Araki, H. Suzuki, S. Horiuchi, T. Kodama, Miyauch and K. Takahashi, Immunohistochemical and ultrastructural detection of advance glycation end products in atherosclerotic lesions of human aorta using a novel specifi monoclonal antibody, *Am. J. Pathol,* 1995, **147**, 654-667.
4. Horiuchi, AGE-modified proteins and their potential relevance to atherosclerosis, *Trends Cardiovascular Med.*, 1996, **6**, 163-168.
5. Miyata, O. Oda, R. Inagi, Y. Iida, N. Araki, N. Yamada, S. Horiuchi, Taniguchi, K. Maeda and T. Kinoshita, β_2-Microglobulin modified with advanced glycation end products is a major component of hemodialysis-associated amyloidosis, *J. Clin. Invest.*, 1993, **92**, 1243-1252.
6. Miyata, S. Taneda, R. Kawai, K. Otani, S. Horiuchi, M. Hara, K. Maeda and V. M. Monnier, Identification of pentosidine as a native structure for advanced glycation end products in β_2-microglobulin forming amyloid fibrils in patients with dialysis-related amyloidosis, *Proc. Natl. Acad. Sci. USA*, 1996, **93**, 2353-2358.
7. Kimura, J. Takamatsu, K. Ikeda, A. Kondo, T. Miyakawa and S. Horiuchi, Accumulation of advanced glycation end products of the Maillard reaction with age in human hippocampal neurons, *Neurosci. Lett,*, 1996, **208**, 53-56.
8. Mizutari, T. Ono, K. Ikeda, K. Kayashima and S. Horiuchi, Photo-enhanced modification of human skin elastin in actinic elastosis by N^ε-(carboxymethyl)lysine, one of the glycoxidation products of the Maillard Reaction, *J. Invest. Dermatol.*, 1997, **108**, 797-802.
9. Vlassara, R. Bucala and L. Striker, Pathogenic effects of advanced glycosylation biochemical, biologic, and clinical implications for diabetes and aging, *Lab. Invest.*, 1994 **70**, 138-151.

0. Vlassara, Y.M. Li, F. Imani, D. Wojciechowicz, Z. Yang, F.T. Liu and A. Cerami Identification of Galectin-3 as a high-affinity binding protein for advanced glycation end products (AGE): a new member of the AGE-receptor complex, *Molecular Med.*, 1995, **1**, 634-646.
1. Schmidt, M. Vianna, M. Gerlach, J. Brett, J. Ryan, J. Kao, C. Esposito, Hegarty, W. Hurley, M. Clauss, F. Wang, Y. E. Pan, T. C. Tsang and D. Stern, Isolation and characterization of two binding proteins for advanced glycosylation end products from bovine lung which are present on the endothelial cell surface, *J. Biol. Chem.*, 1992, **267**, 14987-14997.
2. Araki, T. Higashi, T. Mori, R. Shibayama, Y. Kawabe, T. Kodama, K. Takahashi, M. Shichiri and S. Horiuchi, Macrophage scavenger receptor mediates the endocytic uptake and degradation of advanced glycation end-products of the Maillard reaction, *Eur. J. Biochem.*, 1995, **230**, 408-415.
3. Sakai, A. Miyazaki, H. Hakamata, T. Sasaki, T. Yui, M. Yamazaki, M. Shichiri and S. Horiuchi, Lysophosphatidylcholine plays an essential role in the mitogenic effect of oxidized low density lipoprotein on murine macrophage, *J. Biol. Chem.*, 1994, **269**, 31430-31435.
4. K. Takata, S. Horiuchi, N. Araki, M. Shiga, M. Saitoh and Y. Morino, Endocytic uptake of nonenzymatically glycosylated proteins is mediated by a scavenger receptor for aldehyde-modified proteins, *J. Biol. Chem.*, 1988, **263**, 14819-14825.
5. Pongor, P. C. Ulrich, F. A. Bencsath and A. Cerami, Aging of proteins: Isolation and identification of a fluorescent chromophore from the reaction of polypeptides with glucose, *Proc. Natl. Acad. Sci. USA*, 1984, **81**, 2684-2688.
6. Horiuchi, N. Araki and Y. Morino, Immunochemical approach to characterize advanced glycation end products of Maillard reaction; evidence for the presence of a common structure, *J. Biol. Chem.*, 1991, **266**, 7329-7332.
7. H. Suzuki, Y. Kurihara, M. Takeya, N. Kamada, M. Kataoka, K. Jishage, O. Ueda, Sakaguchi, T. Higashi, T. Suzuki, Y. Takashima, Y. Kawabe, O. Cynshi, Y. Wada, M. Honda, H. Kurihara, T. Doi, A. Matsumoto, S. Azuma, T. Noda, Y. Toyoda, H. Itakura, Y. Yazaki, S. Horiuchi, K. Takahashi, T. J. van Berkel, P. C. Steinbrecher, S. Ishibashi, N. Maeda, S. Gordon and T. Kodama, A role for macrophage scavenger receptors in atherosclerosis and susceptibility to infection, *Nature*, 1997, **386**, 292-296.
8. S. M. Shaw, and J. C. Crabbe, Non-specific binding of advanced-glycosylation end-products to macrophages outweighs receptor-mediated interactions, *Biochem. J.*, 1994, **304**, 121-129.

Insulin Accelerates the Endocytic Uptake and Degradation of Advanced Glycation End-Products Mediated by The Macrophage Scavenger Receptor

Hiroyuki Sano,[1] Takayuki Higashi,[1] Yoshiteru Jinnouchi,[1] Ryoji Nagai,[1] Kenshi Matsumoto,[1] Zhu Wen Qin,[1] Kazuyoshi Ikeda,[1] Yousuke Ebina,[2] Hideichi Makino[3] and Seikoh Horiuchi[1]

[1]DEPARTMENT OF BIOCHEMISTRY, KUMAMOTO UNIVERSITY SCHOOL OF MEDICINE, JAPAN; [2]DEPARTMENT OF ENZYME GENETICS, INSTITUTE FOR ENZYME RESEARCH, THE UNIVERSITY OF TOKUSHIMA, JAPAN; [3]DEPARTMENT OF LABORATORY MEDICINE, EHIME UNIVERSITY SCHOOL OF MEDICINE, JAPAN

Summary

The macrophage scavenger receptor (MSR), one of the receptors for advanced glycation end-products (AGEs), mediates endocytic uptake and degradation of AGE-proteins in several cell types. In the present study, we examined whether MSR function was regulated by insulin signaling. Co-expression of human insulin receptor (IR) with MSR in Chinese hamster ovary (CHO) cells showed that insulin accelerated the degradation of AGE-proteins to 160% of the control. The insulin-enhanced endocytic uptake of AGE-proteins was significantly inhibited by phosphatidylinositol-3-OH kinase (PI(3)K) inhibitors, wortmannin and LY294002. Thus, insulin signaling through the PI(3)K pathway may regulate MSR-mediated endocytic uptake of AGE-proteins.

Introduction

Hyperglycemia accelerates the formation and accumulation of AGEs in plasma and tissue, which might be implicated in diabetic complications or vascular dysfunctions.[1,2] However, with intensive insulin therapy, the elevated plasma levels of early-stage products, such as hemoglobin A_{1c}, as well as that of AGE-proteins are reduced, which also reduces the accumulation of AGEs in tissues. The restoration of glycated proteins to normal levels may delay the onset and slow the progression of diabetic complications.[3] Insulin treatment may be expected to reduce the accumulation of AGEs in tissues by a reduction of AGE formation following a lowering plasma glucose levels, and by enhancing the elimination processes of AGEs from plasma or tissues.

AGE-proteins are known to interact with several types of cells through the AGE-binding proteins, or the AGE-receptors. For instance, AGE-proteins undergo receptor-mediated endocytosis by macrophages or macrophage-derived cells. Several AGE-receptors have been identified, which include MSR,[4,5] RAGE[6] and the receptor complex of AGE-R1, R2 and R3 (R1: p60/OST-48, R2: p90/80K-H and R3: p90/galectin-3).[7,8] Gene-targeting strategies have provided useful information to identify the receptor responsible for the endocytic uptake of the ligands. The endocytic capacity for AGE-bovine serum albumin (BSA) of peritoneal macrophages obtained from MSR knockout mice is reduced to 20 – 30% of the wild-type littermate.[9] Thus, endocytic uptake and subsequent degradation of AGE-proteins by peritoneal macrophages is mediated to a large extent by the MSR. Based on this notion, the present study was undertaken to examine the effect of insulin signaling on MSR-mediated endocytic uptake of AGE-proteins.

Materials and Methods

The MSR-transfected cells (CHO-SRII cells) were isolated according to the method of Freeman et al.[10] Insulin receptor (IR) amino acids were numbered according to the system of Ebina et al.[11] CHO-SRII cells were transfected with IR expression vectors (SRαIR, SRαIR^{1030M}, SRαIR^{972F}) and clones expressing wild- or mutant-type IR were named CHO-SRII-IR, CHO-SRII-IR^{1030M}, and CHO-SRII-IR^{972F}, respectively. CHO-IR cells were also obtained by transfecting SRαIR to parent CHO cells. IR^{1030M} is a kinase-deficient mutant, whereas IR^{972F} is lacking binding site for insulin receptor substrate-1 and -2 (IRS-1 and IRS-2). Clones which expressed ~10^5 wild- or mutant-type IRs per cell were selected. CHO-SRII-IR, CHO-SRII, CHO-IR, and parent CHO cells were cultured on type IV collagen-coated 6-well plates. After serum starvation for 5 h, the cells were incubated for 60 min with 2 μg/mL ^{125}I-AGE-BSA and the indicated concentration of human insulin in KRH buffer (136 mM NaCl, 4.7 mM KCl, 1.25 mM MgSO$_4$, 1.25 mM CaCl$_2$, 20 mM Hepes, 1 mg/mL glucose, and 30 mg/mL BSA). The endocytic activity for the ligands was measured as previously described.[12] We examined the effect of the pp70-S6 kinase (pp70^{S6K}) in addition to PI(3)K, since PI(3)K binds directly to IRS-1 and IRS-2 and pp70^{S6K} is located downstream of PI(3)K. Subsequently CHO-SRII-IR cells were pretreated for 30 min either with 10 nM wortmannin, 100 μM LY294002 (PI(3)K inhibitors), 20 ng/mL rapamycin (pp70^{S6K} inhibitor), or dimethyl sulfoxide (vehicle), followed by 60 min incubation with 2 μg/mL ^{125}I-AGE-BSA in the presence or absence of 10 nM insulin. The endocytic activity for AGE-BSA was determined as the same way described above.

Results and Discussion

To assess the role of insulin in AGE-receptor function, we examined co-expression of MSR and insulin receptor (IR). Co-expression of MSR with IR in CHO cells (CHO-SRII-IR cells) showed an insulin-dependent increase in endocytic capacity for AGE-BSA (Figure 1). Since neither CHO-SRII cells nor CHO-IR cells showed a response to insulin (Figure 1), the insulin sensitivity to endocytic capacity in CHO-SRII-IR cells was probably due to co-expression of MSR with IR. On the other hand, the binding of ^{125}I-AGE-BSA to CHO-SRII-IR cells did not change following incubation with insulin (data not shown), indicating that insulin-induced endocytic uptake of the ligands is not due to the increased number of cell-surface MSR, but probably to post-binding events.

To confirm that IR signaling was necessary for the activation of MSR function, we used CHO-SRII cells stably transfected with mutant IRs (IR^{1030M}, IR^{972F}) containing point mutations at sites known to bind specific signaling molecules, including Lys 1030 (ATP) or Tyr 972 (IRS-1 and IRS-2). Expression of IR^{1030M} or IR^{972F} in CHO-SRII cells did not result in a significant increase in endocytic capacity for AGE-BSA (data not shown). Thus, these results suggest that IR kinase activity, especially the IRS-1 or IRS-2 pathways, plays a

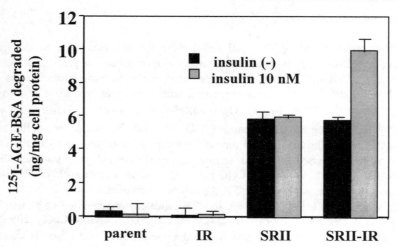

Figure 1. Endocytic degradation of ^{125}I-AGE-BSA by parent CHO (parent), CHO-IR (IR), CHO-SRII (SRII) and CHO-SRII-IR (SRII-IR) cells. Cells were incubated for 60 min with 2 µg/mL of ^{125}I-AGEs-BSA in the absence or presence of 10 nM insulin

key role in the insulin-induced endocytic uptake and degradation of AGE-proteins mediated by MSR.

Insulin fails to activate PI(3)K in CHO cells overexpressed with IR^{1030M} or IR^{972F}, although PI(3)K is activated in response to insulin in CHO-IR cells.[13,14] In this study, we used wortmannin and LY294002, two potent PI(3)K inhibitors. In CHO-SRII-IR cells insulin increased the endocytic degradation of ^{125}I-AGE-BSA to ~160 % of basal level. Both wortmannin and LY294002 inhibited the insulin-induced endocytic uptake of AGE-BSA by CHO-SRII-IR cells (Figure 2).

In common with wortmannin and LY294002, rapamycin blocks the activation of pp70^{S6K} but does not inhibit PI(3)K.[15] Rapamycin did not influence insulin-induced MSR function for endocytic uptake in CHO-SRII-IR cells (Figure 2). Thus, while PI(3)K is necessary for the activation of pp70^{S6K} and the MSR-mediated endocytic uptake of AGE-BSA, it is likely that the pathways leading to such events branch at some point downstream of PI(3)K. These results suggest that PI(3)K activity initiated by IR is coupled to MSR-mediated endocytic uptake of AGE-proteins. However, we cannot exclude a possible role of other signals resulting from the IR acting in concert with PI(3)K to induce the MSR activity. Thus, insulin action may accelerates endocytosis of the ligands, followed by vesicle transport, lysosomal degradation and exocytosis via the classical coated pit/endosomal pathway.

We recently demonstrated that AGEs in plasma were rapidly and effectively eliminated by liver sinusoidal Kupffer and endothelial cells.[16] Since these sinusoidal cells are known to express the MSR and other scavenger receptors, the MSR is thought to play an important role in clearance of plasma AGEs. The present findings provide a novel insight into the action of insulin on the MSR-mediated endocytic uptake of AGE-proteins, suggesting a potential link of the insulin signal to the AGE-elimination system *in vivo*.

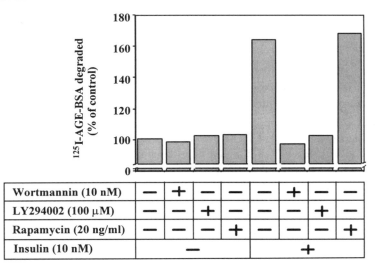

Figure 2. *Effect of insulin, wortmannin, LY294002 and rapamycin on endocytic uptake of ^{125}I-AGE-BSA by CHO-SRII-IR cells*

ACKNOWLEDGMENTS
We are grateful to the following colleagues for helpful discussions: Drs Fumio Shimada, Naokake Hashimoto, Yoshifumi Suzuki at the Department of Internal Medicine, Chiba University School of Medicine, Dr Fumihiko Kanai at the Department of Internal Medicine, Tokyo University School of Medicine, and Dr. Eiichi Araki at the Department of Metabolic Medicine, Kumamoto University School of Medicine. We are also grateful to Dr Tatsuhiko Kodama at the Department of Molecular Biology and Medicine, Tokyo University and Dr Takafumi Doi at the Faculty of Pharmaceutical Science, Osaka University, for providing the MSR expression vector.

References

1. Horiuchi, S., Advanced glycation end products (AGE)-modified proteins and their potential relevance to atherosclerosis, *Trends Cardiovasc. Med.*, 1996, **6**, 163-168.
2. Higashi, T., Sano, H., Saishoji, T., Ikeda, K., Jinnouchi, Y., Kanzaki, T., Morisaki, N., Rauvala, H., Shichiri, M. and Horiuchi, S., The receptor for advanced glycation end products mediates the chemotaxis of rabbit smooth muscle cells, *Diabetes*, 1997, **46**, 463-472.
3. The diabetes control and complications trial research group (DCCT), The effect of intensive treatment of diabetes on the development and progression of long-term complications in insulin-dependent diabetes mellitus, *New Engl. J. Med.*, 1993, **932**, 976-986.
4. Kodama, T., Freeman, M., Rohrer, L., Zabrecky, J., Matsudaira, P. and Krieger, M., Type I macrophage scavenger receptor contains a-helical and collagen-like coiled coils, *Nature*, 1990, **343**, 531-535.
5. Araki, N., Higashi, T., Mori, T., Shibayama, R., Kawabe, Y., Kodama, T., Takahashi, K., Shichiri, M. and Horiuchi, S., Macrophage scavenger receptor mediates the endocytic uptake and degradation of advanced glycation end products of the Maillard reaction, *Eur. J. Biochem.*, 1995, **230**, 408-415.

6. Neeper, M., Schmidt, A. M., Brett, J., Yan, S. D., Wang, F., Pan, Y. C. E., Elliston, K., Stern, D. and Shaw, A., Cloning and expression of a cell surface receptor for advanced glycation end products of proteins, *J. Biol. Chem.*, 1992, **267**, 14998-15004.
7. Li, Y. M., Mitsuhashi, T., Wojciechowicz, D., Shimizu, N., Li, J., Stitt, A., He, C., Banerjee, D. and Vlassara, H., Molecular identity and cellular distribution of advanced glycation endproduct receptors: Relationship of p60 to OST-48 and p90 to 80K-H membrane proteins, *Proc. Natl. Acad. Sci. USA,* 1996, **93**, 11047-11052.
8. Vlassara, H., Li, Y. M., Imani, F., Wojciechowics, D., Yang, Z., Liu, F. T. and Cerami, A, Identification of galectin-3 as a high-affinity binding protein for advanced glycation end products (AGE): a new member of the AGE-receptor complex, *Molec. Med.*, 1995, **1**, 634-646.
9. Suzuki, H., Kurihara, Y., Takeya, M., Kamada, N., Kataoka, M., Jishage, K., Ueda, O., Sakaguchi, H., Higashi, T., Suzuki, T., Takashima, Y., Kawabe, Y., Cynshi, O., Wada, Y., Honda., M., Kurihara, H., Aburatani, H., Doi, T., Matsumoto, A., Azuma, S., Noda, T., Toyoda, Y., Itakura, H., Yazaki, Y., Horiuchi, S., Takahashi, K., Kruijt, J. K., van Berkel, T. J. C., Steinbrecher, U. P., Ishibashi, S., Maeda, N., Gordon, S. and Kodama, T., A role for macrophage scavenger receptors in atherosclerosis and susceptibility to infection, *Nature*, 1997, **386**, 292-296.
10. Freeman, M., Ekkel, Y., Rohrer, L., Penman, M., Freeman, N. J., Chisolm, G. M. and Krieger, M., Expression of type I and type II bovine scavenger receptors in Chinese hamster ovary cells: lipid droplet accumulation and nonreciprocal cross competition by acetylated and oxidized low density lipoprotein, *Proc. Natl. Acad. Sci. USA*, 1991, **88**, 4931-4935.
11. Ebina, Y., Ellis, L., Jarnagin, K., Edery, M., Graf, L., Clauser, E., Ou, J.-h., Masiarz, F., Kan, Y. W., Goldfine, I. D., Roth, R. A. and Rutter, W.J., The human insulin receptor cDNA: the structural basis for hormone-activated transmembrane signalling, *Cell*, 1985, **40**, 747-758.
12. Takata, K., Horiuchi, S., Araki, N., Shiga, M., Saitoh, M. and Morino, Y., Endocytic uptake of nonenzymatically glycosylated proteins is mediated by a scavenger receptor for aldehyde-modified proteins, *J. Biol. Chem.*, 1988, **263**, 14819-4825.
13. Ruderman, N. B., Kapeller, R., White, M. F. and Cantley, L. C., Activation of phosphatidylinositol 3-kinase by insulin, *Proc. Natl. Acad. Sci. USA*, 1990, **87**, 1411-1415.
14. Kanai, F., Ito, K., Todaka, M., Hayashi, H., Kamohara, S., Ishii, K., Okada, T., Hazeki, O., Ui, M. and Ebina, Y., Insulin-stimulated GLUT4 translocation is relevant to the phosphorylation of IRS-1 and the activity of PI3-kinase, *Biochem. Biophys. Res. Commun.*, 1993, **195**, 762-768.
15. Chung, J., Grammer, T. C., Lemon, K. P., Kazlauskas, A. and Blenis, J., PDGF- and insulin-dependent pp70^{S6K} activation mediated by phosphatidylinositol-3-OH kinase, *Nature,* 1994, **370**, 71-75.
16. Smedsrød, B., Melkko, J., Araki, N., Sano, H. and Horiuchi, S., Advanced glycation end products are eliminated by scavenger-receptor mediated endocytosis in hepatic sinusoidal Kupffer and endothelial cells, *Biochem. J.*, 1997, **322**, 567-673.

Endocytic uptake of AGE-modified low-density lipoprotein by macrophages leads to cholesteryl ester accumulation *in vitro*

Yoshiteru Jinnouchi,[1,2] Takayuki Higashi,[1] Kazuyoshi Ikeda,[1,2] Hiroyuki Sano,[1] Ryoji Nagai,[1] Hideki Hakamata,[1] Masaki Yoshida,[2] Shoichi Ueda[2] and Seikoh Horiuchi[1]

[1]DEPARTMENT OF BIOCHEMISTRY AND [2]DEPARTMENT OF UROLOGY, KUMAMOTO UNIVERSITY SCHOOL OF MEDICINE, HONJO 2-2-1, KUMAMOTO 860, JAPAN

Summary
Recent studies disclosed that proteins modified by advanced glycation endproducts (AGE) are taken up by macrophages or macrophage-derived cells by the macrophage scavenger receptor (MSR) *in vitro* and that AGE-proteins are accumulated in foam cells in the human atherosclerotic lesions *in vivo*, suggesting a possibility that AGE-modified low density lipoprotein (AGE-LDL) *in situ* might be involved in atherogenic processes *in vivo*. As a first step, AGE-LDL was prepared by incubating LDL with glycolaldehyde (GA), a highly reactive intermediate of the Maillard reaction. GA-modified LDL (GA-LDL) is characterized by an increase in negative charge, fluorescent intensity and reactivities to anti-AGE antibodies, suggesting that these physicochemical and immunochemical properties of GA-LDL were highly similar to those of AGE-proteins. Furthermore, studies of cellular interaction of GA-LDL with mouse peritoneal macrophages showed that GA-LDL is recognized and endocytosed, followed by intralysosomal degradation by these cells. Endocytic uptake of GA-LDL by these cells was competitively inhibited by acetylated LDL (acetyl-LDL), a representative ligand for MSR. Endocytic degradation of ^{125}I-acetyl-LDL was competed for by GA-LDL. Furthermore, incubation of GA-LDL effectively converted them into foam cells. Similar results were obtained from CHO cells overexpressing MSR. These results suggest that LDL modified by AGE *in situ* is taken up by macrophages mainly via MSR and contributed to foam cell formation in the early atherosclerotic lesions.

Introduction

The incidence of atherosclerotic vascular disease is 3-4 times higher in diabetic patients than in normal subjects. Four mechanisms for pathogenesis of diabetic vascular complications postulated so far include the polyol pathway, oxidative stress, non-enzymatic glycation, and protein kinase C. Among them, hyperglycemia increases the levels of the glycated plasma proteins and enhances subsequent formation of advanced glycation endproducts (AGE) *in vivo*.

Recent studies have emphasized that chemical modification of low density lipoproteins (LDL) *in situ* is a key step for foam cell formation in the early atherosclerotic lesions of arterial walls. As a cause for the higher incidence of atherosclerosis in diabetic patients, it was reported that glycation of plasma LDL results in a marked reduction of its ligand activity for the LDL receptor, which leads to the delay of LDL clearance from circulation and an increase in plasma LDL cholesterol levels, thus enhancing the atherogenic processes in an indirect way.

Recent immunohistochemical studies demonstrated AGE-accumulation in macrophage-derived foam cells in early atherosclerotic lesions.[1] Furthermore, it has recently become clear that AGE-modified proteins are taken up by macrophages or macrophage-derived cells by the macrophage scavenger receptor (MSR).[2,3] These findings suggest another interesting possibility that AGE-modified LDL (AGE-LDL) *in situ* play a direct role in atherogenic processes *in vivo*. As a first step to test this notion, the present study was undertaken to prepare AGE-LDL by incubating LDL with glycolaldehyde (GA), a highly reactive intermediate of the Maillard reaction, and compared its physicochemical as well as biological properties with other AGE-proteins prepared by incubation with glucose.

Materials and Methods

Preparation of lipoproteins

LDL (d=1.019-1.063 g/mL) was isolated by sequential ultracentrifugation from normolipidemic human plasma and dialyzed at 4°C against 0.15 M NaCl and 1 mM EDTA, pH 7.4. To prepare GA-LDL, 2 mg/ml of LDL was incubated with 33 mM glycolaldehyde (GA) for various intervals up to 3 days. Incubation was terminated by dialyzing at 4°C against PBS containing 1 mM EDTA. Acetyl-LDL was prepared by chemical modification of LDL with acetic anhydride.[4]

Physicochemical and immunochemical characterization of GA-LDL

Agarose gel electrophoresis was performed using a Universal Gel/8 electrophoresis kit. SDS-PAGE was done on a 2-15% gradient slab gel system. Lysine modification of LDL was measured by TNBS methods.[5] Amounts of AGE-formation were measured by ELISA using anti-AGE antibodies.[6]

Cellular interaction of GA-LDL

For binding assay, mouse peritoneal macrophages (2×10^6 cells) were incubated at 4°C for 90 min with 0-40 μg/mL of ^{125}I-GA-LDL in the presence or absence of excess unlabeled GA-LDL. The cells were washed and the cell-bound radioactivity was counted. For endocytic uptake and degradation assay, macrophages were incubated at 37°C for 5 h with 0-20 μg/mL of ^{125}I-GA-LDL in the presence or absence of excess unlabeled GA-LDL or ligands to be tested. To determine intracellular cholesteryl esters (CE)-accumulation, macrophages were incubated at 37°C for 18 h with 10 μg/mL of LDL, GA-LDL, acetyl-LDL and oxidized-LDL in the presence of 0.1 mM [^{3}H]oleate conjugated with BSA. The lipids of these cells were extracted and a spot of CE was separated by TLC and measured for the radioactivity. Cellular experiments with CHO-SR cells were also performed in the same way.[1]

Results and Discussion

Physicochemical Properties of GA-LDL

Human LDL was incubated with 33 mM glycolaldehyde (GA) at 37°C for indicated intervals up to day 3. The extents of lysine modification increased to 33% by 4-h incubation and 68.4% by 12-h incubation, reaching a plateau level for 2 days. The electrophoretic mobility was similarly increased with the incubation time and the electrophoretic mobility of GA-LDL obtained by 24-h incubation was almost similar to those of acetyl-LDL and oxidized LDL. GA-LDL showed an excitation maximum at 325 nm and an emission maximum at 400 nm, a fluorescent pattern indistinguishable from those reported with proteins modified by glycolaldehyde.[7,8] These results suggest that physicochemical properties of GA-LDL are closely similar to those of AGE-proteins modified with glucose.

We next determined the reactivity of these GA-LDL preparations to anti-AGE antibodies. AGE-structures so far reported include fluorescent and cross-linked structures such as pentosidine[9] and crosslines[10] and nonfluorescent and non-crosslinked structures such as N^ε-carboxymethyllysine (CML)[11] and pyrraline.[12] Two antibodies used in the present study are the antibody specific for CML (the CML-specific antibody) and the antibody recognizing a structure(s) other than CML, pyrraline, pentosidine and crosslines (non-CML specific antibody). ELISA using the CML specific antibody could detect CML-adducts in these GA-LDL preparations and an increase in their contents with the time of incubation, indicating the presence of CML in these GA-LDL preparations as one of AGE-structures. Moreover, the non-CML specific antibody also showed the positive reaction to these GA-LDL preparations, indicating that in addition to CML, a non-CML structure(s) of AGE was also formed during incubation of LDL with glycolaldehyde. These results indicate that GA-LDL shares several AGE-specific structures in common with AGE-proteins.

Biological Properties of GA-LDL

AGE-proteins are known to be recognized and endocytosed vividly by macrophages and macrophage-derived cells. We examined whether GA-LDL could bind to mouse peritoneal macrophages through the specific binding sites. The total binding of ^{125}I-GA-LDL to these cells at 4°C increased dose-dependently, exhibiting a saturation kinetics. The total binding of ^{125}I-GA-LDL was competed for by >80% by an excess unlabeled GA-LDL. The specific binding also gave a saturation pattern. Scatchard analysis of the specific binding disclosed a high-affinity binding site on these cell surfaces with a Kd value of 2.1 µg/mL and the maximum cell binding of 268 ng/mg cell protein, suggesting the presence of a GA-LDL binding protein(s) on these macrophages.

Experiments at 37°C demonstrated that amounts of ^{125}I-GA-LDL degraded by these cells increased in a dose-dependent manner and effectively inhibited by an excess amount of the same unlabeled ligand by more than 90%. A ligand concentration required for half saturation was calculated as 3-4 µg/mL of GA-LDL, indicating that GA-LDL underwent a receptor-mediated endocytosis as was the case with acetyl-LDL and oxidized LDL.

We next examined whether endocytic degradation of GA-LDL was mediated by the macrophage scavenger receptor (MSR). The endocytic degradation of ^{125}I-GA-LDL (2 µg/mL) was inhibited dose-dependently by unlabeled acetyl-LDL (up to 200 µg/mL) in a manner closely similar or identical to GA-LDL, whereas LDL had no effect. Furthermore, the endocytic degradation of ^{125}I-acetyl-LDL by these cells was effectively competed for either by unlabeled GA-LDL or acetyl-LDL. This cross-competitive effect strongly suggested a major role of MSR in the endocytic degradation of GA-LDL by these macrophages. This is consistent with our recent observations obtained both from CHO cells overexpressing MSR[2] and from peritoneal macrophages prepared from MSR-knockout mice.[3]

Acetyl-LDL and oxidized LDL are known to induce intracellular accumulation of cholesteryl esters (CE), converting macrophages into foam cells in vitro.[13,14] We examined the CE-accumulation capacity of GA-LDL in mouse peritoneal macrophages by incorporation of [^{3}H]oleate into cholesteryl esters at 37°C for 18 h. As shown in Figure 1, GA-LDL obtained from 24-h incubation showed a marked CE-accumulation capacity. Interestingly, the CE-accumulation capacity of GA-LDL was 1.5-fold and 2.8-fold higher than that of acetyl-LDL and oxidized LDL, respectively.

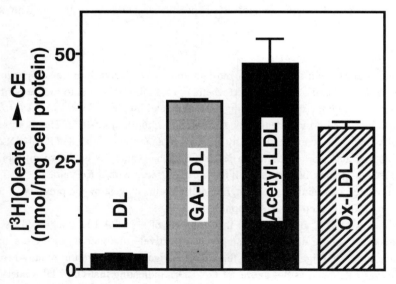

Figure 1 Effects of GA-LDL on the intracellular of cholesteryl esters in mouse peritoneal macrophages. Macrophages (2 x 10^6) were incubated at 37°C for 18 h with 10 μg/ml of LDL, GA-LDL, acetyl-LDL and Ox-LDL in the presence of 0.1 mM [^{3}H]oleate. Cellular lipids were extracted and radioactivity of cholesteryl [^{3}H]oleate. LDL (■), GA-LDL(□), acetyl-LDL (■), oxidized LDL (▨).

Finally, the binding experiments of ^{125}I-GA-LDL to CHO-SR cells overexpressing type II MSR were performed. Whereas the binding of ^{125}I-GA-LDL to wild-type CHO cells was negligible or extremely weak, its binding to CHO-SR cells was significant and increased dose-dependently and at least 5-fold higher than wild-type CHO cells. These results taken together indicate that GA-LDL is actively endocytosed by macrophages via a route identical to MSR, leading to foam cell formation.

Conclusions

In the present study, GA-LDL was prepared as a model of AGE-LDL. The physicochemical properties such as an increase in negative charge and the fluorescent intensity as well as the immunochemical reactivity to anti-AGE antibodies were closely similar between GA-LDL and AGE-proteins prepared by long-term incubation with glucose. In addition, GA-LDL was recognized and endocytosed by macrophages, followed by intralysosomal degradation which was competed for by acetyl-LDL, indicating that GA-LDL undergoes a receptor-mediated endocytosis. Interaction of GA-LDL with macrophages induced intracellular CE-accumulation (foam cells). These biological properties of GA-LDL were also highly similar to those of acetyl-LDL, indicating that endocytic uptake of GA-LDL is mediated mainly by MSR. These *in vitro* data, when combined with the recent *in vivo* data that AGE-modified proteins occur in human atherosclerotic lesions,[1] likely suggest that LDL modified by AGE *in situ* is taken up by macrophages mainly via MSR and contributed to foam cell formation in the early atherosclerotic lesions.

References

1. Kume, S., Takeya, M., Mori, T., Araki, N., Suzuki, H., Horiuchi, S., Kodama, T., Miyauchi, Y., Takahashi, K. (1995) Immunohistochemical and ultrastructural detection of advanced glycation end products in atherosclerotic lesions human aorta with a novel specific monoclonal antibody. *Am. J. Pathol.* 147:654-667.
2. Araki, N., Higashi, T., Mori, T., Shibayama, R., Kawabe, Y., Kodama,T.,Takahashi, K., Shichiri , M., Seikoh, H. (1995) Macrophage scavenger receptor mediates the endocytic uptake and degradation of advanced glycation end products of Maillard reaction. *Eur. J. Biochem.* 230:408-415.
3. Suzuki, H., Kurihara, Y., Takeya, M., Kamada, N., Kataoka, M., Jisage, K., Ueda, O., Sakaguchi, H., Higashi, T., Suzuki, T., Takashima, Y., Kawabe, Y., Cynshi, O., Wada, Y., Honda, M., Kurihara, H., Aburatani, H., Doi, T., Matsumoto, A., Azuma, S., Noda, T., Toyoda, Y., Itakura, H., Kruijt, J. K., van Berkel, T. J. C., Steinbrecher, U. P., Ishibashi, S., Maeda, N., Gordon, S., Kodama, T. (1997) A role for macrophage scavenger receptors in atherosclerosis and susceptibility to infection. *Nature* 386:292-296.
4. Miyazaki, A., Sakuma, S., Morikawa, W., Takiue, T., Miake, F., Terano, T., Sakai, M.,

Hakamata, H., Sakamoto, Y., Naitoh, M., Ruan, Y., Takahashi, K., Hpriuchi, S. (1995) Intravenous injection of rabbit apolipoprotein A-I inhibits the progression of atherosclerosis in cholesterol-fed rabbits. *Artherioscle.r Thoromb. Vasc. Biol.* 15;1882-1988.
5. Habeeb, A. F. S. A. (1966) Determination of free amino groups in proteins by trinitrobenzenesulfonic acid. *Anal. Biochem.* 14:328-336.
6. Ikeda, K., Higashi, T., Sano, H., Jinnouchi, Y., Yoshida, M., Araki, T., Ueda, T., Horiuchi, S. (1996) N$^\varepsilon$-(carboxymethyl)lysine protein adducts is a major immunological epitope in proteins modified with advanced glycation end products of the Maillard reaction. (1996) *Biochemistry* 35:8075-8083.
7. Acharya, A. S., Manning, J. M. (1983) Reaction of glycolaldehyde with proteins: Latentcrosslinking potential of α-hydroxyaldehyde. *Proc. Natl. Acad. Sci. USA* 80:3590-3594.
8. Fong, Y., Edelstein, D., Wang, E. A., Brownlee, M. (1993) Inhibition of matrix-induced bone differentiation by advanced glycation end-products in rats. *Diabetologia* 36:802-807.
9. Sell, D.R., Monnier, V.M. (1989) Structure elucidation of senescence cross-link from human extracellular matrix. *J. Biol. Chem.* 264:21597-21602.
10. Ienaga, K., Nakamura, K., Hochi, T., Nakazawa, Y., Fukunaga, Y., Kakita, H., Nakano, K. (1995) Crosslines, fluorophoes in the AGE-related cross-linked proteins. *Contrib. Nephrol.* 112:42-51.
11. Ahmed, M. U., Thorpe, S. R., Bynes, J. W. (1986) Identification of N$^\varepsilon$-carboxymethyllysine as a degradation product of fructoselysine in glycated proteins. *J. Biol. Chem.* 261:4889-4894.
12. Hayase, F., Nagaraj, R., Miyata, S., Njoroge, F. G., Monnier, V.M. (1989) Aging of proteins: immunological detection of a glucose derived pyrrole formed during Maillard reaction in vivo. *J. Biol. Chem.* 263:3758-3764.
13. Goldstein, J. L., Ho, Y. K., Basu, S. K., Brown, M.S. (1979) Binding site on macrophages that mediates uptake and degradation of acetylated low density lipoprotein, producing massive cholesterol deposition. *Proc. Natl. Acad. Sci. USA* 76:333-337.
14. Henriksen, T., Mahoney, E. M., Steinberg, D. (1981) Enhanced macrophage degradation of low density lipoprotein previously incubated with cultured endothelial cells: recognition by receptor for acetylated low density lipoprotein. *Proc. Natl. Acad. Sci. USA* 78:6499-6503.

Poster Abstracts

Volatiles in Hydrolysed Vegetable Protein

Margit Dall Aaslyng and J Stephen Elmore*; *The Royal Agricultural and Veterinary University of Copenhagen, Chemical Department, Thorvaldsensvej 40, DK-1871 Frederiksberg C, Denmark, and *The University of Reading, Department of Food Science and Technology, Whiteknights, Reading RG6 6AP, UK*

Hydrolysed Vegetable Protein is a savory flavouring product widely used all over the world. It is traditionally produced by acid hydrolysis using HCl and temperatures between 100 and 140°C. Hydrolysis using proteolytic enzymes is an interesting alternative to acid hydrolysis.

Acid hydrolysis was performed using defatted untoasted soygrits and 4 M HCl. The solution was heated to 110°C for 6 hours, followed by neutralization to pH 6.5 with 4 M NaOH. After centrifugation and filtration, the hydrolysate was freeze dried. The product (HVP) was dark brown with a strong meaty savoury flavour.

Enzymic hydrolysis was performed using defatted untoasted soygrits and the protease complexes Alcalase™ and Flavourzyme™ from Novo Nordisk A/S. Before hydrolysis, the soygrits solution (8% protein) was pasteurized at 85°C. The hydrolysis was performed at 50°C for 5 hours at pH 7 followed by 19 hours at pH 5. The enzymes were deactivated at 85°C for 5 minutes and the pH was adjusted to pH 6.5 with 4 M NaOH. After centrifugation and filtration, the hydrolysate was freeze dried. The product (EVP) was yellow-brown with a meaty savoury flavour of much lower intensity than HVP.

A great part of the difference between the two kinds of hydrolysate is probably due to different Maillard reactions having taken place. To characterize the volatiles of the two products, they were analysed by a headspace procedure, followed by GC-MS analysis. This showed that ketones and aldehydes were present in both types of product. Sulfides, furfural, furanones and other furans dominated in HVP, whereas pyrazines and oxazoles dominated in EVP.

The results can be explained by the different reaction conditions, HVP production being characterized by low pH and high temperature, favouring the furfurals and compounds from reactions of furfurals, whereas EVP is characterized by a more neutral pH and a lower temperature, favouring pyrazines and other reaction products derived via the Strecker degradation.

Role of the Maillard Reaction Products in Affecting the Overall Antioxidant Properties of Processed Foods

Monica Anese, Lara Manzocco*, M Cristina Nicoli* and Carlo R Lerici*; *The University of Bari, Istituto di Produzioni e Preparazioni Alimentari,Via Napoli 25, 71100 Foggia, Italy, *The University of Udine, Dipartimento di Scienze degli Alimenti,Via Marangoni 97, 33100 Udine, Italy*

Although the antioxidant properties of the Maillard reaction products (MRPs) have been known for almost thirty years, only in the last decade has considerable attention been paid to the influence of these compounds on food quality, stability and health. Experimental evidence, mostly based on food model systems, demonstrated that certain MRPs can exhibit strong antioxidant activity, functioning mainly as chain breakers and oxygen scavengers. Currently, heat treatments, during which the development of the Maillard reaction is desired, are generally believed to be responsible for the decrease in the overall antioxidant capacity of foods, due to the thermal degradation of the naturally occurring compounds. This may not always be the case, since it is known that, during the development of the Maillard reaction, formation of compounds with new antioxidant properties can occur.[1] Thus, depending on the extent of the applied thermal treatment, the overall antioxidant potential of food products is expected to be maintained or even enhanced as a consequence of the formation of the MRPs.[2]

In this paper, the contribution of the MRPs to the overall antioxidant properties of some thermally processed foods is considered. Reaction conditions which promote the formation of MRPs with antioxidant capacity and their mechanisms of action are discussed.

[1] Nicoli, M.C.; Anese, M.; Parpinel, M.; Franceschi, S.; Lerici, C.R. (1997). *Cancer Letters*, 114, 1-4.
[2] Nicoli, M.C.; Anese, M.; Manzocco, L.; Lerici, C.R. (1997). *Lebensm.-Wiss. u. -Technol.*, 30, 292-297.

Separation of Coloured Components of *Kecap Manis* (an Indonesian Soy Sauce) by HPLC and Capillary Electrophoresis

Anton Apriyantono, Santi Marianti, Louise Royle*, Richard G Bailey* and Jennifer M Ames*; *Bogor Agricultural University, Department of Food Technology and Human Nutrition, Kampus IPB Darmaga, PO Box 220, Bogor 16002, Indonesia, *The University of Reading, Department of Food Science and Technology, Whiteknights, Reading RG6 6AP, UK*

Kecap manis is a typical Indonesian soy sauce and is prepared from black soyabeans, which are subjected to mould and brine fermentation to give *moromi*. The moromi, coconut sugar and spices are cooked for 1-2 h to give *kecap manis*. Coconut sugar represents 45-50 % of the total raw materials before cooking.

Fractions of the coloured components of a commercial *kecap manis* were prepared by extracting with solvents possessing a range of polarity. Based on absorbance of the concentrated extracts diluted in a standard volume of methanol, the amount of coloured material extracted from the *kecap manis* ranged from 0.003% for light petroleum (b.pt. 60-80°C) to 0.179% for acetonitrile, suggesting that the majority of the coloured components of *kecap manis* are polar and, therefore, not extractable.

The ethyl acetate and acetonitrile extracts, as well as the *kecap manis* were analysed by reversed-phase HPLC and by capillary electrophoresis (CE) with a borate buffer, both with diode array detection. Many of the main components of the solvent extracts could not be detected in the original *kecap manis* and are probably only minor contributors to the colour of the soy sauce. CE separated far more components of *kecap manis* than HPLC, but the resolved peaks were observed superimposed on a broad peak or 'hump'. Ultrafiltration, using a membrane with a cut-off of 1000 daltons, gave a fraction which on analysis by CE showed no hump, indicating that the material responsible must possess relatively high molecular mass.

The behaviour of *kecap manis* on HPLC and CE exhibited many similarities with those of an aqueous xylose-glycine model system heated under reflux for 2 h and analysed using exactly the same conditions.[1] This suggests some similarities between the components of *kecap manis* and those of the model Maillard system.

[1] Royle, L.; Bailey, R.G.; Ames, J.M. (1998). *Food Chem.*, in press.

Determination of Dicarbonyl Compounds as Aminotriazines during the Maillard Reaction and *in Vivo* Detection in Aminoguanidine-Treated Rats

Atsushi Araki, Marcus A Glomb, Motoko Takahashi and Vincent M Monnier; *Institute of Pathology, Case Western Reserve University, 2085 Adelbert Road, Cleveland, Ohio 44106, USA*

Aminoguanidine has been proposed as a drug for prophylaxis of diabetic complications in animal experiments. However, the exact mechanism by which aminoguanidine retards or prevents diabetic complications in experimental animals remains unknown. The postulated action of aminoguanidine is to trap dicarbonyl compounds which are formed during the Maillard reaction and which may have deleterious effects on cells or tissues. Dicarbonyl compounds may react with aminoguanidine to form 3-amino-1,2,4-triazines. Therefore, we determined 3-amino-1,2,4-triazines (ATZs) formed in the presence of aminoguanidine at various steps along the Maillard reaction. ATZs were identified and measured using a GC-MS method after extraction and derivatization with a TMS reagent.

1. As a model of glucose autoxidation, glucose (100 mM) was incubated with 2 μM Cu^{2+} in 100 mM sodium phosphate buffer (pH 7.4) at 37°C in the presence of 5 mM aminoguanidine for 8 days. During glucose autoxidation, glucosone-ATZ and glyoxal-ATZ were detected.
2. When glucose (100 mM) was incubated with BSA or lysine in PBS (pH 7.4) at 37°C in the presence of 7 mM aminoguanidine, 3-deoxyglucosone (3-DG)-ATZ and methylglyoxal-ATZ were major ATZs. Glyoxal-ATZ and glucosone-ATZ were also detected.
3. When glycated BSA was incubated with 7 mM aminoguanidine, 1-DG-ATZ was a major ATZ and 3-DG-ATZ, glucosone-ATZ, methylglyoxal-ATZ and glyoxal-ATZ were also detected.
4. In plasma from diabetic rats treated with aminoguanidine (1 g/l drinking water) for several days, 3-DG-ATZ and methylglyoxal-ATZ were about ten times higher than in plasma from non-diabetic rats.

Both in vitro and in vivo, 3-DG-ATZ and methylglyoxal-ATZ were major products formed by the reaction of dicarbonyl compounds with aminoguanidine. The formation of the ATZs in vivo may well contribute to the preventive action of aminoguanidine in diabetic complications.

Glucosone and Glyoxal are Major Dicarbonyl Compounds Formed during Glucose Autoxidation: Role of Hydrogen Peroxide and Hydroxyl Radicals for Dicarbonyl Formation

Atsushi Araki*†, Jaffar Nourooz-Zadeh†, Marcus A Glomb†, Vincent M Monnier† and Simon P Wolff*; *Division of Clinical Pharmacology and Toxicology, Department of Medicine, University College London Medical School, 5 University Street, London WC1E 6JJ, UK, †Institute of Pathology, Case Western Reserve University, 2085 Adelbert Road, Cleveland, Ohio 44106, USA

It has been postulated that glucose autoxidation, a transition metal-catalysed process, generates hydrogen peroxide, hydroxyl radicals and the dicarbonyl compound, glucosone, and that hydroxyl radicals can attack glucose to form dicarbonyl compounds, such as glyoxal, as degradation products of glucose. To clarify the mechanism of dicarbonyl compound formation during glucose autoxidation, glucose (100 mM) was incubated with 2 µM Cu^{2+} in 50 mM sodium phosphate buffer (pH 7.4) at 37°C for up to 3 weeks and dicarbonyl compounds were measured using the Girard T reagent.

Dicarbonyl-compound formation was dependent on glucose, copper and phosphate concentrations in the buffer and increased with time. Hydroxyl radical scavengers (formate, dimethyl sulfoxide, thiourea), a metal-chelating agent, DTPA, and catalase inhibited the formation of dicarbonyl compounds, the net production of hydrogen peroxide as well as benzoate hydroxylation, an indicator of hydroxyl radical formation. These results support the theory that dicarbonyl-compound formation is due to the reaction of hydroxyl radicals derived from hydrogen peroxide in a transition metal-catalysed process with glucose.

In addition, when dicarbonyl compounds were measured over the incubation time specifically as 3-amino-1,2,4-triazines after derivatizing with aminoguanidine using a GC-MS method, the production rate of glucosone and glyoxal was approximately 10 µM and 1 µM per day in the presence of 100 mM glucose and 2 µM Cu^{2+}. The formation of both glucosone and glyoxal in the presence of copper was inhibited by the hydroxyl radical inhibitors by 35% and 55%, respectively. These results suggest that glucosone is formed mainly through enediol oxidation during glucose autoxidation, while glyoxal is produced mainly by the reaction of hydroxyl radicals with glucose.

Commitment of the European Commission to Finance: "Optimisation of the Maillard Reaction - A Way to Improve Quality and Safety of Thermally Processed Foods"

Anna Arnoldi (co-ordinator); Dipartimento di Scienze Molecolari Agroalimentari, Università di Milano, Via Celoria 2, 20133 Milano, Italy

In September 1995, a proposal with the title given above was submitted and the Commission selected it as a three-year shared cost research project, starting in August 1996 (FAIR CT96-1080).

With the general target of improving the competitive position of the food industry throughout Europe, the proposed research aims to understand the factors and inter-relationships between factors that affect flavour, colour and nutritional quality and toxicological safety of all foods that undergo the Maillard reaction (MR) during thermal processing and to obtain better control of the MR through process optimisation.

Control of the reaction may be interpreted in two ways. First, it can mean preventing the reaction, as far as possible, by inhibiting it or by minimising it. Second, control may involve the optimisation of the reaction to achieve optimum flavour and colour development, coupled with reduced levels of toxic compounds and maximised nutritional value. The ability to control the reaction, simultaneously with regard to flavour, colour, nutrition and toxicology, represents an important challenge for the food technologist.

In consequence, the objectives of the project are (a) to obtain greater understanding of (i) the complex chemical processes known as the MR, (ii) how the controlling factors influence the reaction, such information greatly aiding the production of foods with improved safety and sensory quality, and (iii) the kinetics of the processes involved, and (b) to apply the data obtained to important foods, such as milk, milk products, meat and fish, and to processed flavourings.

The project gathers together 13 research groups located in 8 states: Italy (A. Arnoldi, University of Milan, and G. Randazzo, Univ. Naples), Germany (H. Erbersdobler, Univ. Kiel, and L. Kroh and R. Tressl, Tech. Univ. Berlin), Iceland (H. Einarsson, Icelandic Fisheries Laboratories, Reykjavik), Spain (S. Jiménez-Peréz, Istituto del Frio, Madrid), Sweden (M. Jägerstad, Univ. Lund, and M. Johansson, Swedish Meat Research Institute, Kavlinge), The Netherlands (M. van Boekel, Univ. Wageningen, and H. Weenen, Quest International, Bussum), United Kingdom (J. Ames, Univ. Reading, and B. Wedzicha, Univ. Leeds).

Involvement of Free Radicals in Pyrazine Formation in the Maillard Reaction

Anna Arnoldi, Alessandra D'Agostina and Monica Negroni; *Dipartimento di Scienze Molecolari Agroalimentari, Università di Milano, via Celoria 2, 20133 Milano, Italy*

The flavour of processed foods depends on the heat treatments necessary for their preparation. The reaction between amino acids and sugars produces many volatile heterocyclic compounds, whose structures, odour thresholds and concentrations determine the aroma of a particular food. Generally, foods also contain fats which can in part be oxidised to aldehydes and ketones by a radical chain mechanism during storage. These lipid degradation products (aldehydes and ketones) can influence the flavour of food by their presence and through interaction with the Maillard reaction products.

Of the few studies that have focused on the interaction of lipids with the Maillard reaction, most relate to the formation of meat flavour. The subject was extensively reviewed by Whitfield,[1] and recent studies on low-moisture systems have indicated that the effects of phospholipids is increased in the presence of water.[2]

In order to achieve better insight into the effect of vegetable oils on the Maillard reaction, we studied sugar-amino acid model systems with added lipid (corn oil or extra virgin olive oil) and compared them with systems containing methyl stearate or lacking any lipophilic phase. Subsequently, we added to the model systems free radical initiators or scavengers and observed that they influenced the formation of some Maillard reaction products,[3] especially in the control model systems without lipid. For comparison, results were obtained for the same additives in aqueous reaction model systems.

The influence of pH on the Maillard reaction has been widely recognised. To be sure that the effects observed were due to free-radical modulators and not to slight pH differences or the action of buffers, the pH was monitored and kept constant by the addition of base. The model systems investigated were glucose-lysine, fructose-lysine and xylose-lysine. The compounds most sensitive to the modulators were pyrazines.

The work is supported by the European Community, FAIR CT96-1080.

[1] Whitfield, F.B. (1992). *Crit. Rev. Food Sci. Nutr.*, **31**, 1-58.
[2] Mottram, D.S.; Whitfield, F.B. (1995). *J. Agric. Food Chem.*, **43**, 984-988.
[3] Arnoldi, A.; Corain E. (1996). In *Flavour Science: Recent Developments*, Taylor, A.J., Mottram, D.S., Eds, Royal Society of Chemistry: Cambridge, pp. 217-220.

New Coloured Compounds from Model Systems: Xylose-Lysine

Anna Arnoldi, Leonardo Scaglioni and Emanuela Corain; *Dipartimento di Scienze Molecolari Agroalimentari, Università di Milano, Via Celoria 2, 20133 Milano, Italy*

Colour formation is one of the most important effects of the Maillard reaction. Most of the coloured compounds formed are melanoidins, i.e., high molecular mass polymers. Owing to their irregular structures, they are particularly difficult to study and little information is available regarding them. More work has been done on the characterisation of low molecular mass coloured compounds, that could represent some of the substructures incorporated into melanoidins. They were isolated only from model systems, often from simple amine-sugar mixtures. Among amino acids, only glycine and lysine have been investigated. Few of the compounds isolated were completely characterised: in general they have two or three rings. These rings often contain oxygen or nitrogen and are frequently linked by a -CH= group.[1]

Two coloured compounds were separated from a xylose-lysine model system by a complex procedure involving ethyl acetate extraction and TLC and HPLC separation. Compound 1 had already been separated by us some years ago.[2] At that time it was possible to report only limited information regarding its structure. The isolation of a larger sample permitted the collection of MS spectra which gave a relative molecular mass of 482-484, while ^{1}H and ^{13}C NMR gave additional structural information, including evidence of the presence of two furan rings. Although it was impossible to elucidate the structure completely, it is worth stressing that this is the low molecular mass coloured Maillard compound with the highest molecular mass recorded to date.

However, MS and two-dimensional NMR allowed complete elucidation of the structure of Compound 2, the first coloured compound containing three rings to be identified in model xylose-lysine systems.

[1] Ames, J. M.; Nursten, H. E. (1989). In *Trends in Food Science*, Lien W.S., Foo, C.W., Eds, Singapore Institute of Food Science and Technology: Singapore, pp. 8-14.
[2] Ames, J. M.; Apriyantono, A.; Arnoldi, A. (1993). *Food Chem.*, **46**, 121-127.

Nonenzymic Browning Reaction Products Present in Pekmez (Concentrated Grape Syrup)

Neriman Bağdatlioğlu and Yaşar Hişil*; *Celal Bayar University, Faculty of Engineering, Food Engineering Department, Manisa, Turkey, *Ege University, Faculty of Engineering, Food Engineering Department, Izmir, Turkey*

Turkey is the country with the fifth highest grape production in the world. Grapes are processed by drying and by making vinegar, wine and pekmez, in addition to being consumed fresh. Pekmez is made by concentration of grape juice after treatment with $CaCO_3$ to reduce the acidity. Traditionally, grapes were preserved by making concentrated grape syrup for hundreds of years.

The nonenzymic browning reaction provides characteristic taste and odour during concen-tration of grape syrup to obtain pekmez by heat processing and the reaction products cause the colour to darken and the aroma to become more burnt with increasing time and temperature. The major pathways for nonenzymic flavour and aroma formation in syrups are caramelisation and the Maillard reactions.[1-3]

Nonenzymic browning reaction products were investigated by GC-MS. Pekmez samples were treated with pentane:diethyl ether (1:2) in a continuous liquid-liquid extractor. Typical heat-accelerated reaction products of reducing sugars were found, especially dihydro-2-methyl-3(2H)-furanone, furfural, 2-furanmethanol and 2-acetylfuran.

[1] Buera, M.P.; Chirife, J.; Resnik, S.L.; Wetzler,G. (1987). *J. Food Sci.*, **52**, 1063-1067.
[2] Baiser, W.M.; Labuza, T.P. (1992). *J. Agric. Food Chem.*, **40**, 707-713.
[3] Kroh, L.W. (1994). *Food Chem.*, **51**, 373-379.

Effective Inhibition of Free Radical Formation in the Maillard Reaction by Flavonoid Anion-Radicals

Jasna Čanadanović-Brunet, Bozidar Lj Milić and Sonja M Djilas; *University of Novi Sad, Faculty of Technology, Organic Chemistry Department, Bulevar Cara Lazara 1, YU-21000 Novi Sad, Yugoslavia*

In the past few years, considerable attention has been paid to flavonoids, which are present in substantial amounts (0.5-1.5%) in plants, in relation to human nutrition and the quality of food products.[1,2]

The influence of some flavonoids (myricetin, fisetin, rutin, morin, naringenin) on formation of the 1,4-pyrazine cation free radicals in the Maillard reaction between D-(+)-glucose and 3-aminobutanoic acid was studied by Electron Spin Resonance (ESR) spectroscopy.

Under typical reaction conditions (98°C, pH 9.0), flavonoid compounds formed anion radicals, which react directly with the free radicals formed in the Maillard reaction between D-(+)-glucose and 3-aminobutanoic acid (both 1.0 M).

From relative intensity measurements and shape analysis of the ESR spectra, it can be concluded, that the flavonoids which possess a 3',4'-*ortho*-dihydroxy grouping in Ring B effectively inhibit 1,4-pyrazine cation free radicals during Maillard reactions.

[1] Jovanovic, S.V.; Steenken, S.; Tosic, M.; Marjanovic, B.; Simic, M.G. (1994). *J. Amer. Chem. Soc.*, **116**, 4846-4851.
[2] Djilas, S.M.; Milic, B.Lj. (1994). In *Maillard Reactions in Chemistry, Food, and Health*, Labuza, T.P., Reineccius, G.A., Monnier, V.M., O'Brien, J., Baynes, J.W., Eds, Cambridge: Royal Society of Chemistry, pp. 75-81.

Flavour and Oil-Stability Enhancement by Monitoring Maillard-Reaction Precursors in Peanut Kernels and Roasting Environments

Robin Y Y Chiou; *National Chiayi Polytechnic Institute, Department of Food Science and Technology, Chiayi, Taiwan, Republic of China*

A pleasant peanutty flavour, formed mainly through the Maillard reaction, is efficiently produced by roasting peanut kernels at an elevated temperature in a relatively short period of time. When peanut kernels were roasted under various conditions, the best flavour was obtained by roasting under N_2 or CO_2. When raw peanut oil was combined with partially defatted peanut meal at a ratio of 10:1 (w/w) and roasted at 160°C for 30 min in a closed chamber under various atmospheric conditions, the addition of 10-20% moisture to the peanut meal was essential for the formation of peanutty flavour.[2] A unique, pleasant roasted peanut flavour was achieved when roasting was carried out under of CO_2.

Concerning the oxidative stability of the oil, significant antioxidative activity was observed in oil prepared from peanuts subjected to pretreatments consisting of rehydration, blanching and dehydration, followed by roasting at 160°C for 90 min.[1] During roasting, more sucrose, total amino acids and free amino acids were degraded in treated compared to untreated peanuts. At each roasting time, iron, copper and free fattty acid contents in the oils and lipoxygenase activity in the defatted flours prepared from untreated peanuts were higher than in the oils and flours prepared from pretreated peanuts.[3]

In more recent studies, when peanut kernels were subjected to germination, the free amino-acid composition changed significantly as functions of germination time and the nature of the individual amino acids. These changes in relation to flavour contribution by the typical and atypical flavour precursors when the kernels were subjected to roasting will be addressed and discussed.

[1] Chiou, R. Y.-Y. (1992). *J. Agric. Food Chem.*, **40**, 1958-1962.
[2] Chiou, R. Y.-Y.; Cheng, S.-L.; Tseng, C.-Y.; Lin, T.-C. (1993). *J. Agric. Food Chem.*, **41**, 1110-1113.
[3] Chen, M.-J.; Chiou, R. Y.-Y. (1995). *Int. J. Food Sci. Nutr.*, **46**, 145-148.

Effects of Advanced Glycation Endproducts on Vascular Permeability Increase as Induced by Histamine

C M S Conde, F Z G A Cyrino, E Michoud*, D Ruggiero-Lopez*, N Wiernsperger* and E Svensjö; *Laboratório de Pesquisas em Microcirculação, Universidade do Estado do Rio de Janeiro, Rio de Janeiro, Brasil, *LIPHA-INSERM U 352, INSA-Lyon, Building 406, 69621 Villeurbanne Cedex, France*

To test the hypothesis that advanced glycation endproducts (AGEs) may enhance vascular permeability of large molecules in diabetic subjects, AGEs from albumin (AGE-Alb) and control albumin (Alb) were prepared, labelled with fluorescein-isothiocyanate (FITC) and injected intravenously into anaesthetized hamsters at a dose of 4-7 mg/100 g body weight. Two groups of normoglycaemic hamsters were studied; the control group (n = 5) was given FITC-Alb and the experimental group (n = 5) FITC-AGE-Alb. Vascular permeability changes were measured by direct intravital microscopy of hamster cheek pouch preparations in fluorescent light. Increased permeability was recorded as number of sites (leaks) with extravasation of FITC-labelled Alb in postcapillary venules.

No changes were seen during 1 hour after intravenous injection of either FITC-Alb or FITC-AGE-Alb. Local application of 5×10^{-6} M histamine to the cheek pouch for 5 minutes induced reversible increases in vascular permeability in both Alb- and AGE-Alb-injected hamsters. The maximal number of leaks/cm² was 175 ± 5 (SEM) and 295 ± 16 in the Alb- and AGE-Alb-treated groups, respectively ($p < 0.001$). In two streptozotocin-diabetic hamsters (duration: 3 months; glycaemia: 15.6 and 19.9 mM, respectively), the histamine response was 169 leaks/cm² with Alb and 265 leaks/cm² with AGE-Alb, suggesting that the AGE may also augment histamine-induced permeability in hamsters with diabetic microangiopathy to the same degree.

In conclusion, AGEs injected into anaesthetized hamsters augmented the increase in vascular permeability due to histamine by 70%.

Maillard Products: Interference in Carbonyl Assay of Oxidized Protein

Peter J Coussons, Meena Bagga, Stephanie Mulligan and James V Hunt*; *Department of Pathology, University of Cambridge, Tennis Court Road, Cambridge,CB2 1QP, UK, *Molecular Toxicology Group, Glenfield Hospital, Groby Road, Leicester, LE3 9QR, UK*

The carbonyl assay[1,2] measures the production of available ketone and aldehyde groups on proteins. Protein oxidation is believed to be associated with the aetiology of many disease states including Alzheimer's disease and atherosclerosis,[3,4] and carbonyl assay has been used to detect the products of oxidative damage associated with these diseases. However, carbonyl content may also be increased by the non-enzymic attachment of glucose to proteins (termed 'glycation'), which eventually leads to the formation of Maillard products.[3,4] This alternative route to increased carbonyl content of tissues is also associated with many oxidative disease states[3,4] and thus it is important to understand the specificity of assays used to measure protein oxidation and protein glycation.

Here we use regression analysis on measurements of fluorescence,[5] fragmentation[6] and carbonyl content[1,2] of glycated protein to determine which parameters correlate best with protein oxidation or protein glycation. Following glycation of albumin, we find that carbonyl assay is unable to distinguish between glycation and protein oxidation *per se*. Comparing parameters of protein glycation and protein oxidation, the best correlation was between glycation-mediated fluorescence changes and carbonyl content (r values ranged between 0.8 to 0.9). A reduced correlation was observed between fragmentation (diagnostic of protein oxidation) and carbonyl content (r was between 0.5 to 0.8) and no correlation was shown between glycation-mediated fluorescence and fragmentation. Thus, care must be taken when interpreting the carbonyl assay as diagnostic of protein oxidation, particularly when there is the possibility of glycation chemistry affecting biological samples

[1] Levine, R.L.; Williams, J.A.; Stadtman, E.R.; Shacter, E. (1994). *Methods in Enzymology*, **233**, 346-354.
[2] Reznik, A.Z.; Packer, L. (1994). *Methods in Enzymology*, **233**, 357-365.
[3] Brownlee, M.; Vlassara, H.; Cerami, A. (1987). In *Diabetic Complications, Scientific and Clinical Aspects*, Crabbe,M.J.C., Ed., Churchill Livingstone: Edinburgh, pp. 94-139.
[4] Hunt, J.V. (1994). In *Glucose Chemistry and Atherosclerosis in Diabetes Mellitus, Free Radicals in the Environment, Medicine and Toxicology*, Nohl, H., Esterbauer H., Rice-Evans, C., Eds, pp. 137-162.
[5] Hunt, J.V.; Skamarauskas, J.T.; Mitchinson, M. (1994). *Atherosclerosis*, **111**, 255-265.
[6] Skamarauskas, J.T; McKay, A.G.; Hunt J.V. (1996). *Free Radical Biology and Medicine*, **21**, 801-813.

Development of Enzyme-Linked Immunosorbent Assays to Measure Advanced Glycation Endproducts in Human Serum

Catherine A Dorrian, Sylvia Cathcart, Jes Clausen* and Marek H Dominiczak; *Department of Biochemistry, Gartnavel General Hospital, Glasgow G12 0YN, UK, *Novo Nordisk, Novo Alle, DK-2880 Bagsvaerd, Denmark*

Advanced glycation endproducts (AGEs) are formed *in vivo* and increases in their rate of formation and/or accumulation are associated with various pathophysiological conditions. Whilst a number of structures has been described for AGEs (e.g., pentosidine, CML and FFI), it still remains uncertain what the structures of those AGEs found *in vivo* are. Therefore it is important to develop well characterised immunoassays for these compounds using a variety of polyclonal and monoclonal antibodies to permit the measurement and characterisation of the circulating forms.

We have been developing enzyme-linked immunosorbent assays (ELISAs) for AGEs in human serum using polyclonal and monoclonal antibodies raised against KLH-ribose, pentosidine and CML-BSA. The cross-reactivities of various AGEs and related antigens in these assays have been established and the most appropriate standard materials determined. Human serum samples from normal, diabetic and renal failure subjects have been analysed at several dilutions to assess parallelism and recovery experiments have been performed.

These assays should prove useful in studies relating to the development of diabetic complications and provide a useful tool for monitoring diabetic patients.

Antioxidative Activity of Maillard Reaction Products in Lipid Oxidation

Sonja M Djilas, Božidar Lj Milić and Jasna M Čanadanović-Brunet; *University of Novi Sad, Faculty of Technology, Department of Organic Chemistry,Bulevar Cara Lazara 1, YU-21000 Novi Sad, Yugoslavia*

Lipid oxidation is a major determinant of food quality and can directly affect many characteristics, such as flavour, colour, texture and nutritive value.[1] A number of antioxidants, both artificial and natural, has been examined in attempts to control lipid oxidation in food products. Maillard reaction products (MRPs), formed from the interaction of reducing sugars and amino acids, have been found to have antioxidative activity in food products.[2,3]

The antioxidative activity of MRPs, obtained by heating of D-(+)-glucose and 2-, 3- and 4-aminobutanoic acid, respectively, was investigated during lipid oxidation of sunflower oil by a combination of "spin-trapping" and Electron Spin Resonance (ESR) spectroscopy. The effective antioxidative activity of these MRPs was estimated by measuring the relative intensity of signals in the ESR spectra of N-t-butyl-α-phenylnitrone - alkoxyl radical adducts and of N,N'-disubstituted 1,4-pyrazine cation free radicals.

Antioxidative activity of MRPs in lipid oxidation of sunflower oil was changed by altering the level added and the heating time of glucose and aminobutanoic acid in alkaline aqueous solutions (98°C; pH 9.0). MRPs formed by heating for 20 min and added to the sunflower oil at a 1% level had the strongest antioxidative activity. MPRs formed by heating of D-(+)-glucose and 3-aminobutanoic acid had stronger activity than that of MRPs formed by heating of D-(+)-glucose and 4-aminobutanoic acid or D-(+)-glucose and 2-aminobutanoic acid.

The results of this investigation also indicate that the antioxidative activity of MRPs is the result of the formation of free radicals during heating of sugars and amino acids.

[1] Pearson, A.M.; Gray, J.I.; Wolzak, A.M.; Horenstein, N.A. (1983). *Food Technol.*, **37**(7), 121-129.
[2] Bailey, M.E. (1988). *Food Technol.*, **42** (6), 123-126.
[3] Djilas, S.M.; Milic, B.Lj. (1994). In *Maillard Reactions in Chemistry, Food, and Health*, Labuza, T.P., Reineccius, G.A., Monnier, V.M., O'Brien, J., Baynes, J.W., Eds, The Royal Society of Chemistry: Cambridge, pp. 75-81.

The Maillard Reaction in the Formation of Brown Pigments in Preserves

Wael El-Gendy, Hanaa Ismail and Eglal Salem; *Nutrition Department, High Institute of Public Health, Alexandria University, 165 El-Horrea Avenue, El-Hadara, Alexandria, Egypt*

Nonenzymic browning was monitored in canned and glass-packed marmalade and apricot jam at 4, 25 and 40°C for 21 weeks. Browning occurred faster and developed more intensely in canned than in glass-packed preserves. Absorbance at 420 nm indicated less browning in samples stored at 25 and 40°C for preserve packed both in cans and in glass. 4°C was considered best together with the glass containers, because the browning was relatively similar to that of the control (stored at -10°C). The brown pigments resulting from the Maillard reaction have been separated using HPLC, using a reversed-phase column, 5% v/v aqueous acetonitrile and UV detection, and they increased in amount with time of storage.

Sensory evaluation confirmed the similarity between samples stored at 4°C and the controls for both types of packaging. A change in odour was observed for marmalade after storage for 6 weeks at 40°C and after storage for 12 weeks at 25°C. A change in odour of apricot jam was detected after storage for 3 weeks at 40°C in cans, but only after 6 weeks in glass.

The Maillard Reaction in Meat Flavour Formation: The Role of Selected Precursors

Linda J Farmer*[†], Omiros Paraskevas[†] and Terence D J Hagan*; *Food Science Division, Dept. of Agriculture for Northern Ireland and, †The Queen's University of Belfast, Newforge Lane, Belfast BT9 5PX, UK

The volatile compounds which give cooked meat its characteristic aroma and flavour are derived from reactions which occur during cooking. Of these reactions, the Maillard reaction is particularly important for the formation of a range of furanthiols and disulfides which possess distinctive 'meaty' odours and very low odour thresholds.[1,2] Studies have indicated that some potential precursors of these odour-impact compounds have a significant effect on the aroma of cooked beef and on some of the volatile products of the flavour-forming reactions.[3,4] Further investigations of the effects on volatile compounds have now been conducted.

Ribose, ribose-5-phosphate, glucose, glucose-6-phosphate and inosine monophosphate were added to raw homogenised beef (2 mmoles 100g^{-1}) and equilibrated overnight prior to cooking (100°C, 30 min). The volatile compounds were collected on Tenax GC (60°C, 30 min) and analysed by GC-MS. Ribose-5-phosphate and ribose gave the most pronounced effects. However, differences in the effects of each precursor on the various classes of aroma compounds mean that the quantities of these precursors in the meat may affect the balance of flavour compounds formed. For example, ribose and ribose-5-phosphate reduce the headspace concentrations of the C_6 to C_9 n-aldehydes, with inosine monophosphate having a lesser effect yet causing the most suppression of the higher n-aldehydes (e.g., C_{15} and C_{16}). Ribose-5-phosphate and, to a lesser extent, ribose can cause up to 20-fold increases in some of the sulfur-containing compounds believed to contribute to meat flavour, such as bis(2-methyl-3-furyl) disulfide, 2-methyl-3-methyldithiofuran and methional. Thus, the quantities of such sugars in raw beef may be a limiting factor for meat flavour formation.

[1] Gasser, U.; Grosch, W. (1988). *Z. Lebens.-Unters. Forsch.*, **186**, 489.
[2] Farmer, L.J.; Patterson, R.L.S. (1991). *Food Chem.*, **40**, 201-205.
[3] Mottram, D.S.; Madruga, M.S. (1993). In *Trends in Flavour Research*, Maarse, H., van der Heij, D.G., Eds, Elsevier: Amsterdam, pp. 339-344.
[4] Farmer, L.J.; Hagan, T.D.J.; Paraskevas, O. (1996). In *Flavour Science: Recent Developments*, Taylor, A.J., Mottram, D.S., Eds, Royal Society of Chemistry: Cambridge, pp. 225-230.

Identification of the Lactosylation Site of β-Lactoglobulin

Vincenzo Fogliano, Simona Maria Monti, Attilio Visconti, Alberto Ritieni and Giacomino Randazzo; *The University of Naples "Federico II", Dept. of Food Science, Via Università 100, 80055, Portici (NA), Italy*

Different methods have been developed to measure protein-bound lactose; the furosine method, which measures one of the products of the acidic hydrolysis of the Amadori compound, being the most widely adopted. Nevertheless the possibility of developing an immunometrical method able to quantify protein lactosylation with high sensitivity and low cost of analysis suggests that the production of antibodies able to recognize protein-bound lactulosyl-lysine is worth investigating. Although different groups have obtained antibodies which specifically recognize advanced glycation end products deriving from Maillard Reaction[1] or N^ϵ-deoxylactulosyl-lysine,[2] none of these can be used in routine milk quality analysis.

A different immunological approach would be the use of selected lactosylated peptides as antigens. If a significant sequence is injected, the antisera obtained should be able to give a response correlated with the milk's thermal history. To follow this approach, knowledge of the lactosylation sites of milk protein is essential. Although glycation occurs on all milk proteins, whey proteins are the most reactive. Above 60°C, side chains of β-lactoglobulin (βLG), which are normally buried within the native structure, become exposed, with an increase in their reactivity.[3] As few data on lactosylation sites are available,[4] we identified βLG lactosylation sites and raised antibodies against glycated and unglycated peptides containing these sites.

βLG was purified from different heat-treated milks and subjected to partial hydrolysis using different proteases. The resulting peptides were purified by reversed-phase HPLC using an Aquapore RP 300 column. The same procedure was repeated on βLG purified from raw milk and lactosylated by heating with lactose (5 mg/ml) at 40°C for 120 min. By comparison of the chromatograms, peptides corresponding to peaks which were affected by thermal treatment were identified by N-terminal sequence analysis.

1. Matsuda, T.; Ishiguro, H.; Ohkubo, I.; Sasaki, M.; Nakamura, R. (1992). *J. Biochem.*, **111**, 383-387.
2. Fogliano, V.; Monti, S.M.; Ritieni, A.; Marchisano, C.; Peluso, G. *et al.* (1997). *Food Chem.*, **58**, 53-58.
3. Anema, S. and McKenna, A. (1996). *J. Agric. Food Chem.* **44**, 422-428.
4. Otani, H.; Hosono, A. (1987). *Nippon Chikusan Gakkaiho*, **58**, 472-482.

Serum 'Free' Pentosidine Levels and Urinary Excretion Predict Deteriorating Renal Function in Diabetic Nephropathy

Miriam A Friedlander, Roger A Rodby*, Edmund J Lewis* and Donald Hricik; *Division of Nephrology, Case-Western Reserve University and University Hospitals of Cleveland, 11100 Euclid Avenue, Cleveland, Ohio 44106, USA, *Division of Nephrology, Rush University and Rush-Presbyterian-St Luke's Medical Center, 1655 W Congress Parkway, Chicago, Illinois 60612, USA*

Renal function declines progressively in patients who have diabetic nephropathy. Treatment with captopril has been shown to protect against deterioration of renal function in patients with diabetes mellitus and proteinuria in a large multi-centre trial.[1] To identify a role for advanced glycation endproducts (AGEs) in the progression of diabetic nephropathy, 112 sets of baseline blood and urine samples from the trial were examined in a blind fashion. Mean ± SD baseline creatinine was 1.58 mg/100 ml and 24-h protein excretion 4.54 ± 3.1 g/TV. The samples represented 56 patients, in whom creatinine levels doubled over a subsequent three year study period, and 56 controls, matched for serum creatinine, 24 h-urinary protein, age, sex, race and treatment (captopril or placebo), in whom there was no progression of diabetic nephropathy. Pentosidine was measured in serum and urine, in both its protein-bound and free forms, as previously described.[2]

	n	Pentosidine (pmol/mg protein)	'Free' Pentosidin (pmol/ml)	Pentosidine Excretion (pmol/TV)	'Free" Pentosidin Excretion (pmol/TV)
Doublers	56	2.057±1.39	1.356±0.96	4.63±6.4	41.0±28.7
Non-Doublers	56	2.058±1.61	0.947±0.61	4.21±5.3	26.1±24.3
p value (t test)		NS	.0093	NS	.0043

Data expressed as Mean ±SD. Treatment with captopril did not significantly affect levels of 'free' or protein-bound pentosidine in serum or urine (data not shown).

In summary, 'free' pentosidine levels in serum and total urinary excretion of free pentosidine were increased in patients with subsequent progression of diabetic nephropathy. Both of these factors were also significant co-variants in a Cox model of time to doubling of the baseline creatinine. These data suggest that progression of diabetic nephropathy is associated with an increased rate of breakdown and excretion of 'free' pentosidine. Because baseline plasma levels of protein-bound pentosidine did not differ between the two groups, we hypothesize that patients with diabetes who develop progressive nephropathy may respond to protein-bound pentosidine with increased uptake and catabolism of AGE-modified proteins.

[1] Lewis, E.; Hunsicker, L.; Bain, R.; Rohde, R. (1993). *New Engl. J. Med.*, **329**, 1456-1462.
[2] Friedlander, M.; Wu, Y.; Elgawish, A.; Monnier V. (1996). *J. Clin. Invest.*, **97**, 728-735.

Production of an Antibody against Fructated Proteins and its Usefulness to Detect *in Vitro* and *in Vivo* Fructation

Junichi Fujii, Nobuko Miyazawa, Yoshimi Kawasaki, Myint Theingi, Ayumu Hoshi and Naoyuki Taniguchi; *Department of Biochemistry, Osaka University Medical School, 2-2 Yamadaoka, Suita, Osaka 565, Japan*

We have raised an antibody against fructated lysine in proteins by immunizing a rabbit with fructated lysine-conjugated ovalbumin. The antibody specifically recognized proteins incubated with fructose, but not with other reducing sugars. Fructated lysine was able to compete in binding to this antibody, but not glucated lysine, glucose, fructose or lysine. When bovine serum albumin was incubated with various concentrations of fructose, the reactivity of the antibody increased in a dose-dependent manner. When soluble proteins prepared from either normal or streptozotocin-induced diabetic rat tissues were analysed by ELISA using this antibody, increases in the reactive components were observed during the aging process, as well as under diabetic conditions in some tissues. Western blot analysis showed that lens crystallin was highly reactive to this antibody. Fructose is mainly synthesized through the polyol- metabolizing pathway that is enhanced in diabetic conditions and the activity of this pathway in the lens is known to be high. Thus, this antibody promises to be a useful tool in investigating the roles of fructose-mediated glycation reactions and polyol metabolism, both of which are likely candidates for the cause of diabetic complications.

Cross-Linking of Proteins by Maillard Processes: Model Reaction of an Amadori Compound with N^{α}-Acetyl-L-Arginine

Fränzi Gerum, Markus O Lederer* and Theodor Severin; *Institut für Pharmazie und Lebensmittelchemie, Universität München, Sophienstrasse 10, D-80333 München, Germany, *Institut für Lebensmittelchemie, Universität Hohenheim, Garbenstrasse 28, D-70593 Stuttgart, Germany*

Cross-linking of proteins in the course of Maillard reactions is well established. The mechanism of this process is not fully understood, although sugar-derived α-dicarbonyl compounds, such as 3-deoxyhexosuloses, methylglyoxal and glyoxal, are supposed to be involved. Pentosidine, a fluorescent heterocyclic compound linking lysine and arginine by a C_5 moiety, was found in a variety of human tissues, though in very low amounts.[1] Hence, one would expect other structures beside pentosidine to be responsible for the extensive protein cross-linking in certain mammalian tissues.

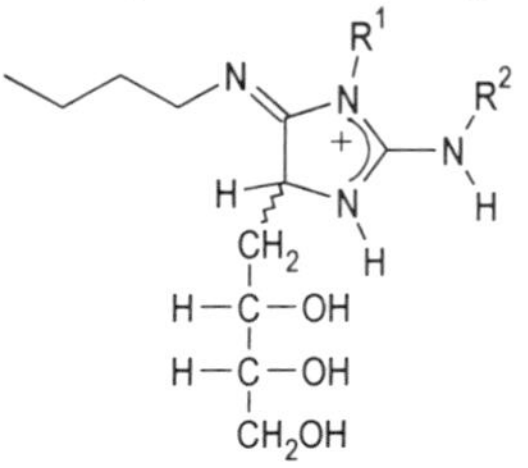

1a: R^1=H, R^2= $-(CH_2)_3-CH(NHAc)-COO^-$

1b: R^1= $-(CH_2)_3-CH(NHAc)-COO^-$, R^2=H

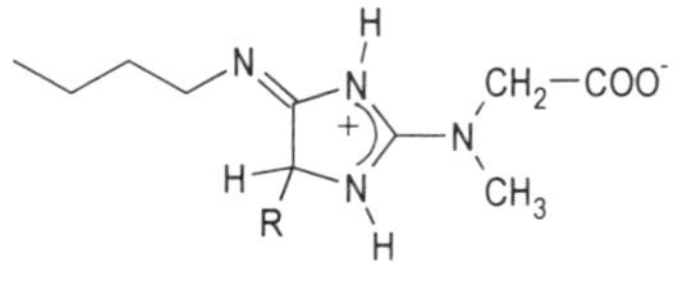

2 a: R=H
b: R=CH_3

From the reaction of 1-deoxy-1-butylamino-D-fructose with N^{α}-acetyl-L-arginine under physiological conditions, two main products with a characteristic UV maximum at 253 nm were isolated. The aminoimidazoline imine structure **1a** or **1b** is proposed on the basis of the spectroscopic data; each of these isomers can exist in two diastereoisomeric forms. Further NMR experiments are required to confirm the postulated structures, but these have been held up, as purification of **1a,b** proved difficult. For an unequivocal structural assignment of the heterocyclic core, we have reacted butylamine and creatine with glyoxal and methylglyoxal, respectively. The resulting two model compounds, 5-butylimino-2-(*N*-carboxymethyl-*N*-methylamino)-2-imidazoline (**2a**) and 5-butylimino-2-(*N*-carboxymethyl-*N*-methylamino)-4-methyl-2-imidazoline (**2b**), were formed in comparatively high yield; their UV spectra are similar to those of **1a,b**. In solution, both **2a** and **2b** exist in the zwitterionic form shown above.

On the basis of these results, it seems justified to expect cross-linking of proteins by the action of reducing carbohydrates to proceed in an analogous manner.

[1] Sell, D. R.; Monnier, V. M. (1989). *J. Biol. Chem.*, **264**, 21597-21602.

Structure and Biological Significance of Pentodilysine, a Novel Fluorescent Advanced Maillard Reaction Protein Crosslink

Lila Graham, R H Nagaraj*, R Peters, L M Sayre* and V M Monnier*; *Departments of Orthopaedic Research, Biological Chemistry & Molecular Pharmacology and Medicine, Harvard University, Cambridge, MA, USA, *Departments of Pathology, Chemistry and Ophthalmology, Case Western Reserve University, Cleveland, OH, USA*

Pentodilysine (penK$_2$) is a fluorescent (exc. 320 and 366 nm/em. 440 nm), acid-stable Maillard product, which was originally isolated from human cataractous lenses and baptized LM$_1$.[1] PenK$_2$ was synthesized both from ribose or ascorbate with serum albumin, whereby dehydroascorbate was a penK$_2$ precursor even under N$_2$ (in contrast to pentosidine which formed only under oxygen). For structure elucidation, penK$_2$ was prepared by the reaction of lysine with ribose and isolated by ion exchange, gel filtration and HPL chromatography.[2]

The following structure was deduced from Electrospray MS and proton NMR analysis:

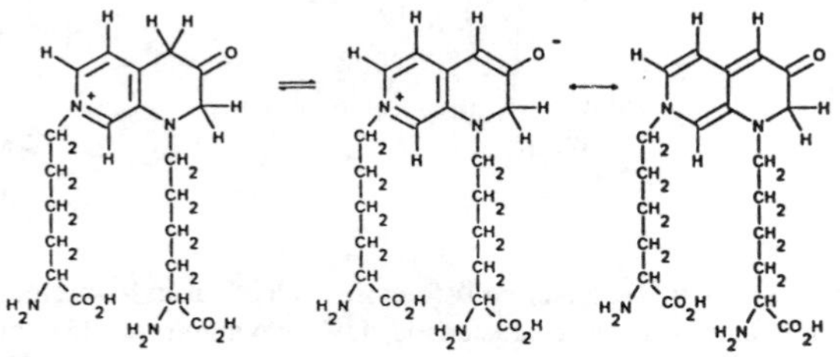

PenK$_2$ levels in human lenses increased with age in the water-insoluble fraction to reach 0.2 mmol/mol protein at 70-80 years. PenK$_2$ levels correlated with severity of diabetes in lenses from diabetic dogs with 5 years of diabetes.[3] Surprisingly, whereas levels were elevated in lenses of dogs with moderate control of diabetes, pentosidine levels were elevated only in lenses from poorly controlled animals. Thus, penK$_2$ is a sensitive marker of metabolic dysfunction by mild levels of hyperglycaemia. Whereas glucose and fructose were not penK$_2$ precursors during a 3-week incubation period, the structure of penK$_2$ suggests that its *in vivo* formation could result from a condensation of two residues of ascorbate oxidation products linked to lysine. Thus, pentodilysine emerges as a unique, potentially highly useful marker of intracellular ascorbate oxidation.

[1] Nagaraj, R.H.; Monnier, V.M. (1992). *Biochim. Biophys. Acta*, **1116**, 34-42.
[2] Graham, L. (1996). *Biochim. Biophys. Acta*, **1297**, 9-16.
[3] Kern, T.; Engerman, R.E.; Larson, M.E. (1990). *Metabolism*, **39**, 638-640.

Poster Abstracts

The Use of Electron Microscopy and X-Ray Diffraction to Study the Effects of Glycation on the Charge Distribution of Collagen and to Measure the Subsequent Swelling Behaviour of Corneal Tissue

J C Hadley, K M Meek and N S Malik; *Biophysics Group, Oxford Research Unit, The Open University, Oxford OX1 5HR, UK*

The purpose of this study was to investigate any changes in the surface charge distribution of collagen fibrils caused by fructation, hence to locate the site at which fructose binds to Type I collagen and to measure the effect of fructation on swelling behaviour of corneal collagen.

Glycation sites were located by analyses of phosphotungstic acid (PTA) stain uptake. PTA is an anionic stain that binds to positively charged amino acid side-chains. This binding of stain gives rise to the banding pattern of collagen observable under an electron microscope. A decrease in stain uptake would be expected upon glycation through a decrease in the number of positively charged side-chains. Banding patterns of glycated and non-glycated collagen were scanned with a laser densitometer and correlated with each other and also with a histogram of the positive charges along the collagen D-period. Five regions of decreased stain uptake were identified, the greatest decrease being apparent at the c1 band. The weightings given to certain lysine and arginine residues were altered to attempt to improve the correlation of the charge histogram with the observed staining pattern and thus to locate which residues are involved in glycation. X-ray diffraction patterns of glycated and non-glycated collagen were also examined and the meridional intensities compared. Meridional reflections appear to be less distinct in glycated samples. Corneal tissue incubated in different buffer and sugar solutions were subsequently hydrated using polyethylene glycol and the interfibrillar spacings were plotted as a function of hydration. No significant difference was noted.

This study suggests that fructose binds to Type I collagen at several sites along the D-period. Not all lysine residues are involved; there appear to be specific residues that favour attachment of sugar. For example, the amino-acid involved in the glycation of the c1 band is postulated to be lysine-479. Sugar-induced crosslinks do not appear to affect the rate of swelling in corneal tissue, but there is evidence that the extent of swelling may be reduced in fructated samples.

This work was supported by Research into Aging and British Telecom.

Increased Formation of the Glycoxidation Product N^ε-Carboxymethyl-Lysine in Retinas from Diabetic Rats

Hans-Peter Hammes, Konrad Federlin, Eva Wagner* and Erwin Schleicher†; *Department for Internal Medicine, Justus-Liebig-Universität Giessen, Rodthohl 6, D-35385 Giessen, Germany, *Institute for Diabetes Research, Kölner Platz 1, D-80804 München, Germany, †Department for Internal Medicine, Division of Endocrinology, Metabolism, Pathobio-Chemistry, Otfried-Müller-Straße 10, D-72076 Tübingen, Germany*

The formation and localization of proteins carrying carboxymethyl-lysine (CML) groups in retinas from diabetic rats and the effect of aminoguanidine, an advanced glycation endproduct-inhibitor, on retinal CML formation was investigated.

Methods: A polyclonal antibody was raised and characterized as specifically recognizing CML-modified proteins. Male Wistar rats were rendered diabetic with streptozotocin (i.v. injection of 65 mg/kg). After 6 months, retinas were obtained from the following groups: 1. normal controls (NC: n = 9; blood glucose (BG) 5.2 ± 0.34 mmol/l; HbA1 4.28 ± 0.76%); 2. untreated diabetics (DC: n = 9; BG 25.77 ± 2.54; HbA1 14.1 ± 1.89); 3. diabetics, treated with aminoguanidine (0.5 g/l drinking water; D-AG; n = 8; BG 26.33 ± 2.45; HbA1 13.0 ± 2.17). Immunohistochemistry of CML was performed on vertical and horizontal cryostat sections. The effect of aminoguanidine on CML-formation was also studied *in vitro* by aerobic incubation of glycated human serum albumin in the presence and absence of 0.1 M aminoguanidine.

Results: *In vitro*, aminoguanidine reduced the formation of CML from glycated human serum albumin by 30%. *In vivo*, CML-immunolabelling was detectable both in vascular and non-vascular parts of the inner retinas of non-diabetic rats. In diabetic rats, a major increase of CML-immunoreactivity was found in all parts of the vascularized retina, predominantly in the inner plexiform layer and the outer capillary network. Aminoguanidine had no effect on both localization and intensity of CML-immunoreactivity of the retina.

Conclusion: This study demonstrates for the first time that diabetes causes an increased formation of CML in retinas, possibly indicating a significantly increased oxidative stress. Aminoguanidine had no discernible effect on CML-formation in diabetic retinas.

Factors Affecting Non-Enzymic Browning Reactions during the Aging of Port

P Ho and M C M Silva; *Escola Superior de Biotecnologia, Universidade Católica Portuguesa, Rua Dr António Bernardino de Almeida, Porto 4200 Portugal*

During the aging of port in wooden barrels, several different chemical reactions occur between its constituents and the environment. Many ports, especially tawny ports, spend many years in barrel exposed to oxidation and numerous temperature changes. A study has shown that higher amounts of hydroxymethylfurfural (HMF) and furfural can be found in older tawny ports.[1] Port contains an excess of reducing sugars over amino acids and this is believed to promote the rate of Maillard browning. However, as port has a pH of about 3.5, amino acids are in their protonated form and hence availability for participating in the Maillard reaction is reduced. Therefore, the role and importance of the Maillard reaction and caramelisation during port aging is not well understood. However, a study with model wines with different types and amount of reactants showed that both the Maillard reaction and caramelisation can produce similar levels of HMF as a madeira wine.[2]

In this study, accelerated aging of port was conducted at different temperatures and initial concentrations of oxygen, with wines containing different combinations and amounts of amino acids and sugars, in order to examine how these reactions affect each other and which pathway predominates. The rate of amino acid and sugar losses, HMF and furfural production and browning of the wines were followed with time and the results will be presented.

[1] Ho, P.; Hogg, T.A.; Silva., M.C.M. (1995). Variation of phenolic compounds and furans in tawny port wines. 2º Encontro de Química de Alimentos, Universidade de Aveiro, Portugal, 19-21 July.

[2] Kroh, L.W. (1994). Caramelisation in food and beverages. *Food Chem.*, **51**, 373-379.

Characterization of Food Melanoidin

Seiichi Homma, Masatsune Murata, Naoko Terasawa* and Young-Soon Lee†; *Department of Nutrition and Food Science, Ochanomizu University, Ohtsuka,Bunkyo-ku, Tokyo 112, Japan, *Faculty of Education, Kanazawa University, Kakuma-cho, Kanazawa-City,Ishikawa 920-11, Japan, †Department of Food Science and Nutrition, Kyung Hee University, Hoegi-dong, Dongdaemoon-ku, Seoul 130-701, Korea*

Brown pigments in foods can be categorized by the major compounds involved in colour formation, such as reducing sugar/amine products, caramel, phenolics or oxidized lipid.

Melanoidin is an amphoteric chelator. The dissociation constant (μM) of the nondialysable melanoidin prepared from glucose and glycine was found to be 1.2-5.6 for Cu^{2+}. That of a melanoidin fraction from instant coffee for Zn^{2+} was about 8. Metal-chelating Sepharose 6B columns separated coloured compounds into groups, as follows:

Brown compounds in foods were hardly retained on the Zn^{2+}-chelated column. Most parts of melanoidins of the model system and of soy and fish sauces were retained on the Cu^{2+} and Fe^{2+} columns and released when the pH of the eluent was decreased. Coloured phenolic compounds, such as those from coffee, were retained on the Fe^{2+} column and most of them compounds were not released from the column when the pH of the eluentwas decreased; they were released with EDTA.

The type of fish sauce pigments was characterized. The pigments were not electrofocused clearly, but were adsorbed on the Cu^{2+}-chelated Sepharose column. Most of the pigments were released by changing the pH of the eluent from 7 to 5 and a part of the pigments was adorbed on a ConA-lectin column. 3-Deoxyglucosone was determined. Fish sauce melanoidin is found to be of the sugar-type.

We isolated from soil two species of microorganisms to decolorize brown pigments: *Streptomyces werraensis* TT14 for the nondialysable model melanoidin and *Paecilomyces canadensis* NC-1 for instant coffee. The incubation of these two microorganisms, as well as *Coriolus versicolor* IFO 30340, with various kinds of browned foods gave different degrees of decoloration depending on the species of microorganisms, thus allowing the type of food melanoidin to be categorized.[1]

[1] Terasawa, T.; Murata, M.; Homma, S. (1996). *J. Food Sci.*, **61**, 669-672.

Immunochemical Detection of Ascorbylated Proteins

Birgit Huber and Monika Pischetsrieder; *Institute of Food Chemistry, Ludwig-Maximilians-Universität Munich, Sophienstrasse 10, 80333 Munich, Germany*

During the incubation of L-ascorbic acid (AA) with proteins, covalent binding of AA or its degradation products occurs in a process called protein ascorbylation. From model studies it can be deduced that proteins are ascorbylated in a Maillard-type reaction.[1] AA itself can react with amino acids,[2] but, more likely, AA is first oxidized to L-dehydroascorbic acid (DHA) and to further degradation products which represent the reactive agents.[3] However, so far almost nothing is known about ascorbylated protein (ASC-protein).

In this study, immunochemical methods were used to characterize ASC-protein. Thus, sheep serum albumin (SSA) was incubated with DHA and a polyclonal anti-ASC-protein antibody was prepared. In a non-competitive ELISA, the antibody reacts with several ascorbylated proteins, such as ASC-lactoglobulin or ASC-human serum albumin (HSA), but not with the native proteins. The signals are dependent on the antiserum dilution and significant signals can be produced with antiserum diluted up to 1:100,000. Since the antibody recognizes ASC-proteins, a competitive ELISA was developed to test cross-reactivity with various modified proteins. Antibody binding can be totally inhibited by the addition of protein which had been incubated with DHA or AA under aerobic conditions, whereas protein incubated with AA under low-oxygen conditions cross-reacts to a lesser extent. ASC-N^α-BOC-L-lysine inhibits the binding almost completely, whereas ASC-N^α-BOC-L-arginine, ASC-N^α-acetyl-L-histidine and the unmodified amino acids do not show significant cross-reactivity under these conditions. These results indicate that lysine residues are the main targets of protein ascorbylation.

Furthermore, proteins were glycosylated with other carbohydrates, such as glucose, glucose-6-phosphate, fructose, ribose, maltose and lactose, and tested for cross-reactivity with anti-ASC-protein antibody. Inhibition was found for various glycosylated proteins, indicating that there exist some common structures of ascorbylated and glycosylated proteins. CML-protein (CML = carboxymethyl-lysine), which is an important glycosylation product and is known to be formed in low amounts from AA, proved to be an inhibitor for the antibody binding, indicating that CML or a similar structure is at least partially responsible for the anti-ASC-protein antibody binding.

[1] Ortwerth, B.; Olesen, P. (1988). *Biochim. Biophys. Acta*, **956**, 10-22.
[2] Pischetsrieder, M.; Larisch, B.; Müller, U.; Severin, Th. (1995). *J. Agric. Food Chem.*, **43**, 3004-3006.
[3] Larisch, B.; Pischetsrieder, M.; Severin, Th. (1996). *J. Agric. Food Chem.*, **44**, 1630-1634.

Novel Observations of the Effect of Aminoguanidine on Previously Glycated Protein: A Pro-Oxidant Activity that Increases Proteolytic Susceptibility

James V Hunt and Donna Schultz; *Department of Pathology, University of Cambridge, Tennis Court Road, Cambridge CB2 1QP, UK*

Proteins altered by Maillard chemistry are typically protease resistant, an observation which has been linked to the accumulation of glycated protein with increasing age and diabetes mellitus. The ability of aminoguanidine to affect Maillard chemistry and the generation of fluorescent advanced glycation endproducts on proteins is assumed to involve its ability to interact with aldehydic products.

However, aminoguanidine has been shown to possess pro-oxidant activity[1] that may affect protein glycation.[2] Here, we show that such pro-oxidant activity of aminoguanidine (0, 0.5, 1.0, 5.0, 10.0 mM) can affect protein-bound Maillard products. Using glycated human serum albumin (HSA), a model for Maillard products in aged and diabetic individuals, we show that aminoguanidine can affect the observed levels of glucose remaining attached to protein and levels of fluorescence attributable to products of glycation. In addition, proteins become fragmented and oxidised, as assessed by the formation of protein fragments soluble in trichloroacetic acid, the carbonyl assay of protein oxidation and SDS PAGE analysis.

Furthermore, studies show that the treatment of native or glycated protein with aminoguanidine increased the protein's susceptibility to several proteases *in vitro* and to degradation by cells in culture. Aminoguanidine's anti-glycation effects seem to be due to its pro-oxidant activity, which appears to contradict the emphasis placed on the adverse effects of oxidative stress in aging and diabetic individuals.[3]

[1] Ou, P.; Wolff, S.P. (1993). *Biochem. Pharmacol.*, **46**, 1139-1144.
[2] Skamarauskas, J.T.; McKay, A.G.; Hunt, J.V. (1996). *Free Rad. Biol. Med.*, **21**, 801-812.
[3] Wolff, S.P.; Jiang, Z.Y.; Hunt, J.V. (1991). *Free Rad. Biol. Med.*, **10**, 339-352.

Effect of Pyrophosphate Buffer on 4-Hydroxy-5-Methyl-3-(2H)-Furanone Formation in Aqueous Model Systems

Alexandra Jacquemier*, Louise Royle, Jane K Parker, Catherine M Radcliffe and Jennifer M Ames; *Department of Food Science and Technology, The University of Reading, Whiteknights, Reading, RG6 6AP, UK, *Permanent address: ENSBANA, 1 Esplanade Erasme, 21000 Dijon, France*

Aqueous solutions of either arabinose on its own or arabinose and alanine (0.25 M with respect to each reagent) were heated under reflux for up to 2 h. The pH was maintained at 5 either by the intermittent addition of sodium hydroxide solution or with 0.8 M pyrophosphate buffer. The rate of browning (absorbance at 420 nm) was similar for arabinose heated with alanine without buffer and for arabinose heated without alanine but in buffer. In contrast, the rate of browning was about 30 times higher for the system containing all three reagents.

Analysis by capillary electrophoresis (CE) facilitated a qualitative and quantitative comparison of the model systems. The profile of reaction products varied. Of particular interest is that the rate of formation of 4-hydroxy-5-methyl-3(2H)-furanone, after heating for 2 h, was about 10 times greater when arabinose was heated without alanine but in buffer, compared with when it was heated with alanine but without buffer. In the presence of both alanine and buffer, the rate of formation of this compound was about 100 times greater than when heating with alanine but without buffer. It is suggested that pyrophosphate catalyses the degradation of arabinose itself and/or the reactions between arabinose and alanine that lead to the formation of 4-hydroxy-5-methyl-3(2H)-furanone.

The Use of an Electronic Nose for Detection of Volatile Aroma in Fried Meat Compared to GC-MS Analysis and Sensory Evaluation

Maria Johansson and Eva Tornberg; *Swedish Meat Research Institute, P.O.Box 504, S-244 24 Kävlinge, Sweden*

To produce a delicious food product with an appetizing appearance is a challenge to the food industry, including the meat industry. The control of the complex Maillard reaction includes minute control of various parameters during food processing as well as analysis of raw materials. Rapid, robust, cheap and reliable methodology for on-line measurements of the "right" taste is lacking. However, recent progress in sensor development may provide a new generation of on-line instruments for improved process and quality control. Several studies have been devoted to the identification of volatiles in cooked meat products and model systems. However, quantitative relationships between the content of volatile flavours and sensory evaluation of these compounds in meat products are still missing.

The aim of the present study was to develop a methodology for the use of an electronic nose as a detector of flavour in cooked meat and yo correlate the results with headspace analysis using GC-MS and with sensory evaluation of the products. Minced meat was fried at different temperatures using a thermostatically controlled frying device. Volatiles from the cooked meat samples were analysed using an electronic nose consisting of 15 non-specific gas sensors combined with a pattern recognition algorithm. For GC-MS analysis, volatiles from the samples were trapped onto Tenax. The fried meat samples were analysed sensorily by a trained panel. Several specific parameters regarding flavour, aroma and appearance of the products were judged by the panel. Aroma profiles obtained from analysis, using the electronic nose, were compared with aroma profiles from headspace analysis, using GC-MS and the sensory panel. The results were evaluated using multivariate data analysis.

Alteration of Skin Surface Protein with Dihydroxyacetone: A Useful Application of the Maillard Browning Reaction

John A Johnson and Ramon M Fusaro; *University of Nebraska Medical Center, Department of Internal Medicine (Dermatology), 600 South 42nd St., Omaha, NE 68198-4360, USA*

Sunless tanning preparations contain dihydroxyacetone (DHA), which undergoes the Maillard reaction with skin protein to produce a brown colour.[1] Since DHA-induced skin protein (DISP) is integral with the skin surface, its durability is limited only by natural sloughing of the skin. Recent studies suggested that DISP affords a Sun Protection Factor (SPF) of three.[2] Such a modest level of day-long protection can be more effective than episodic use of high-SPF sunscreens, and their action can be potentiated by application over DISP.

Dihydroxyacetone is useful in the treatment of photosensitivity (exaggerated response to sunlight). Many photosensitive persons react to UVA radiation (320-400 nm). The brown colour of DISP arises from absorption in the lower end of the visible region (above 400 nm), with overlap into long-wavelength UVA (below 400 nm). This window of vulnerability for photosensitive patients is not shaded by conventional sunscreens. We investigated the mechanism of action of DHA on skin protein and evaluated the efficacy of other potential browning agents.[3] Results obtained with monohydroxyacetone, methylglyoxal, etc., offer unique clues to the mechanism of the Maillard reaction. Since DHA is a triose, it undergoes less complex processes than hexose-initiated reactions.

[1] Johnson, J.A.; Fusaro, R.M. (1987). *Dermatologica*, **175**, 53-57.
[2] Sayre, R. M.; Torode, D.L.; Johnson, J.A.; Fusaro, R. M. (1995). In *Melanin: Its Role in Human Photoprotection*, Zeise, L. et al., Eds, Valdemar Pub. Co: Overland Park, Kansas, pp. 39-47.
[3] Johnson, J.A.; Fusaro, R.M. (1994). In *Maillard Reactions in Chemistry, Food and Health*, Labuza, T.P. et al., Eds, Royal Society of Chemistry: Cambridge, pp. 114-119.

Metabolism of Fructoselysine in the Kidney

Francis Kappler, Shirin Djafroudi*, Holger Kayser*, Bangying Su, Sundeep Lal, William C Randall, Michael A Walker, Anne Taylor, Benjamin S Szwergold, Helmut Erbersdobler* and Truman R Brown; *Fox Chase Cancer Center, Department of NMR and Medical Spectroscopy, Philadelphia, Pennsylvania, 19111, USA, *Christian-Albrechts-Universität Kiel, D-24105 Kiel, Germany*

The mechanisms by which hyperglycaemia causes diabetic complications are poorly understood. However, there is evidence for the involvement of glycated protein in the development of some complications, especially nephropathy.[1]

Protein glycation involves condensation of glucose with protein (usually a lysine residue), followed by rearrangement to a more stable ketoamine. These products can react further via a complex series of rearrangements and dehydrations to produce fluorescent products. These reactions all involve large protein molecules. On the other hand, protein catabolism yields smaller peptides and free amino acids, as well as the glycated amino acid, fructoselysine (FL). This compound has been studied in some detail, since it is also a component of food, especially processed dairy products, the general conclusion being that it is not metabolized by mammals.[2] Contrary to these findings, we report here that FL is metabolized in the kidney by phosphorylation to fructoselysine 3-phosphate (FL3P). This is an extremely reactive compound, rapidly decomposing to 3-deoxyglucosone (3DG), lysine and inorganic phosphate. 3DG is a reactive α-dicarbonyl, which may be able to modify kidney proteins directly. As is to be expected, the kidney has mechanisms to remove such a toxic compound, the primary means of detoxification being by reduction to 3-deoxyfructose (3DF), which is then excreted in the urine.

Studies using a rat model and labelled compounds demonstrate a rapid conversion in the kidney of FL to FL3P. Direct catheterization of a rat's carotid and jugular, as well as cannulating the bladder, gave data immediately after an FL injection, FL, 3DG and 3DF appearing in the urine within one minute. Dietary input of FL as glycated protein in both rats and humans produces excess excretion of FL, 3DG and 3DF.

[1] Cohen, M.P.; Hud, E.; Wu, V.-Y. (1994). *Kidney Int.*, **45**, 1673-1679.
[2] Erbersdobler, H.F.; Brandt, A.; Scharrer, E.; von Wangenheim, B. (1981). *Progr. Food Nutr.Sci.*, **5**, 257-263.

The Effect of Glycated Proteins on the Expression of Cell-Adhesion Molecules of Human Endothelial Cells

Naoki Kashimura, Masaru Kitagawa*, Seiji Noma*, and Shiro Nishikawa; *Department of Bioscience, Faculty of Bioresources, Mie University, Tsu, Japan, *Pharmaceutical Research Center, Toyobo Co Ltd, Katada, Japan*

Glycated proteins and the Amadori rearrangement products are known to generate active oxygen species under physiological conditions.[1] We also reported a systematic survey of the inhibitory effect of substituted sugars on cell adhesion-dependent generation of active oxygen species by human neutrophils.[2] It was further revealed by us that these compounds were also able to inhibit the expression of ICAM-1, ELAM-1 and VCAM-1 of human umbilical vein endothelial cells (HUVEC).

In this report, the promotive effect of some glycated or glycosylated proteins on the CAM expression of HUVEC is described together with their effects on the oxidative burst of neutrophils. The assay system included 5 h of incubation of HUVEC with a glycated protein, which had been prepared from BSA or lysozyme by the *in vitro* reaction with various reducing sugars, in the presence or absence of 100 u/ml of TNF-α, followed by an ELISA. It was found that glycated proteins prepared from D-ribose promoted the expression of adhesion molecules on HUVEC, whereas preparations from D-glucose, D-glucose 6-phosphate, or 3-*O*-methyl-D-glucose did not. The effect was most significant in the promotion of ELAM-1 expression in the absence of TNF-α. A similar promotive effect was observed for commercially available BSA derivatives containing 15-30 D-glucosamine residues linked *via* amide bonds or 28-34 *O*-D-galactosyl residues, whereas D-glucosyl- or D-galactosamide-containing BSA preparations had no activity. Glycated proteins with promoting activity showed a slight accelerating effect on neutrophil adhesion to KLH and an inhibitory effect on the PAF-stimulated generation of active oxygen species by the neutrophils, when the cells were incubated with the sugars for 30 min. In ELAM-1 expression, the promotive effect by the glycated BSA was inhibited by herbimycin A or genistein, tyrosine kinase inhibitors, in a way similar to that observed with the effect of TNF-α.

[1] Zu, J.; Morita, J.; Nishikawa, S.; Kashimura, N. (1996). *Carbohyd. Lett.*, **1**, 457-464.
[2] Kashimura, N.; Kitagawa, M.; Deguchi, S.; Tsuji, J.; *et al.* (1995). *Carbohyd. Lett.*, **1**, 261-268.

Inhibitory Effect of Polei Tea Extract on the Formation of Advanced Glycation Endproducts *in Vivo*

Naohida Kinae, Masanori Matsuda, Mutuo Shigeta and Kayoko Shimoi; *School of Food and Nutritional Sciences, University of Shizuoka, 52-1 Yada, Shizuoka 422, Japan*

We already demonstrated the inhibitory effect of green tea extract on the formation of advanced glycation endproducts (AGEs) in the plasma of streptozotocin-treated rats,[1] as well as in the model system containing D-glucose and bovine serum albumin.[2] In this work, we report the inhibitory effect of the extract of another tea, called Polei tea (PT), produced by fermentation of *Camellia assamica* in China.

Diabetes was induced in 8-week-old male Wistar rats by i.v. injection of 45 mg/kg of streptozotocin. The rats were divided into three groups (A-C). After two weeks, Groups A, B and C were given 0.05% PT extract, 0.1% aminoguanidine or tap water as drinking water, respectively. Nondiabetic Wistar rats were also divided into three parallel groups (D-F). All groups were fed commercial pellets and drinking water *ad libitum*. The AGE content of the plasma was examined by affinity HPLC analysis. After 12 weeks, rat lens crystallins were solubilized and submitted to AGE determination by ELISA, using anti-AGE antibody. Lipid peroxide contents in liver and kidney were determined by the thiobarbituric acid method.

After 6 weeks, the AGE content of the plasma of Group A decreased compared with that of Group C. The AGE value of lens crystallins of Group A also decreased significantly after 12 weeks. The reaction potency of the antibody to the AGE indicated carboxymethyl-lysine to be a key substance and a useful marker in these animals. The chemical components showing high inhibition potency in the aqueous fraction of PT extract are now under investigation.

[1] Kinae, N.; Masumori, S.; Nagai, R.; Shimoi, K. (1994). In *Maillard Reactions in Chemistry, Food, and Health*, Labuza, T.P. *et al.*, Eds, Royal Society of Chemistry: Cambridge, pp. 369-374.
[2] Kinae, N.; Shimoi, K.; Masumori, S.; Harusawa, M.; Furugori, M. (1994). In *Food Phytochemicals for Cancer Prevention II*, ACS Symposium Series **547**, Ho. C-T. *et al.*, Eds, Amer. Chem. Soc.: Washington DC, pp. 68-75.

Meat Surface Effects: Marinating Before Grilling Can Inibit or Promote the Formation of Heterocyclic Amines

Mark G Knize, Cynthia P Salmon and James S Felton; *Biology and Biotechnology Research Program, L-452, Lawrence Livermore National Laboratory, PO Box 808, Livermore, CA 94551-9900, USA*

The formation in muscle meats of heterocyclic amines (HA) that are potent mutagens in bacteria and carcinogens in rodents has been well documented. A carcinogenic effect in humans has not been established, in part because of uncertainty in determining the dietary dose. In addition to the type of meat, cooking temperature and cooking time, meat marinating greatly affects formation of 2-amino-3,8-dimethylimidazo[4,5-f]quinoxaline (MeIQx) and 2-amino-1-methyl-6-phenylimidazo[4,5-b]pyridine (PhIP).

We compared MeIQx and PhIP levels in marinated and unmarinated chicken breast meat flame-grilled on a propane grill. We marinated chicken prior to grilling and, using solid-phase extraction and HPLC,[1] determined the levels of several HA formed during cooking. Compared with unmarinated controls, a 92 to 99% decrease in PhIP was observed in whole chicken breast marinated with a mixture of brown sugar (sucrose), olive oil, cider vinegar, garlic, mustard, lemon juice and salt and then grilled for 10, 20, 30, or 40 minutes.

Conversely, MeIQx increased over 10-fold with marinating, but only at the 30 and 40 minute cooking times. Despite the increase in MeIQx, marinating greatly reduced the total amount of detectable HA.

A change in free amino acids, which are known precursors of HA, might explain the decrease in PhIP and increase in MeIQx, yet we were unable to detect any change in free amino acids. Marinating chicken in each ingredient individually showed that sucrose was involved in the increased MeIQx. The involvement of glucose with MeIQx formation has been shown in a model system.[2] We have no explanation for the decrease in PhIP which occurred after marinating with most of the individual ingredients. This work shows that marinating can significantly affect HA formation conditions.

This work was performed under the auspices of the U.S. DOE by LLNL under contract W-7405-Eng-48 and supported by NCI grant CA55861 and NCI agreement YO1CP2-0523-01.

[1] Gross G. A.; Grüter A. (1992). *J. Chromatog.*, **592**, 271-278.
[2] Skog K.; Jägerstad, M. (1992). *Carcinogenesis*, **14**, 2017-2031.

Flavour Formation in Soy Hydrolysates Prepared Using Enzymes

Lene V Koťod, Morten Fischer and Donald S Mottram*; *Novo Nordisk A/S, Enzyme Development and Application, Bagsværd, DK-2880, Denmark, * The University of Reading, Department of Food Science and Technology, Whiteknights, Reading RG6 6AP, UK*

Hydrolysed vegetable protein (HVP) is widely used as a flavouring ingredient in different food products, such as sauces, bouillons and snacks. The simplest way of producing a protein hydrolysate is by acid hydrolysis, in which the proteinaceous material is hydrolysed in 4M HCl at 100-140°C for many hours. Hydrolysis of residual sugars also occurs. The resulting amino acids and reducing sugars participate in Maillard reactions and undergo pyrolysis at the elevated temperatures. The volatile reaction products, thus formed, are the main contributors to the flavour of the hydrolysate. A typical proteinaceous material for HVP is defatted soy bean meal.[1]

The hydrochloric acid hydrolysis has the major disadvantage of producing mono- and dichloropropanols (MCP and DCP), which are considered to be potential health hazards. Therefore, an alternative is needed. Due to the development of new powerful enzyme systems, hydrolysis of the proteinaceous material by enzymes has become a realistic possibility for the industry.[2] The enzyme treatment takes place under much more gentle conditions and, therefore, no DCP or MCP is produced. However, the formation of flavour volatiles is greatly reduced. The flavour intensity of EVP can be enhanced by a maturation step comprising a prolonged heat treatment, at moderate temperature and pH, after the hydrolysis, during which Maillard reactions occur.

Compared with acid hydrolysis, enzyme treatment results in lower degrees of hydrolysis of the protein. However, the lower degree of hydrolysis potentially allows differential release of amino acids from the soy meal. Thus, by varying the mix of enzymes used for hydrolysis and also the physical and chemical parameters for the maturation step, hydrolysates with very different flavour profiles can be obtained.

Examples showing the use of enzymes for the production of soy meal hydrolysates with different flavour characteristics will be presented, as well as the GC-MS analysis of the flavour volatiles.

[1] Nagodawithana, T.W. (1995). *Savory Flavors.* Esteekay Associates: Milwaukee.
[2] Pommer, K. (1995). *Cereal Foods World*, 745-748.

Role of Oligo- and Polymeric Carbohydrates in the Maillard Reaction

Lothar W Kroh, Jörg Häseler and Anke Hollnagel; *Institute of Food Chemistry, Technical University Berlin, Gustav-Meyer-Allee 25,D-13355 Berlin, Germany*

With regard to the carbohydrate component of the Maillard reaction, investigations have centred largely on the use of monosaccharides, either hexoses or pentoses, in model systems as well as in food. Up to now, relatively little attention has been paid to the role of oligo- and polymeric carbohydrates, even though such compounds are the predominating carbohydrates in food.

According to current opinion, polymeric sugars are degraded to monomers or at least to dimers, before they interact with amino compounds in the Maillard reaction. There is no clear idea about the mechanism of the degradation reactions or about the nature of the reactive non-volatile intermediates.

The non-volatile reaction products of the thermolysis of maltodextrins and soluble starch in a closed system at 180°C are now analysed and compared with those of the thermolysis of D-glucose. Without derivatisation, the methods of AMD/HPTLC and HPAEC/PAD are suitable for the quantitative determination of mono- and oligomeric saccharides. Under the reaction conditions used the thermolysis of polymeric α-glucans first leads to the formation of D-glucose and predominantly to 1,4-linked maltooligosaccharides. The latter intermediates undergo further reactions, such as hydrothermolysis, transglycosylation, dehydration and isomerisation. As a result of these reactions, non-volatile degradation products, such as D-glucose, anhydrosugars, fructose-containing sugars and short-chain, partially branched, maltodextrins, are formed. These reaction products can act as starting material for non-enzymic browning reactions and produce either typical caramelisation products, connected with changes of the sugar structure, or Maillard reaction products in presence of amino compounds, such as glycine.[1]

Recent work on the structure and the reaction pathways to different reaction products after thermolysis as a result of the Maillard reaction of α-glucans are to be discussed.

[1] Kroh, L.W.; Jalyschko, W.; Häseler, J. (1996). *Starch/Stärke*, **48**, 426-433.

Thermal Degradation of Lachrymatory Precursor of Onion

Roman Kubec and Jan Velíšek; *Institute of Chemical Technology, Faculty of Food and Biochemical Technology, Department of Food Chemistry and Analysis, Technická 1905, 166 28 Prague, Czech Republic*

Important naturally occurring constituents of many plants belonging to the *Liliaceae* and *Brassicaceae* families are *S*-alk(en)yl-L-cysteines and their sulfoxides. These unique sulfur-containing, non-protein amino acids are important precursors of characteristic flavours, enzymically developed on disruption of their cellular tissue. However, culinary processing (boiling, frying, baking etc.) causes inactivation of alliinase and thus considerable amounts of aroma precursors remain to participate in flavour generation of thermally processed vegetables.[1-3]

The lachrymatory precursor of onion, *S-trans*-1-propenyl-L-cysteine sulfoxide (isolated from onion), and its stereoisomer *S-cis*-1-propenyl-L-cysteine sulfoxide (synthesized), were treated under different conditions at temperatures exceeding 80°C and the volatiles released were evaluated by sensory analysis and identified by means of GC-MS and GC-FTIR. Formation of volatile compounds was observed in dependence on temperature (80-200°C), time of heating (1-60 min), content of water (0-98%), and polarity of reaction medium (water-oil).

The aim of this work is focused on the sulfur-containing volatiles arising from *S-trans*-1-propenyl-L-cysteine sulfoxide and its stereoisomer, their flavour characteristics and mechanisms of their formation as influenced by reaction conditions.

[1] Yu, T.-H.; Wu, C.-M.; Ho, C.-T. (1994). *J. Agric. Food Chem.*, **42**, 1005-1009.
[2] Yu, T.-H.; Wu, C.-M.; Rosen, R.T.; Hartman, T.G.; Ho, C.-T. (1994). *J. Agric. Food Chem.*, **42**, 146-153.
[3] Yu, T.-H.; Wu, C.-M.; Ho, C.-T. (1994). *Food Chem.*, **51**, 281-286.

Isolation and Characterisation of Melanoidins from Heat-Treated Fish Meat

Margarita J Kuntcheva, Sonia M Rogacheva and Tzvetan D Obretenov; *Higher Institute of Food and Flavour Industries, Department of Organic Chemistry, 26 Maritza Boulevard, 4002 Plovdiv, Bulgaria*

The present study is an extension of our investigation on the processes of melanoidin formation and the methods of isolation of melanoidins from foodstuffs.[1-3]

The influence of the temperature and time of the baking of fish meat on the formation of water-soluble melanoidins has been reviewed. The heat treatment was performed at 180°C for 30, 60, 90, 120 and 150 min and at 100, 150, 180, 220 and 250°C for 60 min. The melanoidins formed were isolated by water extraction and purified by chromatography on DEAEC-anionite. It was established that, in general, increasing temperature and time of roasting favoured the formation of higher molecular mass products, but their quantity decreased at the higher roasting temperatures, because of the formation of water-insoluble polymers and partial melanoidin depolymerization. The unfractionated melanoidins were characterised by gel-permeation chromatography, molecular spectra and elementary analyses. Fractions with different molecular mass had similar UV characteristics. The similarity with some model and food melanoidins has been pointed out. A significant feature observed was a high nitrogen content of up to 16%.

The method used is applicable to the isolation of water-soluble melanoidins from heat-processed meat products and for their quality assessment.

[1] Obretenov, T.D.; Ivanova, S.D.; Kuntcheva, M.J.; Somov, G.T. (1993). *J. Agric.Food Chem.*, **41**, 653-658.
[2] Kuntcheva, M.J.; Panchev, I.N.; Obretenov, T.D. (1994). *J. Food Proc. Pres.*, **18**, 9-21.
[3] Kuntcheva, M.J.; Obretenov, T.D. (1996). *Z. Lebensm. Unters. Forsch.*, **202**, 238-243.

Autoxidation Mechanism of Reductones and its Significance in the Maillard Reaction

Tadao Kurata, Noriko Miyake and Yuzuru Otsuka*; *Institute of Environmental Science for Human Life, Ochanomizu University, Tokyo 112, Japan, *Faculty of Education, Tottori University, Tottori 680, Japan*

Various types of reductones, including highly reactive amino-reductones, are known to play a key role in Maillard reactions in food and biological systems. For instance, effects of oxygen on the reaction rates and products of the Maillard reaction are well known, and, in most cases, the reductones are considered to be involved. Since most of the reductones are unstable and difficult to isolate, their exact structural characteristics and chemical reactivities, especially the mechanisms of their auoxidation, are still not clear. Quite recently, the reaction mechanism of autoxidation of L-ascorbic acid (ASA), which is a typical aci-reductone, was investigated. A new autoxidation mechanism, including the formation of a C2-oxygen adduct of ASA as the reaction intermediate, was proposed on the basis of both experimental results and those of semi-empirical molecular orbital (MO) calculations.[1,2] Formation of the superoxide anion and C2-C3 cleavage products in the autoxidation of ASA was also considered to occur through this C2-oxygen adduct. A similar reaction mechanism was also confirmed to be operative in the autoxidation of triose-reductone. On the other hand, autoxidation of amino-reductones, such as the enaminol form of fructose-amino acids, such as fructoselysine, is considered to be important as the precursor of some bio-marker compound of advanced glycation endproducts, and especially glycoxidation products, *in vivo*, but, again, the detailed reaction mechanism is not known.

Here, the autoxidation reactions of ene-diol aci-reductones, as well as enaminol aminoreductones, have been investigated by using the semi-empirical MO method, and their mechanisms, including that of the formation of *N*-carboxymethyl-lysine, is to be described. A common feature of the autoxidation mechanisms of these reductones and their significance in Maillard reactions will also be discussed.

[1] Kurata, T.; Miyake, N.; Suzuki, E.; Otsuka, Y. (1996). In *Chemical Markers for the Quality of Processed and Stored Foods*, Lee, T.-C., Kim, H.-J., Eds, American Chemical Society Symposium Series: Washington, DC, pp. 137-145.
[2] Kurata, T.; Miyake, N.; Otsuka, Y. (1996). *Biosci. Biotech. Biochem.*, **60**, 1212-1214.

Ascorbic Acid as the Principal Reactant Causing Browning in an Orange Juice Model System

Theodore P Labuza, Amar Kaanane* and Catherine Davies; *Department of Food Science and Nutrition, University of Minnesota,St Paul, MN 55108, USA, *Départment de Chimie-Biochimie Alimentaire, Institut Agronomique et Vétérinaire Hassan II, Rabat, Morocco*

The loss of vitamin C, degree of nonenzymic browning and furfural build-up in model systems simulating orange juice were studied. Model systems containing amino acids, minerals and the sugars found in orange juice were prepared, one with vitamin C and one without. These were evaluated during storage at 21°C and 45°C, conditions simulating distribution of UHT pasteurized juice in a typical punch box. The results showed that furfural production and browning development were absent in the system without vitamin C, even though ample sugars were present (sucrose, fructose and glucose). In addition, the solutions were absolutely clear, i.e., no colour developed for up to 57 days storage at either temperature. However, in the system with vitamin C, degradation of vitamin C, furfural accumulation and browning occurred in a linear manner. This indicates that the browning in citrus fruit juices is most likely to be due to ascorbic acid degradation and not to a sugar-amine (Maillard) reaction, probably because of the low pH.

[1] Handwerk, R.L.; Coleman, R. L. (1988). Approaches to citrus browning problem: A review. *J. Agric.Food Chem.*, **36**, 231-236.
[2] Kaanane, A.; Kane, D.; Labuza, T.P. (1988). Time and temperature effect on stability of Moroccan processed orange juice during storage. *J.Food Sci.*, **53**, 1470-1473.
[3] Trammell, D.J.; Dalsis, DE.; Malone, C.T. (1986). Effect of oxygen on taste, ascorbic acid loss and browning for HTST-pasteurized single-strength orange juice. *J. Food Sci.*, **51**, 1021-1023.

Maldi Mass Spectrometry in the Evaluation of Glycation Level of γ-Globulins in Healthy and Diabetic Subjects

Annunziata Lapolla, Rosaria Aronica, Michele Battaglia, Massimo Garbeglio, Domenico Fedele, Martina D'Alpaos, Roberta Seraglia* and Pietro Traldi†; *Istituto di Medicina Interna, Malattie del Metabolismo, Via Giustiniani 2, I-35100 Padova, Italy, *CNR, Area di Ricerca, Corso Stati Uniti 4, I-35100 Padova, Italy, †CNR, Centro di Studio sulla Stabilità e Reattività dei Composti di Coordinazione, Via Marzolo 1, I-35100 Padova, Italy*

It has been suggested that relationships between inflammatory reactions or immune disorders in diabetics may be connected with alterations of biological activity in glycated γ-globulins, themselves related to modification of antigen-binding capacity or to modifications of their ability to activate the complement system. Consequently, analytical methods able to give effective information on the glycation level of γ-globulins are of great interest. Recently a new mass spectrometric method, matrix-assisted laser-desorption/ionisation mass spectrometry (MALDI), was shown to be very powerful in the protein glycation field, being able to determine, with an accuracy of 0.1%, the molecular mass of intact proteins, without any need of digestion procedures. Such a method has been applied in the determination of glycation levels of plasma proteins, e.g., human serum albumin and β-globins, providing evidence of clear differences between healthy subjects and well and badly controlled diabetic patients. In particular, in the case of β-globins, the technique allowed glycated and glyco-oxidised products to be distinguished, leading to the design of specific therapies.

Here we report the results obtained by the same analytical approach in the determination of the glycation level of γ-globulins in 10 healthy subjects and 10 well controlled and 20 badly controlled type 2 diabetic patients. While in the case of healthy subjects, the mean molecular mass of the major component of the γ-globulins was found to lie in the range 148,400-148,700 Da, in the case of well controlled diabetic patients, it increased to 148,800-150,100 Da. This difference becomes greater in the case of badly controlled diabetic patients, where the range was 150,100-152,600 Da. Such results provide good experimental evidence of the occurrence of extensive γ-globulin glycation in the case of diabetic subjects and thus give the physician a parameter to relate to possible immune disorders.

The Maillard Reaction of Glucose and Ascorbic Acid with Guanosine under oxidative stress

Bernd Larisch, Monika Pischetsrieder and Theodor Severin; *Institut für Pharmazie und Lebensmittelchemie der Universität München, Sophienstraße 10, D-80333 München, Germany*

In long-living cells, DNA is susceptible to chemical modifications, such as oxidation, deamination, depurination or alkylation, resulting in mutations and, as a consequence, in oncogenesis and genetic dysfunctions. Glucose, glucose-6-phosphate or other sugar-derived substances can also react with DNA, causing strand-breaks and cross-links (e.g., Ref. 1). *In vitro* and *in vivo* studies have shown that DNA nucleotides are glycated in a process similar to the well known Maillard reaction. From reaction mixtures of glucose and guanosine, the diastereomeric guanosine derivatives, N^2-(1-carboxy-4,5,6-trihydroxypentyl)guanosine and N^2-(1-carboxyethyl)guanosine (CEG), have been isolated as main products. L-Ascorbic acid (AA) can also undergo Maillard reactions, resulting in the formation of brown and fluorescent products and protein cross-links, and some reaction products have been isolated.[2] Especially α-dicarbonyl compounds, which are formed during the oxidative breakdown of carbohydrates or AA, are powerful glycating agents. For example, oxidation of glucose yields glucosone and DHA results from AA under aerobic conditions. Since both products have higher glycation activity than unmodified sugars, they can easily bind to amino groups of proteins or amino acids, initiating Maillard reactions. Assuming that these dicarbonyl compounds are also more reactive towards nucleotides, we incubated guanosine with AA, DHA or xylosone and the mixtures were investigated by HPLC. Two products were identified as diastereomers of CEG. It can be assumed that, in parallel to the formation of CEG from glucose, AA first breaks down to give methylglyoxal, which reacts further with guanosine, resulting in the products. The structures of the main products were elucidated as diastereomers of N^2-(1-carboxy-3-hydroxypropyl)-guanosine (CHPG). CHPGs are formed from AA and guanosine under oxidative conditions, but also when dehydroascorbic acid and xylosone are used as educts. The fact that xylosone is a reactive intermediate of this reaction leads to the assumption that other sugar osones, such as glucosone, react in a similar manner, resulting in the formation of analogous products. The reaction products, CEG and CHPG, can be detected after heating at higher temperatures or incubating at 37°C and pH 7.4. This study shows that L-ascorbic acid and dicarbonyl compounds derived from carbohydrates or AA can be responsible for guanosine modification.

[1] Bucala, R.; Model, P.; Cerami, A. (1984). *Proc. Natl. Acad. Sci. U.S.A.*, **81**, 105–109.
[2] Larisch, B.; Pischetsrieder, M.; Severin, Th. (1996). *J. Agric. Food Chem.*, **44**, 1630–1634.

Amadori Products and a Pyrrole Derivative from Model Reactions of D-Glucose/3-Deoxyglucosone with Phosphatidylethanolamine

Markus O Lederer; *Institut für Lebensmittelchemie, Universität Hohenheim, Garbenstr. 28, 70593 Stuttgart, Germany*

Nonenzymic glycosylation of aminophospholipids is supposed to play an important role in lipid oxidation *in vivo* (1). There is a significant correlation between lipid-linked advanced glycosylation endproducts (AGEs) and oxidised low-density lipoprotein (LDL). Knowledge about both primary and secondary products from the reaction of D-glucose with aminophospholipids could provide a deeper insight into the mechanism of AGE-initiated lipid oxidation. We here report on how 1-deoxy-1-[2-(1,2-dialkanoyl-*sn*-glycero-3-phosphooxy)ethylamino]-D-fructose (**1,2**) and 1-[2-(1,2-ditetradecanoyl-*sn*-glycero-3-phosphooxy)ethyl]-5-hydroxymethylpyrrole-2-carbaldehyde (**3**) can be identified in the products of model reactions of D-glucose/3-deoxyglucosone (3-DG) and the respective phosphatidylethanolamine (PE).

$$1, 3: R^1 = C_{13}H_{27}$$
$$2: R^1 = C_{15}H_{31}$$

Independent syntheses and unequivocal structural characterisation are given for peracetylated E/Z-1-deoxy-1-(2-hydroxyethylamino)-D-fructose O-methyloxime (**4a,b**), E/Z-1-deoxy-1-(2-hydroxyethylamino)-D-fructose (3-methylbenzothiazolin-2-ylidene) hydrazone (**5a,b**) and 1-(2-hydroxyethyl)-5-hydroxymethylpyrrole-2-carbaldehyde (**6**). Chromatographic and spectroscopic data for these compounds were established by either GLC/MS or HPLC with diode array detection (DAD). PE and D-glucose/3-DG were incubated at 37°C and pH=7.4 in neat buffer or ethanol-buffer mixtures for four weeks, and the phospholipid fraction was purified on a C18 solid-phase extraction column. The phosphatidic acid was cleaved either with phospholipase D or 1 N ethanolic KOH to form the free Amadori product and the free pyrrole, respectively. The Amadori product was derivatised to give **4a,b** or **5a,b**. Both these derivatives and **6** were identified from the respective incubations by GLC/MS and HPLC/DAD analyses. With Amadori product and pyrrole formation from D-glucose/3-DG and phosphatidylethanolamine now established definitively, further investigations are in progress on whether and how these compounds influence lipid oxidation.

[1] Bucala, R.; Makita, Z.; Koschinsky, T.; Cerami, A.; Vlassara, H. (1993). *Proc. Natl. Acad. Sci.*, **90**, 6434-6438.

Adverse Effect of Advanced Glycation Endproduct-Modified Laminin on Neurite Outgrowth and its Implications for Brain Aging

Jenny J Li*; *Glaxo Institute for Molecular Biology, 14 Chemin des Aulx, 1228 Geneva, Switzerland,*
Current address: The Picower Institute for Medical Research, Laboratory of Diabetes and Aging, 350 Community Drive, Manhasset, NY 11030, USA

Glucose reacts nonenzymically with a wide range of proteins and lipids to form advanced glycosylation endproducts (AGEs) that accumulate in the tissues. It is well known that tissue accumulation of AGEs is increased in diabetes and aging. Long-lived, extracellular matrix proteins of peripheral tissues are particularly susceptible to AGE modification. The mechanisms which contribute to cognitive function decline, including decreased learning and memory capacity in aged humans, remain unknown. Recently, the normal aging human brain was shown to accumulate AGEs in neurons and glial cells.[1] Laminin, a major extracellular matrix component in brain parenchyma, plays a crucial role in neurite outgrowth and plasticity.[2] Since memory formation and learning involve constant remodelling of nerve terminals (synapses), we examined the effect of AGE-modified laminin on neurite growth of both differentiated pheochromocytoma PC12 cells and primary rat cortical neurons. Laminin (EHS sarcoma; 5 mg/ml) was incubated in the presence or absence of glucose (0.4 g/ml) in phosphate buffer (pH 7.2) under sterile conditions at 37°C for 10 weeks. At the end of the incubation, laminin was dialysed extensively to remove excess glucose, and AGE-modification of laminin was quantified by AGE-specific ELISA and Western blot analysis. PC12 cells were incubated on AGE-modified laminin (330 AGE units/mg laminin) or control laminin (26 AGE units/mg) pre-coated 8-chamber slides (1×10^4 cells/chamber). After overnight incubation, cells were induced to differentiate in Dulbecco's modified Eagle medium (DMEM), containing 10% FCS, 80 ng/ml nerve growth factor (NGF) and 1 mM cAMP. The extent of neurite outgrowth was assessed at 5 and 7 days with an inverted microscope (Axiovert 10, Zeiss) and semi-quantified with a computer imaging system (Vidas, Zeiss). Our results showed that PC12 cells grown on AGE-modified laminin had significantly shorter neurites (2459 ± 1604, N=34, perikarya counted) when compared with cells incubated on control laminin (4573 ± 2300, N=36). Our results also showed that this effect was not a result of the AGE-laminin affecting PC12 cell differentiation into mature neurons. Similar results were also obtained with primary rat cortical neuronal cultures, in that arborization of primary cortical neurons grown on AGE-modified laminin was impaired.

In summary, our results suggest that accumulation of AGE-modified extracellular matrix proteins such as laminin in brain parenchyma could have adverse effects on neurite outgrowth, and thus on synapse remodelling. These effects could contribute to the impairments in learning and short-term memory formation associated with brain aging.

[1] Li, J.J.; Surini, M.; Catsicas, S.; Kawashima, E.; Bouras, C. (1995). *Neurobiol. Aging*, **16**, 69-76.

[2] Murtomäki, S.; Risteli, J.; Risteli, L.; Koivisto, U.-M.; Johansson, S.; Liesi, P. (1992). *J Neuroscience Res.*, **32**, 261-273.

Effect of Low Concentrations of Aminoguanidine on Formation of Advanced Glycation Endproducts *in Vitro*

Jason Liggins, Nicola Rodda, Viki Burnage*, Jim Iley* and Anna Furth; *The Open University, Oxford Research Unit, Boars Hill, Oxford OX1 5HR, UK, *The Open University, Department of Chemistry, Milton Keynes, MK7 6AA, UK*

Despite current interest in the therapeutic potential of aminoguanidine (AG), there is little information on its mode of action at the low concentrations likely to occur *in vivo*. Data from patients with end-stage renal disease, for example, suggest that oral doses of 1.2 g give plasma levels of the order of 0.13 mM.[1] We therefore investigated the lowest effective concentrations at which AG inhibits AGE formation *in vitro*, looking in particular at fluorophores, crosslinks, protein-bound carbonyls and carboxymethyl-lysine (ML).

When 0.6 mM BSA was incubated in 0.1 M glucose for 7 days, 1 mM AG blocked AGE crosslinking between pre-glycated and native protein[2] by 100%. The formation of AGE fluorophores was blocked by 60% and of protein-bound carbonyls by 10%. In this low millimolar range, output of fluorophores was very sensitive to further changes in AG concentration, being fully inhibited in only 4 mM AG. Output of CML was similarly affected.

However, some carbonyls could be detected even after glycation in the presence of high AG concentrations. The plot of AG concentration against output of protein-bound carbonyls flattened off, with 70% inhibition in 25 mM AG; higher AG concentrations had no further inhibitory effect. This suggests that not all the protein-bound carbonyls produced by glycoxidation are fully accessible to AG. (The carbonyls of sugar derivatives, for example, may be involved in ring closure; slow acyclisation rates would then effectively protect these groups from AG.) Hence our results support suggestions[3] that the main targets for AG are free, small-molecule carbonyl derivatives postulated in Maillard reaction schemes.

[1] Buccala, R.; Makita, Z.; Vega, G.; Grundy, S.; Kochinosky, T.; Cerami, A.; Vlassara, H. (1994). *Proc. Nat. Acad. Sci.*, **91**, 9441-9445.
[2] Liggins, J.; Furth, A.J. (1996). *Biochem. Biophys. Res. Comm.*, **219**, 186-190.
[3] Requena, J.R.; Vidal, P.; Cabezas-Cerrata, J. (1993). *Diabetes Res. Clin. Prac.*, **19**, 23-30.

Maillard Reaction in Glucose-Glycine Systems Studied by Differential Scanning Calorimetry

Lara Manzocco, Paola Pittia and Enrico Maltini; *The University of Udine, Dipartimento di Scienze degli Alimenti, Via Marangoni 97, 33100 Udine, Italy*

The Maillard reaction (MR) is a complex heat-inducted network of chemical reactions, occurring during food preparation, preservation and storage. It is well known that the development of this reaction can strongly affect colour and flavour, as well as the nutritional properties and the stability of foods. Because the MR is often undesirable, great attention has always been paid to the search for suitable indicators able to detect even the early stages of the reaction. Currently the extent of Maillard reaction is generally followed through some chemical and physical indicators, such as hydroxymethylfurfural, furosin, isomaltoglycoside, carbon dioxide and colour.[1] As heat plays an important role in the MR pathway, thermal analysis techniques, because of their high sensitivity, should be an interesting tool for evaluating the extent of the reaction, but, so far, few literature data on the thermal behaviour of the MR are available.[2]

In this preliminary study differential scanning calorimetry (DSC) was used to analyse the thermal behaviour of glucose, glycine and glucose-glycine mixtures subjected to different heat treatments. The relative importance of the thermal phenomena observed and the temperature range in which they appear are discussed.

[1] Lerici C.R.; Barbanti D.; Manzano M.; Cherubin S. (1990). *Lebensm. Wiss. u. Tech.*, **23**, 289-294.
[2] Raemy A.; Hurrel R.F.; Loelinger J. (1983). *Thermochem. Acta*, **65**, 81-92.

Interaction between Maillard Reaction Products and Lipid Oxidation in Intermediate-Moisture Model Systems

Dino Mastrocola, Marina Munari, Maria Cioroi* and Carlo R Lerici; *The University of Udine, Department of Food Science, Via Marangoni 97, 33100 Udine, Italy, *The University "Dunarea de Jos" of Galati, Department of Chemistry, Domneasca St. 47, 6200 Galati, Romania*

The effect of Maillard reaction products (MRP) on the kinetics of lipid oxidation in model systems, containing pregelatinized starch, glucose, lysine and soya oil, has been studied.

The samples, either containing all the components or excluding one or more of them, were heated at 100°C for different periods. Lipid oxidation and browning indices were determined and the results confirmed the ability of MRP to retard peroxide formation. Under the conditions adopted, the rate of the Maillard reaction seems to be increased by the presence of the oil and its oxidation products.

The antioxidant action of the MRP was also evaluated using a peroxide scavenging test based on crocin bleaching.[1] The results demonstrated that antioxidant activity developed with increased browning of the samples.

[1] Bressa, F.; Tesson, N.; Dalla Rosa, M.; Sensidoni, A.; Tubaro, F. (1996). *J. Agric. Food Chem.*, **44**, 692-695.

Suppressive Effect of 4-Hydroxyanisole on Pyrazine Free Radical Formation in the Maillard Reaction

Bozidar Lj Milić, Sonja M Djilas, Jasna Čanadanović-Brunet and Natasa B Milić; *University of Novi Sad, Faculty of Technology, Department of Organic Chemistry, Bulevar Cara Lazara 1, YU-21000 Novi Sad, Yugoslavia*

The effect of phenolic compounds on the Maillard pathways responsible for the formation of free radicals, as precursors of aroma compounds as well as aminoimidazoazarenes, has received little attention. Studies on related topics suggest several routes by which phenolic compounds may interact with the Maillard reaction.[1]

4-Hydroxyanisole as suppressant of pyrazine cation radical formation in the Maillard reaction has been studied by Electron Spin Resonance (ESR). The model system consisted of 2-aminobutanoic acid and D-(+)-glucose (both 1.0 M), in aqueous solution at pH 9.0, heated at 98°C for 10 minutes, with and without 4-hydroxyanisole (0.15 to 0.45 M). The reaction mixture after heating was transferred to a Bruker quartz cell 4121, and spectra were recorded on a Bruker 300E ESR spectrometer, using settings as follows: field sweep width, 50 G; receiver gain, 1.021 G; microwave power, 6.32×10^{-1} mW; center field, 3440 G; time constant, 2.56 ms; sweep time, 335.544 s; microwave frequency, 9.640 G; and modulation amplitude, 0.204 G.

The results of the investigation by ESR spectroscopy, including the measured relative intensity of ESR spectra, showed that 4-hydroxyanisole has a suppressive effect on the formation of the pyrazine cation free radical during the Maillard reaction.

4-Hydroxyanisole, under reaction conditions characteristic for the Maillard reaction, has shown its semiquinone free radical to be unreactive towards oxygen ($k \ll 10^5$ M^{-1} s^{-1}), although the reaction with reactive oxygen species to form the quinone free radical is very rapid ($k = 8.7 \times 10^8$ M^{-1} s^{-1}).[2] This distinctive trait of 4-hydroxyanisole to form sequinone/quinone radicals has been applied to the Maillard reaction as a suppressant of the formation of pyrazine cation radicals, produced during the induction period of the reaction between carbonyl and amino compounds.

[1] Milić, B.Lj.; Djilas, S.M.; Čanadanović-Brunet, J.M. (1996). *J. Serb. Chem. Soc.*, **61**, 797-801.
[2] Cooksey, C.J.; Land, E.J.; Riley, P.A.; Sarna, T.; Truscot, T.G. (1987). *Free Rad. Res. Comm.*, **4**, 131-138.

Characterisation of the Fluorescence Associated with Human Serum Albumin Incubated with Glucose and Icodextrin-Based Peritoneal Dialysis Fluids

David J Millar, Samra Turajlić, Thomas Henle* and Anne Dawnay; *Renal Research Laboratory, G3 Dominion House, 59 Bartholomew Close, St Bartholomews Medical College, London EC1A 7BE, UK, *Lehrstuhl für Milchwiss., Techn. Universität, München, D85350 Freising-Weihenstephan, Germany*

Glycation and subsequent advanced glycation endproduct (AGE) formation, arising from the nonenzymic interaction of reducing sugars with amino groups of biomolecules, is a major contributing factor to diabetic complications and aging. Furthermore, it is to be expected that the high glucose concentration required in glucose-based peritoneal dialysis (PD) fluids may be responsible for the reduction in peritoneal membrane ultrafiltration capacity suffered by many long-term PD users. Use of icodextrin (a glucose polymer) as the osmotic agent in PD fluids has the potential to reduce glycation and hence AGE formation, due to the lower molar concentration of icodextrin required for successful ultrafiltration. Recent work demonstrated that commercial PD fluids inhibit the growth of a variety of tissue culture cells.[1] Analysis PD fluids[2] identified a number of reactive aldehydes arising from glucose breakdown during heat sterilisation. We demonstrated that there is a rapidly reacting component(s) in heat-sterilised PD fluids which contributes to fluorescence generation.[3]

Here, we compares the generation of fluorescence by glucose and icodextrin in heat- and filter-sterilised solutions and the ability of aminoguanidine, an AGE inhibitor, to quench the fluorescence generated by both sugars and their degradation products. The fluorescence arising from glucose and icodextrin incubated with human serum albumin was scanned for excitation and emission. Pyralline, a nonfluorescent AGE, was determined to ascertain whether a characterised AGE correlated with the fluorescence.

[1] Wieslander, A.P.; Nordin, M.K.; Martinson, E.; et al. (1993). *Perit. Dial. Int.*, **13**, 208-213.
[2] Nilsson-Thorell, C.B.; Muscalu, N.; Andren, A.H.G.; et al. (1993). *Clin. Nephrol.*, **39**, 343-348.
[3] Millar, D.J.; Dawnay, A.B.S. (1995). *J. Amer. Soc. Nephrol.*, **6**, 551.

Advanced Glycation in Diabetic Embryopathy: Increase in 3-Deoxyglucosone in Rat Embryos in Hyperglycaemia *in Vitro*

Harjit S Minhas, Paul J Thornalley, Parri Wentzel* and Ulf J Eriksson*; *University of Essex, Department of Biological and Chemical Sciences, Central Campus, Wivenhoe Park, Colchester CO4 3SQ, UK, *Department of Medical Cell Biology, University of Uppsala, Biomedicum, S-751 23, Uppsala, Sweden*

Diabetic pregnancy is associated with increased risk of foetal malformation. The causative agent(s) of embryo dysmorphogenesis in diabetes are not known, but hyperglycaemia and advanced glycation may be involved.[1] We examined the concentrations of the potent α-oxoaldehyde glycating agents, glyoxal, methylglyoxal and 3-deoxyglucosone (3-DG), in embryos (4-5 per culture), membranes (yolk sac and amnion) and medium of rat embryos in culture medium containing 10, 30 and 50 mM glucose after 48 h, changing the medium after 24 h. α-Oxoaldehydes were assayed by derivatization with 1,2-diamino-4,5-dimethoxybenzene and HPLC of adducts.[2]

For cultures with 10 mM glucose, the concentrations of glyoxal and methylglyoxal in the culture medium were 0.1-0.3 μM, in embryos 0.1-0.3 nmol and in membranes 0.2-0.8 nmol, but did not increase significantly with increased glucose concentration. After 24 h, the concentration of 3-DG in the culture medium with 10 mM glucose was 0.4 μM, and increased markedly to 2.4 μM and 5.6 μM in cultures with 30 mM and 50 mM glucose, respectively (n = 3, P < 0.01). Similar increases were found in the medium at 48 h, and also in tissues. In embryos (n = 3): 10 mM glucose, 0.3 nmol 3-DG; 20 mM glucose, 1.7 nmol 3-DG (P < 0.05) and 50 mM glucose, 5.5 nmol 3-DG (P < 0.001). In membranes: 10 mM glucose, 0.8 nmol 3-DG; 20 mM glucose, 1.7 nmol 3-DG (P < 0.05) and 50 mM glucose, 6.5 nmol 3-DG (P < 0.001). 3-DG in embryos therefore increased 18-fold in model hyperglycaemia, with similar increases in the culture medium and membranes. 3-DG is formed by degradation of fructosamines and fructose-3-phosphate. It reacts with proteins to form cysteinyl hemithioacetal adducts and advanced glycation endproducts, pyrraline and imidazolones,[3] and induced apoptosis.[4] 3-DG may be a prospective novel initiator of diabetic embryopathy.

[1] Stryrud, J.; Thunber, L.; Nybacka, O.; Eriksson, U. (1995). *Pediatr. Res.*, **37**, 343-353.
[2] McLellan, A.C.; Phillips, S.A.; Thornalley, P.J. (1992). *Anal. Biochem.*, **206**, 17-23.
[3] Thornalley, P.J. (1996). *Endocrinol. and Metab.*, **3**, 149-166.
[4] Okado, A.; Kawasaki, Y.; Hasuike, Y.; *et al.* (1996). *Biochem. Biophys. Res. Commun.*, **225**, 219-224.

Reduced Susceptibility of Pyrraline-Modified Albumin to Lysosomal Proteolytic Enzymes

Satoshi Miyata, Bing-Fen Liu, Hiroyuki Shoda, Takeshi Ohara, Hiroyuki Yamada, Kotaro Suzuki and Masato Kasuga: *Kobe University School of Medicine, The Second Department of Internal Medicine, Kobe 650, Japan*

The Maillard reaction is thought to play a role in the pathogenesis of angiopathy in diabetes and in the aging process. Plasma levels of pyrraline, one of the advanced glycation endproducts (AGEs), are elevated in diabetic rats and humans as determined by ELISA using antibodies to pyrraline.[1,2] Pyrraline has also been demonstrated to be present in vascular lesions of diabetic and elderly subjects using immuno-histochemical techniques.[2] Among the several pathways leading to pyrraline, highly reactive dicarbonyl compounds, such as 3-deoxyglucosone (3-DG), were identified as precursors, which react with free amino groups to form pyrraline.[1,2] We recently reported that plasma 3-DG levels were elevated in diabetic rats.[3]

AGEs are also known to alter the structural and functional properties of proteins. Furthermore, interaction between proteins modified by AGEs and cells, such as macrophages, may be involved in diabetic complications. In the present study, we examined whether modification of albumin by pyrraline influences its degradation by phagocytes. Degradation of pyrraline-modified albumin by these cells was diminished, causing accumulation of the albumin in the cells. Our findings suggested that the accumulation of pyrraline-modified albumin in the cells was due to the reduced susceptibility of the protein to lysosomal enzymic degradation. Such alterations in the interaction between AGE-modified protein and phagocytes may contribute to angiopathy in elderly subjects and patients with diabetes.

[1] Hayase, F.; Nagaraj, R.H.; Miyata, S.; Njoroge, F.G.; Monnier, V.M. (1989). *J. Biol. Chem.*, **264**, 3758-3764.
[2] Miyata, S.; Monnier, V.M. (1992). *J. Clin. Invest.*, **89**, 1102-1112.
[3] Yamada, H.; Miyata, S.; Igaki, N.; Yatabe, H.; Miyauchi, Y.; Ohara, T.; Sakai, M.; Shoda, H.; Oimomi, M.; Kasuga, M. (1994). *J. Biol. Chem.*, **269**, 20275-20280.

Photo-Enhanced Modification of Human Skin Elastin in Actinic Elastosis by N^ε-(Carboxymethyl)lysine, one of the Glycoxidation Products of the Maillard Reaction

Kumiko Mizutari, Tomomichi Ono, Kazuyoshi Ikeda*, Ken-ichi Kayashima and Seikoh Horiuchi*; *Department of Dermatology and *Department of Biochemistry, Kumamoto University School of Medicine, Kumamoto 860, Japan*

In the skin following chronic exposure to sunlight, especially in persons with a fair complexion, hyperplasia of the elastic tissue is usually evident in the upper dermis by the age of 30. These changes found in the photoaged skin constitute actinic elastosis. Although it is not clear by what mechanism chronic solar exposure can induce these accumulations of the elastic fibres, deleterious effects of sunlight can often be attributed to oxidation by free radical intermediates. It is therefore possible that ultraviolet-induced oxidative stress may have some role in the pathogenesis of actinic elastosis through the formation of oxidation products. Since N^ε-(carboxymethyl)lysine (CML) has been proposed as a potential biomarker of oxidative damage of tissue proteins *in vivo*, we have attempted in the present study to determine whether CML-modification could occur in the lesions of actinic elastosis. Immunohistochemical and immuno-electron microscope exami-nation with a monoclonal anti-AGE antibody (6D12), whose epitope is CML,[1,2] demonstrated CML-accumulation predominantly in elastic fibres, especially in the amorphous electron-dense material corresponding to photo-induced degeneration rather than electron-lucent regions. Immunochemical analyses with enzyme-linked immuno-sorbent assay of elastase-soluble fractions demonstrated that the CML-levels of the sun-exposed area were significantly higher than those of the unexposed area. The lesions of actinic elastosis were similarly stained by our polyclonal anti-AGE antibody, whose epitope is not known, but differs from CML or pentosidine. In addition, these lesions were also stained positively by the polyclonal anti-pentosidine antibody, indicating the presence of pentosidine as well as other AGE-structure(s). We conclude that ultraviolet-induced oxidation may accelerate formation of CML in actinic elastosis of the photoaged skin.

[1] Horiuchi, S.; Araki, N.; Morino, Y. (1991). *J. Biol. Chem.*, **266**, 7329-7332.
[2] Ikeda, K.; Higashi, T.; Sano, H.; Jinnouchi, Y.; Yoshida, M.; Araki, T.; Ueda, S.; Horiuchi, S. (1996). *Biochemistry*, **35**, 8075-8083.

Identification and Antioxidative Activity of the Main Compounds from a Lactose-Lysine Maillard Model System

Simona M Monti, Vincenzo Fogliano, Alberto Ritieni and Giacomino Randazzo; *Università di Napoli "Federico II", Dipartimento di Scienza degli Alimenti, Via Università 100, 80055, Portici (NA), Italy*

    The reducing disaccharide lactose occurs in abundance in milk and reacts extensively by the Maillard reaction with the ε-amino group of lysine This reaction is responsible for the marked development of off flavour, colour and/or instability. The Maillard reaction leads to a very complex mixture of products, including substances of low, intermediate and high molecular mass, the last presumably melanoidins.[1]
    The objective of this study was to investigate the products of the thermal lactose-lysine Maillard reaction with particular attention to their antioxidative activity. Aqueous molal solutions of the lactose-lysine model system were refluxed for 1, 2, 3 or 4 h without pH control. Samples were collected after each hour and the reaction monitored by measuring the final pH, the absorbance and the antioxidative activity.[2] Clear differences were evident with increased time of heating, the mixture heated for 4 h showing the greatest absorbance and antioxidant power. This suggests that the coloured intermediates of the Maillard reaction are the main contributors to the antioxidative activity of Maillard reaction products. Next, this solution was separated into the three fractions of different molecular mass, which were analysed directly by HPLC with diode-array detection.[3] The main resolved peaks were collected and analysed by NMR and their antioxidative activity was determined.

[1] Hodge, J. E. (1953). *J. Agric. Food Chem.*, **1**, 928-943.
[2] Pryor, W.A.; Cornicelli, J.A.; Devall, L.J.; Tait, B.; Trivedi, B.K.; Witiak, D.T.; Wu, M. (1995). *J. Org. Chem.*, **58**, 3521-3532.
[3] Bailey, R.G.; Ames, J.M.; Monti, S.M (1996). *J. Sci. Food Agric.*, **72**, 97-103.

Effect of Model Melanoidins on Oxidative Stress of Neural Cells

G S Moon, W Y Lim and J S Kim; *Department of Food Science and Nutrition, Inje University, Kimhae, Korea*

    Melanoidins, the final products of the Maillard reaction, are well known for their antioxidant activity. Neural cells are readily damaged by oxidative stress. In this study, the antioxidant activity of melanoidins on neural cells under oxidative stress, provoked by paraquat or H_2O, was determined. For this study, three kinds of model melanoidins were made, from glucose-glycine, glucose-lysine and xylose-arginine. They were made to have the same, brown colour intensity. The SK-N-SH cell line was used to provide the neural cells.
    When the comparative antioxidant activity in linoleic acid emulsions of three kinds of model melanoidins was determined by the ferric thiocyanate method, peroxide value and conjugated diene content, the xylose-arginine melanoidin showed the strongest activity, despite being at the same colour intensity. The xylose-arginine melanoidin also showed the strongest electron-donating ability, implying that the antioxidant activity of melanoidin is due to its electron-donating activity.
    To characterise the melanoidins, each was separated on a Sephadex G-50 column and, for each fraction, the brown colour intensity, reducing power, ninhydrin-positive reaction and antioxidant activity were determined. The antioxidant activity of the melanoidin fractions was closely related to their reducing power and colour intensity.
    Model melanoidins were effective in protecting the neural cells from oxidative damage by paraquat or H_2O. Xylose-arginine melanoidin was the most effective in protecting from oxidative damage by paraquat. Xylose-arginine and glucose-lysine melanoidins also prevented oxidative damage to neural cells by H_2O.
    Model melanoidins increased the enzyme activity of QR but not SOD activity.

Analysis of Fluorescent Compounds Bound to Protein in Casein-Lactose Systems

Francisco J Morales, Carmen Romero and Salvio Jiménez-Pérez; *Instituto del Frío (CSIC), Department of Dairy Products, Ciudad Universitaria, s/n, Madrid 28040, Spain*

The Maillard reaction is a complex network of chemical reactions which usually occur during processing and storage of foods. The extent of the reaction is affected by many factors, such as temperature, water activity, pH, moisture content and chemical composition of the food system. The initial stage, in which Amadori compounds are formed, does not give rise to colour, fluorescence or absorbance in the near UV. Upon prolonged heating, the Amadori rearrangement products undergo dehydration and fission and yield colourless reductones as well as fluorescent substances, some of which may also be pigmented. At this point, an increasing amount of unsaturated carbonyl compounds is observed. The fluorescence as well as the light-absorbing properties of the system increases due to formation of smaller molecules. In the advanced stage of the Maillard reaction, proteins are modified, giving coloured, fluorescent and cross-linked molecules. In the process, many reactive compounds are formed as intermediates, such as keto and aldosamines, deoxyosones, furans, pyrroles, pyrazines, pyranones and imidazoles. In stages prior to the formation of brown pigments, fluorescent compounds are formed. Several authors have pointed out that the coloured and fluorescent compounds were not identical and that the fluorogens may be precursors of the brown pigments.[1]

Previously, we have studied the accumulation of free fluorescent compounds in milk and milk-resembling systems (sodium caseinate or concentrate of whey proteins with lactose), which seem to take two different reaction routes: (a) the more important route is the Maillard reaction, but free fluorescent compounds are also formed in trace amounts by (b) Lobry de Bruyn-Alberda van Ekenstein transformation.[2]

The aim of this work is to study the development of fluorescence associated with proteins in casein-lactose systems. The heated systems are to be digested under controlled conditions using pronase. Subsequently, a clarified solution (solid-phase step) of the peptides is to be analysed by HPLC, assessing their elution time in relation to their polarity. Different column types are to be used to extend the study.

[1] Ames, J.M. (1992). In *Biochemistry of Food Proteins*, Hudson, B.J.F., Ed., Elsevier Applied Science: London, pp. 99-153.
[2] Morales, F.J.; Romero, C.; Jiménez-Pérez, S. (1996). *Food Chem.*, 57, 423-428.

Metal-Binding Properties of Glycated Proteins and Amino Acids

Valeri V Mossine and Milton S Feather; *University of Missouri, Department of Biochemistry, Columbia, MO 65211, USA*

Recent reports suggest that oxidative reactions may be important factors in the overall Maillard reaction, and that metal ions may function as redox catalysts for these degradations. Metal ions, such as copper(II) and iron(III) could, therefore, be involved in oxidative pathways in both *in vivo* and *in vitro* Maillard reactions. Zinc, calcium and iron have been reported to be excreted or to accumulate in mammalian excretory organs in elevated amounts when feed contains Maillard reaction products (MRP); cell aggregation and apoptosis, which are believed to depend on calcium(II) and zinc(II), respectively, are strongly affected by the presence of MRP. Collectively these data suggest that glycated amino acids, peptides and proteins possess metal-binding properties different from those of their non-glycated analogues.

This was investigated using ultrafiltration and potentiometric techniques. The data collected to date indicate that glycated bovine serum albumin (BSA) possesses an increased capacity to bind non-specifically doubly and triply charged metal ions. At physiological pH, glycated BSA contains at least one additional high-affinity center, which was able to compete with citrate for copper(II) and aluminium(III). Parallel studies were also carried out using Amadori compounds (fructose amino acids) and N^ε-carboxymethyl-lysine as low-molecular-mass models for glycated proteins. For these compounds, conditional binding (affinity) constants, derived from thermodynamic formation constants determined for zinc(II) complexes, were at least ten-fold higher for the glycated amino acids than for the non-glycated analogues, and, for the case of copper(II) complexes, the binding constants were hundreds of times higher for the glycated materials vis-a-vis the non-glycated analogues.

Cytotoxicity of Advanced Glycation Endproducts

Gerald Münch, Birgit Geiger, Claudia Loske, Andreas Simm and Reinhard Schinzel; *University of Würzburg, Physiological Chemistry I, Biocenter, Am Hubland, 97074 Würzburg, Germany*

In Alzheimer's disease and haemodialysis-associated amyloidosis, protein modified by advanced glycation endproducts (AGEs) is deposited and accumulates in tissue over time.[1,2] Although AGEs were previously considered to be inert components, there is considerable evidence that they interact with cells.[3] In this study, we investigated the direct toxic effects of AGEs to baby hamster kidney (BHK) cells.

For this purpose we incubated the cells for 16 h with different concentrations of ovalbumin-AGE, BSA-AGE and carnosine-AGE. Cell viability was assessed by counting the cell number and by the MTT (metabolic activity) assay.

All AGEs tested inhibited cell growth in a dose-dependent manner; in high concentrations (e.g., about 20 µM chicken egg albumin-AGE), they caused significant cell death. However, AGE toxicity was attenuated by the addition of intermediates of the citric acid cycle, such as ketoglutarate or pyruvate, and intracellularly acting antioxidants, such as thioctic acid. To investigate further the mechanism of cell death, changes in cell morphology were followed for several hours. The cells began to shrink to about 25% of their original cell volume 4 hours after the addition of AGEs, but retained an intact cell membrane which disintegrated in the later stages. The chronology of events leading to cell death suggests initially apoptosis, but later exhibits characteristics similar to necrotic cell death.

[1] Smith, M.A.; Taneda, S.; Richey, P.L.; Miyata, S.; Yan, S.D.; Stern, D.; Sayre, L.M.; Monnier, V.M.; Perry, G. (1994). *Proc. Natl. Acad. Sci. USA*, **91**, 5710-5714.

[2] Vitek, M.P.; Bhattacharya, K.; Glendening, J.M.; Stopa, E.; Vlassara, H.; Bucala, R.; Manogue, K.; Cerami, A. (1994). *Proc. Natl. Acad. Sci. USA*, **91**, 4766-4770.

[3] Wautier, J.L.; Wautier, M.P.; Schmidt, A.M.; Anderson, G.M.; Hori, O.; Zoukourian, C.; Capron, L.; Chappey, O.; Yan, S.D.; Brett, J. (1994). *Proc. Natl. Acad. Sci. USA*, **91**, 7742-6.

Determination of Advanced Glycation Endproducts in Serum by Fluorescence Spectroscopy and Competitive ELISA

Gerald Münch, Regina Keis, Ullrich Bahner*, August Heidland*, Horst-Dieter Lemke†, Holger Kayser† and Reinhard Schinzel; *University of Würzburg, Physiological Chemistry I, Biocenter, Am Hubland, 97074 Würzburg, Germany, *Curatorium for Dialysis and Transplantation, Hans-Brandmann-Weg 1, 97078 Würzburg, Germany, †Central Research RMB, AKZO-Nobel AG, 63784 Obernburg, Germany*

Advanced glycation endproducts (AGEs) play a significant role in the evolution of complications in degenerative and chronic diseases, especially in diabetic complications (renal failure) and dialysis-associated amyloidosis.[1] AGEs accumulate in maintenance haemodialysis (MHD) patients by the insufficient removal of low molecular mass AGE peptides. Determination of AGE levels in the serum is mainly achieved by competitive ELISAs, but this method suffers from several drawbacks, for example, low sensitivity and long analysis time.[2,3]

In this study, the measurement of the intrinsic fluorescence of AGEs is introduced as a second independent method of quantifying AGEs in the serum. Fluorescence emission spectra between 400 and 480 nm (with a peak at 440 nm) at the AGE excitation wavelength of 370 nm are almost identical for synthetic AGEs and the serum. The AGE serum level measured by competitive ELISA was approximately zero for the controls and 0.32 ± 0.16 AGE units for the MHD patients. AGE serum levels measured by fluorescence also increase in MHD patients in comparison to healthy controls ($1.11 \pm 0.27 \times 10^5$ AU, n = 158, vs $0.31 \pm 0.06 \times 10^5$ AU, n = 19). Comparison of these assays demonstrated a significant correlation (r = 0.53, p < 0.001). The difference of AGE values in each assay suggests a distinct distribution of antigenic vs fluorescent AGEs for each individual patient. Preliminary results show that only low molecular mass AGE levels can be reduced *in vivo* by dialysis, high-flux membranes being more efficient in this than conventional membranes. Both methods will be employed to monitor the serum AGE level of dialysis patients in order to determine the best membrane type for this purpose.

[1] Buccala, R.; Cerami, A. (1992). *Adv. Pharmacol.*, **23**, 1-34.

[2] Friedlander, M.A.; Wu, Y.C.; Schulak, J.A.; Monnier, V.M.; Hricik, D.E. (1995). *Am. J. Kidney Dis.*, **25**, 445-451.

[3] Papanastasiou, P.; Grass, L.; Rodela, H.; Patrikarea, A.; *et al.* (1994). *Kidney Int.*, **46**, 216-222.

Advanced Glycation Endproducts as "Pacemakers" of β-Amyloid Deposition in Alzheimer's Disease

Gerald Münch, Claudia Loske, Arne Neumann, Johannes Thome*, Reinhard Schinzel and Peter Riederer*; *University of Würzburg, Physiological Chemistry I, Biocentre, Am Hubland, 97074 Würzburg, Germany,* *University of Würzburg, Clinical Neurochemistry, Psychiatric Hospital, Füchsleinstrasse 15, 97080 Würzburg, Germany*

Advanced glycation endproducts (AGEs) are structural components of neurofibrillary tangles and amyloid plaques in Alzheimer's disease (AD) patients.[1] We have shown that nucleation-dependent polymerization of β-amyloid peptide is significantly accelerated by AGE-mediated crosslinking.[2] This process is further accelerated by the addition of μM amounts of transition metal ions such as Cu and Fe. These data suggest that the reported increase of free transition metal in AD may be also responsible for the accelerated formation of amyloid deposits *in vivo*.

With the exception of ascorbic acid, formation of the AGE-crosslinked β-amyloid aggregates can be attenuated by metal chelators, antioxidants and inhibitors of AGE formation, suggesting innovative pharmacological strategies for the treatment of AD.[3] Circumstantial evidence for this hypothesis comes from clinical trials with tenilsetam, where a significant improvement in cognition and memory in up to 87% of AD patients was observed.[4] These data support the rationale that drugs which interfere with AGE formation, subsequently attenuating oxidative stress, may become promising therapeutic agents in the prevention or treatment of AD.[5]

[1] Harrington, C.R.; Colaco, C.A.L.S. (1994). *Nature*, **370**, 247-248.
[2] Münch, G.; Mayer, S.; Michaelis, J.; Hipkiss, A.R.; et al. (1997). *Biochem. Biophys. Acta*, **1360**, 17-29.
[3] Münch, G.; Taneli, Y.; Schraven, E.; et al. (1994). *J. Neural Transm. [P-D-Sect.]*, **8**,193-208.
[4] Ihl, R.; Perisic, I.; Maurer, K.; Dierks, T. (1989). *J. Neural Transm. [P-D-Sect.]*, **1**, 84-85.
[5] Münch, G.; Thome, J.; Double, K.; Riederer, P. (1996). *Alzheimer's Dis. Rev.*, **1**, 71-74.

Browning; Does the Matrix Matter?

Wan A W Mustapha, Sandra E Hill, John M V Blanshard and William Derbyshire; *The University of Nottingham, Department of Applied Biochemistry and Food Science, Sutton Bonington, Loughborough LE12 5RD, UK*

Browning of foods has been studied extensively due to the importance of colour in consumer perception of quality. Melanoidins are the coloured final products of the Maillard reaction. Their rate of production is dependent on concentration and type of reactants (amino and carbonyl groups), pH, temperature and water activity (a_w). The Maillard reaction is important in solid foods of limited moisture content and, therefore, the solubilities and mobilities of the reactants have been used to explain the difference in the browning rate observed. The bell-shaped curve relating reaction rate to a_w has been explained in terms of solubility and hence the concentration in solution of the reactants.[1] The mobilities of the reactants within the solid matrix have been considered in terms of the flexibility of the molecular structure of the matrix. Karel and Buera[2] demonstrated that different rates of browning were dependent on the glass-transition temperature of the system. Although these concepts provide some understanding and therefore ability to predict the likely degree of browning in food systems, they are not a total explanation.

Recent work has shown that the physical state of the matrix, although important, is not the only relevant parameter in governing the rate of the Maillard reaction. The chemical composition of the matrix is also important, as is its ability to form different phases. Studies have been carried out using different liquid matrices (miscible and non-miscible). Solids plus liquids have also provided matrices and the studies using starch as the solid component have shown unexpected results. These may have been due to different macro and micro environments existing within the system, reflecting the structure of many real foodstuffs, which are also not homogeneous and contain regions that have different concentrations of materials and show a range of physical properties. In this type of matrix, an equilibrium state may not be achievable due to movement and generation of the different materials. The rates of Maillard reactions are to be demonstrated in a range of matrices and then related to the initial and changing composition of the matrix.

[1] Labuza, T.P. (1980). *Food Technology*, **34** (4), 36-41, 59.
[2] Karel, M.; Buera, M.P. (1994). In *Maillard Reactions in Chemistry, Food and Health*, Labuza, T.P. *et al.*, Eds, Royal Society of Chemistry: Cambridge, pp. 164-169.

N^ε-(Carboxymethyl)lysine Formation from Amadori Products by Hydroxyl Radicals

Ryoji Nagai, Kazuyoshi Ikeda, Takayuki Higashi, Hiroyuki Sano, Yoshiteru Jinnouchi, Kenshi Matsumoto and Seikoh Horiuchi; *Department of Biochemistry, Kumamoto University School of Medicine, Honjo 2-2-1, Kumamoto 860, Japan*

Recent immunological studies demonstrated the presence of N^ε-(carboxymethyl)lysine (CML)-modified proteins in several tissues and their increase on aging and diseases such as diabetic complications and atherosclerosis. Since CML formation from glucose-protein adducts or fatty acids needs oxidation,[1] it is considered to be a biomarker of glycoxidation or lipid oxidation products *in vivo*. Takanashi et al.[2] showed that addition of Fe^{3+} to glycated human serum albumin (glycated HSA) leads to CML formation and proposed that enediol chelated with Fe^{3+} might be oxidized to the 2,3-dicarbonyl. However, the mechanism of the subsequent decomposition of the 2,3-dicarbonyl into CML and erythronic acid is not well understood. In the present study, to examine the effects of radical oxygen species on CML formation from an Amadori product, CML generated from glycated HSA was measured by the immunochemical method using anti-CML antibody.[3] Our glycated HSA prepared by 7 days' incubation with 1.6 M glucose and DETAPAC did not contain sufficient CML to be detectable. Incubation of the glycated HSA with 0.4 mM $FeCl_2$ for 1 hour led to CML formation at a detectable level. CML formation was enhanced dose-dependently by addition of hydrogen peroxide, but was significantly inhibited by catalase or mannitol. Superoxide anion, generated by the xanthine/xanthine oxidase system or hydrogen peroxide itself, failed to produce CML from glycated HSA. In addition, superoxide dismutase had no effect on this process. These data indicate that hydroxyl radicals generated by the Fenton reaction between Fe^{2+} and hydrogen peroxide might play a major role in the decomposition of the 2,3-dicarbonyl into CML and erythronic acid.

[1] Fu, M.; Requena, J.R.; Jenkins, A.J.; Lyons, T.J.; *et al.* (1996). *J. Biol. Chem.*, **271**, 9982-9986.
[2] Takanashi, M.; Sakural, T.; Tsuchiya, S. (1992). *Chem. Pharm. Bull.*, **40**, 705-708.
[3] Ikeda, K.; Higashi, T.; Sano, H.; Jinnouchi, Y.; Yoshida, M.; *et al.* (1996). *Biochemistry*, **35**, 8075-8083.

Effects of an Organo-Germanium Compound (Ge-132) on Naturally Occurring Diabetes Mellitus Rats "OLETF"

Kunie Nakamura, Yoshikuni Fujita*, Toshihiko Osawa† and Norihiro Kakimoto‡; *Molecular Biology Laboratory and *Department of Internal Medicine, Kitasato University School of Medicine, 1-15-1 Kitasato, Sagamihara, Kanagawa, 228 Japan, †Faculty of Agriculture, Nagoya University, Furo-cho, Chikusaku, Nagoya ,464 Japan, ‡Asai Germanium Research Institute, 1-6-4 Izumi Honcho, Komae, 201 Tokyo, Japan*

An organo-germanium compound [bis(2-carboxyethylgermanium)trioxide, Ge-132] induced the reversible solubilization of advanced glycation endproducts (AGEs) *in vitro* and *in vivo*.[1] In diabetes mellitus (DM) induced in rats by streptozotocin, Ge-132 suppressed the progress of DM-complications by peroral administration in aqueous form.[2-4] The OLETF (Otsuka Long Evans Tokushima Fatty) rat is the model for naturally occurring human Type II DM. In the present study, 60 OLETF rats were observed for 72 weeks with/without administering Ge-132 at 100 mg/kg/d, performing clinical examinations and immuno-pathological observations on kidney and brain, in comparison with 20 normal LETO (Long Evans Tokushima Otsuka) rats without DM. Though plasma glucose concentration was not reduced by Ge-132, urinary volume, urinary glucose, creatinine, β-D-N-acetylglucosaminidase (NAG), glycated albumin and fructoseamine showed significant differences between OLETF and those treated with Ge-132. Immuno-pathological staining with anti-AGEs antibodies (Abs) demonstrated the retention of AGEs in the proximal tubules and the basement membrane of glomeruli in OLETF-kidneys, and Ge-132 caused weakening of the stain by the Abs. Anti-NO_2-Tyr Abs reacted with endothelial cells in OLETF rat kidney, but only gave a weak reaction with OLETF kidney treated with Ge-132. Other clinical parameters for DM are under investigation. Thus, it may be concluded that organo-germanium compounds such as Ge-132 can affect the progress of organ damage due to AGEs by preventing the formation of AGEs in OLETF rats, resulting in the suppression of DM complications that usually cause serious clinical problems in the treatment of DM.

[1] Nakamura K.; Uga, S.; Miyata, M.; Kakimoto, N. (1990). In *Maillard Reaction in Food Processing, Human Nutrition and Physiology,* Finot, P.A. *et al.*, Eds, Basel: Birkhäuser, pp. 461-468.
[2] Nakamura K.; Nomoto, K.; Kriya, K.; Nakajima, Y.; *et al.* (1991). *Amino Acids*, **1**, 263-278.
[3] Nakamura K.; Osawa, T.; Kawakishi, S.; *et al.* (1996). *Main Group Metal Chemistry*, **19**, 301-306.
[4] Nakamura K.; Miyata, M., Shimojo, N.; Naka, K.; *et al.* (1996). *J. Subtle Energy Researches*, **1**, 43-50.

Influence of pH and Metal-Ion Concentration on the Kinetics of Nonenzymic Browning

John O'Brien; *The University of Surrey, School of Biological Sciences, Guildford GU2 5XH, UK*

pH has a significant influence on the mechanism and kinetics of nonenzymic browning reactions. Reaction rates generally increase with increasing pH above pH 7 up to a maximum at pH ~9-10. Reaction rates also increase, though less dramatically, at very low pH values, resulting in the characteristic 'U'-shaped curve for browning versus pH. During processing and storage, the pH of foods may change, adversely affecting their stability to nonenzymic browning.

The overall Maillard reaction and the rate of browning of Amadori products is accelerated by Fe(II), Fe(III) and Cu(II) ions.[1] However, the extent to which the effect of metal ion may be influenced by pH effects has not been investigated. Furthermore, there is evidence that several products of the Maillard reaction may form complexes with metal ions.[2] It is still not clear how and to what extent complex formation may influence the kinetics and mechanism of the Maillard reaction. However, it has been reported that complex formation is a prerequisite for the promoting activity of Cu and Fe on the rate of browning.[3] This is the first study to examine systematically the influence on nonenzymic browning of the interaction between pH and metal-ion concentration.

The study involved the incubation of a model system at pH 2-10 for up to 750 h in the presence of various concentrations of Mn(II), Mg, Pb, K, Sn, Cu(II), Zn(II), Na, Ni(II), Sr, Fe(II), Fe(III), Al, Ca or Co(II). The results suggest that the effect of metal ions on browning rates depends on the initial pH of the system. Browning rates were accelerated dramatically in the presence of Cu or Fe. Several pH-browning curves showed evidence of complex formation. Paradoxically, some ions such as Mn appear to promote the rate of browning at one pH while inhibiting it at another.

[1] O'Brien, J. (1997). In *Advanced Dairy Chemistry, Vol.3*, Fox, P.F., Ed., Chapman and Hall: London, pp. 155-231.
[2] O'Brien, J.; Morrissey, P.A. (1997). *Food Chem.*, **58**, 17-27.
[3] Kato, H.; Matsumura, M.; Hayase, F. (1981). *Food Chem.*, **7**, 159-168.

First Evidence for Accumulation of Protein-Bound and Protein-Free Pyrraline in Human Uraemic Plasma by Mass Spectrometry

Hiroko Odani, Toru Shinzato, Yoshihiro Matsumoto, Jun Usami and Kenji Maeda; *Department of Internal Medicine, Daiko Medical Center, Nagoya University, Nagoya, Japan*

Glucose-derived advanced glycation endproducts (AGEs) cross-link proteins and cause various types of biological tissue damage.[1] One of the crosslinks, pyrraline [ε-(2-formyl-5-hydroxymethylpyrrol-1-yl)-L-norleucine], has been demonstrated by using antibodies to accumulate in plasma and sclerosed matrix of diabetic individuals, which suggests their responsibility for diabetic complications. To elucidate the involvement of pyrraline in uraemia, we examined the pyrraline levels in patients with chronic renal failure by a mass spectrometric approach. We show that protein-free pyrraline, as well as pyrraline bound to protein, is significantly increased in non-diabetic uraemic plasma compared to healthy subjects. Our results suggest that circulating pyrraline could be a substance contributing to complications in uraemia.[2]

[1] Monnier, V.M.; Cerami, A. (1981). *Science*, **211**, 491-494.
[2] Odani, H; Maeda, K. (1996). *Biochem. Biophys. Res. Comm.*, **224**, 237-241.

Oxidative Damage to Proteins and Abnormal Concentration of Pentosidine during Haemodyalysis and after Renal Transplantation

Patrizio Odetti, Giovanna Gurreri, Silvano Garibaldi, Luana Cosso*, Irene Aragno, Sabina Valentini, Maria Adelaide Pronzato* and Umberto Maria Marinari*; *Department of Internal Medicine (DIMI) and *Institute of General Pathology, University of Genoa, Italy*

Nonenzymic glycoxidation and oxidative damage to proteins have been reported to be significantly increased during uraemia. In twelve uraemic subjects, undergoing chronic haemodialysis (for at least 3 months) and in six uraemic patients, who had undergone renal transplantation, endproducts of those reactions have been determined in plasma and urine in order to evaluate the mean level during substitutive therapy and the trend after renal replacement. Plasma pentosidine, a marker of advanced glycation endproducts (28 ± 3 pmol/mg protein) and plasma carbonyl content (1.6 ± 0.1 nmol/mg protein) were high during haemodialysis, but dropped after transplantation. In the months following the transplantation, pentosidine and carbonyl level continued to decrease, but after 24 months the first had not yet reached the normal range (4.1 ± 0.8 pmol/mg protein), whereas the carbonyl groups were within the normal range (1.2 ± 0.2 nmol/mg protein). In transplant patients, 24-hour pentosidine excretion was not significantly different from that of controls (50-200 nmol/24h) and pentosidine clearance increased toward the normal range (0.7-1.6 ml/min).

To elucidate the structure of the peptide containing pentosidine removed by kidney, we pooled the urines of uraemic patients on haemodialysis, of those who had undergone renal transplantation and of control subjects, respectively. The samples were loaded onto a 1.6/70 Superdex 30 column (Pharmacia) and the peptide identified. The molecular mass was about 1000 Da and similar for the three groups.

These findings suggest that both oxidative stress and glycation damage are elevated in uraemia and require a long time to restore the redox balance and to improve the removal system, which is likely to have been compromised by uraemia. A negative influence of the immunological therapy cannot be ruled out. The pentosidine-containing peptide is likely to be a degradative product of circulating and tissue glycated protein.

A Novel Type of Advanced Glycation Endproduct Found in Diabetic Rats

Toshihiko Osawa, Tomoko Oya, Harue Kumon, Yasujiro Morimitsu, Hiroyuki Kobayashi*, Mitsuo Akiba* and Norihiro Kakimoto*; *Department of Applied Biological Sciences, Nagoya University, Chikusa, Nagoya 464-01, Japan, *ASAI Germanium Institute, Komae, Tokyo 201, Japan*

Much attention has been focused on the formation of advanced glycation endproducts (AGEs), because the accumulation of AGEs has been taken to represent an integrated measure of exposure to glucose over time in diabetic patients. Until now, several different types of AGEs, such as pyrraline, crossline and pentosidine, have been isolated and identified as fluorescent protein cross-link and glycation markers. Recently, we have succeeded in detecting a novel fluorescent product (Em 340 nm, Ex 402 nm), named MRX, in the acid hydrolysates of rabbits' erythrocyte membrane ghosts after the reaction with glucose for two weeks. This has prompted us to make a chemical investigation of MRX. The model reactions of HSA, BSA, lysozyme and ribonuclease with glucose have also been found to produce MRX.

In order to isolate and identify MRX, a large scale preparation and isolation of MRX has been carried out from the reaction of a mixture of BSA with glucose. Pure MRX was obtained. By spectroscopic analyses, such as those using 2-dimensional NMR, FAB-MS, FT-IR and UV, the chemical structure of MRX has been confirmed as the conjugate of the oxidation product of glucose with a cysteine residue of BSA (molecular mass 162 Da), formed in the presence of arginine (or a guanidino-type compound). On detailed examination, a large quantity of MRX was detected in urine excreted by rats with streptozotocin-induced diabetes. MRX was also detected in urine excreted by OLETF, the model of naturally occurring human diabetes mellitus (DM) type II.

MRX is supposed to be produced from the precursor bound to protein, and this paper discusses recent work examining the formation mechanisms of MRX both in vitro and in vivo systems.

[1] Kawakishi, S.; Nasu, R.; Cheng, R.Z.; Osawa, T. (1996). In *Chemical Markers for Processed and Stored Foods*, Lee, T-C., Kim, H-J., Eds, American Chemical Society Symposium Series: Washington DC, pp. 77-84.

[2] Kato, Y.; Tokunaga, K.; Osawa, T. (1996). *Biochem. Biophys. Res. Commun.*, **226**, 923-927.

Effect of L-Ascorbate on the Oxidative Reaction of Lysine in the Formation of Collagen Cross-Links

Meguni Otsuka, Miwa Kuroyanagi, Eriko Shimamura and Nobuhiko Arakawa; *Ochanomizu University, Department of Nutrition and Food Science, 2-1-1, Otsuka, Bunkyo-ku, Tokyo 112, Japan*

Pyridinoline (Pyr), a mature cross-link of collagen fibrils, is especially abundant in collagen of cartilage and bone. Data from the cartilage of guinea pigs, as well as that from humans, indicated the remarkable increase with growth.[1] We also found that depletion of ascorbate in guinea pigs during growth resulted in an increase of Pyr in the cartilage. The purpose of this study was to clarify the role of ascorbate in the formation of Pyr *in vitro*. The initial step in the formation of Pyr is a deamination of lysyl and hydroxylysyl residues, catalysed by lysyl oxidase, yielding reactive aldehydes which in turn form Schiff bases with the ε-amino groups of hydroxylysyl residues to provide dehydrodihydroxylysinonorleucine (deDHLNL). Pyr is formed by a non-enzymic condensation of hydroxylysine and deDHLNL.

In a model reaction, using partially purified lysyl oxidase from bovine aorta and soluble collagen prepared from bovine cartilage, we examined changes of deDHLNL and Pyr as a function of time. The content of deDHLNL, which normally increases with incubation time, decreased remarkably on addition of ascorbate. However, the content of Pyr increased when the initial incubation of the reaction mixture was carried out in the absence of ascorbate. These results indicate that ascorbate is involved in the formation of Pyr at the stage of the initial enzymic reaction. Enzyme activity of lysyl oxidase, which was highly purified from bovine aorta according to the method of Shackleton, was determined using polylysine as the substrate. Addition of ascorbate to the reaction mixture under aerobic conditions led to decreasing enzymic activity. Thiols such as glutathione had no effect on the activity of lysyl oxidase, whereas addition of erythorbate and 3,4-dihydroxybenzoate, which are structural analogues of ascorbate, lowered the activity in a similar manner to ascorbate. This implies that the inhibitory effect of ascorbate on the activity of lysyl oxidase is a characteristic of its structure. These findings suggest that ascorbate has an important regulatory role in the oxidative reaction of lysine in the formation of collagen cross-links.

[1] Kim, M.; Otsuka, M.; Arakawa, N. (1994). *J. Nutr. Sci. Vitaminol.*, **40**, 95-103.
[2] Otsuka, M.; Kim, M.; Shimamura, E.; Arakawa, N. (1994). In *Maillard Reactions in Chemistry, Food, and Health*, Labuza, T.P., Reineccius, G.A., Monnier, V., O'Brien, J., Baynes, J., Eds, Royal Society of Chemistry: Cambridge, p. 431.

Different Responses of Retinal Microvascular Cells to Advanced Glycation Endproducts

C Paget*, N Rellier*, D Ruggiero-Lopez, M Lecomte, M Lagarde and N Wiernsperger; *LIPHA-INSERM U 352, INSA-Lyon, Building 406, 69621 Villeurbanne Cedex, France (*Contributed equally to this work)*

Diabetic retinopathy is a microangiopathy involving changes in the thickness of the vascular basement membrane, as well as modifications of the biology of microvascular cells, such as endothelial cells and pericytes. A mechanism that could be involved is the accelerated Maillard reaction of glucose with cellular and extracellular proteins, which leads to the formation of advanced glycation endproducts (AGE). Since retinal endothelial cells and pericytes seem to behave differently during the evolution of diabetic retinopathy, we investigated whether AGE could affect parameters, such as cell proliferation, enzymic glycosylation and enzymic antioxidant defences, in a different way according to the type of cell.

Accordingly, pure cultures of endothelial cells and pericytes from bovine retinal microvessels were incubated in the presence of AGE from bovine serum albumin. Cell proliferation was assessed by cell counting and DNA measurement. Enzymic glycosylation was studied by analysis of the sugar composition of cell glycoproteins using lectin affinoblotting. Enzymic antioxidant defences were evaluated by measurement of catalase and glutathione peroxidase activities. The results showed that, for each parameter studied, endothelial cell and pericyte responses to AGE were markedly different. Indeed, in pericytes, AGE decreased proliferation and increased catalase activity without any significant changes in the cell glycoprotein composition. On the contrary, in endothelial cells, proliferation was increased by AGE, catalase activity remained unchanged and significant modifications in endothelial cell glycoproteins were observed.

The different responses of endothelial cells and pericytes to AGE raise the question of the pathways involved in the cellular effects of AGE and whether these underlie the well known differences in behaviour of both constitutive cell types in diabetic retinopathy.

Mitochondrial Aminophospholipid Modification by Advanced Maillard Reaction is Related to the Longevity of Mammalian Species

Reinald Pamplona, Manuel Portero-Otín, Maria Josep Bellmunt, Joan Prat and Jesús R Requena*; *Department of Basic Medical Sciences, University of Lleida, 25198 Lleida, Spain, *Department of Chemistry and Biochemistry, University of South Carolina, Columbia, SC 29208, USA*

Among mammals, as the specific metabolic rate (vO_2, SMR) decreases, the rate of aging decreases and the maximum life span (MLSP) increases. This "rate of living" phenomenon may be explained by the increases in mitochondrial free radical production as mitochondrial vO_2 increases. Accordingly, it has been observed that the rate of mitochondrial O_2^- and H_2O_2 generation is inversely correlated to MLSP and directly related to SMR. Moreover, there is an inverse relationship between oxygen consumption rate and protein half-life. Oxidation is also implicated in the *in vivo* Maillard reaction (glycoxidation).

We evaluated the levels of carboxymethylethanolamine (CME) in liver mitochondria, as a biomarker of mitochondrial aminophospholipid damage by oxidative processes, from adult specimens of ten species, each at an age ≈30% of its MLSP (total n = 45), namely, mouse, Sprague-Dawley rat, guinea pig, rabbit, sheep, dog, pig, cow, horse and man. Their MLSP values (3 to 90 years) were obtained from the literature. The SMR of the animals was calculated using Kleiber's equation: SMR (cal g^{-1} day^{-1}) = 393 (g body weight)$^{-0.25}$. Mitochondrial fractions were isolated by differential centrifugation, their lipids were extracted with a chloroform: methanol:water (2:1:1 v/v) system, and CME was determined by means of GC/MS.

The results show that CME correlates logarithmically with MLSP (CME (mmol/mol ethanolamine) = 0.50 + 0.22 ln MLSP (yr), r = 0.81, p < 0.005, n = 45) and inversely in a linear fashion with SMR (CME (mmol/mol ethanolamine) = 1.38 - 0.004 SMR, r = -0.79, p < 0.007, n = 45). The present data show unexpectedly that longer-living mammals display higher levels of oxidative damage in terms of CME in mitochondrial membrane aminophospholipids. This trend could result from differences in rates of phospholipid turnover, as proposed for pentosidine accumulation in the extracellular matrix proteins.

The Inclusion of 5-Hydroxymethylfurfural in the Sulfite-Inhibited Maillard Reaction Scheme

Jane K Parker, Jennifer M Ames and Douglas B MacDougall; *The University of Reading, Department of Food Science and Technology, Whiteknights, Reading, RG6 6AP, UK*

The inhibition of the Maillard reaction by sulfite is well documented. We show here, however, that 5-hydroxymethylfurfural (HMF), which is often used as a marker for the Maillard reaction, is not necessarily inhibited to the same extent as the development of brown colour. We suggest an explanation in terms of the scheme proposed by Wedzicha.[1] An aqueous model fruit system, comprising 30% sugars (glucose and fructose) and 1% amino acids (glutamic acid, aspartic acid, glutamine and γ-aminobutyric acid), was buffered at pH 4. Different amounts of sodium bisulfite (2-20000 ppm) were added to inhibit the formation of brown pigments. The solutions were heated at 97°C for 3 h and the absorbance at 420 nm was obtained at 15 min intervals. The amount of HMF was determined by HPLC after 3 h.

At high levels of sulfite (>2000 ppm), both the formation of pigments and HMF are inhibited, because of the competing irreversible reaction of the precursor, 3,4-dideoxyhexosulos-3-ene (DDH), with sulfite. Below 50 ppm, there is very little effect on the level of either pigments or HMF. However, at intermediate levels of sulfite, our results show that, although the formation of pigments has been substantially inhibited, the levels of HMF have not been reduced to the same extent. The two dose-response curves for sulfite show that, for example, at 700 ppm sulfite, inhibition of pigments is >85%, whereas levels of HMF have been reduced by <20%. Thus, of the 3 competing reactions of DDH (formation of brown pigment, formation of HMF and addition of sulfite), addition of sulfite is still important, but the levels of DDH are high enough to give significant amounts of HMF. One might have expected the third reaction (the formation of brown colour) to compete for the DDH, but its contribution turns out to be small. We suggest that, since the formation of pigments consists of a series of polymerisation reactions involving other reactive carbonyl species, the low levels of DDH (and other carbonyls tied up as hydroxysulfonates) become the limiting factor in the development of Maillard browning. In the absence of competition from the polymerisation reactions, the formation of HMF is relatively fast. HMF is still often used to monitor the Maillard reaction, but our results clearly show that, in this case, HMF is not a good marker of Maillard browning.

[1] Wedzicha, B.L. (1982). *Food Chem.*, **14**, 173-187.

Identification and Inhibition of Glycation Cross-Links Impairing the Function of Collagenous Tissues

R G Paul, T J Sims, N C Avery and A J Bailey; *Collagen Research Group, Division of Molecular and Cellular Biology, University of Bristol, Langford, Bristol, BS18 7DY, UK*

Glucose-mediated tissue damage has been implicated in the development of disease such as diabetes mellitus and in the progressive loss of tissue function with age. One of the most damaging effects is the formation of intermolecular cross-links with a concurrent stiffening of the tissue. The mechanism by which these cross-links form has been shown to be an extremely complex one, involving metal-catalysed oxidation.

Recent work in our laboratory has identified a non-fluorescent cross-link, NFC-1, which, in contrast to pentosidine, is present in molar concentrations sufficient to account for the changes in tissue properties both *in vitro*[1] and *in vivo*.[2,3] Consequently we are interested in the mechanism of formation of these cross-links and, ultimately, its inhibition.

Both NFC-1 and pentosidine are found during incubation of collagen with either glucose or ribose. However, each of these sugars produces additional but different components, indicating a different reaction mechanism. We have shown that NFC-1 is detectable during the early stages of the changes in physical properties of collagen, whilst pentosidine appears later. Both cross-links are inhibited by antioxidants, sulfhydryl compounds and chelating agents. Surprisingly, pentosidine formation is increased in the presence of DTPA and Tiron during incubation with ribose, but not glucose, supporting the concept of different reaction mechanisms for these two sugars. Ribose is often used for rapid glycation, but it is clearly not a substitute for the reactions of glucose.

Aminoguanidine is one of the few inhibitors that can be used *in vivo* to alleviate the deleterious effects of glycation cross-linking, but can also be studied *in vitro*. A study of the hydrothermal isometric tension generated on heating of tendons as a measure of stable intermolecular cross-links revealed obvious differences between tendons incubated in the presence of either sugar and aminoguanidine. Tendons, in which the formation of known cross-links had been completely inhibited, surprisingly still showed marked changes in physico-chemical properties, which were consistent with the formation of additional unknown cross-links. The use of such inhibitors must, therefore, be treated with caution until all aspects of the mechanism have been investigated.

[1] Bailey, A.J.; Sims, T.J.; Avery, N.C.; Halligan, E.P. (1995). *Biochem. J.*, **305**, 385-390.
[2] Sims, T.J.; Rasmusson, L.M.; Oxlund, H.; Bailey, A.J. (1996). *Diabetologia*, **39**, 946-951.
[3] Paul, R.G.; Bailey, A.J. (1996). *Int. J. Biochem. Cell Biol.*, **27**, 1-14.

Structural Modifications and Bioavailablity of Starch Components upon the Extent of the Maillard Reaction: An Enzymic Degradation and Solid State ^{13}C CP MAS NMR Study

Laura Pizzoferrato*, Maurizio Paci and Giuseppe Rotilio*†; *University of Rome, Tor Vergata, Department of Chemical Science and Technology, Rome, Italy, *National Institute of Nutrition, Via Ardeatina, Rome, Italy, †University of Rome, Tor Vergata, Department of Biology, Rome, Italy*

The modification of the starch components, amylose and amylopectin, from different sources has been studied from a structural point of view in order to understand the modifications induced by different levels of the Maillard reaction occurring in the presence of amino acids and of proteinaceous materials. In fact, this reaction produces changes in the starch components able to induce the formation of different allomorphs in the macromolecular structure. These change are clearly evident for starch components in the hydrated solid state.

The differences induced in parallel in the bioavailability of these compounds by these modifications were studied by enzymic degradation kinetics, the glucose released upon digestion being monitored. Structural characterization was obtained by solid-state ^{13}C CP MAS NMR, which has the potential to monitor solid-state phase changes. Changes in the crystalline state give rise to modifications of the chemical shifts in the NMR spectrum.

The results revealed that changes are induced by the Maillard reaction in the macromolecular structure of amidaceous materials. The purpose here is to correlate these modifications with changes in the digestibility of the starch and starch components.

Presence of Pyrraline in Human Urine and its Relationship with Glycaemic Control

Manuel Portero-Otín, Reinald Pamplona, Maria Josep Bellmunt and Joan Prat; *Metabolic Physiopathology Research Group, Department of Basic Medical Sciences, University of Lleida, Lleida, E-25198 Spain*

Oxidatively derived Maillard reaction products, such as pentosidine and carboxymethyl-lysine (CML), have been found in urine, probably as a result of protein catabolism. In contrast, the urinary elimination of pyrraline, a non-oxidative product, which has been found in plasma and in extracellular matrix proteins, has not been explored. Here, we seek to establish whether pyrraline is found in urine. Urine was submitted sequentially to two reversed-phase HPLC systems with diode-array detection. The material obtained was analysed by ES-MS. Pyrraline levels were evaluated in urine from insulin-treated diabetes-mellitus patients without complications and a healthy population. The presence of pyrraline in urine was established by means of HPLC, UV spectrometry and ES-MS, as well as by its sensitivity to acid and alkaline hydrolysis. Free urinary pyrraline mean values (1.61 ± 0.4 µg/mg creatinine, mean ± SEM) greatly exceed those of urinary pentosidine and approach those of urinary CML. It was observed that urinary pyrraline correlated inversely with plasma urea concentration ($p = 0.05$, $r = 0.21$, pyrraline (pmol/µl) = -0.07 x plasma urea (mmol/L) + 9.93), supporting the relationship between renal function and pyrraline urinary elimination. Pyrraline mean values were not significantly higher in the overall diabetic group, but its correlation with glucaemia ($p < 0.008$, $r = 0.38$, pyrraline (µg/mg creatinine) = 0.0065 x glucaemia (mg/dL) + 0.7), HbA$_{1c}$ ($p < 0.02$, $r = 0.36$, pyrraline (µg/mg creatinine) = 0.235 x HbA$_{1c}$ (%) - 0.35), and last year mean HbA$_{1c}$ ($p < 0.02$, $r = 0.4$, pyrraline (µg/mg creatinine) = 0.238 x HbA$_{1c}$ (%) - 0.41) prove its relationship with hyperglycaemia. Moreover, the negative correlation between urinary pyrraline and plasma urea concentrations supports the notion that urinary pyrraline is related to plasma pyrraline concentration, as reinforced by recent data.[1] The data presented here prove the relationship between sustained hyperglycaemia and urinary pyrraline elimination, suggesting that pyrraline might be related to the development of diabetic complications.

[1] Odani, H.; Shinyato, T.; Matsumoto, Y.; *et al.* (1996). *Biochem. Biophys. Res. Commun.*, **224**, 237-241.

Fluorescent Products from Aminophospholipids and Glucose

Joan Prat, Maria Josep Bellmunt, Reinald Pamplona and Manuel Portero-Otín; *Metabolic Physiopathology Research Group, Department of Basic Medical Sciences, University of Lleida, Lleida, E-25198 Spain*

The free amino group of phospholipids can react with glucose. Some of the products known at present seem to include oxidative steps in their synthetic pathway. On the other hand, oxidation of fatty acids is a well established process in which peroxidative breaking of aliphatic double bonds leads to reactions that generate very reactive aldehydes. The grafting of these aldehydes onto major biological molecules produces very complex structures resembling glycation products. In order to understand how the two reactions interact in the formation of fluorophores in a lipid-glucose mixture, we prepared several buffered (PBS pH 7.4) tubes containing glucose and lipids and incubated them at 37°C for 15 days. The following variables were taken into consideration: (a) Reactive or unreactive amino group in the phospholipid molecule (phosphatidylethanolamine or phosphatidylcholine); (b) Saturated or unsaturated fatty acids (synthetic 16:0 or a soybean mix-ture with more than 50% of 18:2); (c) Oxidative or non-oxidative conditions (PBS or PBS+BHT+EDTA+N$_2$). Samples treated under the same conditions but without glucose and/or lipids were used as controls. The lipids were extracted with chloroform. Fluorescence of the organic media was measured. The results show: (1) A reactive amino group is necessary to obtain fluorescent products. (2) Unsaturated lipids are necessary to obtain fluorescent products. (3) Under non-oxidative conditions, only glucose incubated with unsaturated phosphatidylethanolamine displays fluorescence, but phosphatidylethanolamine incubated alone does not. (4) Under oxidative conditions, unsaturated phosphatidylethanolamine displays fluorescence on incubation with or without glucose, the intensity being greater in the former situation. We conclude that: (I) In all situations, fluorescent-product formation requires the presence of a reactive amino group and unsaturated fatty acids, which, at least in part, contain oxidised fatty acids that react with the amine. (II) The presence of glucose increases the fluorescence obtained in phosphatidylethanolamine incubations. This could have two explanations: (1) the reaction of glucose with the amino group must start and/or accelerate some reactions leading to the formation of fluorescent lipid derivatives, probably through the generation of oxidised species, or (2) unsaturated fatty acid oxidation products can accelerate the formation of glyco-oxidative fluorescent products.

Diabetes and Effects of Collagen Glycation in Heart

Timothy J Regan; *UMD-New Jersey Medical School, 185 South Orange Avenue, University Heights, Newark, NJ 071043-2714, USA*

Collagen accumulation in myocardial interstitium in diabetes is considered to promote diastolic stiffness. Whether advanced glycosylation is responsible for the glycoprotein increment and the mechanical alteration is the subject of this pharmacological study in a canine alloxan-diabetic model of six-month duration.

Left ventricular chamber stiffness, derived from pressure-volume measurements *in vivo*, was increased in untreated diabetic (Gr. II) compared to normals (Gr. I). Basal heart rate, aortic pressure and ejection fraction were comparable. There was a rise in collagen concentration and Sirius Red stain revealed an increased interstitial distribution, but there was no change in collagen phenotype by gel electrophoresis. To assess the role of collagen glycation, the insoluble fraction was extracted, disgested with collagenase and measured for fluorescence by spectrofluorometry. Advanced glycation endproducts (AGEs) increased in Gr. II to 10.6 ± 1.6 FU/mg collagen *versus* 6.9 ± 0.7 FU/mg in Gr. I ($p<0.01$). Immunochemical assay of AGEs did not differ. Since angiotensin-converting enzyme inhibition (ACEI) modifies collagen in other heart disease models, enalapril was chronically administered to diabetics (Gr. III). Although glycation was normal, the chamber stiffness increment was reduced by 50%, but collagen concentrations remained elevated. Aminoguanidine was given daily at 20 mg/kg p.o. for 6 months (Gr. IV); glycation and chamber stiffness were normal. However, collagen concentrations were elevated. Thus, fluorescent collagen-linked glycosylation appeared to be a major but not exclusive process affecting diastolic function in untreated diabetics, but collagen increments were independent of glycosylation and chamber stiffness.

Detection of Age-Lipids *in Vivo*: Carboxymethylation of Aminophospholipids in LDL and Red Cell Membranes

Jesús R Requena, Mahtab U Ahmed, Sharanya Reddy, C Wesley Fountain, Thorsten P Degenhardt, Alicia J Jenkins*, Cliff Perez*, Timothy J Lyons*, John W Baynes and Suzanne R Thorpe; *The University of South Carolina, Department of Chemistry and Biochemistry, Columbia, SC 29208, USA, *The Medical University of South Carolina, Divison of Endocrinology, Charleston, SC 29425, USA*

Carbohydrate-dependent modification of amino groups in protein has been a major focus of studies on the Maillard reaction *in vivo*. Aminophospholipids also have free amino groups and are also susceptible to structural and functional damage resulting from the Maillard reaction. N^ε-(Carboxymethyl)lysine (CML) is the major glycoxidation product formed during browning of protein *in vitro* with glucose under aerobic conditions. At the same time, we have recently shown that CML is readily formed during the peroxidation of polyunsaturated fatty acids (PUFAs) in the presence of protein. Therefore, we speculated that carboxymethylation may be a likely modification of aminophospholipids *in vivo*, derived from oxidation of either Amadori adducts or PUFAs. We developed GC/MS assays to measure CM-ethanolamine (CME) and CM-serine (CMS), following hydrolysis of aminophospholipids. *In vitro*, CME was formed during glycation of dioleylphosphatidylethanolamine under air, but not under anaerobic conditions (N_2, DTPA). These results demonstrated the formation of CME via glycoxidation *in vitro*. *In vivo*, CME and CMS were detected in LDL isolated from plasma, red blood cell (RBC) membranes and urine of healthy individuals. In RBC ghosts (n=8), CME was present at 0.13 ± 0.06 mmol/mol ethanolamine and was comparable to levels of CML, 0.21 ± 0.07 mmol/mol lysine in these same proteins. At the same time, like CML, CME is not quantitatively a major source of lipid modification, but rather a marker of oxidative Maillard-reaction damage. The presence of CME in urine suggests active metabolism of CM-lipids.

We also hypothesized that CME, because of its structural homology with CML, would be detected by anti-AGE-protein antibodies, which are CML specific. CME effectively inhibited the detection of AGE-protein in a competitive ELISA using an anti-AGE antibody prepared in our laboratory. Therefore, previous reports of the detection of AGE-lipid in AGE-LDL, using anti-AGE-protein antibodies, may have been detecting CME. Further, the detection of AGE epitopes in lipid-rich atheromatous plaque may also be explained in part by the presence of CM-aminophospholipids in lesions. Comparison of levels of CM-aminophospholipids in blood, tissues and urine may prove useful for assessing the role of the lipids modified by the Maillard reaction in aging, diabetes, atherosclerosis and inflammatory disease.

Acknowledgements: Research grant AG11472 from the National Institutes on Aging. JRR is the recipient of a Juvenile Diabetes Foundation Post-doctoral Fellowship.

Studies of the Formation of 2-Acetyl-1-Pyrroline in Model Systems and in Bread

Lauren L Rogers, Colin G Chappell and Jerzy A Mlotkiewicz; *Dalgety PLC, Dalgety Food Technology Centre, Station Road, Cambridge CB1 2JN, UK*

2-Acetyl-1-pyrroline is a potent odour compound which plays an important role in the aroma of rice[1] and wheat bread crust.[2] Its formation has been studied in this laboratory using isotopically labelled standards and mass spectrometry. Studies elsewhere, based upon original proposals by Hodge,[3] suggest that 2-acetyl-1-pyrroline is formed from the acetylation of 1-pyrroline (from proline decarboxylation) by 2-oxopropanal.[4,5] Using proline labelled with a single ^{13}C at the carboxylic acid group, it was shown that the label is lost when 2-acetyl-1-pyrroline is formed. This was enforced by the study of model systems using U-^{13}C-labelled glucose and unlabelled proline, where the 2-acetyl-1-pyrroline formed is labelled in both positions of the acetyl moiety only. This suggests that the acetyl group is sugar-derived, and the ring system is proline-derived. Experiments in this laboratory with U-^{13}C-labelled glucose also support this, although the use of glucose does not prove the eventual involvement of 2-oxopropanal.

Further experiments have shown that U-^{13}C-proline, incubated with unlabelled 2-oxopropanal in a phosphate buffer model system, produces 2-acetyl-1-pyrroline containing only three labels, not four, demonstrating that the mechanism involves a ring-opening step, allowing the loss of an additional ^{13}C label besides that of the carboxylic acid. However, in further experiments with U-^{13}C proline and unlabelled 2-oxopropanal in bread, as opposed to model systems, 2-acetyl-1-pyroline containing four ^{13}C labels predominated, though some contained three ^{13}C labels. Some unlabelled product was also formed from endogenous substrate. These findings support the view that, predominantly, 2-acetyl-1-pyrroline can be formed in bread from the acetylation of 1-pyrroline, but that the acetylating agent is not 2-oxopropanal directly. A different mechanism is proposed by which 2-oxopropanal and 1-pyrroline interact *via* a ring-opening step. Further experiments with ornithine have assisted in understanding the mechanisms involved.

[1] Buttery, R.G.; Ling, L.C. (1982). *Chemistry and Industry*, **23**, 958-959.
[2] Schieberle, P.; Grosch, W. (1985). *Z. Lebensm. Unters. Forsch.*, **80**, 474-478.
[3] Hodge, J.E.; Mills, F.D.; Fisher, B.E. (1972. *Cereal Science Today*, **17**, 34-38, 40.
[4] Schieberle, P.(1989). In *Thermal Generation of Aromas*, ACS Symposium Series **409**, Washington DC, pp. 268-275.
[5] Grosch, W. and Schieberle, P. (1991). In *Volatile Compounds in Foods and Beverages*, Maarse, H., Ed., Marcel Dekker: New York, pp. 41-47.

Identification and Quantification of Class IV Caramels in Soft Drinks

Louise Royle, Jennifer M Ames, Laurence Castle*, Harry E Nursten and Catherine M Radcliffe;*The University of Reading, Department of Food Science and Technology, Whiteknights, Reading RG6 6AP, UK, *CSL Food Science Laboratory, Norwich Research Park, Colney, Norwich NR4 7UP, UK*

Class IV caramels are used to add colour to soft drinks. The colour intensity of a caramel is related to its nitrogen and sulfur content, which itself is determined by the amount of ammonium sulfite added to the sugar during the production process. Currently the method used to determine the amount of caramel in soft drinks is relatively unspecific, consisting simply of measurement of colour intensity.

Capillary electrophoresis (CE) in phosphate buffer at pH 9.5 is used here to analyse a range of Class IV caramels with different nitrogen and sulfur contents. A direct relationship between the migration time of the principal caramel peak and the sulfur content of the caramel has been found. This enables the type of Class IV caramel to be indicated by capillary electrophoresis. There is also a linear relationship between caramel concentration and peak area in the range 0.1 to 10 mg/ml, allowing quantification.

A range of soft drinks was analysed by the CE method developed, where the only sample pretreatment was with a 0.2 μm filter to remove particulates. The type of Class IV caramel in all of the samples analysed was indicated to be a high-nitrogen, high-sulfur content caramel (*ca* 5% N, 9% S). The concentration of this caramel was between 0.3 and 0.6 mg/ml in the cola drinks, and *ca* 1.3 mg/ml in Dandelion and Burdock.

Capillary electrophoresis has been shown to provide a fast and reproducible method for the direct determination of Class IV caramel type and concentration in soft drinks.

Reaction of Metformin with Reducing Sugars and Dicarbonyl Compounds

D Ruggiero-Lopez, M Lecomte, N Rellier, M Lagarde and N Wiernsperger; *LIPHA-INSERM U352, INSA-Lyon, Building 406, 69621 Villeurbanne Cedex, France*

The reaction between reducing sugars, such as glucose, and amino groups in proteins, which leads to the formation of advanced glycation endproducts (AGEs), has been shown to play a role in the development of the characteristic tissue pathology of diabetes. Guanidine derivatives, such as aminoguanidine, inhibit this process by blocking the reaction of amino groups with glucose or the dicarbonyl compounds derived from glucose. Despite the fact that the antidiabetic metformin (1,1-dimethylbiguanide) is a guanidine-like compound, it has not sofar been investigated for possible inhibitory effects on AGE formation.

Metformin was incubated at pH 7.4 and 37°C, in the presence of various reducing sugars and dicarbonyl compounds (glyoxal and methylglyoxal), which are more reactive with proteins than glucose. Reaction kinetics were assessed by analysis of the spectral changes and the reaction products were analysed by TLC. The results showed that metformin reacts with glucose, fructose and glucose 6-phosphate. The reactivity of metformin is 400-times higher with glyoxal (one of the main intermerdiates of the Maillard reaction), and even greater with methylglyoxal, which increases in diabetes mellitus. Reaction of metfor-min with glucose was oxygen dependent, whereas that with glyoxal was not. To study whether metformin can inhibit AGE production, albumin (6 mg/ml) was incubated in the presence of 1 mM glyoxal or methylglyoxal, with or without metformin. After isolation by ultrafiltration, the albumin-AGEs were assessed by measuring the fluorescence (370 nm/440 nm). We observed that albumin glycoxidation was decreased by glyoxal or methylglyoxal by about 30% or 50%, respectively, in the presence of 1 mM metformin.

These results suggest that, besides its known antihyperglycaemic effect, metformin is also able to decrease AGE formation by reaction with reducing sugars or glycoxidation intermediates.

Determination of Heat Damage in Foods by Analysing N^ε-Carboxymethyl-Lysine together with Available Lysine (Homoarginine Method)

Axel Ruttkat, Anke Steuernagel and Helmut F Erbersdobler; *Institut für Humanernährung und Lebensmittelkunde, University of Kiel, Düsternbrooker Weg 17, D-24105 Kiel, Germany*

Furosine [N^ε-(2-furoylmethyl)-L-lysine] has been established as an ideal indicator for early Maillard reactions in moderately heated foods, especially in milk products. However, furosine may lead to an underestimation of heat damage in more severely heat-treated food and, in fructose-rich foods, uncertain outcomes by the furosine method must be expected, since glucoselysine, the reaction product from fructose and lysine, does not form furosine. As an alternative, heat damage can be evaluated by measuring the losses in available lysine, e.g., by guanidination of the ε-amino group with *O*-methylisourea, thus forming homoarginine, which is detected by IEC.[1] For several years, *N*-carboxymethyl-lysine (CML) has been analysed by HPLC as a marker of heat damage,[2] indicating Amadori products as well as Heyns products,[3] and also ε-threuloselysine, a reaction product of ascorbic acid and lysine. CML is more stable during the advanced Maillard reaction and thus a useful indicator in more severely heated food samples.

Furosine, CML and available lysine were determined in model mixtures of casein with glucose or fructose and in 19 food items, which had been severely heated or were rich in fructose (products for diabetics or honey-sweetened food items). The results show that, in the casein mixtures, CML was formed earlier by the reaction with glucose than with fructose, while, after longer heating, the mixture with fructose yielded higher CML values. On the other hand, the concentration of available lysine decreased very rapidly in the casein-glucose mixture and remained always below the casein-fructose model.

The results of the analyses of the commercial food items show that heat damage was underestimated by the furosine method, especially in severely heat-treated or fructose-rich products. However, the determination of available lysine to assess the heat damage is of less value for food items with protein sources of different origin or of unknown processing. Although, in contrast to furosine, CML does not enable one to calculate the available lysine content, it gives in combination with the homoarginine method more precise and reliable information about the protein quality, especially for severely heat-damaged or fructose-rich products.

[1] Mauron, J.; Bujard, E. (1964). *Proc. 6th Int. Nutr. Congr.*, Mills, A.R., Passmore, R. Eds, Livingstone: London, pp. 489-490.
[2] Hartkopf, J.; Pahlke, C.; Lüdemann, G.; Erbersdobler, H.F. (1994). *J Chromatogr.*, **672**, 242-246.
[3] Ruttkat, A.; Erbersdobler, H.F. (1995). *J. Sci. Food Agric.*, **68**, 261-263.

Browning Reactions of Dehydroascorbic acid with Aspartame in Aqueous Solutions and Beverages

Hidetoshi Sakurai, Kenji Ishii, Huong T T Nguyen*, Zuzana Réblová* and Jan Pokorný*; *Nihon University, College of Bioresources Sciences, Department of Agricultural and Biological Chemistry, Shimouma, Setagaya-ku, Tokyo 154, Japan, *Prague Institute of Chemical Technology, Department of Food Chemistry and Analysis, Technická St. 5, CZ-166 28 Prague 6, Czech Republic*

In aqueous solutions, L-dehydroascorbic acid (DHA), formed by oxidation of L-ascorbic acid, undergoes various condensation and caramelization reactions. The formation of brown products was measured by UV/VIS spectrophotometry and the decomposition of DHA by high-performance liquid chromatography (HPLC). The degradation of DHA follows approximately first-order kinetics. The reaction products possess caramel and burnt flavour notes; whereas the former develop from the beginning of heating and even very slowly at room temperature, the latter do so only in the advanced stages of the reaction.

In presence of amines, Maillard reactions, including the Strecker degradation, compete with the caramelization. In soft drinks, sugar is partially or completely replaced by synthetic sweeteners, often by aspartame (A). A contains an amino group, which enters the Maillard reactions. The decomposition of DHA is about the same in model solutions or beverages containing A, but the reaction course is not the same. The reaction was studied at 80-100°C, as it proceeds only very slowly at room temperature. The browning in mixtures with A proceeds at about the same rate as in its absence, and the degradation rate of DHA is also of the same order. The decomposition of A (determined by HPLC) was much slower than the decomposition of DHA. Caramel and roasted flavour notes were much weaker in solutions containing A, and the burnt flavour note was not detected. On the contrary, weak floral and fruity notes were perceived, accompanied by musty, cardboard and rotten off-flavours. They could be attributed to N-(2-pyron-3-yl)aspartame, which was identified among the secondary products, and which has similar flavour in aqueous solutions. In lemon, orange and grape juices, fortified with L-ascorbic acid and containing A instead of sugar, deterioration may be observed after a storage of a month or more, accompanied by weak off-odours and off-flavours, and by a brownish hue. These adverse changes may be attributed to the above reactions.

Effect of High Glucose Concentrations on Soluble Proteins from Mesophilic and Thermophilic Bacteria

Reinhard Schinzel and Gerald Münch; *Universität Würzburg, Theodor-Boveri-Institut, Physiologische Chemie I, Am Hubland, 97074 Würzburg, Germany*

Although nonenzymically glycated proteins[1] and nucleic acids,[2] as well as proteins able to bind advanced glycated endproducts (AGEs),[3] have been observed in prokaryotes, the turnover of macromolecules in mesophilic bacteria seems too rapid to allow accumulation of glycated proteins and the subsequent formation of AGEs. Since formation of AGEs is strongly accelerated by high temperatures, nonenzymic glycation might have more relevance to thermophilic bacteria. Therefore the formation of AGEs was examined in the thermophilic eubacterium, *Thermus thermophilus* in comparison with the mesophilic *E. coli*. Indeed, a higher content of Maillard products was found in cell-free protein extracts of *Th. thermophilus* grown at 70°C by immunological and spectroscopic methods. However, when extracts of *Th. thermophilus* and *E. coli* were incubated at 37°C in presence of glucose or fructose the *E. coli* proteins precipitated, while *Th. thermophilus* proteins remained soluble. In *E. coli*, this precipitation could be partially inhibited by acetylation of the proteins and the presence of metal chelators, such as DTPA, while addition of 5 μM Fe^{2+} and Cu^{2+} enhanced the precipitation. This suggests that glycoxidation processes are involved in the formation of the precipitate. None of these effects was observed for the *Th. thermophilus* proteins, although the concentration of accessible lysine residues seems to be similar to that of *E. coli* extracts. Dialysis and gel filtration of *Th. thermophilus* extracts did not enhance generation of protein precipitates. This indicates, that the resistance of *Th. thermophilus* proteins does not depend on the presence of small scavenger molecules, but is most likely due to evolutionary adaptation of the protein structures to avoid crosslinking at high temperatures.

[1] Bhattacharya, M.; Plantz, B.A.; Kobler, J.D.; Swanson, J.D.; Nickerson, K.D. (1993). *Appl. Environ. Microbiol.*, **59**, 2666-2672.

[2] Lee, A.T.; Cerami, A. (1990). *Mutat. Res.*, **238**, 185-192.

[3] Gerhardinger, C.; Taneda, S.; Marion, M.S.; Monnier, V.M. (1994). *J. Biol. Chem.*, **269**, 27297-27302.

Studies on the Reaction of Glyoxal with Protein-Bound Arginine

Uwe Schwarzenbolz, Thomas Henle and Henning Klostermeyer; *Lehrstuhl für Milchwissenschaft, Technische Universität München, Vöttinger Straße 45, D-85350 Freising-Weihenstephan, Germany*

Advanced Maillard reactions as well as sugar degradation under acid and alkaline conditions give rise to the formation of various α–dicarbonyl compounds. Little is known about the reaction of these sugar degradation products with proteins. Up to now only few protein-bound reaction products of α-dicarbonyles with amino acids have been found. The ornithinoimidazolinone 1, formed in the reaction between methylglyoxal and arginine residues, and pyrraline, originating from lysine and 3-deoxyhexosulose, were found in food samples.[1,2] In model systems consisting of bovine serum albumin and glyoxal, N^ε-carboxymethyl-lysine and C2-diimines could be identified as lysine derivatives.[3]

Now, after incubation of arginine and glyoxal and subsequent ion-exchange chromato-graphy, we were able to detect with ninhydrin a new amino acid, designated as "Glarg". After preparative isolation, the structure of "Glarg" could be established, using ^{1}H-, ^{15}N- and ^{13}C-NMR spectroscopy, as an acid-labile hydantoin derivative, 1-(4-amino-4-carboxybutyl)-2-imino-5-oxo-imidazolidine 2. Studies on the formation of "Glarg" showed that glyoxal reacts very rapidly with arginine under physiological conditions, arginine derivatisation of over 91% being achieved in only one hour (molar ratios N^α-acetylarginine/glyoxal 1:1). After incubation of β-casein with different quantities of glyoxal under physiological conditions, "Glarg" was detected in the chromatograms of enzymic hydrolysates. The peak areas of the new compound correlated directly, those of arginine inversely, with the amount of glyoxal added.

"Glarg" represents a new form of post-translational protein modification in the course of the advanced Maillard reaction and thus a possible component of foods and biological systems.

[1] Henle T.; Walter A.; Haeßner R.; Klostermeyer H. (1994). *Z. Lebensm. Unters. Forsch.*, **199**, 55-58.
[2] Henle T.; Klostermeyer H. (1993). *Z. Lebensm. Unters. Forsch.*, **196**, 1-4.
[3] Glomb M.; Monnier V. (1995). *J. Biol. Chem.*, **270**, 10017-10026.

Degradation Products Formed from Glucosamine in Water

Chik-Kuen Shu; R.J. Reynolds Tobacco Company, Bowman Gray Technical Center, 950 Reynolds Blvd, Winston-Salem, NC 27105, USA

It is well known that the first step of the amino acid-reducing sugar reaction is a sugar-amine condensation, leading to an N-substituted Amadori or Heyns intermediate, from which the flavour compounds are generated via subsequent rearrangement and/or degradation. Similarly, when the amino acid is replaced with ammonia, the sugar-amine condensation also occurs initially, leading to the simple sugar amine instead of the Amadori or Heyns intermediate. Previous studies on flavour formation from Amadori and Heyns compounds are abundant, but information on flavour formation from simple sugar amines is limited. The objective of this study was to identify the degradation products formed from glucosamine in water, and to propose the formation mechanism of some of the products identified.

An aqueous solution of glucosamine was heated to 150°C in an enclosed system at different pH values. The product was extracted with ethyl acetate and the concentrated extract was analysed by GC/MS. It was found that the major products formed were furfurals, especially at pH 4 or 7. At pH 8.5, substantial amounts of additional flavour components were generated. Components identified included pyrazines, 3-hydroxypyridines, 1*H*-pyrrole-2-carboxaldehydes, furanones and hydroxyketones. Of these components, it seems that 3-hydroxypyridines and 1*H*-pyrrole-2-carboxaldehydes are derived from furfurals. The ammonia, liberated from glucosamine, can initiate the ring-opening of furfurals to form 5-amino-2-keto-3-pentenals. Intramolecular condensation in these intermediates between the amino and aldehyde groups would lead to the formation of 3-hydroxypyridines, while a similar condensation between the amino and keto groups would lead to the formation of 1*H*-pyrrole-2-carboxaldehydes.

Screening for Toxic Maillard Reaction Products in Meat Flavours and Bouillons

Kerstin Skog, Alexey Solyakov, Patrik Arvidsson and Margaretha Jägerstad; *Department of Applied Nutrition and Food Chemistry, Centre for Chemistry and Chemical Engineering, Lund University, PO Box 124, S-221 00 Lund, Sweden*

Participation of the Maillard reaction in the formation of mutagenic heterocyclic amines was proposed more than 10 years ago. Since then, the support for this route has increased, especially from a study where carbon atoms from ^{14}C-labelled glucose were shown to be incorporated into IQx, MeIQx and DiMeIQx. According to our previous studies, different amino acids produce different species of heterocyclic amines. In a recent study, 20 amino acids were individually heated with glucose and creatinine in an aqueous model system, and the presence of IQx, MeIQx and DiMeIQx was established in most of the samples, with addition of PhIP, Trp-P-1 and Trp-P-2 in some samples. Trp-P-1 and Trp-P-2 have not hitherto received much attention, as they had been thought to be formed exclusively under extreme cooking conditions and consequently were not expected to be found in the Western diet. However, recently these compounds were identified in foods cooked at moderate temperatures.

In the production of meat flavours and bouillons, the ingredients are heated to different temperatures and for various lengths of time depending on the manufacturer's production processes. The presence of creatinine and free amino acids and glucose during the boiling and the final drying step could result in the formation of heterocyclic amines to various extents. To minimise the levels of harmful compounds in meat extracts and bouillon cubes, there is a need to know the extent of the Maillard reaction under different sets of processing conditions.

The discussion covers recent work examining the formation of mutagenic and/or carcinogenic heterocyclic amines, including Trp-P-1 and Trp-P-2, from naturally occurring components in meat and fish, and also the presence of these mutagenic compounds in pan residues, meat extracts and bouillon cubes.

Characterisation of a Novel AGE-Epitope Derived from Lysine and 3-Deoxyglucosone

Ian C Skovsted, Mogens Christensen and Steen B Mortensen; *Health Care Discovery, Novo Nordisk A/S, DK-2880 Bagsvaerd, Denmark*

Nonenzymatic glycation of macromolecules is hypothesised to play a significant role in the development of late diabetic complications. The initial event is the direct modification of sidechains in basic amino acids leading to the formation of advanced glycation endproducts.

3-Deoxyglucosone is a dicarbonyl sugar formed from the fragmentation of the glucose derived Amadori adduct or directly from fructose.[1,2] It is a strong glycation agent, and has been shown to induce protein cross-linking and formation of fluorescent compounds upon incubation with proteins.[2]

Dicarbonyl sugars are generally strong glycating agents and the formation of the compounds, GOLD and MOLD, resulting from incubation of glyoxal and methylglyoxal with lysine, has recently been demonstrated.[3,4]

In this paper we identify a new compound called DOLD, the major product resulting from the reaction between hippuryl-lysine and 3-deoxyglucosone. It has been isolated and analysed by MS and NMR.

We show that the major product from the reaction between hippuryl-lysine and the diketones, glyoxal, methylglyoxal and 3-deoxyglucosone, share the same core structure, indicating a common Maillard reaction pathway.

[1] Knecht, K.J.; Feather, M.S., Baynes, J.W. (1992). *Arch. Biochem. Biophys.*, **294**, 130-137.
[2] Ledl, F.; Schleicher, E. (1990). *Angew. Chem.*, **29**, 565-594.
[3] Wells-Knecht, K.J.; Brinkmann, E.; Baynes, J.W. (1995). *J. Org. Chem.*, **60**, 6246-6247.
[4] Brinkmann, E.; Wells-Knecht, K.J.; Thorpe, S.R.; Baynes, J.W. (1995). *J. Chem. Soc., Perkin Trans. 1*, 2817-2818.

Malondialdehyde Reactions with Collagen with Related Chemical Studies on Model Systems

David A Slatter and Allen J Bailey; *Collagen Research Group, Division of Molecular and Cellular Biology, University of Bristol, Langford, Bristol BS18 7DY, UK*

One feature of diabetes is the random accretion of glucose onto the ε-amino groups of exposed lysine (and arginine) residues of proteins via Schiff base formation and Amadori rearrangement. Subsequent reactions eventually yield stable advanced glycation endproducts (AGEs). Long-lived matrix proteins, such as collagen and crystallin in particular, accrue AGEs with impairment of function whenever a cross-link occurs between the proteins or an important lysine is modified. An added complication is that the oxidised state of these glycated proteins has been shown to catalyse the oxidation of polyunsaturated fatty acids, to give a variety of reactive aldehydes, a major one being malondialdehyde (MDA, propandial). Excess formation of MDA from low-density lipoprotein in the cardiovascular system leads to damaging effects, in particular stiffening of the collagen and elastin by intermolecular cross-linking. We have therefore embarked on an investigation of the effects of MDA on collagen.

MDA is relatively long-lived as an aldehyde; over 99.9% of it is in an unreactive enol form at physiological pH, allowing it to build up to roughly 50 nM in healthy serum. It has been shown to cross-link bovine serum albumin, polylysine and γ-crystallin and to alter collagen properties markedly. Chemical studies show that MDA could cross-link two lysine residues via their ε-NH_2 groups.[1] However, studies of the MDA reaction with CBZ-lysine at physiological pH give several products and are complicated by a slow internal equilibrium as MDA itself cleaves to the more reactive ethanal and formic acid. MDA incubated with collagen from rat-tail tendon for different periods has been shown to generate several potential borohydride-reducible cross-links stable to acid hydrolysis, as verified by new components seen by ion-exchange and gel-filtration during the early stages of the reaction. Enzymic hydrolysis of the MDA-cross-linked collagen is being undertaken to allow studies on non-reducible but acid-labile products.

Associated model chemical studies imply that these MDA/collagen reactions are unlikely to be affected by the presence of glucose or ribose, being much faster with the reagent CBZ-lysine. MDA also appears to either inhibit or modify acid-labile pyrrole products generated in CBZ-lysine and ribose incubations, as detected by dimethylaminobenzaldehyde, demonstrating that it may interfere with glycation reactions *in vivo*. Its reaction with CBZ-arginine is at least 2 orders of magnitude slower, and so is unlikely to be significant.

[1] Chio, K.S.; Tappel, A.L. (1969). *Biochemistry*, **8**, 2821-2832.

Influence of Maillard Browning on Cow and Buffalo Milk and Milk Proteins - A Comparative Study

A Srinivasan, A Gopalan and S Ramabadran; *Protein Research Unit, Department of Chemistry, Loyola College, University of Madras, Madras-600034, India*

Systematic investigations have been undertaken to evaluate the intensities of browning in cow and buffalo milk samples, heated at different temperatures and for different periods in the presence of sweetening agents, such as glucose and sucrose of known concentrations. Attempts have been made to study the influence of browning on the physico-chemical properties of the heated milk samples and also on the nature, composition and nutritional status of isolated milk proteins, such as casein and coprecipitates of milk and whey proteins.

Results of the investigations are of special interest to Asian countries, such as India, where milk is consumed only after boiling for flavour and safety reasons, when it is used directly or with coffee or tea in the presence of sweetening agents. Further indigenous dairy products, such as koah and therattupal, are prepared by continuously boiling and evaporating milk with constant stirring with sweetening agents. Thermal processing under such conditions induces nonenzymic browning, affecting protein quality and shelf life. The intensity of browning increases with temperature, time of heating and concentration of sweetening agent. Buffalo milk and the presence of glucose led to greater intensity of browning. Browning was found to affect the digestibility and nutritional value of milk proteins. Milk melanoidins with a deep brown colour and a pleasant flavour, having antioxidant and metal-ion absorbing properties, were obtained on roasting with sucrose or glucose both casein and coprecipitates of milk. Oxidised milk proteins were isolated from hydrogen peroxide-treated milk, casein and coprecipitates of milk and, on roasting with sweetening agents, produced attractive pigments of deep red or bluish violet colour.

Nonenzymic Glycation of Phosphatidylethanolamine in Vivo

Kyozo Suyama and Kazutoshi Watanabe; *Tohoku University, Department of Applied Biological Chemistry, Faculty of Agriculture, Aoba-ku, Sendai 981, Japan*

It has been reported that nonenzymic glycation of protein is a common post-translational modification of body protein to yield ketoamine derivatives as stable covalent adducts.[1] The reaction proceeds through a Schiff base, followed by Amadori rearrangement. The present research aims to elucidate the parallel non-enzymic glycation of the amino group of phosphatidylethanolamine (PE) by reducing sugars *in vivo*. First, a model reaction between dilauroyl-PE and glucose to form the ketoamine derivative of PE (KA) was carried out under physiological conditions. For the analysis of KA by HPLC, the furosyl derivative[2] of KA, named furosylethanolamine (FE), was synthesised. Second, the increase of KA (both fructosyl- and tagatosyl-phosphatidylethanolamine) in the phospholipid fraction of hyperglycaemic rat tissues (streptozotocin-diabetic and galactose-administered galactosaemic after 4 weeks) was measured and compared with that for control rats. KA in the phospholipid fraction of human blood before and after administration of glucose was also determined.

Dilauroyl-KA was separated from the model reaction products and gave a relative molecular mass of 742 on FAB/MS. The structure of FE obtained from dilauroyl-KA was established by NMR spectra. KA was detected in the phospholipid fraction of porcine liver and KA was detected as FE in every tissue phospholipid fraction analysed. When glucose was orally administered to normal adult volunteers, the KA level of blood phospholipids increased markedly as compared with that before administration. KA levels of the phospholipids of diabetic rat tissues were very much higher than those of controls (almost 2, 5 and 10 times higher in blood, brain and kidney, respectively). Those of galactosaemic rats were lower than those of diabetic rats, except for brain phospholipids (almost 10 times higher than those of control rats).

[1] Thorpe, S.R.; Baynes, J.W. (1982). In *The Glycoconjugates*, Horowitz, M.I., Ed., Vol. III, Academic Press: New York, pp. 113-132.
[2] Schleicher, E.; Scheller, L.; Wieland, O.H. (1981). *Biochem. Biophys. Res. Comm.*, 99, 1011-1019.

Overexpression of Aldehyde Reductase Will Protect PC 12 Cells from Cytotoxicity of Methylglyoxal or 3-Deoxyglucosone

Keiichiro Suzuki and Naoyuki Taniguchi; *Osaka University Medical School, Department of Biochemistry, 2-2 Yamadaoka, Suita, 565, Japan*

Intermediate products in the glycation reactions are responsible for modification and cross-linking of long-lived proteins. In hyperglycaemic conditions, the level of cytotoxic aldehydes such as methylglyoxal (MG) and 3-deoxyglucosone (3-DG), increases. 3-DG, a 2-oxoaldehyde, is produced through the degradation of Amadori compounds and MG is a physiological metabolite. Both compounds are elevated during hyperglycaemia and accelerate glycation reactions. We have recently reported that aldehyde reductase (ALR) is one of the oxide reductases that protects against the cytotoxicity of such aldehydes.

To examine the intracellular role of ALR in the diabetic complications of neural cells, its gene was overexpressed in rat phaeochromocytoma PC 12 cells, which normally express a low level of ALR. Cytotoxicity, including apoptotic cell death, which was determined by fluorescent microscopy using a fluorescent DNA-binding dye, Hoechst 33258, was observed with 100 µM MG in normal PC12 cells. Western blot analysis showed that ALR protein in the ALR gene-transfected cells was more than twice as much as in the control (mock) cells. In the ALR gene-transfected cells, the cytotoxicity of MG and 3-DG was decreased and apoptotic cell death decreased also. This suggests that intracellular ALR protects neural cells from the cytotoxicity of 3-DG or MG, and that neural cells, which normally express a low level of ALR, should be susceptible to diabetic complications evoked by intermediate products of the Maillard reaction, such as 3-DG and MG.

Isolation, Purification and Characterization of Amadoriase Isoenzymes (Fructosylamine:Oxygen Oxidoreductase EC 1.5.3) from *Aspergillus* Sp.

Motoka Takahashi, Monika Pischetsrieder and Vincent M Monnier; *Institute of Pathology, Case Western Reserve University, Cleveland, Ohio, USA*

Four "Amadoriase" enzyme fractions, which oxidatively degrade glycated low molecular mass amines and amino acids with formation of hydrogen peroxide and glucosone, were isolated from an *Aspergillus* sp. soil strain, selected on fructosyladamantanamine as sole carbon source. The enzymes were purified to homogeneity using a combination of ion exchange, hydroxyapatite, gel filtration and Mono Q column chromatography. Molecular masses of Amadoriase enzymes Ia-c were 51 kDa, and 49 kDa for Amadoriase II. Apparent kinetic constants for N^ϵ-fructosyl-N^α-t-Boc-lysine and fructosyladamantanamine were almost identical for enzymes Ia-c, but corresponding values for enzyme II were significantly different. FAD was identified in all enzymes based on its typical absorption spectrum. The N-terminal sequence was identical for enzymes Ia and Ib (Ala-Pro-Ser-Ile-Leu-Ser-Thr-Glu-Ser-Ser-Ile-Ile-Val-Ile-Gly-Ala-Gly-Thr-Trp-Gly-) and for Ic except that the first 5 amino acids were truncated for it. The sequence of enzyme II was different (Ala-Val-Thr-Lys-Ser-Ser-Ser-Leu-Leu-Ile-Val-Gly-Ala-Gly-Thr-Trp-Gly-Thr-Ser-Thr-). All enzymes had the FAD cofactor-binding consensus sequence Gly-X-Gly-X-X-Gly within the N-terminal sequence.

In summary, these data demonstrate the presence of two distinct Amadoriases in the *Aspergillus* sp. soil strain selected on fructosyladamantanamine and induced by fructosylpropylamine. In contrast to previous described enzymes, these novel Amadoriases can deglycate both glycated amines and amino acids. Molecular cloning and expression studies are in progress and the results of these studies will be presented.

Nonenzymic Glycation as Risk Factor in Osteoarthritis

Johan M teKoppele, Jeroen de Groot, Nicole Verzijl and Ruud A Bank; *Gaubius Laboratory, TNO Prevention and Health, P.O. Box 2215, 2301 CE Leiden, The Netherlands*

Osteoarthritis (OA, cartilage degeneration) occurs in the majority of the elderly, and age is the major 'risk factor'. How aging contributes to this process is largely unknown. Collagen proteins in articular cartilage with a half-life of 200-400 years are especially prone to accumulate nonenzymic glycation (NEG) products with age. This is supported by the colour change of articular cartilage from bluish in youth to yellow/brownish in old age and aged cartilage collagen contains a 7-fold higher NEG level (pentosidine per triple helix) than collagen from other tissues.[1] Here the role of NEG in the occurrence of OA is explored by investigating the accumulation of pentosidine in human articular cartilage with age, its relation to tissue remodelling and its role in tissue function (i.e., stiffness).

In healthy human cartilage (knee condyle; age range: 2.5-92 yr, n = 62), pentosidine levels were low (about 0.001 pentosidine/triple helix) up to age 20. After age 20 yr, pentosidine levels increased linearly with age (r = 0.93, p < 0.001), up to 70-fold. This increase is the largest age-related change in articular cartilage of any of the parameters studied so far. The findings indicate tissue remodelling up to age 20 (no accumulation of pentosidine due to relatively high turnover of collagen), and a low rate of remodelling thereafter. This was supported by levels of the lysyl oxidase-mediated collagen crosslink, lysylpyridinoline (LP), as a function of the distance from the articular surface. In mature cartilage (30 and 43 yr), LP levels were highest at the surface. In young tissue (9 yr), the opposite was seen. Tissue sections at age 14 yr presented an intermediate situation: in the top layer LP levels are high (as in mature tissue), the lowest levels being reached halfway through the tissue, followed by an increase towards the bone (as in young tissue). Thus, the extensive remodelling of articular cartilage during the first two decades seems zonally regulated: maturation starts in the upper half of the tissue and occurs last in the lower half of the tissue.

To study the hypothesis that NEG is a key process in the known age-related stiffening of articular cartilage, 25 yr-old human cartilage was treated with ribose (0.6 M, 37°C, 4 days), resulting in yellowish cartilage with a 6-fold increase in pentosidine level. Stiffness of the collagen network, determined as 'instantaneous deformation',[2] was significantly increased by ribosylation (p < 0.001, n = 7).

Altogether, our findings indicate that articular cartilage undergoes extensive NEG after maturity, and that the known age-related stiffening of articular cartilage can - at least partly - be accounted for by NEG. As such, NEG presents a plausible molecular mechanism for the factor most predisposing for OA, age.

[1] Sell, D.R.; Monnier, VM. (1989). *J. Biol. Chem.*, **264**, 21597-21602.

[2] Mizrahi, J.; Maroudas, A.; Lanir, Y.; Ziv, Y.; Webber, T.J. (1986). *Biorheology*, **23**, 311-330.

A Mass Spectrometric Investigation on the Products Arising from the Reaction of Glucose with Nucleotides

Pietro Traldi, Roberta Seraglia*, Martina D'Alpaos†, Annunziata Lapolla†, Rosaria Aronica†, Michele Battaglia†, Massimo Garbeglio† and Domenico Fedele†; *CNR, Centro di Studio sulla Stabilità e Reattività dei Composti di Coordinazione, Via Marzolo 1, I-35100 Padova, Italy, *CNR, Area di Ricerca, Corso Stati Uniti 4, I-35100 Padova, Italy, †Istituto di Medicina Interna, Malattie del Metabolismo, Via Giustiniani 2, I-35100 Padova, Italy*

Half of the perinatal deaths in infants of insulin-dependent diabetic mothers are due to malformations affecting multiple organ systems. A *rationale* of these malformations has been proposed in terms of DNA mutations, which play an important role in embryopathy and teratogenesis. Some studies demonstrated that high glucose levels promote the formation of advanced glycation endproducts (AGE), exerting mutagenic effects. Recently, evidence was obtained that nucleic acids, like proteins, can react with sugars, supporting direct glycation as responsible for DNA modification. To investigate this aspect further, a series of *in vitro* incubations were performed of adenine, guanine and cytosine as free bases, as well as adenosine, guanosine and cytidine, the disodium salts of adenosine 5'-monophosphate, guanosine 5'-monophosphate and cytidine 5'-monophosphate, of adenosine 3',5'-cyclic monophosphate, guanosine 3',5'-cyclic monophosphate and cytidine 3',5'-cyclic monophosphate (each at 0.02 M), with glucose (0.02 and 0.1 M) in sodium phosphate buffer (0.05 M) pH 7.5, in sterile conditions, for 21 days. The products were analysed by HPLC/Electrospray/Mass Spectrometry. The formation of glycated products was observed from the first day of incubation and showed an increase directly related to incubation time, reaching a steady state after 14 days. The structure of such products has been determined from their molecular mass and MS/MS experiments performed on the molecular species. Multiple tandem mass spectrometry experiments (MS^3) allowed the structure modifications induced by the glycation processes to be investigated.

Antibody Titres against Oxidatively Modified Proteins Reveal an Increased Oxidative Stress in Diabetic Rats

N Traverso, S Menini, L Cosso, P Odetti*, D Cottalasso, M A Pronzato, E Albano† and U M Marinari; *Institute of General Pathology and *Department of Internal Medicine, University of Genova, Via L B Alberti 2, 16132, Genova, Italy; †Department of Medical Sciences-Novara, University of Turin, Via Solaroli 17, 28100, Novara, Italy*

Several authors maintain that the Maillard reaction can be a source of oxygen free radicals and that the interaction between advanced glycation endproducts (AGE) and RAGE, a recently characterized receptor for Maillard products, is able to trigger oxidation inside the cells. Consequently, the Maillard reaction could be responsible for increased oxidative stress in diabetes. An important consequence of oxidative stress is lipoperoxidation with generation of reactive aldehydes, such as malondialdehyde (MDA) and hydroxynonenal (HNE), which are able to modify proteins, generating fluorescent adducts. In rat skin collagen during aging, the fluorescence related to protein adducts with MDA and with HNE and the well known glycation-derived fluorescence accumulate in a parallel way.[1] In BB rats, spontaneously diabetic, faster rates of accumulation have been observed.[2] In order to verify whether HNE- and MDA-protein adducts actually develop *in vivo*, sera of diabetic and non-diabetic rats have been evaluated by ELISA[3] for the level of antibodies against HNE- and MDA-modified rat serum albumin (RSA). Diabetic rats exhibited a higher mean level, with a wider distribution, of both the antibodies tested.

Protein can also be modified by direct action of reactive oxygen species (ROS). If oxidative stress is increased in diabetes and ROS-modified proteins occur *in vivo*, diabetic rats should develop antibodies against them. Sera of diabetic rats have been tested for the level of antibodies against ROS-modified RSA (by γ-irradiation) and showed a higher mean level, with a wider distribution, than non-diabetic rats.

Our data point to a stronger immune response of diabetic rats against proteins modified by lipoperoxidative aldehydes and oxygen free radicals and support the hypothesis of an increased oxidative stress in diabetes.

Supported by grants Progetto Strategico No 96.04995. ST74 and MURST 40% and 60%.

[1] Odetti, P.; Pronzato, M.A.; Noberasco, G.; Cosso, L.; et al. (1994). *Lab. Invest.*, **70** (1), 61-67.
[2] Odetti, P.; Traverso, N.; Cosso, L.; Noberasco, G.; et al. (1996). *Diabetologia*, **39**, 1440-1447.
[3] Bellomo, G.; Maggi, E.; Poli, M.; Agosta, F.A.; Bollati, P.; Finardi, G. (1995). *Diabetes*, **44**, 60-66.

Antioxidant Reduction of Lens Damage in Model Diabetic Cataract

John R Trevithick, Fusan Kilic and Jon Caulfield; *The University of Western Ontario, Department of Biochemistry, Faculty of Medicine, London, Ontario, Canada N6A 5C1*

Diabetic cataract may be modelled by incubation of rat eye lenses in tissue culture medium containing 55.6 mM glucose (ten times the normal medium concentration). Damage to the lens cells and cell death result in the release of proteins, including lactate dehydrogenase (LDH). Since it has been suggested that nonenzymic glycation is accompanied by free-radical generation, several natural antioxidants, such as vitamins E and C and taurine, were tested against protein/LDH leakage, coupled to the putative increase in free-radical production. The diabetic lens loses endogenous osmolyte antioxidants, such as taurine, glutathione, cysteine and ascorbate, to compensate for the increased sorbitol concentration. Addition of antioxidants to the medium, to counter this loss, significantly reduces lens damage and protein/LDH loss. Resupply of lens taurine, for instance, appears to prevent the damage. Luminescent tests show taurine can scavenge reactive oxygen species, supporting the idea that the lens maintains high levels of this important osmolyte, because of its antioxidant activity. The concentration of taurine in the lens is almost identical to the concentration which completely scavenges reactive oxygen species *in vitro*.

Studies of the Mutagenic Action of Maillard Reaction Products from Triose Reductone and Amino Acids

Yaw-Kun Tseng and Hirohisa Omura*; *Department of Food Science and Technology, National Chia-Yi Institute of Technology, Chia-Yi, Taiwan, Republic of China, *Food Chemistry Institute, Faculty of Agriculture, Kyushu University, Fukuoka, Japan*

The Maillard reaction products of triose reductone (TR) and amino acids (AA) were obtained by heating reaction mixtures, which were prepared by first dissolving completely 0.5 M amino acid in 0.2 N HCl, adding 0.5 M TR and heating at 80-100°C for about 1 hour under a stream of nitrogen. The yields of the condensation products, the aminoreductones (TR-AA), were 33-40%.

The mutagenic action of the TR-AA was studied by the Ames test with *Salmonella typhimurium* TA 100 and the Rec-assay with *Bacillus subtilis* strains H 17 Rec$^+$ and M 45 Rec$^-$. It was confirmed that all the TR-AA induced high mutagenic activity. The activity depended on concentration, TR-Met, TR-Phe, TR-Gly and TR-Ile showing maximum activity at 20, 40, 60 and 80 mM, respectively. The activity of TR-Trp was constant from 20 to 100 mM. In general, the mutagenic activity of TR-AA was decreased by Cu^{2+}, but TR-Phe was activated by a very small amount of Cu^{2+}.

Digestibility and the Peptide Patterns of Lysozyme Modified by Glucose

Hiromi Umetsu and Nguyen van Chuyen; *Department of Food and Nutrition, Japan Women's University, Tokyo 112, Japan*

Proteins modified by glucose are known to have decreased digestive value, since crosslinks are formed in the process of the reaction with reducing sugars. However, few studies have been carried out on the peptides generated during the digestion process and on the cleavage pattern of the peptide bonds.

In this study,[1] lysozyme was modified by glucose in the dry state (50°C, 75% r.h.) and polymerization, changes in amino acid composition and peptide patterns after hydrolysis by pepsin-pancreatin were investigated using HPLC. Furthermore, the amino-acid sequence of peptides generated by the enzymic digestion was determined by protein sequencer and mass spectrometer. Modified lysozyme was also administered to rats to make clear the cleavage of peptide bonds *in vivo*.

The following became apparent: Digestibility and the pattern of peptides of molecular mass below 3,000 Da of the hydrolysate of the modified lysozyme (loss of Lys was about 50%) were similar to those of native lysozyme. The pattern of peptides of molecular mass between 3,000 and 10,000 Da of the hydrolysate of the modified lysozyme was also similar to that of native lysozyme. High molecular mass peptides of over 10,000 Da were detected in the hydrolysate of the modified lysozyme. The amino-acid sequences of peptides with molecular mass below 3,000 Da from native or modified lysozyme were: (a) H_2N-Lys-Val-Phe-Gly-COOH (b) H_2N-Ser-Leu-Gly-Asn-Try-Val-COOH (c) H_2N-Ser-Leu-Gly-Asn~Try-Val-Cys-COOH. These peptides correspond to residues 1-4, 24-30 and 24-31 of lysozyme, counting from the N-terminal. Thus, these results indicate that native and modified lysozyme were cleaved similarly by enzymic digestion even though 50% Lys residues had been modified. Peptide patterns of native or modified lysozyme in the rat small intestine were the same as those of the control. Therefore, even *in vivo*, it seems that modified lysozyme was digested as easily as the native form.

[1] Umetsu, H.; van Chuyen, N. (1996). Presented at the Annual Meeting of the Japan Society for Bioscience, Biotechnology, and Agrochemistry.

The Study of Renin-Angiotensin-Aldosterone in Experimental Diabetes Mellitus Rats

B Üstündağ, M Cay*, M Nazirŏglu*, N Ilhan, N Dilsiz†, and M J C Crabbe‡; *Department of Biochemistry, Medical School; *Department of Physiology, Veterinary Faculty, †Department of Biology, Faculty of Science; Firat University, Elazig, 23119, Turkey, ‡School of Animal and Microbial Sciences, University of Reading, Whiteknights,PO Box 228, Reading, RG6 6AJ, UK*

It is generally accepted that hypertension and other vascular pathologies increase in diabetes mellitus patients as a result of the angiotensin-aldosterone system. In this study, changes in the level of the renin-angiotensin-aldosterone system (RAS) were determined in streptozotocin-injected rats. In total, 46 female Wistar albino rats (180-220 g body weight) were examined.

The rats were injected intraperitoneally with streptozotocin (60 mg/kg body weight), dissolved in 0.1 M citrate buffer, pH 4-5. The non-diabetic rats were injected with sterilized buffer alone as a control group. Blood glucose level was 398±8.2, 488±11.75 and 658±29.6 mg/dl at days 3, 12 and 30, respectively. The level of plasma renin activity (PRA) was measured as 7.69± 1.07, 1.82±0.22 and 0.67±0.12 ng/ml/h at days 3, 12 and 30, respectively. These values showed that the PRA level decreases with time. The serum angiotensin-converting enzyme (ACE) level increased at days 12 and 30 ($P<0.05$ and $P<0.005$, respectively), whereas serum aldosterone increased at days 3 and 12 ($P<0.05$). The level of urea and creatine increased at days 12 and 30 ($P<0.005$ and $P<0.005$, respectively) when compared with the control group.

The evidence from these experiments is that the PRA level decreased, whereas ACE increased in diabetic rats compared with the control. The aldosterone level increased in the first step of experiment, but subsequently decreased as a result of the changes in renin and ACE levels.

Kinetic Modelling of the Maillard Reaction between Glucose and Glycine

Martinus A J S van Boekel; *Wageningen Agricultural University, Department of Food Science, PO Box 8129, 6700 EV, Wageningen, The Netherlands*

The Maillard reaction between glucose and glycine has been the subject of many studies; for an overview, see.[1] However, many papers pay attention only to one or two reaction steps or reaction products. This limits the possible kinetic analysis of the complete reaction network. We therefore attempted to follow the various reaction steps as closely as possible, by analysing isomers of the sugar, the Amadori product and degradation products, both of the sugar and the Amadori product.

Analyses were as follows: sugars and organic acids by HPLC, glycine and the Amadori product, fructosyl-lysine, by amino acid analyser, brown colour formation by spectrometry and fluorescent compounds by fluorometer. pH was measured as a function of reaction time. Experiments were done in an aqueous system, buffered to pH 7, over the temperature range 40-120°C with reaction times varying from hours to days, depending on the temperature. The Amadori product, fructosyl-lysine, was synthesized and also heated in the absence of sugar, in order to be able to study the degradation of the Amadori product without interference of the sugar.

The reaction products identified sofar are, besides glucose and glycine, fructose, fructosyl-lysine, acetic acid and formic acid. Formation of organic acids was especially marked at the higher temperatures, and causes, of course, a pH decrease. As the pH strongly affects the Maillard reaction, as well as isomerization, this pH decrease complicates the reactions even further.

Research is underway to propose a kinetic model based on the main reaction products identified and the mass balance. The model proposed will be tested, and, if necessary, adjusted by multiresponse modelling,[2] which also yields the relevant kinetic parameters (rate constants, activation enthalpies and entropies for every reaction step).

[1] Labuza, T.P.; Baisier, W.M. (1992). In *Physical Chemistry of Foods*, Schwartzberg, H.G., Hartel, R.W., IFT Basic Symposium Series 7, Marcel Dekker, New York, pp. 595-649.
[2] Van Boekel, M.A.J.S. (1996). *Neth. Milk Dairy J.*, 50, 245-266.

Improvement of Diabetes Mellitus Complications by Dietary Antioxidants

Nguyen van Chuyen, Jimaima Veisikiaki Jale, Keiko Shinada, Nami Kemmotsu and Hanae Arai; *Department of Food and Nutrition, Japan Women's University, Tokyo 112, Japan*

It is known that the reaction between glucose and protein contributes to the increase of chemical modification and cross-linking of long-lived tissue proteins in diabetes mellitus. Levels of glycated collagen and glycohaemoglobin ($GHbA_{1C}$) are higher in diabetic patients than in healthy subjects. The Amadori rearrangement products have been reported to produce reactive oxygen species in the presence of transition metal under physiological conditions. Thus, in diabetes mellitus, lipid peroxidation is considered to be accelerated. The complications of diabetes mellitus, such as cataract, are considered to be mostly related to the glycation of lens crystallin. However, the mechanism is still in obscure.

In this study, dietary antioxidants, such as vitamin E, β-carotene, astaxanthin and catechin, were administered to STZ diabetic rats to elucidate whether antioxidants could affect the glycation and complications of diabetes or not. The rats were fed short term (2 or 4 months) or long term (7 months) with various antioxidants. Lipid oxidation, glycohaemoglobin, SOD GST activity, ketoamine, solubility of skin collagen, pentosidine and degree of cataract formation were determined in rats.

The following became apparent: (a) Strong inhibition of lipid peroxidation was observed when antioxidant was administered. (b) Inhibition of glycohaemoglobin was observed only when catechin was added to the diet. (c) Inhibition of ketoamine formation and of cross-linking of collagen was observed only to a small extent. (d) Formation of pentosidine was also inhibited only to a small extent. (e) Cataract formation was drastically inhibited by vitamin E and astaxanthin.

From these results, it can be deduced that lipid peroxidation does not contribute to the accumulation of crosslinks, but it does aggravate the formation of cataract. Consequently, dietary antioxidants are useful for the improvement of some diabetes mellitus complications.

[1] Shinada, K.; Kemmotsu, N.; Kamimura, K.; van Chuyen, N. (1995). Effect of antioxidants on the improvement of diabetes mellitus. Presented at the Annual Meeting of the Japan Society for Bioscience, Biotechnology, and Agrochemistry.

α-Acetyl-N-Heterocycles from Glutamic Acid and Carbohydrates

Jos G M van der Ven and Hugo Weenen; *Bio-Organic Chemistry Department, Quest International, PO Box 2, 1400 CA Bussum, The Netherlands*

2-Acetyl-1,4,5,6-tetrahydropyridine (ACTP) and 2-acetyl-1-pyrroline (ACP) are strongly smelling compounds that are responsible *inter alia* for the impression given by wheat bread crust. Their odour has been described as roasty, popcorn-like and cracker-like (odour thresholds in water: ACTP 1.0 ppb, ACP 0.1 ppb). They belong to the class of α-acetyl-N-heterocycles which have a tautomeric α-alkanoyl-imine (I) or α-alkanoyl-enamine (II) structure as part of a ring system and which all elicit a roasty and cracker-like aroma.

The formation of ACTP and ACP from amino acids and carbohydrates in model experiments was first investigated by Hunter *et al.*[1] and Hodge *et al.*,[2] and later in more detail by Schieberle,[3] who hypothesised that, in bread, these roasty and cracker-like aroma compounds are formed from components present in the low molecular mass fraction of baker's yeast *(Saccharomyces cerevisiae)*. It was shown that amino acids that are only present in yeast, like ornithine and citrulline, are precursors for ACP, whereas ACTP is exclusively formed from proline.

The presence of glutamic acid in yeast, and especially in gluten proteins, prompted us to undertake model experiments with glutamic acid and carbohydrates. We observed the unexpected formation of ACP in our model experiments and propose a mechanism for its formation from glutamic acid and carbohydrates.

[1] Hunter, I.R.; Walden, M.K.; Scherer, I.R.; Lundin, RE. (1969). *Cereal Chem.*, **46**, 189-195.
[2] Hodge, J.F.; Mills, F.D.; Fischer, B. F. (1972). *Cereal Sci. Today*, **17**, 176-182.
[3] Schieberle, P. (1990). *Z. Lebensm. Unters. Forsch.*, **191**, 206-209.

Maillard Reaction of free and Nucleic Acid-Bound 2-Deoxy-D-Ribose and D-Ribose with ω-Amino Acids

Georg Wondrak, Roland Tressl and Dieter Rewicki*; *Technische Universität Berlin, Institut für Biotechnologie, Seestrasse 13, 13353 Berlin, Germany, *Freie Universität Berlin, Institut für Organische Chemie, Takustrasse 3, 14195 Berlin, Germany*

Nucleic acids, both DNA and RNA, are an abundant source of intracellular polymer-bound sugars, as well as of bases bearing an amino group. It has been argued that the amino groups of DNA bases may serve as a crucial target for covalent modification by the Maillard reaction.[1,2] The following facts, (a) frequent base loss from nucleic acids *in vitro* as well as *in vivo*, forming polymer-bound open-chain sugar phosphates, (b) accelerated Maillard reactivity of sugar phosphates and (c) the abundant complexation of nucleic acids by polyamines and histones (containing up to 30% lysine residues), led us to investigate the Maillard reactivity of nucleic acids as sugar-type components.

The Maillard reaction of free and nucleic acid-bound 2-deoxy-D-ribose and D-ribose with ω-amino acids (4-aminobutyric acid and 6-aminocaproic acid) was studied under both stringent and mild conditions. With amines, 2-deoxyribose displays stronger browning activity than ribose and, accordingly, DNA is much more reactive than RNA, even in the absence of amines. From reaction under stringent conditions (160°C) between 2-deoxy-ribose (or DNA) and methyl 4-aminobutyrate, methyl 4-(2-pyrrolidonylmethyl-1-pyrrolyl)butyrate was identified as a new 2-deoxy-D-ribose-specific key compound trapped by pyrrolidone formation. Levulinic acid-related *N*-substituted lactams were identified as the predominant products from incubations of DNA with amino acids, whereas RNA paralleled the reaction with D-ribose, but in lower yield. The polyphosphate backbone of nucleic acids may be involved in catalysing the reaction.

[1] Ochs, S.; Severin, T. (1994). *Liebigs Ann. Chem.*, 851-853.
[2] Papoulis, A.; Al-Abed, Y.; Bucala, R. (1995). *Biochemistry*, **34**, 648-655.

Modification of Protein Structure following Reaction with Epoxyalkenals

Rosario Zamora and Francisco J Hidalgo; *Instituto de la Grasa, CSIC, Avda. Padre García Tejero, 4, 41012-Sevilla, Spain*

Some short-chain aldehydes have been implicated in the nonenzymic browning reactions involving lipids. Among them, epoxyalkenals have been shown to modify very rapidly reactive groups in amino acids, leading to colour and fluorescence.[1] As a continuation of these studies, this investigation was undertaken to determine the modifications induced in the protein structure as a consequence of its reaction with epoxyalkenals. Bovine serum albumin (BSA), as a model protein, was incubated with eight concentrations (10 μM-10 mM) of 4,5(E)-epoxy-2(E)-heptenal (EH) at pH 7.4 and 37°C for seven periods of time (0-24 h). Changes produced in the primary structure of the protein were determined by analysing the losses produced in basic amino acid residues, the formation of N^{ε}-pyrrolylnorleucine (Pnl) and the generation of protein carbonyl derivatives. The oxidative-stress product Pnl is produced as a consequence of the reaction between EH and the ε-amino group of lysine residues, and the generation of protein carbonyl derivatives results from the reaction of EH with the imidazole ring of histidine residues. Changes produced in the secondary and tertiary structures were analysed by studying protein denaturation and polymerization. Protein polymerization was determined by both gel permeation chromatography and high-performance capillary electrophoresis. The presence of EH produced changes in the primary structure of BSA, which were a function of EH concentration and the incubation time. These changes in the BSA primary structure also modified the secondary and tertiary structure. In addition, all these changes occurred parallel to the development of colour and fluorescence in the protein, which was a consequence of protein polymerization produced by a mechanism which had been characterized previously.[2] The above results indicate that epoxyalkenals are able to modify protein structure, contributing to the formation of colour and fluorescence. A failure in the degradation of these modified proteins might induce their accumulation, leading to participation in lipofuscin or formation of AGE pigments.

[1] Zamora, R.; Hidalgo, F.J. (1995). *Biochim. Biophys. Acta*, **1258**, 319-327.
[2] Hidalgo, F.J.; Zamora, R. (1993). *J. Biol. Chem.*, **268**, 16190-16197.

Radiation-Induced Maillard Reactions

Alicja Zegota and Stefania Bachman; *Institute of Applied Radiation Technique, Technical University of Lódz, Poland*

The above concept covers not only the complexity of chemical composition of natural food products, but also the reaction of components of natural food products with the products of water radiolysis. Preliminary studies were mainly conducted on model systems, composed of reducing sugars and amino acids. The products of these reactions have a critical influence on the quality of food products and their acceptance by consumers. There are prospects that in the near future irradiation techniques will be accepted by the Joint FAO/WHO/IAEA (UN organizations) and consequently introduced on a commercial scale.

In this paper, the results are presented of investigations of the effect of ionizing radiation from ^{60}Co on the acceleration of Maillard reactions in model system containing an aqueous solution of fructose (F, 0.003 mol/dm^3) and alanine (Ala, 0.001 mol/dm^3). Solutions of F/Ala, irradiated in the range of doses 5 to 30 kGy at a dose rate 1.4 Gy/s, were then heated for some hours at different temperatures, 40°, 60°, 80° and 100°C. To assess their colour intensity, the absorbance at 450 nm in 10 mm cuvettes was measured. The colour intensity of irradiated solutions was found to be dependent on the dose of radiation, the temperature and the heating time. The reaction constants estimated for the system studied increased with temperature from 60°C to 100°C, but decreased with dose from 70.6 kJ/mole for 5 kGy to 60.7 kJ/mole for 30 kGy.

The results of spectrophotometric and chromatographic studies confirmed the formation of carbonyl products fructose radiolysis and their participation in the acceleration of non-enzymic browning reactions..

The changes in the concentration of fructose and alanine during irradiation of the solutions under study were proportional to the dose of radiation. The radiation yield of fructose decomposition was equal to G = 2.6 and that for alanine to G = 0.22. In the irradiated solutions of F/Ala, serine has been found, which so far has not been mentioned as a product of alanine radiolysis.

The studies performed demonstrate the influence of radiation on the initiation and acceleration of the Maillard reaction during subsequent heating at 40°C up to 100°C of systems containing reducing sugars and amino acids and this needs to be taken into consideration when introducing radiation technology to food products to be exposed to further thermal treatment.

Subject Index

Acetic acid 156, 158
Acetol 65
2-Acetylfuran 403
2-Acetyl-5-hydroxymethyl-4,5-dihydropyridin-4-one 76
2-Acetylpyrrolidine 213
2-Acetyl-1-pyrroline 209, 440, 452
2-Acetyl-1,4,5,6-tetrahydropyridine (ACTP) 209, 452
Actinic elastosis 427
Acyclic form 37, 166
Advanced glycation endproducts (AGEs) 3, 239, 262, 279, 285, 310, 322, 327, 339, 357, 363, 374, 380, 386, 404, 405, 408, 416, 423, 424, 426, 427, 430-435, 441
Ageing 3, 316, 423, 436, 439, 447
Alanine 82, 143
Aldehyde reductase (AR) 446
Aldehydes 198
Aldol condensation 65
Aldose reductase pathway 296
Alkylamines 109, 204
Alkylamino ascorbic acid 107
Alzheimer's disease 405, 430, 431
Amadori compounds 28, 37, 323, 415, 429
Amadori product dione (1,4-dideoxy-1-alkylamino-2,3-hexodiulose) 242-243
Amadori product, ene-dione (1,4,5-trideoxy-1-alkylamino-2,3-hexulos-4-ene) 242-243
Amadoriase 28, 447
4-Aminobutyric acid (GABA) 69
2-Amino-3,4-dimethylimidazo[4,5-f]quinoline (MeIQ) 13-15
2-Amino-3,8-dimethylimidazo[4,5-f]quinoxaline (MeIQx) 417, 444
Aminoguanidine 239, 279, 304, 327, 341, 400, 401, 411, 413, 424, 426, 437
2-Amino-3-methyl-3H-imidazo[4,5-f] quinoline (IQ) 13-15
2-Amino-3-methyl imidazo [4,5-f]quinoxaline (IQx) 219
2-Amino-1-methyl-6-phenylimidazo[4,5-b]pyridine (PhIP) 417, 444
Aminophospholipids 245, 363
2-Amino-3,4,8-trimethylimidazo[4,5-f]quinoxaline (DiMeIQx) 13-15
Ammonia 189
Analytical methods 76, 121, 160, 178
Angiotensin 450
Anomeric forms 40
Antibodies, AGE 310
Antioxidant products of the Maillard reaction 228, 231, 399, 404, 406, 425, 428
Antioxidants 4-5
Antioxidants, dietary 451
Apoptosis 345

Apple juice 91 - 92
Arachidonic acid 316
Arginine 104, 109, 143-155, 197, 262
Argpyrimidine 250
Aroma compounds 190, 198, 204, 209
Ascorbic acid 3, 89, 107, 127, 256, 419-421, 435, 442
Ascorbylation 107, 413
Aspartame 442
Aspartic acid 143
Aspergillus fumigatus 28
Atherosclerosis 285, 374, 391, 405, 439

Baking 22
Barley 92
Basement membrane, capillary 369
Beef 92
Beer 92
Beta-cells, pancreatic 345
Bis(2-methyl-3-furyl)disulphide 407
Bovine serum albumin (BSA) 134-136, 429, 434, 453
Bovine serum albumin, glycated 429, 430, 434
Brain 423, 446
Bread flavour 209, 440, 452
Browning 19, 62, 82, 113, 121, 147, 160, 166, 172
2,3-Butanedione 53

Canning 22
Cannizzaro reaction 205, 364
Capillary electrophoresis (CE) 121, 400, 414, 440
Caramelization 172
Caramels 440
Carbenium ions 147
Carbonyl assay 405, 413
Carbonyls 3, 327-331
Carboxyethyllysine 358
Carboxymethylethanolamine (CME) 363, 436, 439
Carboxymethyllysine (CML) 3, 94, 310, 316, 339, 351, 405, 411, 413, 416, 419, 424, 427, 429, 432, 438, 441
Carboxymethylserine 363, 439
Carcinogens 11, 219, 449
Cartilage 447
Casein 154
    β–Casein 100
Catalase 341
Cataract 256, 271, 273, 304
Cell adhesion 416
Cell death 430
Chaperone 273, 304
Chelating agents 258
    in dialysis 341

Chemiluminescence 113
Chemotaxis 374
Cholesterylester synthesis 351
Chromophore, red 85, 130
Chromophore, yellow 84
CIELAB 160
Coffee 92, 181
Collagen 411, 435, 437, 445, 447
Colour 19, 76, 82, 160
Coloured compounds 84, 85, 130, 160, 400, 402
Configuration, effect of 172
Cooked food mutagens 11
Corn meal 188
Cracker-like aroma 191, 209
Crosslines 8
Crosslinking 6-8, 100, 109, 127, 133, 239, 250, 258, 322, 409, 410
Crystallin 271, 273, 304
Cyclization 166
Cycloalliin 194
Cyclotene (3-methylcyclopent-2-en-2-olone) 65
Cytokines 357, 371

Deamidation 187
Deglycating enzymes 28
Dehydroascorbic acid (DHA) 127, 442
3-Deoxyfructose 291, 298
Deoxyfructosyl glycine (see also fructosyl glycine) 43
2-Deoxyguanosine 234-235
3-Deoxyglucosone (3-DG) 3, 57, 141, 262, 291, 298, 333, 400, 426, 427, 444, 446
3-Deoxyglycosones 57
3-Deoxyhexosulose (see 3-deoxyglucosone)
3-Deoxy-2-keto-gluconic acid 298
3-Deoxypentosone 57
Deoxyribose 233-235
3-Deoxytetrosone 57
3-DG (see 3-deoxyglucosone)
Diabetic cataract 449, 451
Diabetic complications 322, 405, 408, 430, 432
Diabetic embryopathy 426
Diabetic nephropathy 408, 415
Diabetic retinopathy 279, 411, 435
Dialysis 333
Dialysis, peritoneal 339
Dicarbonyl compounds 5, 51, 322, 400, 401, 441, 446
Dietary antioxidants 451
Differential scanning calorimetry (DSC) 424
Difructosyl lysine 38-41
Digestibility 450
2,3-Dihydro-3,5-dihydroxy-6-methyl-4H-pyran-4-one 76
Dihydro-2-methyl-3(2H)-furanone 403

Dihydroxyacetone (DHA) 415
DiMeIQx (2-amino-3,4,8-trimethylimidazo[4,5-*f*]quinoxaline) 13-15
1,1-Dimethylbiguanide (metformin) 441
Disaccharide hydrolysis 147

Electronic effects 147
Electronic nose 414
ELISA (see enzyme-linked immunosorbent assay)
Enalapril 439
Endocytosis 380, 386, 391
End-stage renal disease 339
Enolization 37
Enzyme-linked immunosorbent assay (ELISA) 110, 243, 306, 313, 405, 408, 413, 430, 448
Enzymes, amadoriase 28
Epitopes 310
4,5 (*E*)-Epoxy-2(*E*)-heptenal (EH) 453
Erythrocytes 363
Extrusion 21, 137, 187

FFI 405
Fish 419
Flavonoids 403
Flavour 19, 190, 204, 209
    Meat 198, 407, 414, 444
Fluorescence 333, 426, 430, 434, 438, 453
Fluorescent compounds 262, 329
Foam cells 316, 374, 391
Food processing 19, 127, 178
Food quality 19, 160
Formaldehyde 204
Formic acid 156, 158
Fragmentation 57, 69
Free radicals 402, 403, 406, 425, 449
Fructose 327
Fructose-3-phosphate 291
Fructose-3-phosphokinase pathway 291
Fructoselysine (see also fructosyl lysine) 415
Fructosyl amine: oxygen oxidoreductase 28
Fructosyl alanine 177
Fructosyl glycine 39-41
Fructosyl lysine 39-41, 100, 369, 415
2-Furaldehyde (see furfural)
Furan-2-carbaldehyde (see furfural)
Furaneol 43
2-Furanmethanol 403
Furanones 57, 76
Furans 187, 190
Furfural 82, 172, 403, 412
Furosine 339, 441

Furosylethanolamine 446

Galactose 154
Gelation 133
Germanium 431
Glass-transition temperature 431
γ–Globulins 420
β–Glucans 418
Glucosamine 443
Glucose 76, 333
Glucosone 28
Glutamic acid 143-145, 187
Glutamine 189
Glutathione 259
Glycine 51, 143-145, 204
Glycolaldehyde 391
Glycosyl cation 172
Glycoxidation 3, 310, 316
Glyoxal 6-7, 94, 204, 250
GOLD (glyoxal-lysine dimer) 6-8
Grape syrup 403
Guanosine 421

Haemodialysis 339
Hamburgers 11
Harman 219
Hemithioacetal 358
Heterocyclic amines, mutagenic 11, 219
Heterocyclic compounds 89, 198
Heyns product 329
High pressure 121
Hippuryl lysine 444
HMF (see 5-hydroxymethyl furfural)
Hue 160
Human serum albumin (HSA), glycated 431
Human serum albumin 357
Hydrazine 331
Hydrazone 331
Hydrogen peroxide 28, 401
Hydroimidazolium crosslink 358
Hydroimidazolone 358
Hydrolysed vegetable protein 399, 417
Hydroxyacetone 65
4-Hydroxyanisole 425
4-Hydroxy-2,5-dimethyl-3(2H)-furanone (furaneol) 43, 57
Hydroxyketones 65, 191, 448
Hydroxyl radicals 401, 431
4-Hydroxy-3(2H)-furanone 57
4-Hydroxy-2-(hydroxymethyl)-5-methyl-3(2H)-furanone 76
4-Hydroxy-5-methyl-3(2H)-furanone 76, 414

5-Hydroxymethyl furfural (HMF) 172, 412, 436
Hydroxynonenal (HNE) 448
2-(1-Hydroxy-2-oxo-1-propyl) pyrrolidine 212
Hyperglycaemia 438
Hypertension 450

Imidazole 204
Imidazolium crosslinks 8
Imidazolone 262, 357
Immunochemical characterization 110, 310
Inflammation 316, 357
Insulin 262, 275, 386
Interleukin 1β (IL–1β) 360
IQ (2-amino-3-methyl-3H-imidazo[4,5-f] quinoline) 13-15
IQx (2-amino-3-methyl imidazo [4,5-f]quinoxaline) 219
Irradiation 453
Isomerization 154

Jam 406

2-Keto-3-deoxygluconate 291
Kinetics 141, 147, 154, 166, 172, 219, 433, 451

Lachrymatory precursor (in onion) 193
β–Lactoglobulin, lactosylation site of 407
Laminin 423
LDL (see low-density lipoprotein)
Lens 257, 258, 273, 304
Lipid 198
Lipid hydroperoxides 355
Lipid peroxidation 7-8, 225, 228, 316, 345, 356
Lipids 327, 363
L-NAME (nitro-L-arginine methyl ester) 279
Low-density lipoprotein (LDL) 285, 391
    Carboxymethylated 351
    Oxidized 354
    Succinyl, acetyl 356
Lysine 94, 100, 143, 231, 262
Lysozyme 262
Lysyl pyrraline 100

Macrophages 285, 316, 351, 369, 380, 386, 391
Macrophage colony stimulating factor (M-CSF) 357
Macrophage scavenger receptor 285, 356, 369, 374, 380, 386, 391
Maillard reaction, control of 401
Malondialdehyde (MDA) 445, 448
Maltol 57
Maltose 144-145
Marinating 14, 417
Marmalade 406

Mass spectrometry 80, 117, 366, 420, 448
Matrix-assisted laser-desorption/ionisation mass spectrometry (MALDI) 420
Meat 219, 417
Meat flavour 198, 407, 414, 444
MeIQ (2-amino-3,4-dimethyl-imidazo[4,5-*f*]quinoline)13-15
MeIQx (2-amino-3,8-dimethylimidazo[4,5-*f*]quinoxaline) 13-15, 219, 417, 444
Melanoidins 89, 412, 419
Mesangial cells 270
Metal binding/interaction 412, 429, 432, 433, 442
Metal ions 258, 351
Metformin 441
Methional 407
3-Methylcyclopent-2-en-2-olone (cyclotene) 65
Methylglyoxal 3, 178, 250, 285, 333
4-Methylimidazolium crosslink 357
2-Methyl-3-methyldithiofuran 407
Microangiopathy 298, 369
Microorganisms 412, 442, 447
Microsomes 225
Microwave cooking 21-22
Milk 178, 445
MIP26   273
Model systems
        Aldose - amino acid – sulphite 141
        Aminophospholipids 245
        Arabinose-alanine 414
        Ascorbic acid – amino acid 89, 107
        Benzylamine – phenylglyoxal 113
        Bovine serum albumin 327
        Carbohydrate-glutamic acid 452
        Casein – sugar 154
        Creatinine-glucose-amino acid 219
        Dehydroascorbic acid 127
        N-(1-deoxy-D-fructos-1-yl)-glycine 43
        3-Deoxyglycosones 57, 262
        2-Deoxyribose-methyl-4-aminobutanoate 452
        Disaccharides 147
        Fructose-alanine 453
        Fructose-lysine 402
        Glucose-aminobutanoic acid 406
        Glucose – arg-lys 239
        Glucose-glycine 76, 160, 424, 428, 451
        Glucose-lysine 121, 160, 402, 428
        Glucose-lysozyme 450
        Glucose-phosphatidylethanolamine 422
        Glucose-proline 210
        Glycolaldehyde-lysine 250
        Glyoxal-arginine 443
        Glyoxal-glycine 204
        Glyoxal-lysine 94, 250

Labelled amino acids 51
Labelled sugars 69
β–Lactoglobulin 327
Lactose-casein 429
Lactose-lysine 428
Lipid 198
Lysozyme 327
S-(1-Propenyl)-L-cysteine sulphoxide 193
Xylose-arginine 428
Xylose-glycine 76, 160
Xylose-lysine 160, 231, 402
MOLD (methylglyoxal-lysine dimer) 6-8
Molecular modelling 147
Monocytes 357, 369
MSR (see macrophage scavenger receptor)
MSR-knockout mouse 380
Mutagenicity 11, 219, 449
Mutarotation 38-41, 166

NADPHd (NADPH diaphorase) 279
Nephropathy 268, 298
Neural cells 428
Nitric oxide 279
NMR data 37, 60-61, 69, 76, 82
    ^{13}C CP MAS NMR 437
Non-fluorescent cross-link (NFC-1) 437
Norharman 219
N-Phenacylthiazolium bromide 323
Nucleotides 448

Oligosaccharides 141
Onion 193, 418
Orange juice 420, 442
Ornithinoimidazolinone 178
Ortho-tyrosine 6
Osteoarthritis (OA) 447
Oxalic acid alkylamides 109
Oxidation 3, 37, 225, 259
Oxidative stress 3, 225, 316, 345
Oxygen, effect of 113

Pancreatic β–cells 345
Peanuts 404
Pentodilysine (penK$_2$) 410
Pentosidine 6, 8, 100, 178, 266, 339, 405, 408, 409, 427, 434, 438, 447
Peptides 245
Peritoneal dialysis 333, 339, 426
Peroxidation lipid 7-8, 225, 228, 316, 345, 356
pH 399, 402, 433
PhIP (2-amino-1-methyl-6-phenylimidazo[4,5-*b*]pyridine) 11, 219, 417, 444

Subject Index

Phosphatidylethanolamine 422, 438, 446
Phosphatidylinositol-3-OH kinase 386
Phospholipid 198, 245, 363
Polymers 69
Popcorn flavour 211
Port 412
Potatoes, sugar spot in 24
Pretzel 182
Processing, food 19, 127, 178
Product quality 19, 160
Prooxidants 231
Propylamine 108
Protein crosslinking 258, 409, 410
Protein gels 133
Pyralline 426, 427, 433, 438
Pyranones 76
Pyrazines 51, 187, 190, 204, 402, 425, 443
Pyrazinone 51
Pyridinoline (Pyr) 435
Pyrophosphate 414
Pyrraline 8, 178
Pyrroles 204
3(2H)-Pyrrolinone 82
Pyrropyridine 265
Pyruvaldehyde 51

Rapamycin 389
Reactive oxygen species (ROS) 229, 345
Receptor, macrophage scavenger 285, 356, 369, 374, 380, 386, 391
Receptors, AGE 374, 380
Red cells (see erythrocytes)
Red pigment 85, 130
Reductones 419, 449
Renal disease, end-stage 339
Retinopathy 279, 298
ROS (see Reactive Oxygen Species)
Rye 92

Scavenger Receptor, Macrophage 285, 356, 369, 374, 380, 386, 391
Schiff's base 94, 239
Serine 143
Shelf-life, food 23
Singlet oxygen 259
Smooth muscle cells 374
Soft drinks 440
Solvent, effect on mutarotation 166
Soy meal 417
Soy sauce, Indonesian 400
Soya 136-138

Spectra 39-40, 77-80, 116-117, 160, 168, 247
Starch 437
Sterilization, thermal 333
S-*trans*-1-Propenyl-L-cysteine sulphoxide 418
Strecker degradation 205
Streptozotocin 268, 279, 304, 416, 450
Sucrose 147
Sugar configuration, effect of 172
Sugar spot, in potatoes 24
Sulphite 141, 436
Sulphur compounds 193
Sun protection 415
Superoxide radical 41

Taurine 449
Tautomerism 38-41, 65
Tea 416
Tenilsetam 431
Thermal degradation 43, 89
Thermal degradation products 333
Thermal processing 19, 401, 441
Thiazoles 193, 198
Thiazolium salts 322
Thiobarbituric-acid reactive substances 226-228, 233-235
Thiophenes 193
Toxicology 11, 219
Transforming growth factor–β (TGF–β) 374
Trehalose 147
Trout muscle 225
Tumour necrosis factor-alpha (TNF-α) 359

Uraemia 339, 433, 434
UV radiation 260
UV spectra 160

Vascular complications 369
Vascular permeability 404
Visible spectra 160
Vitamin E 268
Volatiles 190, 198, 204, 209

Water activity (a_w) 431
Wortmannin 386

X-ray diffraction 38
Xylose 76, 231
Xylosone 421

Yellow chromophore 84